Linear Circuits and Computation

Linear Circuits and Computation

Bharat Kinariwala

Franklin F. Kuo

Nai-kuan Tsao

Department of Electrical Engineering
University of Hawaii

John Wiley & Sons New York · London · Sydney · Toronto

Library of Congress Cataloging in Publication Data:

Kinariwala, B. K.
Linear circuits and computation.

Includes bibliographical references.
1. Electric networks. 2. Electric circuits.
I. Kuo, Franklin F., 1934- joint author. II. Tsao, Nai-kuan, joint author. III. Title.
TK454.2.K55 621.319'2 72-13821
ISBN 0-471-47750-8

Printed in the United States of America

10 9 8 7 6 5 4 3 2 1

Preface

This text is intended for use in a first course in electrical circuits normally taught at the sophomore or junior level. Most engineering students are required to take this kind of course at some time during their studies but, for electrical engineering students, it is usually their first course in electrical engineering. A first course with a broad scope should concentrate on motivation and specifics to create interest and to build confidence; it should contain a minimum of abstraction and generalities. This was our guiding principle in writing this textbook.

The scope of the book is linear circuits and their analysis. Formal analytical techniques and computational techniques are stressed without compromise. The theory that forms the basis of the formal analysis is part of engineering thought embodied in linear analysis. It is used repeatedly, almost in an intuitive manner, in a range of subjects such as communication theory, signal theory, control theory, and system theory. Thus the basic techniques of linear circuit analysis are covered in a thorough manner.

Digital computation, on the other hand, has made analysis much more than just theory. It allows us to find precise behavior of complicated circuits—a feat that was not possible 10 years ago. Circuit design of an impressive nature can now be performed with precision by using digital computation. Computational techniques of analysis are therefore included at every stage of our discussion. The computational techniques that are useful in circuit analysis are some of the most basic techniques that are useful in most engineering analysis. They are therefore given distinct emphasis by treating each new computational technique in a separate section instead of interspersing them in the text. This approach creates a consistent format: each chapter contains a section dealing with a major computational procedure.

We emphasize and advocate the use of the nodal analysis. It is direct and straightforward—a major asset! Loop and state variable approaches are developed in the discussion but their use is discouraged. Graph theory is avoided because it is unnecessary and too abstract for the sophomore or junior student.

Chapter 1 gives the essentials for circuit analysis and material on computer number systems. Chapter 2 deals with first-order equations and introduces important ideas on numerical solutions, which will be useful for courses that follow. Chapter 3 is a continuation of Chapter 2 for higher-order circuits. Runge-Kutta and predictor-corrector methods of solving differential equations are discussed in Chapters 2 and 3. Chapter 4 treats convolution as related to step and impulse response. This serves as a background for introducing methods of numerical integration.

Chapter 5 discusses the analysis of resistive networks and the important topic of Gaussian elimination and its computer implementation.

Chapter 6 covers the sinusoidal, steady-state analysis, and the related computer procedures deal with complex number operations. Network functions are introduced in Chapter 7 in terms of the sinusoidal steady-state descriptions, which is the way they are usually used. The computation of network functions accompanies the discussion.

Chapter 8 is on Fourier analysis with a thorough but concise treatment of the discrete transform and the Fast Fourier Transform (FFT). This is an important subject that is not usually dealt with at this level.

Chapters 9 and 10 discuss the traditional Laplace transform and two-port analysis; the computational material covers root-finding procedures and partial-fraction expansion techniques.

This textbook contains much more material than can be reasonably taught in a one-semester, three-credit hour course. At the University of Hawaii we cover the material in two semesters. Even then, we do not give substantial attention to computational details, especially the more complex algorithms and error analysis. The amount of material available in the text gives the teacher considerable flexibility in the choice of subjects to be covered. For example, if he skips all of the numerical computation sections, the result would be a standard modern circuits course similar to the courses taught in other U.S. engineering schools. On the other hand, there is enough material in the numerical methods and computation sections alone for a brief course in numerical analysis. We feel that most teachers should adopt a middle course between the two extremes. Since the numerical methods sections are independent of the circuit theory parts, the teacher has a choice of which algorithms to cover. We emphasize, however, that the computer programs can be used easily, even though the student is unfamiliar with the details of the numerical algorithms on which they are based. In fact, a thorough knowledge of FORTRAN IV is not needed, although it is certainly helpful.

We especially thank M. E. (Mac) Van Valkenburg for his constant encouragement of this project; Richard H. Yamamoto, Bhadra K. Patel, and Cecil E. Williams for many helpful comments on the manuscript; and Jeanne Moriyasu, Susan Ong, and Margo Takata for careful typing and preparation of the manuscript. We also thank our students who contributed useful suggestions. Finally, we express our gratitude to our wives for their help and encouragement during the writing of this book.

Honolulu, Hawaii, 1972

BHARAT KINARIWALA
FRANKLIN F. KUO
NAI-KUAN TSAO

Contents

CHAPTER 1
Introductory Concepts

1.1 INTRODUCTION

Perhaps the primary function of an electrical engineer is to design circuits in an efficient and intelligent manner. While practical considerations of cost, size, and ease of fabrication are important factors in good circuit design, the desired circuit performance is always the major consideration. *Circuit analysis*, which is the study of the behavior of electrical circuits under various types of excitations, provides a precise and thorough knowledge of circuit performance. It is one of the basic tools of the electrical engineer.

Electrical circuits serve to process signals. The circuit modifies the energizing signals or *excitations* into other desired signals or *responses*. For example, in many applications, such as in the transmission of radio or TV signals, magnification of the signals propagated through the atmosphere is necessary. The circuit that magnifies the signals is called an amplifier. Another common circuit, found in radio and TV receivers, is the tuner. TV and radio receivers receive many stations (signals) at the same time, but by the use of this selective circuit, the tuner, they eliminate all but the desired station. Both the amplifier and the tuner function to process the exciting signals into desired responses. The term *signal* denotes either a voltage or a current described as a function of time.

Circuit analysis includes an understanding of *theory* or the principles governing the behavior of circuits and an understanding of *computation* or the determination of the actual numerical values describing the circuit behavior. Our discussion of circuits will emphasize both theory and computation at every stage.

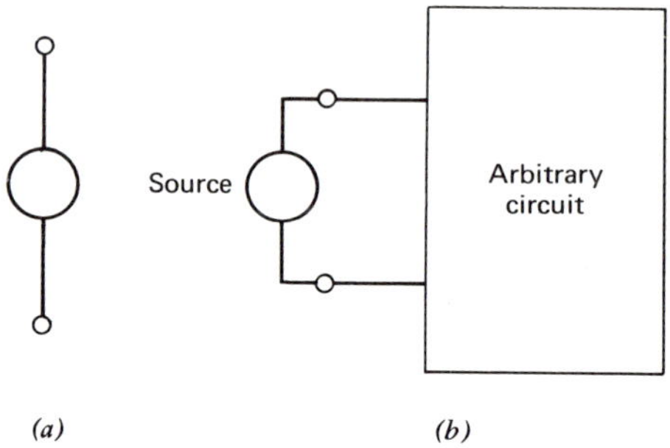

Fig. 1.1

Circuit Analysis

To clarify the concept of analysis, let us introduce some standard symbols and notations. An excitation signal *source* is usually denoted by a circle with two terminals, as shown in Fig. 1.1. This source is connected to some arbitrary circuit. The terminals or *nodes* represent the junction points between two or more circuit elements, for example, the source and the arbitrary circuit. The response signal may be a voltage measured across some two points in the circuit or a current flowing between some two points in the circuit.

An example of a circuit diagram is shown in Fig. 1.2. The circuit elements, connected in the manner shown, represent the *circuit* that is excited by the source. The response signal is observed as either the voltage or current at the desired place in the circuit. For example, the response may be the voltage across points **c** and **d**, or the current flowing between points **b** and **c**, and so forth.

Very often, a simplified diagram is utilized in which the excitation signal is written as $e(t)$ and the response signal as $r(t)$. The circuit or *network* or *device*, referred to as the *system*, is represented as a black box, as shown in Fig. 1.3. The excitation signal is called the *input* signal and the

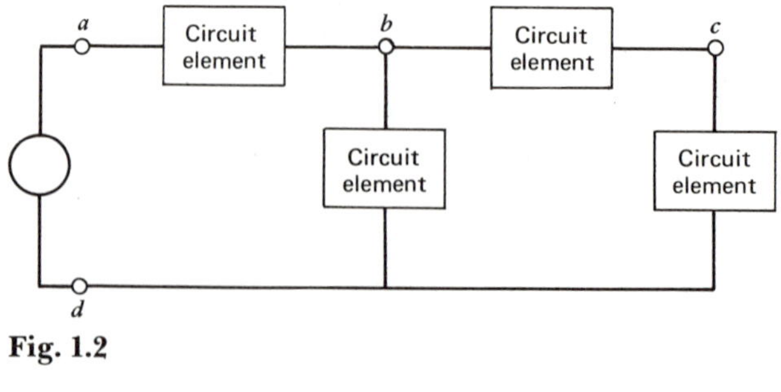

Fig. 1.2

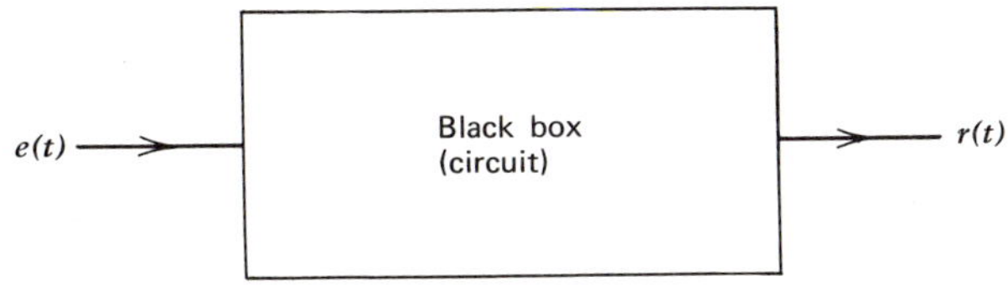

Fig. 1.3 Simplified diagram.

response, the *output* signal. The problem of analysis is to find the response $r(t)$, given the excitation $e(t)$.

The essence of circuit analysis resides in the fundamental laws and relationships that govern the behavior of all circuits. These laws have been established after much observation and study of circuits of all types. The laws that govern the behavior of individual circuit elements are known as *branch constraints.* The most frequently used circuit elements represent separate *branches* of the circuit since they have two terminals, as in Fig. 1.2. The branch constraint expresses the relation between the voltage across and the current through such a *two-terminal element.* Circuit elements with more than two terminals are viewed as two or more branches coupled together in specific ways. The behavior of individual circuit elements is thus described by the branch constraints. For example, the branch constraint governing a resistor is the well-known *Ohm's law*, which states that the ratio of the voltage across the resistor to the current through the resistor is given by a constant defined as the resistance. The laws that define the branch constraints for all types of circuit elements are discussed later at length.

The set of *equilibrium constraints* known as *Kirchhoff's laws* gives us the equilibrium relationships between the voltages or currents pertaining to groups of component branches of the circuit. For example, if several elements are connected together at a junction (node) of a circuit, the current flowing into the node must equal the current flowing out of the node. We will study these equilibrium constraints in a later section.

The voltages and currents are the variables in a circuit. When the equilibrium constraints or equations are written in a mathematical form, and the appropriate branch constraints used, we obtain certain mathematical equations in the circuit variables. It is shown later that it is not possible to solve or analyze these equations without one more piece of information. We must also know the *state of the circuit* at some instant of time. The state of the circuit is given by the energy storage or, more conveniently, the corresponding voltage or current variables in the individual reactive elements of the circuit. Inductors and capacitors, called the *reactive elements*, can store energy. Let us use a simple circuit as an illustration. If a capacitor is connected to a resistor and there is no excitation source, it is not possible to state whether there is any current flowing in the circuit. The answer will depend on the state of the circuit at the time of the connection. If the capacitor is uncharged initially (the state is zero), there is no current in the circuit. If, on the other hand, the initial state is not zero (the capacitor is charged), there is some current in the circuit. This simple example serves to show that the knowledge of the state (actual value) at the initial instant of interconnection is needed to determine the behavior of the circuit.

Scope

The techniques of analysis that we shall develop in this text are founded on the physical properties of the elements (that is, the branch constraints), the equilibrium constraints (Kirchhoff's laws), and the initial state of the circuit. Computation of the numerical values of the circuit variables can be easily performed for very simple circuits. For most practical circuits, on the other hand, computation becomes difficult without the use of a digital computer. Therefore, it is essential to learn the numerical methods for solving circuit equations on a digital computer. Such computational methods are an integral part of our study.

Physical circuit elements exhibit variations in their behavior. Imperfections in fabrication, in environment of operation, as well as in specific applications, all contribute to the variance in behavior of circuit elements. To develop the most effective tools of circuit analysis it is necessary to represent the physical elements by their *idealized* versions. The elements that we will consider in this text will therefore be the *mathematical models* or *abstractions* of their physical counterparts. Physical circuits only approximately resemble the idealized models we will use.

The great bulk of electrical circuits use elements known as lumped elements. The magnetic and electric phenomena associated with *lumped elements* are localized in space. In other words, the actual voltage or current associated with such a circuit element in a complicated interconnection depends solely on the electrical interconnections and not on the physical location (distance) of the element relative to the source. The behavior of a circuit of such lumped elements, therefore, is governed purely by the electrical interconnections. Circuits of lumped elements are also the only ones for which we have a well-developed theory. We will confine ourselves to the discussion of *lumped circuits*—that is, circuits composed entirely of lumped elements.

We will also limit our discussion to circuits excited by external sources and consisting entirely of *linear, time-invariant* elements. General definitions of these terms are given in Sec. 1.7. Specifically, linearity implies homogeneous, straight-line relationships between current and voltage in resistors, charge and voltage in capacitors, flux and current in inductors, and so forth. Time-invariance implies no change with time in these relationships. For most applications, physical elements can be approximated by linear, time-invariant models.

1.2 POLARITIES AND SIGN CONVENTIONS

Throughout our discussion we need to represent the polarities of voltages and the directions of current flow. We shall use the standard conventions shown in Fig. 1.4.

The voltage (in volts) as a function of time t (in seconds) between any two points **a** and **b** in the circuit is given by two quantities: the function $v(t)$ and the polarity indicated by the (+) and (−) signs. The polarity markings indicate that the voltage of point **a** with respect to point **b** (i.e., with point **b** as the reference) is given by the function $v(t)$. This means that if the negative or ground terminal of a voltmeter is connected to the negative polarity, point **b**, and the positive terminal of the voltmeter is connected to the positive polarity, point **a**, then the voltmeter will read the

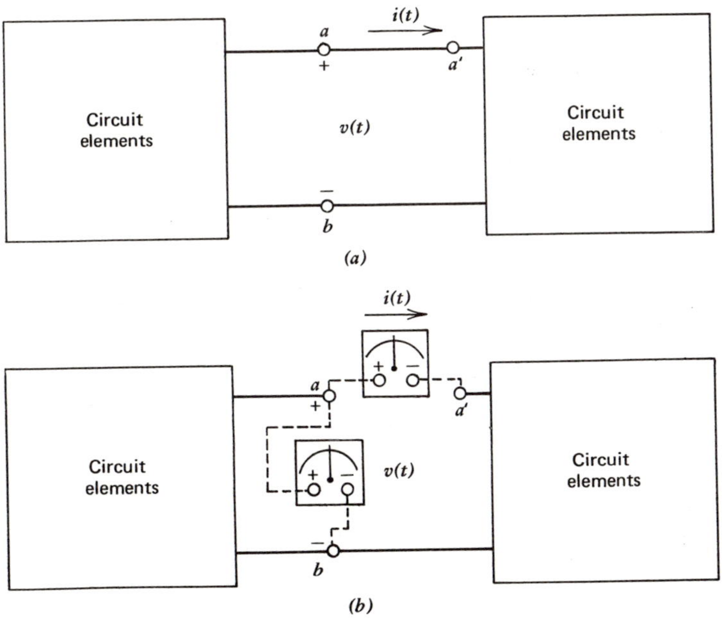

Fig. 1.4 (*a*) Standard sign conventions. (*b*) Sign conventions and measurements.

voltage $v(t)$. However, if the polarity markings were reversed, the voltage at point **b** with point **a** as reference would be $-v(t)$. Finally, with the polarity signs as shown, the voltage $v(t)$ is called a voltage drop from **a** to **b** or a voltage rise from **b** to **a**.

The positive direction of current flow is shown by the arrow in the figure. The value of the current (in amperes) as a function of time is $i(t)$ and the direction of the flow is in the direction of the arrowhead. If we connect the positive terminal of an ammeter to point **a** (towards the tail of the arrow) and the negative terminal of the meter to point **a**′ (towards the head of the arrow), the current measured by the meter would be $i(t)$. As before, reversing the arrow is equivalent to changing the sign of $i(t)$. Finally, we say that the current $i(t)$ flows out of point **a** and into the circuit at point **a**′. Frequently, v and i are used to indicate the voltage and the current functions without explicitly specifying the time dependence.

The above conventions describe combinations of currents and voltages. Currents add if they are in the same direction. Voltages add together if they have the same polarity directions. This is shown in Fig. 1.5.

These properties combine the voltages and currents in circuits. For example, in the circuit element shown in Fig. 1.6, there are two currents i_1 and i_2 flowing in the directions given. The resulting current is an appropriate combination of these currents in the desired direction. Observe that the current i_1 flows towards **b** or $(-i_1)$ flows towards **a**; and i_2 flows towards **a** or $(-i_2)$ towards **b**. The resultant or net current flowing towards point **a** is given by (i_2-i_1); equivalently, the net current flowing towards point **b** is (i_1-i_2).

Similarly, the voltages v_1 and v_2 are given with the indicated polarities in Fig. 1.7. Observe that the voltage of the point **b** with respect to point **c**

Fig. 1.5

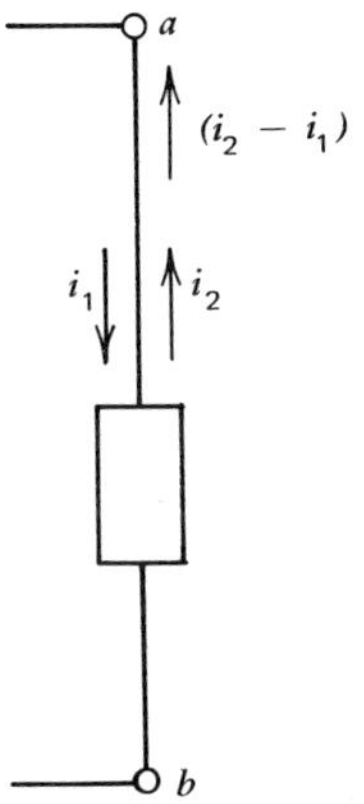

Fig. 1.6

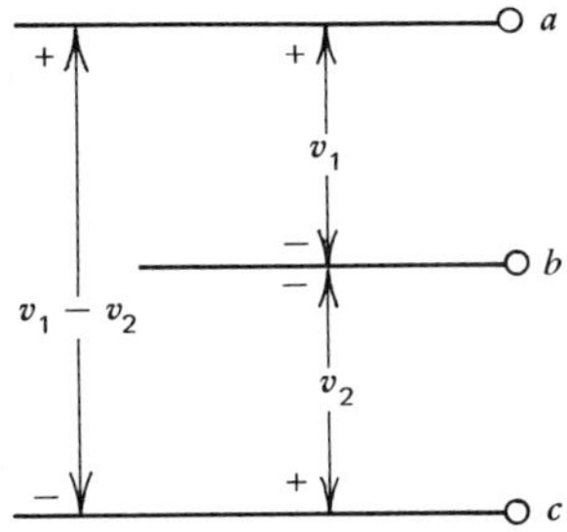

Fig. 1.7

is $(-v_2)$ and the voltage of the point **b** with respect to **a** is $(-v_1)$. Then the voltage of **a** with respect to **c** is $(v_1 - v_2)$, and that of **c** with respect to **a** is $(v_2 - v_1)$.

1.3 TWO-TERMINAL ELEMENTS

The branch constraints for two-terminal elements are defined for the polarity markings as shown in Fig. 1.8. The constraint is defined by the relationship between *the current in the element and the voltage drop across the element in the same direction as the current flow*. Elements with more than two terminals are described similarly. These will be discussed later.

The power delivered to a circuit or a circuit element, in terms of the sign conventions of Fig. 1.8, is the product of the voltage across and the current through it. The energy delivered to the circuit (element) is the

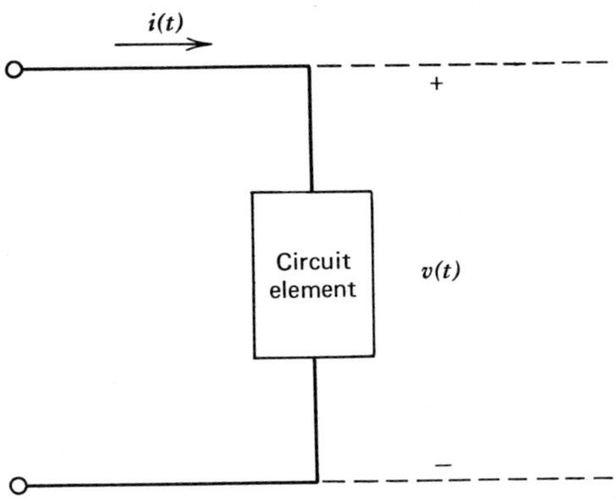

Fig. 1.8

integral of this power over the pertinent time interval. With the voltage and current conventions as shown in Fig. 1.8, the power delivered is given by

$$p(t) = v(t)i(t) \quad \text{watts} \tag{1.1}$$

The energy delivered over an interval of time in seconds from $t = t_1$ to $t = t_2$ is given by

$$w = \int_{t_1}^{t_2} v(t)i(t)\,dt \quad \text{watt-secs} \tag{1.2}$$

When the power and energy are positive quantities, they represent actual delivery; however, when they are negative, the power and energy are flowing out of the circuit.

Short-circuit and open-circuit between any two points, terminals, or nodes in a circuit can also be defined in terms of Fig. 1.8. If the circuit element is replaced by an ideal wire, the terminals are short-circuited. The *voltage across the short-circuit is always zero* regardless of the current through it. The current in a short-circuit is arbitrary and depends on the rest of the circuit connected to the terminals.

The terminals in Fig. 1.8 are open-circuited when the circuit element is removed, so that there is no electrical path between the terminals. *The current in an open-circuit is always zero* regardless of the voltage across it. The voltage across an open-circuit depends on the rest of the circuit connected to the terminals.

We now undertake the description of the important circuit elements and the branch constraints for these elements. We start first with the idealized versions of the sources.

Independent Sources

The exciting signals, of course, can be either voltage or current signals. Thus, we have two ideal sources: *Ideal Voltage Source* and *Ideal Current Source*. In an ideal voltage source the terminals of the source always exhibit the voltage specified by the function $v(t)$ with the polarities as shown in Fig. 1.9. The current flowing in this ideal voltage source has no constraint whatever. Such a source, therefore, can deliver any current desired (required by the rest of the circuit) while maintaining the specified voltage $v(t)$ at its terminals. In other words, an ideal voltage source does not constrain the current through it but constrains only the voltage across the terminals to a specified value. Since the voltage of the source does not depend in any manner on the rest of the circuit, it is also termed an independent voltage source.

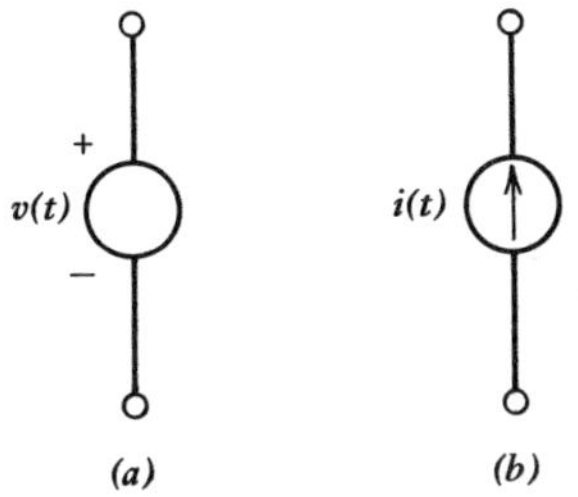

Fig. 1.9 (*a*) Ideal voltage source. (*b*) Ideal current source.

An ideal current source similarly maintains a specified current $i(t)$ through the source terminals in the direction of the arrow. In this case, the voltage across the terminals is completely unconstrained and determined by the rest of the circuit. Again, this is also called an independent current source.

There are many occasions when the source signals are zero or set to zero. An ideal voltage source with the voltage signal zero maintains the voltage across the source terminals at zero independent of the current flowing through the source. Zero voltage across a pair of terminals with the current unconstrained defines a short-circuit. Thus, a zero signal ideal voltage source represents a short-circuit.

A zero signal ideal current source similarly represents an open-circuit since the source always maintains zero current through it regardless of the voltage across its terminals.

The independent sources are commonly viewed as external elements that excite the circuit proper which consists of the physical elements described below. References to circuits as a rule assume that the independent sources are not present within the circuits.

The elements that we consider below represent idealized versions of real physical passive circuit elements. They differ from the ideal sources defined above in that both the voltage and the current are constrained by a mutual relationship for each element.

Resistors

The voltage-current relationship for a typical resistor is shown in Fig. 1.10. The relationship is nonlinear in general, and is expressed as

$$v_R = f_R(i_R) \tag{1.3}$$

where v_R is the voltage across the resistor and i_R is the current through it, with the polarity and sign conventions discussed in Fig. 1.8. The relationship may also be time-varying, that is, it changes with time. For many applications the relationship for a resistor may be approximated by a

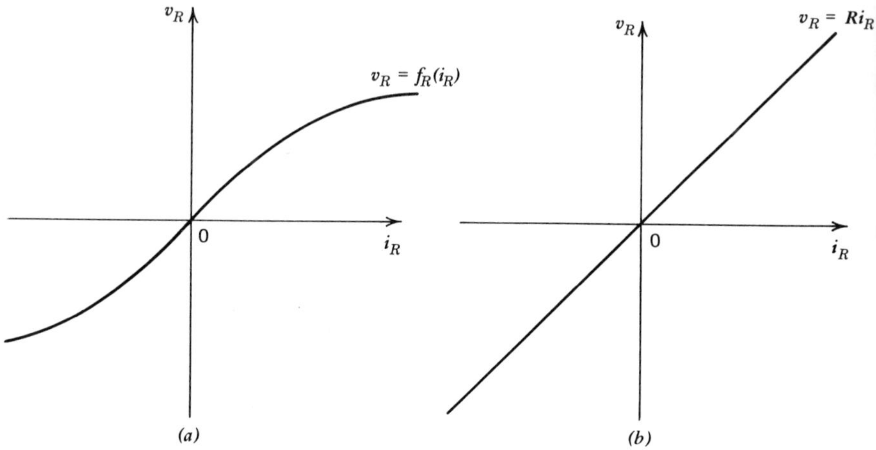

Fig. 1.10 (*a*) Voltage-current relationship for a resistor. (*b*) Linear, time-invariant model of a resistor.

constant (time-invariant) linear curve (straight line) through the origin in which case

$$v_R(t) = Ri_R(t) \tag{1.4}$$

where R is a positive constant called the *resistance*. The units are volts for voltage v_R, amperes for current i_R, and *ohms* (Ω) for resistance R. Equation 1.4 is the well-known *Ohm's law*. A resistor is indicated by the symbol shown in Fig. 1.11.

Quite often the reciprocal of R defined as *conductance* G, ($G = 1/R$) is used. The units of conductance are in *mhos*.

Inductors

Faraday's law of induced voltage for inductors or coils states that the induced voltage, v_L, in a coil is proportional to the rate of change of flux, ϕ_L, linking that coil.

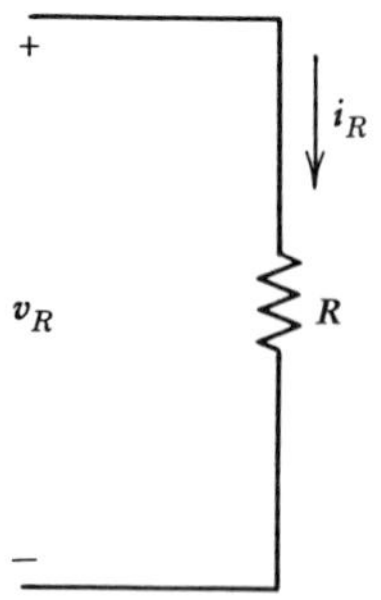

Fig. 1.11 Resistor (R ohms).

$$v_L = \frac{d\phi_L}{dt} \tag{1.5}$$

In Fig. 1.12*a*, we show a coil with the flux ϕ_L linking the coil produced by the current i_L flowing in the coil.

The flux linkage and the current may have a nonlinear relationship $\phi_L = f_L(i_L)$ as in Fig. 1.12*b*. But under linear approximation and with no time variation in the relationship, we can express

$$\phi_L = Li_L \tag{1.6}$$

where L is a positive constant called the *inductance* or *self-inductance* of the

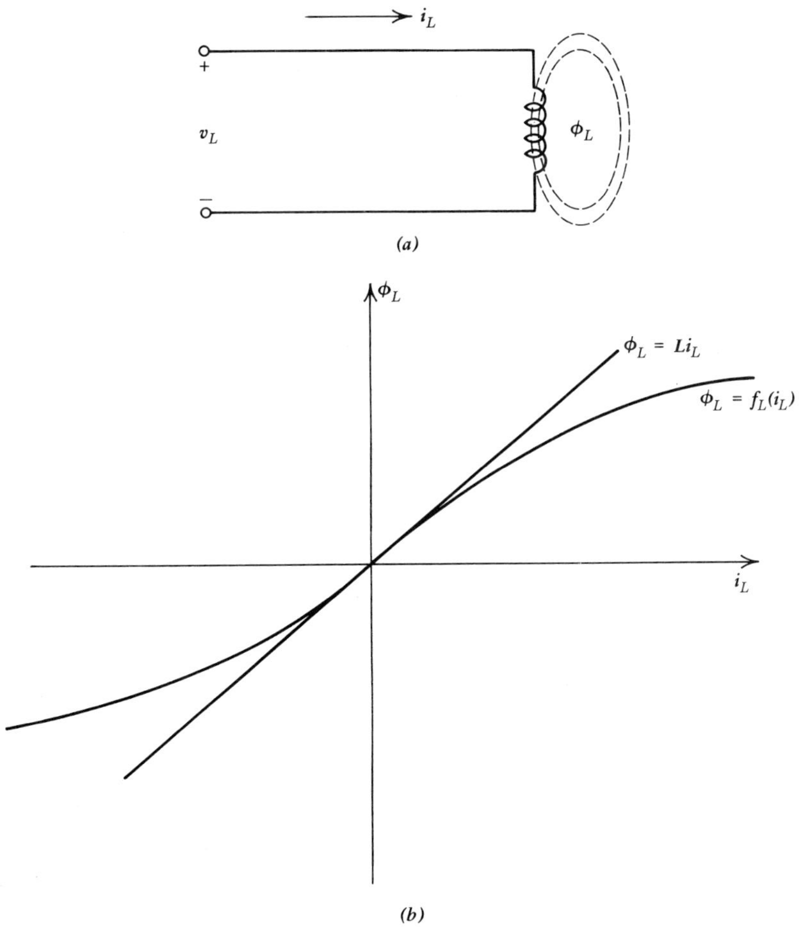

Fig. 1.12 (*a*) Flux linking the coil. (*b*) Flux-current relationship in inductors.

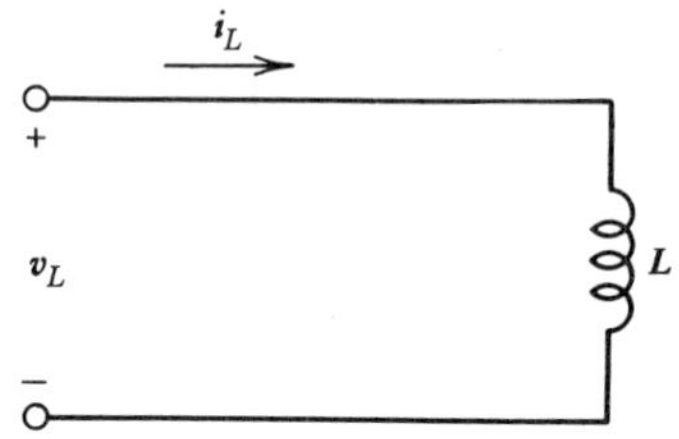

Fig. 1.13 Inductor (L henries).

coil measured in *henries*. Then

$$v_L = L\frac{di_L}{dt} \tag{1.7}$$

Integrating both sides of the last equation, we find

$$i_L(t) = \frac{1}{L}\int_{t_0}^{t} v_L(\tau)\, d\tau + i_{L0} \tag{1.8}$$

where i_{L0} is the initial current through the coil at time $t = t_0$. The representation of coils is given in Fig. 1.13.

Capacitors

For capacitors, we use *Coulomb's law*, which states that the current in any circuit element is the rate of change of charge flowing in the element. The charge q_C in a capacitor is a function of the voltage across it; that is, $q_C = f_C(v_C)$. For our purposes, we assume that the relationship is linear and time-invariant. Under this assumption $q_C = Cv_C$, where C is the *capacitance*, a positive constant, measured in *farads*. Then

$$i_C = \frac{dq_C}{dt} = C\frac{dv_C}{dt} \tag{1.9}$$

Integrating both sides of the above equation, we get

$$v_C(t) = \frac{1}{C}\int_{t_0}^{t} i_C(\tau)\, d\tau + v_{C0} \tag{1.10}$$

where v_{C0} is the initial voltage across the capacitor at time $t = t_0$.

The symbol for the capacitor and the polarities are shown in Fig. 1.14. From Eqs. 1.8 and 1.10, we observe that the initial current i_{L0} in an inductor and the initial voltage v_{C0} across a capacitor do not change in zero time as long as the signals are bounded. The integrals are zero if

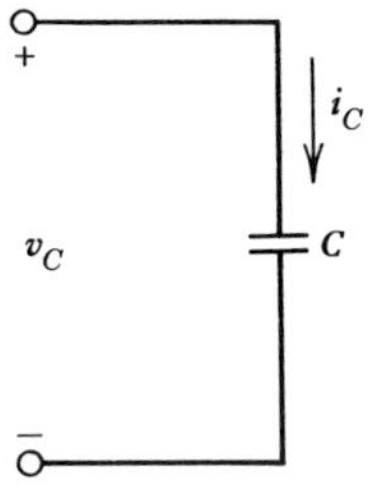

Fig. 1.14

both the limits (upper and lower) are the same. The integrals are nonzero only if $(t - t_0)$ is finite and nonzero. Thus the current in an inductor and the voltage across a capacitor cannot change instantly. If the initial current in an inductor and the initial voltage across a capacitor are given at an instant prior to excitation, say t_0, then the instant after the excitation, say t_{0+}, they will still have the same values. We shall discuss this property of the reactive elements later in Sec. 1.6.

The above property also leads to a useful intuitive understanding of circuit behavior. For example, if the initial voltage of a capacitor were zero, it will still be zero momentarily after excitation. Thus, *at the instant of excitation the uncharged capacitor behaves as if it is a short-circuit* (zero voltage, arbitrary current). Similarly, *a zero initial current inductor behaves like an open-circuit at the instant of excitation.*

1.4 TWO-PORT ELEMENTS

So far we have discussed only two-terminal elements with the voltage and current measured at these terminals as depicted in Fig. 1.15*a*. There are other circuit elements of importance that have four terminals with two of the terminals acting as the input terminals and the other two as the output pair. The relationships between the voltage and current signals

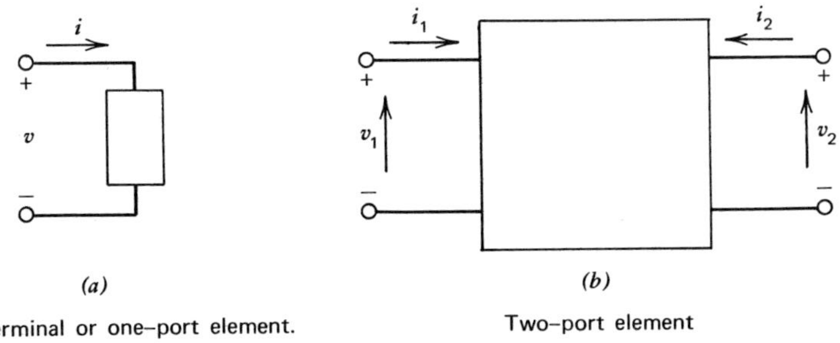

Fig. 1.15 (*a*) Two-terminal or one-port element. (*b*) Two-port element.

at both the input and the output terminal pairs define such a two-terminal pair or a two-port element as shown in Fig. 1.15*b*. In this section we introduce several of the important two-port elements. One important way in which two-port elements differ from circuits involving only one-port elements (i.e., two-terminal elements) is that energy transfer may take place without actual electrical connection between the input and output terminals.

A good example of a two-port element is a pair of coupled coils placed close to each other. They are not electrically connected together. The energy transfer takes place through the medium of the magnetic flux linking both the coils. The current flowing in any one of the coils produces a magnetic flux that links with the other coil. The rate of change of this flux linkage generates induced voltages in both the coils. This is also an example of Faraday's law of induction. The flux linkage between the two coils depends on the relative positions of the two coils. As we shall see, ideal transformers are idealizations of coupled coils.

Other two-port elements are obtained when we model active circuits employing devices such as transistors or vacuum tubes. For example, amplifiers used in many electronic applications (radio, TV, etc.) are modeled as two-port elements.

It should be observed that Fig. 1.15*b* shows power going into the two-port element at both ports. It is, of course, possible to view power going into the element at one port, say $v_1 i_1$, and going out of the element at the other end, say $-v_2 i_2$. The difference in the two quantities may be dissipated, stored, or even generated in the element depending on the sign of this power.

We will now discuss many of the important two-port elements.

Coupled Coils

We show in Fig. 1.16 a coil with the flux linkage ϕ_1 produced by the current i_1 flowing in the coil. We put a second coil near the first coil. At first, let us assume there is no current in this second coil. But the flux produced by i_1 in the first coil or some part of that flux links the second coil as well as the first. Let ϕ_2 be the total flux linkage of the second coil due to i_1. The voltage induced in the first coil is denoted v_1, and the voltage induced in the second is denoted v_2.

Under linear and time-invariant flux-current relationship, we have shown, in Eq. 1.6,

$$\phi_1 = L_1 i_1 \tag{1.11}$$

where L_1 is the self-inductance of the first coil. The self-induced voltage

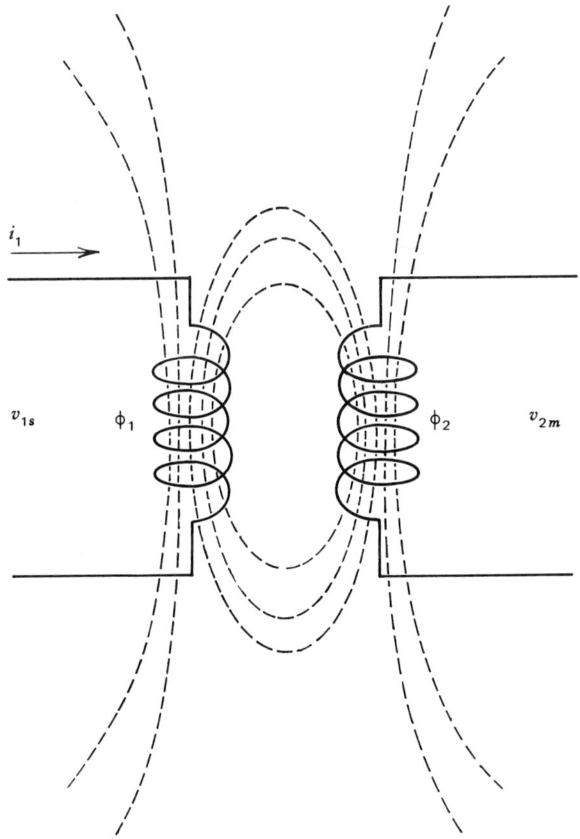

Fig. 1.16 Flux linkages due to i_1 only.

in the first coil due to the flux produced by the current in the same coil is, by Eq. 1.7,

$$v_{1s} = \frac{d\phi_1}{dt} = L_1 \frac{di_1}{dt} \tag{1.12}$$

The flux linkage ϕ_2 is also proportional to i_1 and expressed by

$$\phi_2 = M_{21} i_1 \tag{1.13}$$

where M_{21} is a constant and called the *mutual inductance* (again in henries). The subscript 21 represents that the voltage is induced in coil two due to current in coil one. Now we have the following relationship for the voltage induced in coil two by the mutual coupling with coil one:

$$v_{2m} = \frac{d\phi_2}{dt} = M_{21} \frac{di_1}{dt} \tag{1.14}$$

We could also have current i_2 in coil two producing self-induced voltage in coil two

$$v_{2s} = L_2 \frac{di_2}{dt} \tag{1.15}$$

where L_2 is the self-inductance of the second coil. There is also a voltage induced in coil one due to current in coil two. It is given by

$$v_{1m} = M_{12} \frac{di_2}{dt} \tag{1.16}$$

Note that

$$M_{12} = M_{21} = M \tag{1.17}$$

The representation of the coupled coils is shown in Fig. 1.17.

The terminal voltages v_1 and v_2 are the net induced voltages due to the two currents. The polarities of the mutually induced voltages v_{1m} and v_{2m} relative to the polarities of the self-induced voltages v_{1s} and v_{2s} will determine v_1 and v_2. The polarities of the mutually induced voltages depend upon the relative directions of the two coil windings. The terminal voltages are thus either sums or differences of the corresponding self and mutually induced voltages.

$$\begin{aligned} v_1 &= v_{1s} \pm v_{1m} = L_1 \frac{di_1}{dt} \pm M_{12} \frac{di_2}{dt} \\ v_2 &= v_{2s} \pm v_{2m} = \pm M_{21} \frac{di_1}{dt} + L_2 \frac{di_2}{dt} \end{aligned} \tag{1.18}$$

The representation of coupled coils in Fig. 1.17 is thus not complete and requires some polarity markings to show the polarities of the mutually induced voltages. (The self-induced voltage polarities follow the conventions of Fig. 1.15*a*.) The necessary polarity markings are shown by dots in Fig. 1.18. The dots signify that if a current flows into a coil at the dot end, then the dot end of the other coil has positive polarity (for the mutually induced voltage) with respect to the other end of the coil. Conversely, if the current flows out of a coil at the dot end, then the dot end of the other coil has negative polarity for the induced voltage due to this current.

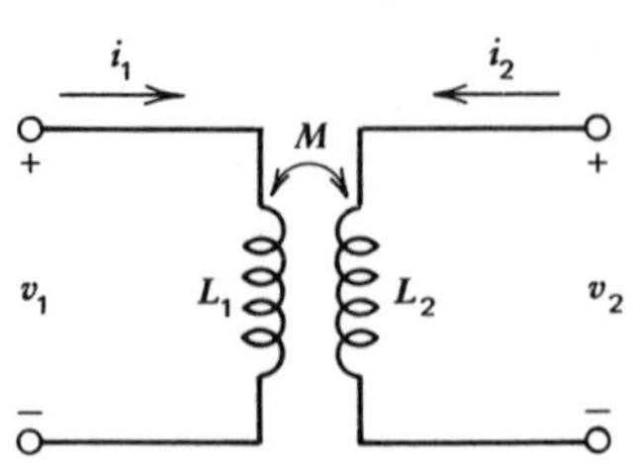

Fig. 1.17 Coupled coils (self and mutual inductances).

Consider the two possible cases in Fig. 1.18. In (*a*), the current i_1 flows into the coil at the dot end; so the induced voltage v_{2m} has positive polarity at the dot end of coil two. Similarly the current i_2 flows into the second coil at the dot end; so the induced voltage v_{1m} has positive polarity at the dot end of the first coil. Thus for this case

$$
\begin{aligned}
v_1 &= v_{1s} + v_{1m} = L_1 \frac{di_1}{dt} + M \frac{di_2}{dt} \\
v_2 &= v_{2s} + v_{2m} = M \frac{di_1}{dt} + L_2 \frac{di_1}{dt}
\end{aligned}
\tag{1.19}
$$

Next, let us consider the case of Fig. 1.18*b*. Here, i_1 flows into the first coil at the dot end; so the dot end of the second coil has positive polarity for v_{2m}. However, the current i_2 flows out of the second coil at the dot end and so the dot end of the first coil has negative polarity for v_{1m}. Thus for this case

$$
\begin{aligned}
v_1 &= v_{1s} - v_{1m} = L_1 \frac{di_1}{dt} - M \frac{di_2}{dt} \\
v_2 &= v_{2s} - v_{2m} = -M \frac{di_1}{dt} + L_2 \frac{di_2}{dt}
\end{aligned}
\tag{1.20}
$$

We will use expressions in Eq. 1.18 when the dot markings are unspecified as in a general discussion of the coupled coils. The expressions for currents in terms of the voltages can be obtained by integrating Eq. 1.18.

$$
\begin{aligned}
L_1 i_1(t) \pm M i_2(t) &= \int_{t_0}^{t} v_1(\tau)\, d\tau + L_1 i_{10} \pm M i_{20} \\
\pm M i_1(t) + L_2 i_2(t) &= \int_{t_0}^{t} v_2(\tau)\, d\tau \pm M i_{10} + L_2 i_{20}
\end{aligned}
\tag{1.21}
$$

where i_{10} and i_{20} are the initial currents at $t = t_0$ in the two coils. The two equations must be solved simultaneously to obtain expressions for $i_1(t)$

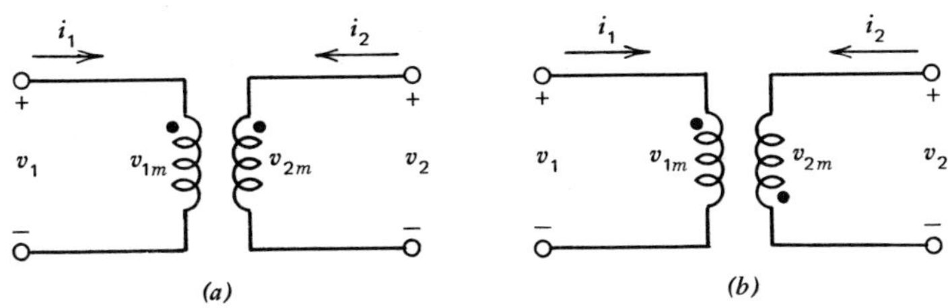

Fig. 1.18 Dot conventions.

and $i_2(t)$. For simplicity, let $i_{10} = i_{20} = 0$. Then

$$i_1(t) = \frac{L_2}{L_1L_2 - M^2}\int_{t_0}^{t} v_1(\tau)\,d\tau \mp \frac{M}{L_1L_2 - M^2}\int_{t_0}^{t} v_2(\tau)\,d\tau$$

$$i_2(t) = \mp\frac{M}{L_1L_2 - M^2}\int_{t_0}^{t} v_1(\tau)\,d\tau + \frac{L_1}{L_1L_2 - M^2}\int_{t_0}^{t} v_2(\tau)\,d\tau \tag{1.22}$$

provided $L_1L_2 - M^2 \neq 0$.

For physical elements $L_1L_2 - M^2 \geqslant 0$. This can be seen by letting $v_2 = 0$, and considering $L_1(di_1/dt)$ using Eqs. 1.22:

$$L_1\frac{di_1}{dt} = \frac{L_1L_2}{L_1L_2 - M^2}v_1 \tag{1.23}$$

If $L_1L_2 - M^2 < 0$, then the induced voltage $L_1(di_1/dt)$ has negative value relative to v_1. This cannot happen for physical situations and so $L_1L_2 - M^2 \geqslant 0$. The limiting case occurs when

$$L_1L_2 = M^2 \tag{1.24}$$

and the Eqs. 1.22 are not valid in this case. The limiting case is called ideal coupling since it represents the maximum coupling value for M. Physically, it implies that all the flux produced links both the coils with no loss of flux linkage.

Ideally Coupled Coils

The ideally coupled coils represent the maximum possible magnetic coupling between two coils. A measure of coupling can be represented as a fraction of the limiting case. Let

$$M^2 = k^2L_1L_2 \tag{1.25}$$

or

$$k = \frac{M}{\sqrt{L_1L_2}} \leqslant 1 \tag{1.26}$$

where k is called the coefficient of coupling. The ideally coupled coils correspond to a value of $k = 1$ and they are therefore known as unity coupled coils.

We have seen that Eqs. 1.22 are not valid for unity coupled coils. In such cases, the two equations in Eq. 1.18 or in Eq. 1.21 show that v_1 and

v_2 are proportional:

$$\frac{v_1}{v_2} = \frac{\pm M}{L_2} = \frac{L_1}{\pm M} \tag{1.27}$$

The two equations in Eq. 1.21 are thus one and the same except for a constant multiplier and so it is not possible to solve for i_1 and i_2. An additional equation constraining i_1 and i_2 is needed in order to solve for the currents. This is usually supplied by the circuit connections.

Ideal Transformers

Further idealizations of unity coupled coils occur when we let the inductances L_1 and L_2 approach infinity. This implies that negligible current is needed to provide the necessary magnetic flux. Such coupled coils are idealizations of physical transformers. Transformers are coupled coils designed to minimize loss of flux linkages and the current necessary to provide the required flux, and are widely used in many electronic circuits as well as in power systems.

Let us examine Eq. 1.21 with $i_{10} = i_{20} = 0$. Divide the first equation by L_1 and the second by L_2. Then

$$\begin{aligned} i_1(t) \pm \frac{M}{L_1} i_2(t) &= \frac{1}{L_1} \int_{t_0}^{t} v_1(\tau)\,d\tau \\ \pm \frac{M}{L_2} i_1(t) + i_2(t) &= \frac{1}{L_2} \int_{t_0}^{t} v_2(\tau)\,d\tau \end{aligned} \tag{1.28}$$

Now, we let $L_1 \to \infty$ and $L_2 \to \infty$ to obtain

$$\begin{aligned} i_1 \pm \frac{M}{L_1} i_2 &= 0 \\ \pm \frac{M}{L_2} i_1 + i_2 &= 0 \end{aligned} \tag{1.29}$$

or,

$$\frac{i_1}{i_2} = \frac{\mp M}{L_1} = \frac{L_2}{\mp M} \tag{1.30}$$

We thus have a constraint on the current ratio in addition to Eq. 1.27 constraining the voltage ratio for unity coupled coils. These two relations of Eqs. 1.27 and 1.30 define an ideal transformer. These relations can now be related to the number of turns of the two coils.

The flux produced by a current in a coil is proportional to the number of turns in the coil, and the flux linkage of a coil is the flux times the number of turns of the coil. It follows that the self-inductance of a coil is proportional to the square of the number of turns of the coil and the mutual inductance is proportional to the product of the number of turns of the two coils involved. So,

$$\begin{aligned} L_1 &= KN_1^2 \\ L_2 &= KN_2^2 \\ M &= KN_1N_2 \end{aligned} \tag{1.31}$$

where N_1 and N_2 are the number of turns of the two coils and K is the proportionality constant.

In the case of ideal transformers, the coupling is assumed ideal and the constant of proportionality is the same for both the self and mutual inductances. From Eqs. 1.27, 1.30, and 1.31, we have

$$\begin{aligned} \frac{v_1}{v_2} &= \pm\frac{N_1}{N_2} = \pm n \\ \frac{i_1}{i_2} &= \mp\frac{N_2}{N_1} = \mp n \end{aligned} \tag{1.32}$$

where n is the ratio of the turns in the first coil (primary) to the turns in the second coil (secondary). The representation for an ideal transformer is shown in Fig. 1.19.

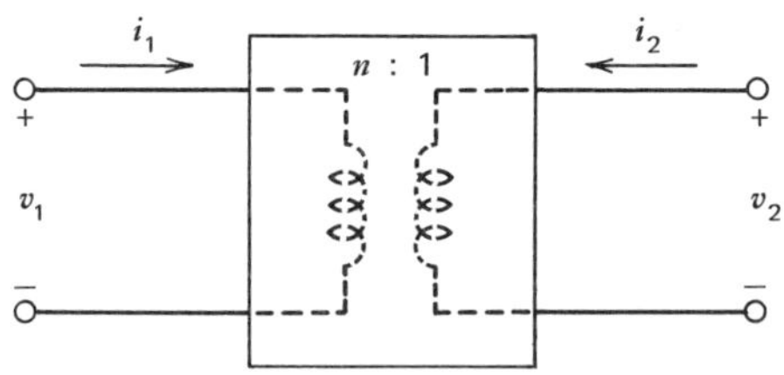

Fig. 1.19 An ideal transformer.

Controlled Sources

We have introduced coupled coils and ideal transformers as two-port elements. Next, we introduce another type of two-port element, the *controlled source*. The controlled source may be a voltage source or a current source controlled by an input current or voltage.

An idealized amplifier is an example of a controlled source. The amplifier is used to obtain a strengthened replica of a weak input signal. That is, the output signal is a constant times the input signal in an ideal

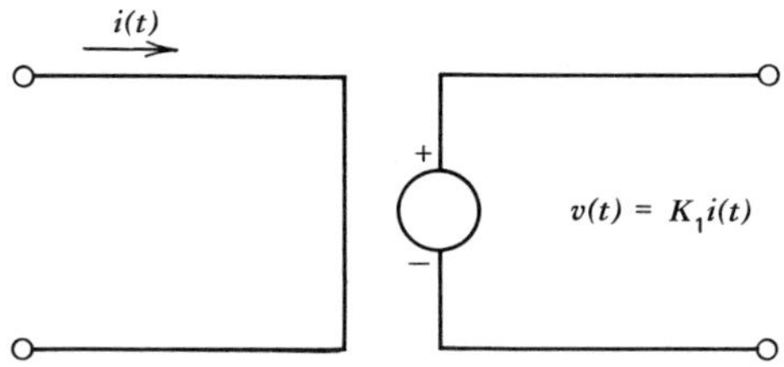

Fig. 1.20 A current controlled voltage source.

amplifier. For analysis purposes, amplifiers may thus be idealized to controlled sources.

We shall now define four ideal controlled sources. In a *current controlled voltage source*, the current, $i(t)$, controls an ideal voltage source and the value of the voltage as a function of time is some constant K_1 times that of the controlling current; this is shown in Fig. 1.20. We denote a current controlled voltage source by the letters CVS.

If the voltage source is controlled by a voltage, the controlled source is represented as shown in Fig. 1.21. The output voltage, $v_2(t)$, is a constant K_2 times the input voltage, $v_1(t)$. Such a two-port element is called a *voltage controlled voltage source* and is denoted by VVS.

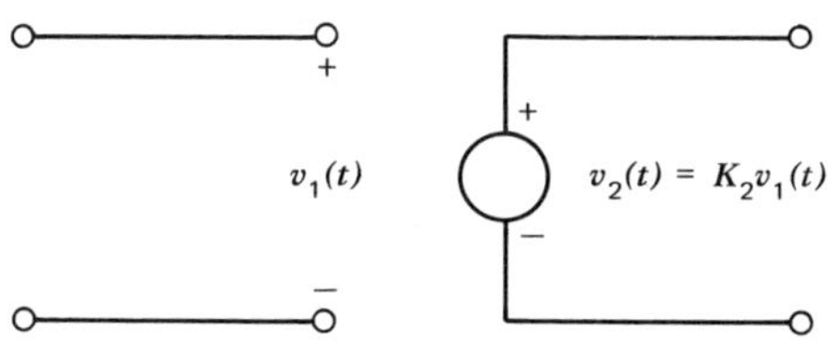

Fig. 1.21 A voltage controlled voltage source.

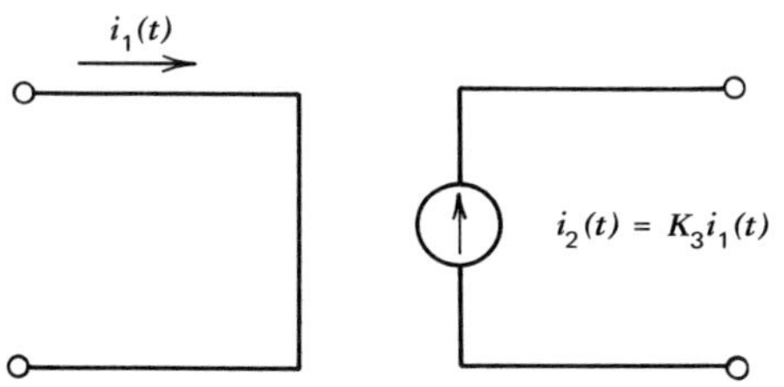

Fig. 1.22 A current controlled current source.

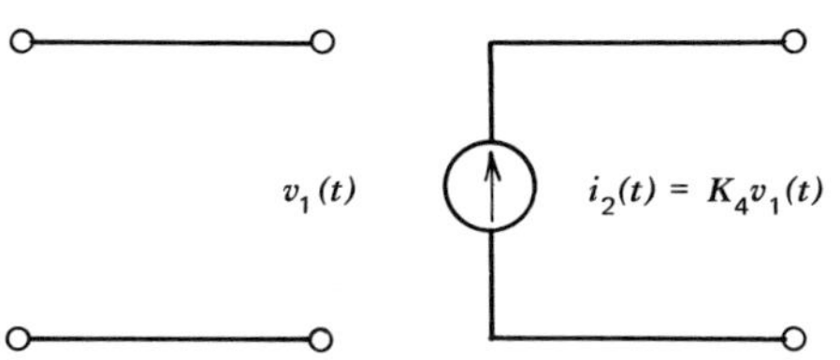

Fig. 1.23 A voltage controlled current source.

Another controlled source is the *current controlled current source* (CCS), in which the output current, $i_2(t)$, is a constant, K_3, times the input current, $i_1(t)$. This is shown in Fig. 1.22.

The *voltage controlled current source* (VCS) is a current source controlled by a voltage. This is shown in Fig. 1.23. The output current, $i_2(t)$, is K_4 times the input voltage $v_1(t)$.

These controlled sources are used as idealizations of many physical devices or circuits. For example, a current controlled voltage source can represent a vacuum tube pentode, and a current controlled current source can represent a transistor, and so on.

1.5 KIRCHHOFF'S LAWS

So far we have discussed the branch constraints for circuit elements. We now proceed to a discussion of the equilibrium equations for circuits or interconnections of the elements. There are two laws known as Kirchhoff's laws that define either the equilibrium relationships among certain voltages or the relationships among certain currents in the circuit.

Kirchhoff's voltage law (KVL) states that *the sum of the voltage drops around each and every closed loop in the circuit must be zero*. These *drops* must be considered across *all* elements of a loop in the *same* direction around the loop.

Let there be N elements, numbered $1, 2, \ldots, N$, in a given loop of a circuit. Let the voltage drops across each of these elements in the same direction around the loop be $v_1(t), v_2(t), \ldots, v_N(t)$. Then KVL states that

$$v_1(t) + v_2(t) + v_3(t) + \cdots + v_N(t) \equiv 0 \tag{1.33}$$

for all time t. This is symbolically written as

$$\sum_{n=1}^{N} v_n^{(d)} = 0 \tag{1.34}$$

where the superscript d implies (voltage) drops and the subscript n refers to the corresponding numbered element in the given loop. Such relationships must hold true for each and every loop of the circuit.

Since voltage rises are the negatives of voltage drops, the KVL also states that the voltage rises across all elements in a loop in a consistent direction sum to zero. If in a given direction around the loop some elements are considered to have voltage drops and others voltage rises, the KVL states that the sum of the drops must equal the sum of the rises.

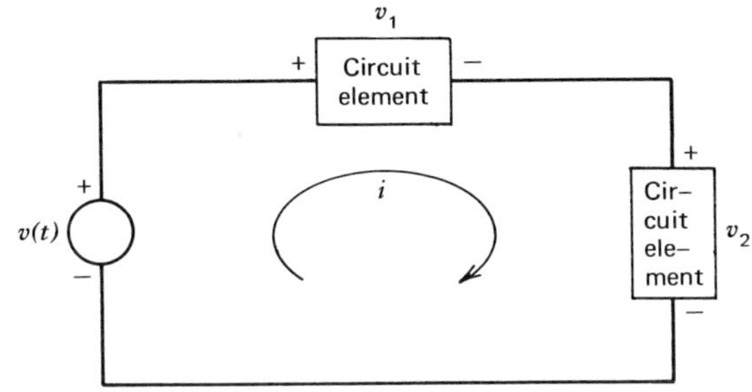

Fig. 1.24

Let us consider a few examples to illustrate the use of KVL. Consider the circuit in Fig. 1.24 consisting of only one loop. Let us sum the voltage drops in a clockwise direction. The voltage drops $v_1(t)$ and $v_2(t)$ plus the voltage drop across the source must all sum to zero. In the clockwise direction, the source voltage $v(t)$ represents a voltage rise; so, the voltage drop across the source is $-v(t)$. Then

$$\text{KVL:} \qquad v_1(t) + v_2(t) - v(t) = 0 \tag{1.35}$$

Consider another circuit as shown in Fig. 1.25. In this circuit, there are three possible loops as shown by assumed currents, i_1, i_2, and i_3, in these loops. Applying KVL to each loop in the direction of the current flow gives the following:

$$\text{KVL-1} \qquad v_1(t) + v_2(t) - v(t) = 0 \tag{1.36}$$

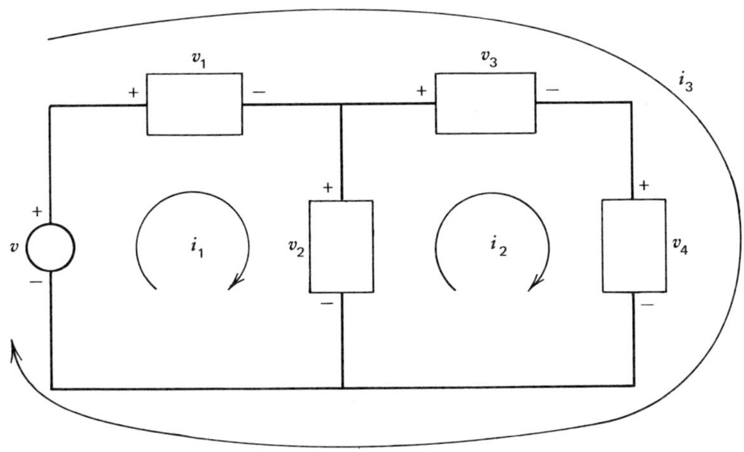

Fig. 1.25

$$\text{KVL-2} \qquad v_3(t) + v_4(t) - v_2(t) = 0 \tag{1.37}$$

$$\text{KVL-3} \qquad v_1(t) + v_3(t) + v_4(t) - v(t) = 0 \tag{1.38}$$

All the above equations represent valid relationships. It is easy to see, however, that they are *not* all *independent* equations. That is, any one of the equations can be obtained by a linear combination of the other two. *Linear combination* of any two or more equations is obtained by multiplying each equation by some constant and then adding these scaled equations together.

A set of equations is *linearly independent* if no equation can be obtained by a linear combination of the others. Otherwise the set is linearly *dependent*.

The Eqs. 1.36–1.38, above, are linearly dependent. For example, the third equation is the sum of the first two. The first equation is obtained by subtracting the second from the third, and so on. It follows that if any two of the equations are satisfied, the third will be also. Therefore, we do not need all three equations to define the circuit behavior. Any two equations will suffice. The two, say Eqs. 1.36 and 1.37, will also be (linearly) independent equations; that is, neither of them can be obtained by a linear combination of the other (i.e., by a multiplicative constant) because each involves voltage drops across elements not present in the other. Both the equations are thus necessary as well as sufficient to completely define the circuit behavior.

From the above discussion we see that the KVL need be applied only to those loops that yield (linearly) independent equations and that such equations result if *each of these loops contains at least one element not contained in the other loops*. Furthermore, the equations must be *mathematically coupled* together; that is, at least one of the voltage drops in a loop equation must be identical to or somehow related to a voltage drop in one of the other loop equations. The circuit is a single unit connected together electrically or magnetically, and the circuit equations must reflect this *interrelationship*.

It follows that once all the elements in the circuit have been included in these independent but interrelated circuit equations, no further loops need be considered.

Let us illustrate our procedure by some examples.

Example 1.5.1 For the circuit of Fig. 1.26 the KVL equations for loops 1 and 2 are:

$$\text{KVL-1} \qquad v_{R1} + v_{R2} - v = 0 \tag{1.39}$$

$$\text{KVL-2} \qquad -v_{R3} - v_L + v_C + v_{R4} = 0 \tag{1.40}$$

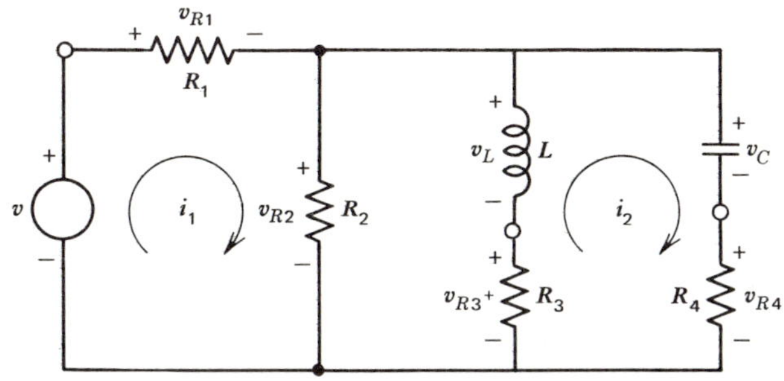

Fig. 1.26

These equations are linearly independent since each loop involves elements not contained in the other. The two loops also include all the circuit elements. The set of equations is not complete, however, since the two equations are not interrelated through a common voltage drop. The loop equation

$$\text{KVL-3} \qquad -v_{R2}+v_L+v_{R3}=0 \tag{1.41}$$

for the loop containing R_2; L and R_3 can complete the interrelated set. The three equations are linearly independent since KVL-2 contains element(s) not contained in KVL-1, and KVL-3 contains elements not contained in either KVL-1 or KVL-2. It is also coupled to KVL-1 through the term v_{R2} and to KVL-2 through v_L and v_{R3}. All the elements are also included in the equations and, therefore, the circuit equations are complete.

Example 1.5.2 For the circuit of Fig. 1.27

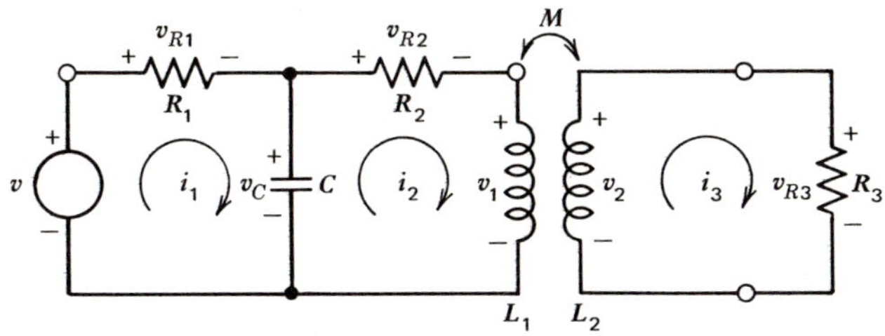

Fig. 1.27

let us apply our procedure to obtain for loop 1

$$\text{KVL-1} \qquad v_{R1}+v_C-v=0 \tag{1.42}$$

Now loop 2 contains R_2 not contained in loop 1, so the equation for loop

2 will be an independent one,

$$\text{KVL-2} \qquad v_{R2} + v_1 - v_C = 0 \tag{1.43}$$

Once again loop 3 contains R_3 not contained in the previous loops, so we have another independent equation,

$$\text{KVL-3} \qquad v_{R3} - v_2 = 0 \tag{1.44}$$

Let us examine the equations to see if they are mathematically interrelated. KVL-1 and KVL-2 are coupled together by the term v_C. KVL-2 and KVL-3 are coupled together by the terms v_1 and v_2, which are related to each other because the two coils are magnetically coupled. The set of circuit equations is thus interrelated.

All the circuit elements are included in the above linearly independent but interrelated equilibrium equations, and so they completely define the circuit behavior.

The total number of linearly independent equations depends on the number of branches, b, the number of nodes, n, and the number of electrically separate parts, k, of the circuit. *Branches* are used synonymously as two terminal circuit elements. *Nodes* are the junctions between two or more branches. Each one of the coupled coils is a separate branch, and thus is a controlled source. Electrically *separate parts* are those in which the elements are electrically connected together. Coupled coils or controlled sources give rise to more than one electrical part in a circuit. While the procedure described above always yields all the independent equations, it can be shown that the total number of such independent loop equations is

$$l = b - n + k \tag{1.45}$$

Example 1.5.3 For Example 1.5.1

$$b = 7$$
$$n = 5$$
$$k = 1$$

The number of independent loop equations needed is

$$l = 7 - 5 + 1$$
$$= 3$$

For Example 1.5.2

$$b = 7$$
$$n = 6$$
$$k = 2$$

and the number of independent loops is

$$\begin{aligned} l &= 7-6+2 \\ &= 3 \end{aligned}$$

This verifies our statements before.

Systematic procedures for writing the independent loop equations for very complex circuits can be developed. They can be quite unwieldy, however, and not really necessary. The KCL procedures described next are by far the simplest ones and just as generally applicable.

Kirchhoff's current law (KCL) states that the sum of all the currents flowing into each and every junction or node must be zero. *All* currents at each node must be considered as flowing *into* the node with appropriate algebraic signs.

The negative of a current flowing *into* a node may be viewed as flowing *out* of the node; the KCL also states that the sum of all the currents viewed as flowing out of a node must sum to zero. Alternatively, if, at a node, some currents are viewed as flowing into the node and others as flowing out of the node, the KCL states that the sum of the currents flowing into a node equals the sum of the currents flowing out of that node.

Let there be N elements connected together at a node and let the corresponding currents through each element flowing into the node be $i_1(t), i_2(t), \ldots, i_N(t)$. Then KCL states that

$$i_1(t) + i_2(t) + \cdots + i_N(t) \equiv 0 \tag{1.46}$$

for all time t. This may also be written as

$$\sum_{n=1}^{N} i_n^{(i)}(t) = 0 \tag{1.47}$$

where the superscript implies current flowing into the node. Such relationships must hold true for every node in the circuit.

Let us consider some examples to illustrate the use of KCL. There are only two nodes in the circuit shown in Fig. 1.28. The KCL must be valid at each node. At node 1, the current $i(t)$ and the currents $-i_1(t)$ and $-i_2(t)$ must all sum to zero.

$$\text{KCL-1} \qquad i(t) - i_1(t) - i_2(t) = 0 \tag{1.48}$$

At node 2, the current $-i(t)$ and the current $i_1(t)$ and $i_2(t)$ must sum to zero.

$$\text{KCL-2} \qquad -i(t) + i_1(t) + i_2(t) = 0 \tag{1.49}$$

The above two equations are not linearly independent. One is the

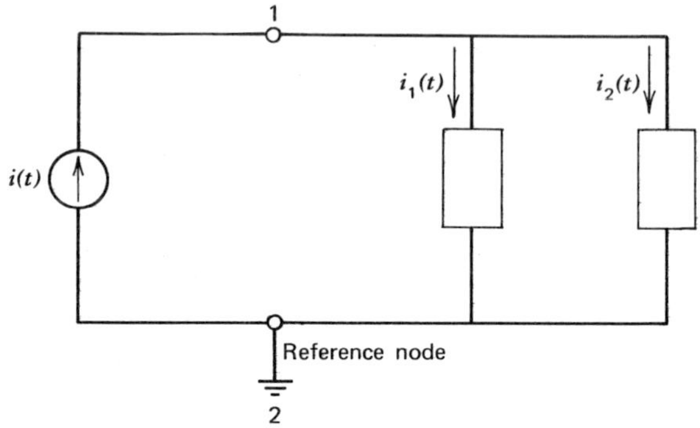

Fig. 1.28

negative of the other. So, either equation will suffice to describe the circuit behavior. The choice of the node, for which the KCL equation is used, is arbitrary. However, as we shall see in later chapters, the currents are expressed in terms of the branch voltages (through the branch constraints); and these voltages must be measured with respect to some reference node in the circuit. The procedure, thus, is to make one of the circuit nodes, say node 2, as the reference node and write the KCL for the other node, that is, node 1.

For the circuit of Fig. 1.29, let node 3 be the reference node. The KCL equations for nodes 1 and 2 are:

$$\text{KCL-1} \qquad i(t) - i_1(t) - i_2(t) = 0 \tag{1.50}$$

$$\text{KCL-2} \qquad i_2(t) - i_3(t) = 0 \tag{1.51}$$

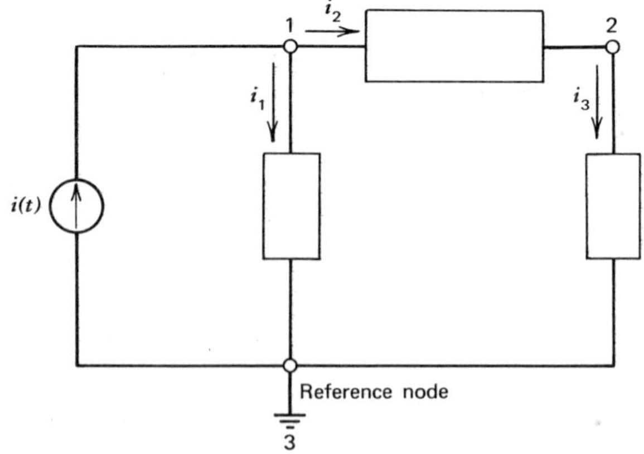

Fig. 1.29

The two equations are both independent and interrelated. The KCL for node 3 is

$$i_1(t)+i_3(t)-i(t)=0 \tag{1.52}$$

which is the negative sum of KCL-1 and KCL-2 and, therefore, not necessary. The first two equations completely define the circuit behavior.

As shown below, for an electrically connected circuit ($k=1$) with n nodes, we specify one node as the reference node. We write the KCL equations for the remaining $(n-1)$ nodes to yield the desired independent and interrelated complete set of circuit equations. If the circuit has k separate parts, each part must be assigned a reference node and the KCL equations written for the remaining $(n–k)$ nodes. Each reference node is the reference point for voltages in the corresponding separate part of the circuit. The KCL equation for each reference node is always a linear combination of the equations for the other remaining nodes in that part of the circuit.

The justification for the above statement is simply demonstrated. We have seen the truth of it for $n=2$ and $n=3$ for an electrically connected circuit ($k=1$). We will now show that if a node is added to any circuit (with $k=1$), one more KCL equation must be included in the circuit description.

In Fig. 1.30, a new node **p** is added to the original circuit. At least two elements are required for **p** to be connected to the circuit. Let A be connected to some node **a**. Then B is either connected to the reference node or to some other circuit node **b**. If B is connected to the reference, an independent KCL for **p** exists, since it includes the current through B that is not included in other KCL equations. Such a KCL equation is

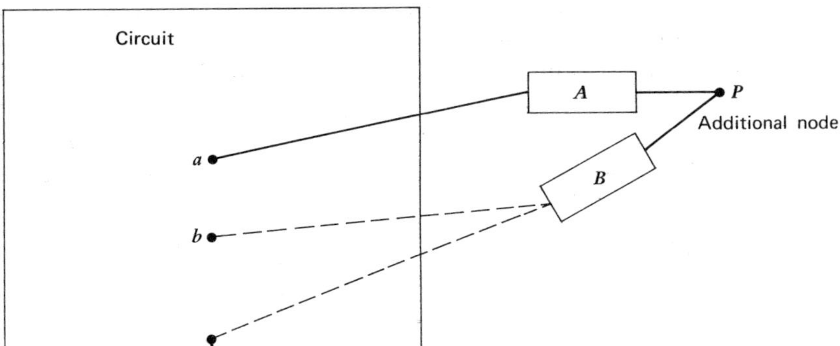

Fig. 1.30

required in order to include all circuit elements since only a KCL node equation at node p can include the current through B.

If B is connected to the node **b**, let us examine the relevant set of KCL equations. The KCL at node **a** includes the current through A. The KCL at node **p** includes the current through A as well as that through B and is therefore independent of the KCL at node **a**. The KCL at node **b** includes the current through B as well as some other current not included in the KCL at node **a**. It is thus independent of the KCL at node **p**. The KCL equations at nodes **a**, **b**, and **p** are thus independent. They are also interrelated through the currents in A and B and they exhaust all elements. The set of equations is now complete with the addition of the KCL at node **p**.

Thus, the addition of a node implies one more KCL equation in the set of equations needed to completely describe the circuit. Each node with the exception of the reference node gives rise to an independent circuit equation. Similar arguments can be developed for each separate part of the circuit. The above general statement for writing the complete set of $(n-k)$ KCL equations is thus justified. The procedure outlined is straightforward and can be applied to any circuit with ease. We conclude this discussion with an illustrative example.

There are two separate parts in the circuit of Fig. 1.31. Let nodes 2 and 5 be the reference nodes. We then write the KCL's for the remaining (3) nodes #1, 3, 4.

$$\text{KCL-1} \qquad i - i_{C1} - i_{R1} = 0 \tag{1.53}$$

$$\text{KCL-3} \qquad Ki_{C1} - i_{R2} - i_{C2} = 0 \tag{1.54}$$

$$\text{KCL-4} \qquad i_{R2} - i_L = 0 \tag{1.55}$$

It is easy to see that the equations are independent, interrelated, and they completely define the circuit.

It should be quite clear that writing the KCL equations for any circuit

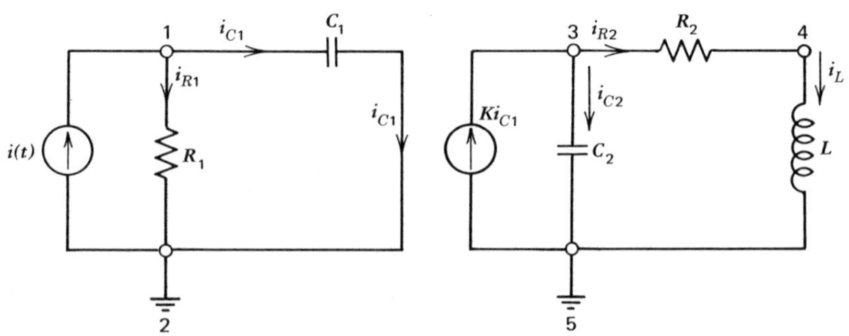

Fig. 1.31

is simple and straightforward and is to be preferred over writing the KVL equations in general. Of course, when the loops are obvious (e.g., as in the case of a single loop), it is sometimes more convenient to write the KVL description. For any complex circuit, on the other hand, the KCL description is the best!

The circuit equilibrium equations are expressed in terms of the unknown loop current or the node voltage variables using the branch constraints. The resulting equations must then be solved for the unknown variables. As mentioned earlier, it is not possible to solve these equations without knowing the (initial) state of the circuit. We shall proceed to a discussion of the state of the circuit, but first we present some simple equivalence relationships for series and parallel combinations of resistors, inductors, and capacitors.

Example 1.5.5 Consider two resistors in series as shown in Fig. 1.32*a*. Such a combination is *equivalent* to a single resistor. Equivalent means that if the series combination and the single equivalent resistor were each placed in a box with only the terminals accessible, it would not be possible to distinguish the two by electrical measurements at the terminals.

In Fig. 1.32*a* we have two resistors in series with the same current i through them. The voltage across R_1 is v_1 and that across R_2 is v_2. The total voltage across the series combination is $v = (v_1 + v_2)$. By Ohm's law, we have

$$v_1 = R_1 i$$
$$v_2 = R_2 i$$

It follows that

$$\begin{aligned} v = v_1 + v_2 &= R_1 i + R_2 i \\ &= (R_1 + R_2) i \\ &= Ri \end{aligned}$$

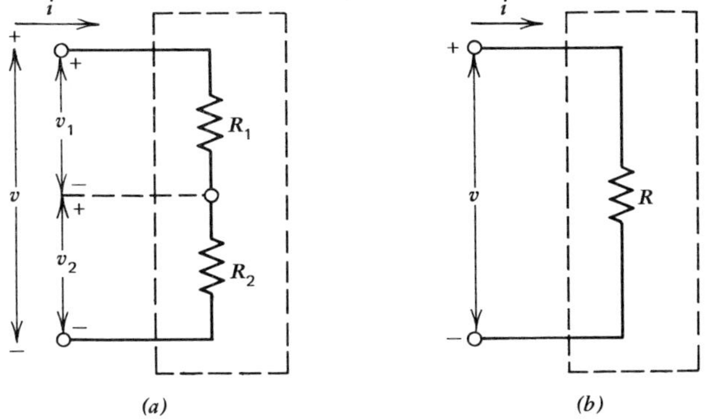

Fig. 1.32

We thus have a voltage-current relationship at the external terminals. An equivalent resistor $R = (R_1 + R_2)$ of Fig. 1.32*b* would exhibit the same terminal relationship. Thus the series combination of two resistors is equivalent to a resistor whose resistance is given by the sum of the two resistances:

$$R = R_1 + R_2$$

Next, let us examine a parallel combination of two resistors as shown in Fig. 1.33*a*. Ohm's law gives us

$$v = R_1 i_1$$
$$v = R_2 i_2$$

By KCL the total current entering the terminals is given by the sum of i_1 and i_2:

$$\begin{aligned} i &= i_1 + i_2 \\ &= \frac{v}{R_1} + \frac{v}{R_2} \\ i &= \left(\frac{1}{R_1} + \frac{1}{R_2}\right) v \\ &= \frac{1}{R} v \end{aligned}$$

where

$$\frac{1}{R} = \frac{1}{R_1} + \frac{1}{R_2}$$

or

$$G = G_1 + G_2$$

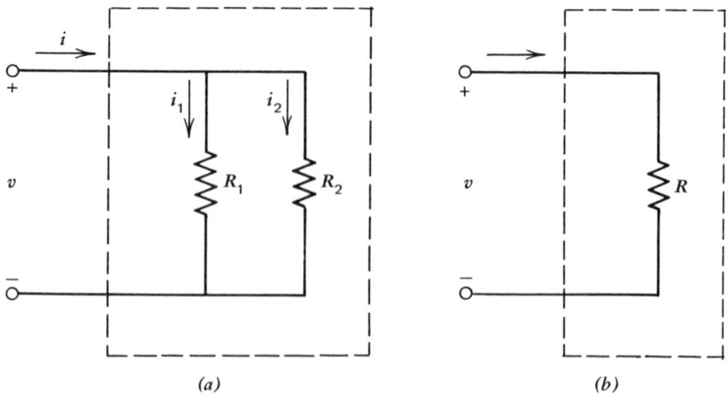

Fig. 1.33

Thus a parallel combination of two resistors is equivalent to a single resistor whose conductance is the sum of the two conductances. One may also express the equivalent resistance as

$$\frac{1}{R} = \frac{R_1 + R_2}{R_1 R_2}$$

or

$$R = \frac{R_1 R_2}{R_1 + R_2}$$

Similarly, series and parallel combinations of inductors and capacitors can be represented by equivalent elements. For inductors L_1 and L_2 in *series*, the equivalent inductance is

$$L = L_1 + L_2$$

The equivalent inductance of L_1 and L_2 in *parallel* is given by

$$\frac{1}{L} = \frac{1}{L_1} + \frac{1}{L_2}$$

or

$$L = \frac{L_1 L_2}{L_1 + L_2}$$

Two capacitors in *series* have an equivalent capacitance given by

$$\frac{1}{C} = \frac{1}{C_1} + \frac{1}{C_2}$$

or

$$C = \frac{C_1 C_2}{C_1 + C_2}$$

The equivalent capacitance of a *parallel* combination is given by

$$C = C_1 + C_2$$

Note that it is the *parallel* combination of capacitors whose equivalent capacitance is a sum of the capacitances. The series combination equivalent has a capacitance whose reciprocal is a sum of the reciprocal capacitances.

1.6 STATE OF THE CIRCUIT

The state of the circuit at any instant of time is defined in terms of the energy storage variables for the individual reactive elements.

The energy stored in a capacitor is obtained by using Eqs. 1.2 and 1.9,

$$\begin{aligned} w_C(t) &= \int_{-\infty}^{t} v_C(\tau) i_C(\tau)\, d\tau \\ &= \int_{-\infty}^{t} C v_C \frac{dv_C}{d\tau}\, d\tau \\ &= \tfrac{1}{2} C v_C^2(t) \end{aligned} \tag{1.56}$$

where it is assumed that at $t = -\infty$, the capacitor is uncharged.

The *state* of the capacitor is then defined by the *value of the voltage variable* v_C at the instant of interest.

Similarly, the energy stored in an inductor at any instant is given by Eqs. 1.2 and 1.7,

$$\begin{aligned} w_L(t) &= \int_{-\infty}^{t} i_L(\tau) v_L(\tau)\, d\tau \\ &= \int_{-\infty}^{t} L i_L \frac{di_L}{d\tau}\, d\tau \\ &= \tfrac{1}{2} L i_L^2(t) \end{aligned} \tag{1.57}$$

and *the state of the inductor is given by* $i_L(t)$.

When mutual inductance is present, the energy stored in one of the coupled coils is given by Eqs. 1.2 and 1.18

$$\begin{aligned} w_{\text{c.c.}}(t) &= \int_{-\infty}^{t} i_1 \left(L_1 \frac{di_1}{d\tau} \pm M \frac{di_2}{d\tau} \right) d\tau \\ &= \tfrac{1}{2} L_1 i_1^2(t) \pm M i_1(t) i_2(t) \end{aligned} \tag{1.58}$$

A similar expression is obtained for the second coil. The values of the two currents in the separate coils thus define the state in the case of coupled coils.

The inductors and capacitors are the only energy storage elements. Resistors do not store energy; they dissipate energy. The branch constraints for the elements show that the currents and the voltages in resistors can change instantly but the same is not true for the reactive elements.

The voltage across individual capacitors and the currents through the individual inductors completely specify the energy in a circuit. The state of the circuit is given by the values of these voltage and current variables. If there are p capacitors, numbered $C_1, C_2, \ldots, C_p$, and q inductors, L_1,

$L_2, \ldots, L_q$, in a circuit, *the state of the circuit* is given by the values of

$$\mathbf{X}(t) = \begin{Bmatrix} v_{C1}(t) \\ v_{C2}(t) \\ \cdot \\ \cdot \\ \cdot \\ v_{Cp}(t) \\ i_{L1}(t) \\ i_{L2}(t) \\ \cdot \\ \cdot \\ \cdot \\ i_{Lq}(t) \end{Bmatrix} \tag{1.59}$$

at the desired instant of time. The array in Eq. 1.59 is called the state vector, and this vector is sufficient to specify *all* the voltages and currents in the circuit for $t > t_0$, where t_0 is the initial instant of time.

Equilibrium equations for a circuit may be written, by using appropriate combinations of KVL and KCL, to yield the circuit description in terms of the state vector. Such equations are known as State-Variable Equations and will be discussed in Chapter 3.

The main purpose of our discussion here is to establish that the state in physical circuits cannot change in zero time. Energy stored in physical systems cannot be changed in zero time. Therefore, charge in capacitors and flux in inductors or the corresponding energy variables in the reactive elements cannot be changed instantaneously. Mathematically, the branch constraints for the elements show that instantaneous changes in the state variables require infinite values for the currents in capacitors and voltages across inductors. For all physical situations, then, the state of a circuit cannot change instantaneously.

Let us examine the implications of the above fact. Let the circuit be connected or put together at time $t = t_0$. We usually say that the circuit is connected together by means of a switch at time $t = t_0$. The instant just before the connection is denoted by t_{0-} and the instant just after is denoted by t_{0+}. Since the value of the state cannot change instantaneously, its value at t_{0+} must be the same as at t_{0-}. That is, the initial state

$$\mathbf{X}(t_0) = \mathbf{X}(t_{0+}) = \mathbf{X}(t_{0-}) \tag{1.60}$$

or,

$$v_{Ck}(t_{0-}) = v_{Ck}(t_{0+}) = v_{Ck}(t_0) \tag{1.61}$$

for all capacitors, and

$$i_{Lk}(t_{0-}) = i_{Lk}(t_{0+}) = i_{Lk}(t_0) \tag{1.62}$$

for all inductors.

We will repeatedly use, in analyzing our circuits, the fact that the state of the circuit immediately after the closing of the switch must be the same as just before the closing of the switch. The term *initial state* is commonly used to denote these initial values of the capacitor voltages and the inductor currents. The initial state values are used to obtain the initial conditions or values of the (current or voltage) variables in the circuit equations. The initial state values should not be confused with the initial conditions for mathematical equations of the circuit.

We will say the state vector has a value zero if all the state variables are zero. In a purely resistive circuit or network, the state has no meaning since there are no reactive elements in the circuit.

1.7 SOME BASIC DEFINITIONS

All the circuits that we shall discuss will consist of interconnections of circuit elements excited by independent sources. The responses will be certain voltages or currents in the circuits. Certain common properties govern these excitation-response pairs *when the initial states of the circuits are zero.* In turn, they describe certain properties of these *zero-state circuits* (networks or systems). For example, all our zero-state circuits are *linear, time-invariant,* and *causal* systems. In addition, if the controlled sources are not included, then the zero-state circuits or systems are also *passive* and *reciprocal.* In this section, we examine what these words actually mean in precise terms. A word of caution! These terms describe circuits of interest to us only when they are initially in zero-state.

Furthermore, the properties are also valid if *the input signals are zero* and the initial state-response pairs are considered as the excitation-response pairs. For example, all our circuits are *zero-input linear* and *time-invariant.*

Linear

A system (network) is linear if (a) the principle of *superposition* and (b) the principle of *proportionality* hold.

By the superposition principle, if, for a given network, $[e_1(t), r_1(t)]$ and $[e_2(t), r_2(t)]$ are excitation-response pairs, then if the excitation were $e(t) = e_1(t) + e_2(t)$, the response would be $r(t) = r_1(t) + r_2(t)$. By the proportionality principle, if the excitation were $C_1e_1(t)$, where C_1 is a

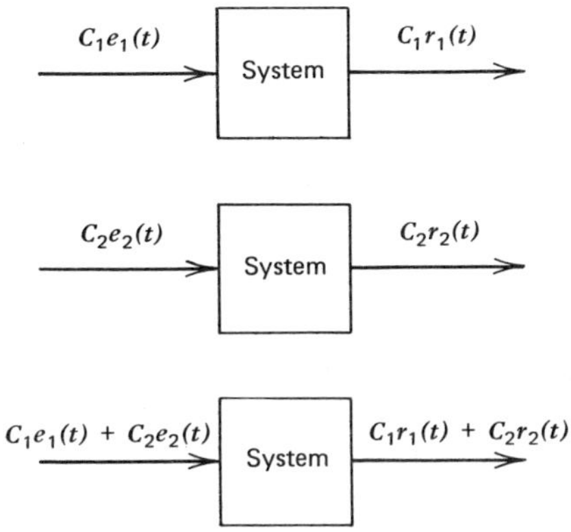

Fig. 1.34 Linear system.

constant, then the response would be $C_1r_1(t)$—that is, the constant of proportionality C_1 is preserved by the linear network. The two conditions of linearity are summarized in Fig. 1.34.

Another definition of a linear network is that the excitation and response of the network are related by a linear differential equation.

Passive

A linear network is *passive** if (a) the energy delivered to the network is non-negative for any arbitrary excitation and (b) if no voltages or currents appear between any two terminals before an excitation is applied.

Reciprocal

A network is said to be *reciprocal* if when the points of excitation and measurement of response are interchanged, the relationship between excitation and response remains the same. This must be true for any choice of points of excitation and response.

Causal

We say a system is *causal* if its response is nonanticipatory, that is, if

$$e(t) = 0 \qquad t < T$$

then

$$r(t) = 0 \qquad t < T \tag{1.63}$$

*G. Raisbeck, A Definition of Passive Linear Networks in Terms of Time and Energy," *J. Appl. Phys.*, **25** (Dec. 1954), 1510–1514.

In other words, a system is causal if before an excitation is applied at $t = T$, the response is zero for $-\infty < t < T$, assuming the system is zero-state, linear.

Time Invariant

A system is *time invariant* if $e(t) \rightarrow r(t)$ implies that $e(t \pm T) \rightarrow r(t \pm T)$, for all t and T, where the symbol $\rightarrow$ means "gives rise to." To understand the concept of time invariance in a linear system, let us suppose that initially the excitation is introduced at $t = 0$, which gives rise to a response $r(t)$. If the excitation were introduced at $t = T_1$ and if the shape of the response waveform were the same as in the first case, but delayed by a time T_1 (Fig. 1.35), then we could say the system is time invariant. It should be mentioned here that linear systems need not be time invariant.

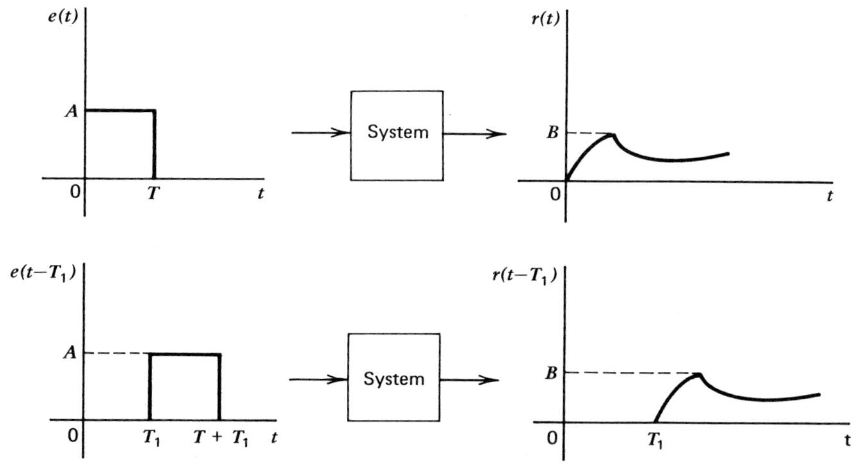

Fig. 1.35 Time-invariant system.

Derivative Property

From the time-invariant and linearity properties we can show that, if $e(t)$ at the input gives rise to $r(t)$ at the output (Fig. 1.36), then, if the input were $e'(t)$, that is, the derivative of $e(t)$, the response would be $r'(t)$. The proof is quite simple. Consider an excitation $e(t+\epsilon)$ where ϵ is a real quantity. By the time-invariant property, the response would be $r(t+\epsilon)$. Now suppose the excitation were

$$e_1(t) = \frac{1}{\epsilon}[e(t+\epsilon) - e(t)] \tag{1.64}$$

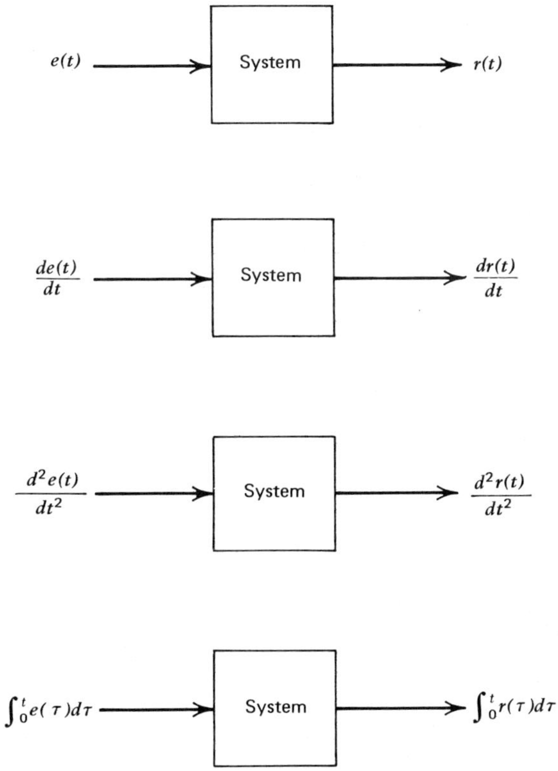

Fig. 1.36

then according to the linearity and time-invariant properties, the response would be

$$r_1(t) = \frac{1}{\epsilon}[r(t+\epsilon) - r(t)] \tag{1.65}$$

Taking the limit as $\epsilon \to 0$, we see that

$$\begin{aligned} \lim_{\epsilon \to 0} e_1(t) &= \frac{d}{dt} e(t) \\ \lim_{\epsilon \to 0} r_1(t) &= \frac{d}{dt} r(t) \end{aligned} \tag{1.66}$$

We can extend this idea to higher derivatives as well as to the integrals of $e(t)$ and $r(t)$, as shown in Fig. 1.36.

1.8 SIGNAL DESCRIPTIONS

In the previous section we considered certain basic properties of circuits. In this section let us examine how we characterize some of the signals that are commonly used as excitation sources in circuit analysis.

The signals represented as functions of time are also called the waveforms. The signals described mathematically are idealizations of the actual signals. The most important signals occurring as excitations, as responses, or as building blocks of excitations and responses, are the exponentials, sinusoids, unit steps, and unit impulses. We now describe these functions and some of their properties. The complex exponential, a very important but nonreal signal function, is described at the end of the section. The complex exponential is a very useful function and considerably simplifies circuit analysis.

The Exponential

The function

$$f(t) = Ke^{at} \tag{1.67}$$

is known as the exponential function. When a is a negative number, it is known as the exponentially decreasing or decaying function; and when a is a positive number, it is known as the exponentially increasing function. The exponential function is shown in Fig. 1.37.

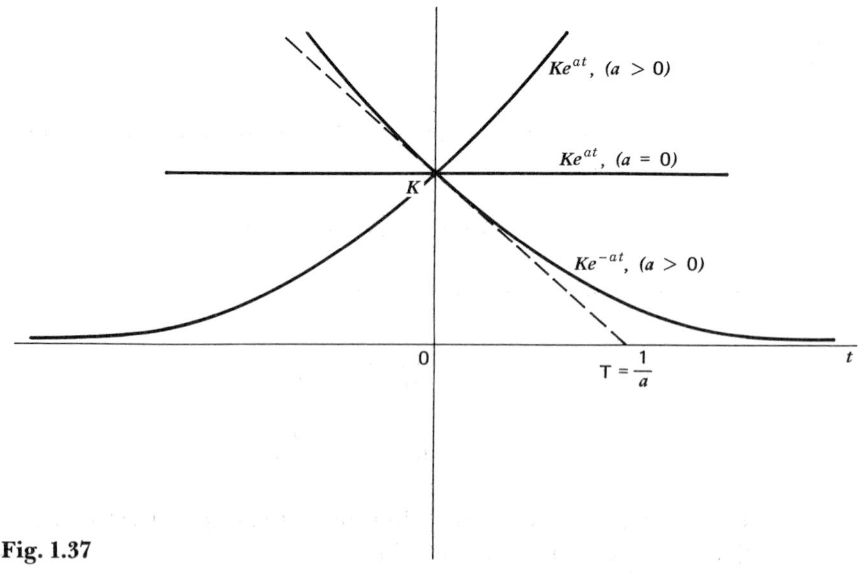

Fig. 1.37

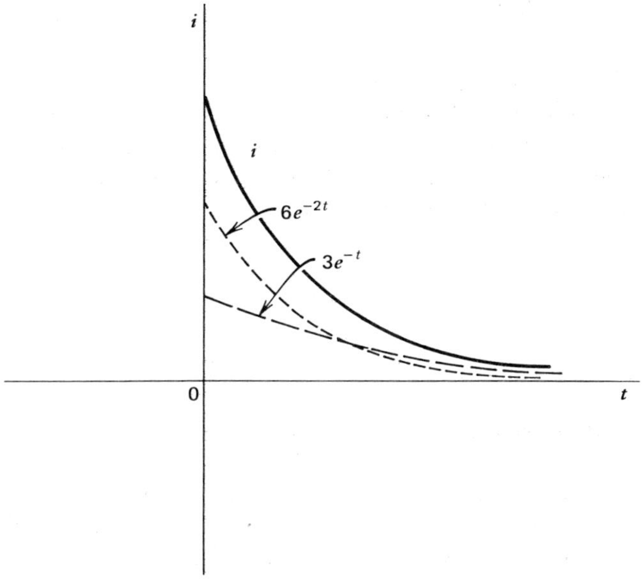

Fig. 1.38

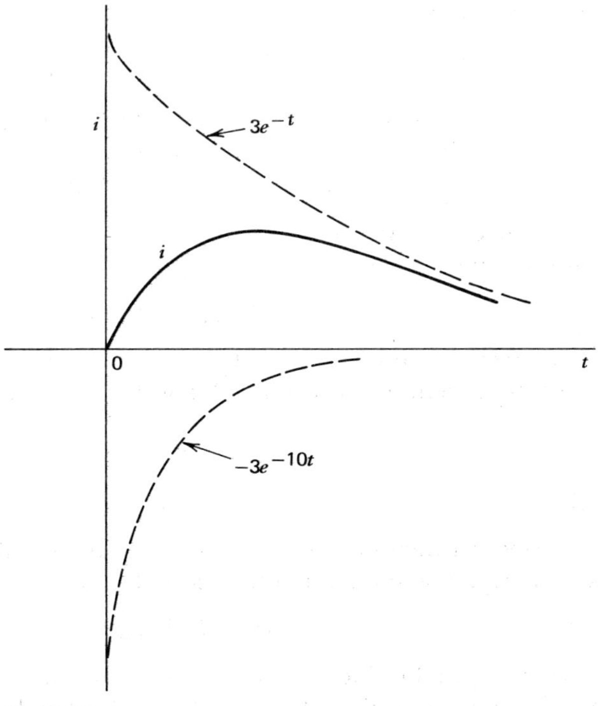

Fig. 1.39

The decaying exponential function takes on the value $(1/e)K$ at time $t = T = (1/a)$. This value $T = (1/a)$ is called *the time constant.* In one time constant, the exponential decreases to 37 percent of its value at $t = 0$.

Derivatives and integrals of exponential functions are also exponentials except for added constants of integration.

$$\frac{d}{dt}Ke^{at} = aKe^{at} \tag{1.68}$$

$$\int_{t_0}^{t} Ke^{a\tau}\,d\tau = \frac{1}{a}Ke^{at} + K_2 \tag{1.69}$$

where K_2 is an arbitrary constant, and τ is a dummy variable of integration.

Circuits that approximate differentiation and integration leave exponential signals unchanged except for some multiplicative or additive constants. Observe also that the slope (derivative) of an exponential at $t = 0$ is given by aK. A tangent at $t = 0$ thus intercepts the t-axis at $(1/a) = T$, the time constant.

As we shall see later, responses of circuits involve linear combinations of exponentials, such as

$$i = c_1e^{-a_1t} + c_2e^{-a_2t} + c_3e^{-a_3t} + \cdots + c_ne^{-a_nt},\ t \geqslant 0 \tag{1.70}$$

are defined only for $t \geqslant 0$ where c_k's and a_k's are real constants. Consider the response

$$i = 3e^{-t} + 6e^{-2t}, \quad t \geqslant 0 \tag{1.71}$$

Its plot is given in Fig. 1.38.

Consider another example,

$$i = 3e^{-t} - 3e^{-10t}, \quad t \geqslant 0 \tag{1.72}$$

The first term decays ten times slower compared to the second term; the value of the function is zero at $t = 0$. Fig. 1.39 shows the function.

The Sinusoid

The periodic functions, sine and cosine, are familiar functions from trigonometry. A sine wave as a function of time

$$v(t) = V \sin \omega t \tag{1.73}$$

is shown in Fig. 1.40. The value of ωt represents the angular displacement of a point moving uniformly around a circle of radius V. The value of v

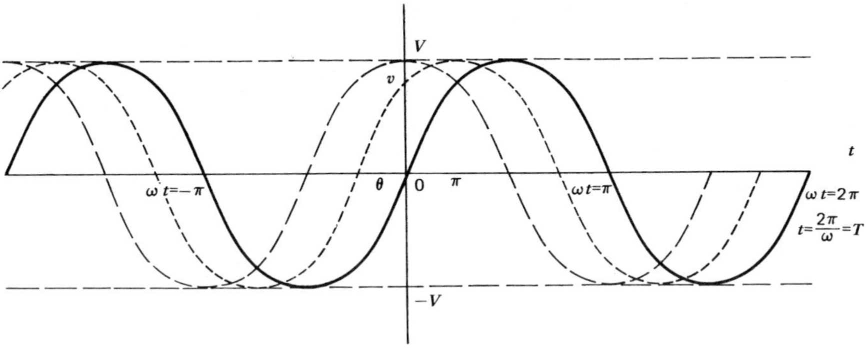

Fig. 1.40

is the vertical projection of the point at time t seconds. The constant ω is called the angular frequency in radians per second. The maximum magnitude of the function is V. Observe that this sine wave has value zero at zero time.

If the sine wave is shifted by a *phase angle*, say θ, so that it is not zero at zero time, the function is

$$v_1(t) = V \sin(\omega t + \theta) \tag{1.74}$$

The cosine function is given by a 90° shift of the sine function

$$v_2(t) = V \cos \omega t = V \sin\left(\omega t + \frac{\pi}{2}\right) \tag{1.75}$$

and

$$v_3(t) = V \sin(\omega t + \theta) = V \cos(\omega t + \phi) \tag{1.76}$$

where ϕ and θ differ by 90°.

We see from the above that a sine function can be represented as a cosine function with appropriate change in phase angle and vice versa. We will use the term *sinusoid* to refer to either a sine or a cosine function.

Finally, we say that v_1 *leads* v if θ is positive and that v_1 *lags* v if θ is negative.

In Fig. 1.40, we see that the waveform repeats itself every 2π radians. For this reason, the function is called *periodic*. The time T when

$$\omega T = 2\pi \tag{1.77}$$

that is

$$T = \frac{2\pi}{\omega} \tag{1.78}$$

is called the *period* of the periodic function. This is expressed as

$$v(t) = v(t + T) \tag{1.79}$$

The function in any interval of T seconds is called a cycle. If the function is known in any one cycle, it is known completely since we merely repeat the same cycle over and over again. The number of cycles in one second is called the frequency and given by

$$f = \frac{1}{T} \tag{1.80}$$

The unit of frequency in cycles per second is hertz (Hz). Kilohertz (kHz $= 10^3$ Hz) and megahertz (MHz $= 10^6$ Hz) are also commonly used.

We will have occasions to refer to derivatives and integrals of sinusoids that are also sinusoids but with different magnitudes and phase.

$$\frac{d}{dt}\sin(\omega t) = \omega \cos \omega t = \omega \sin\left(\omega t + \frac{\pi}{2}\right) \tag{1.81}$$

$$\int_{t_0}^{t} \sin \omega\tau \, d\tau = -\frac{1}{\omega}\cos \omega t + \frac{1}{\omega}\cos \omega t_0$$

$$= \frac{1}{\omega}\sin\left(\omega t - \frac{\pi}{2}\right) + K \tag{1.82}$$

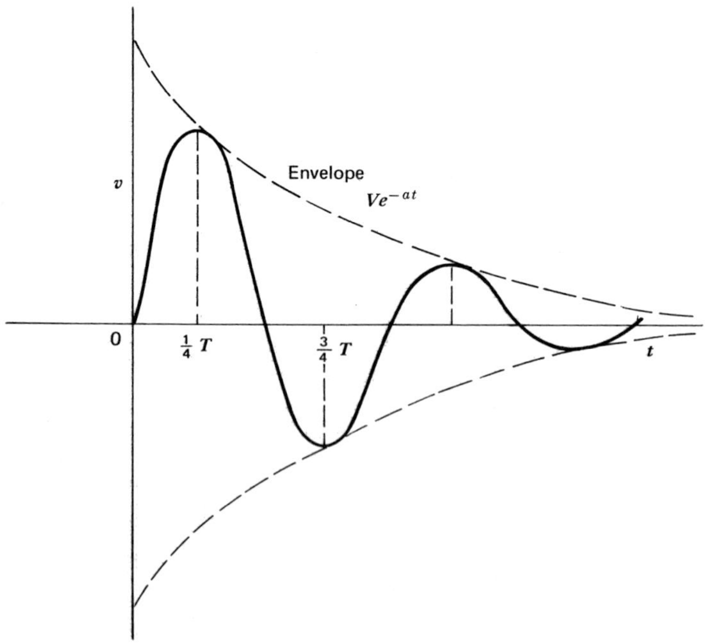

Fig. 1.41

A *damped sinusoid*, often encountered in circuit analysis, is described by

$$v(t) = Ve^{-at} \sin \omega t \qquad a > 0 \qquad t \geqslant 0 \tag{1.83}$$

The waveform is shown in Fig. 1.41. Recall that the maximum values of the sinusoid are ± 1. It follows that at intervals of $\frac{1}{2}T$ seconds, the function $v(t)$ takes on the values $\pm Ve^{-at}$. This function that bounds the oscillations of the whole function is called the *envelope* of the damped sinusoid.

The Unit Step Function

Another basic signal that we will deal with is the unit step function $u(t)$ shown in Fig. 1.42, and defined as

$$\begin{aligned} u(t) &= 0 \qquad t < 0 \\ &= 1 \qquad t \geqslant 0 \end{aligned} \tag{1.84}$$

The physical analogy of a unit step excitation corresponds to a switch S, which closes at $t = 0$ and connects a battery of 1 volt to a given zero-state circuit, as shown in Fig. 1.43. Note that the physical analogy is valid only for a zero-state circuit (i.e., a circuit whose initial state is zero). If the initial state is not zero, then the voltage at the circuit terminals for $t < 0$ may not be zero and the unit step analogy breaks down. For a zero-state

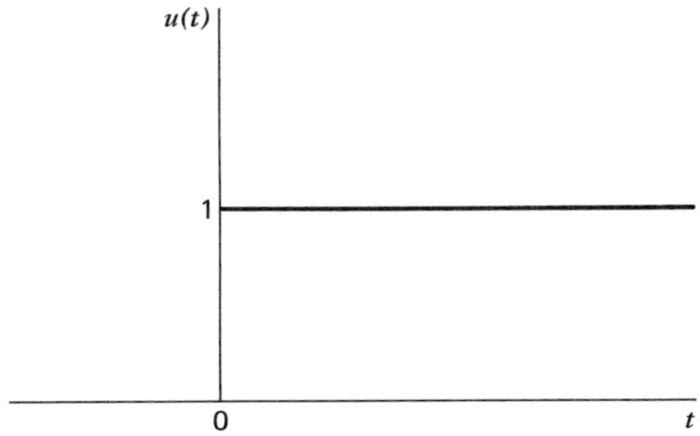

Fig. 1.42 Unit step function.

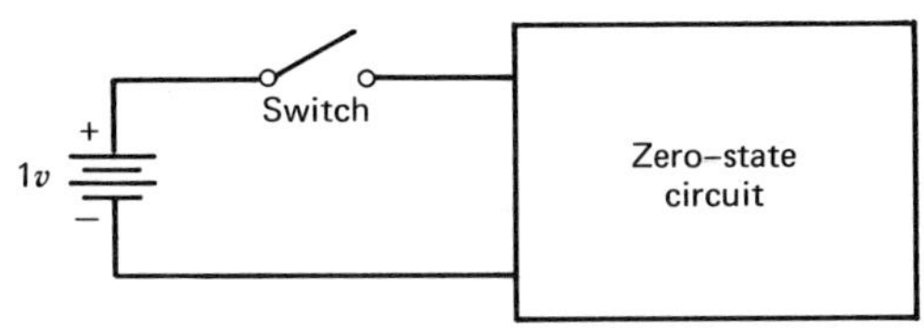

Fig. 1.43 Network analog of unit step.

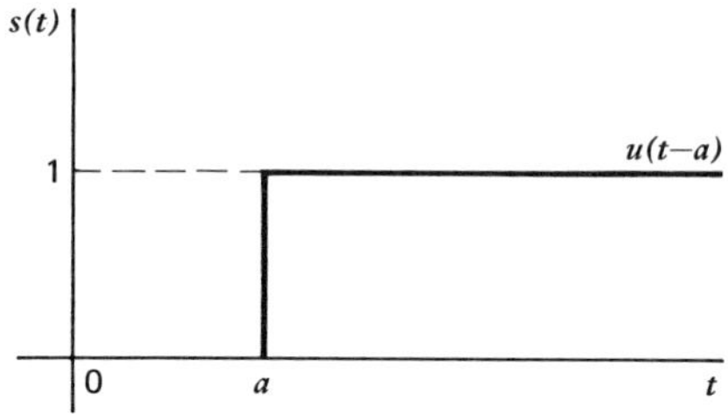

Fig. 1.44 Shifted step function.

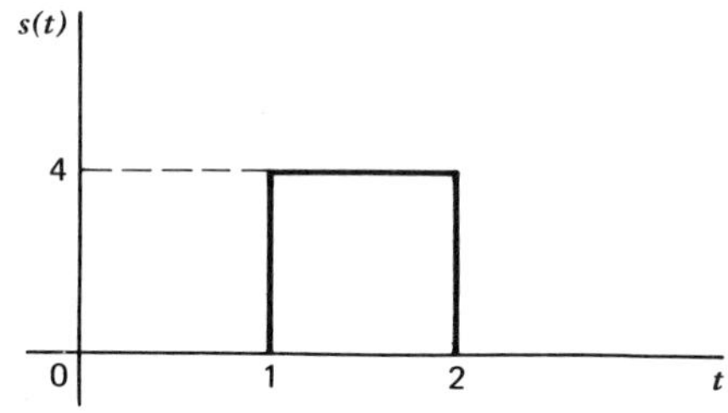

Fig. 1.45 Square pulse.

circuit, the voltage at the circuit terminals for $t < 0$ is zero, and the unit step analogy is valid. Note that the unit step is zero whenever the argument (t) within the parenthesis is negative, and is unity when the argument (t) is greater than zero. Thus the function $u(t-a)$, where $a > 0$, is defined by

$$\begin{aligned} u(t-a) &= 0 \qquad t < a \\ &= 1 \qquad t \geqslant a \end{aligned} \tag{1.85}$$

and is shown in Fig. 1.44. Note that the jump discontinuity of the step occurs when the argument within the parentheses is zero. This forms the basis of the *shifting* property of the step function. Also, the height of the jump discontinuity of the step can be scaled up or down by the multiplication of a constant K.

With the use of the change of amplitude and the shifting properties of the step function, we can proceed to construct a family of pulse waveforms. For example, the square pulse in Fig. 1.45 can be constructed by the sum of two step functions

$$s(t) = 4u(t-1) + (-4)u(t-2) \tag{1.86}$$

as given in Fig. 1.46. The "staircase" function, shown in Fig. 1.47 is

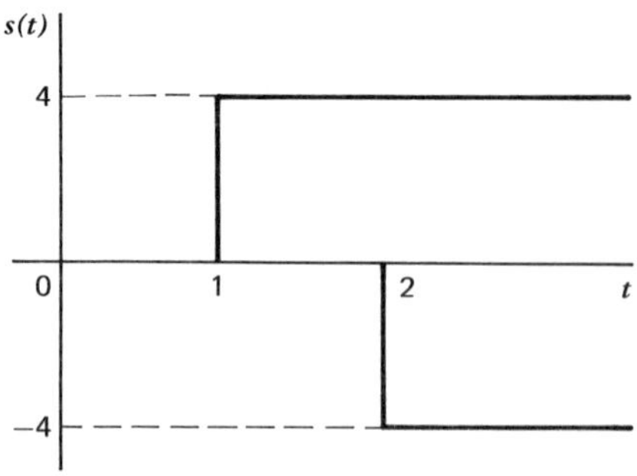

Fig. 1.46 Construction of square pulse by step function.

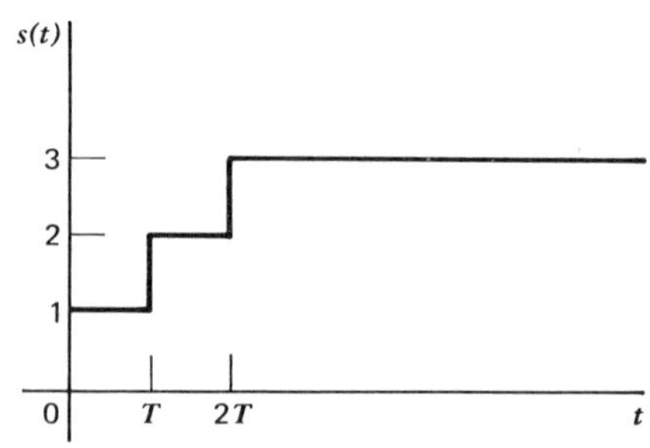

Fig. 1.47 Staircase function.

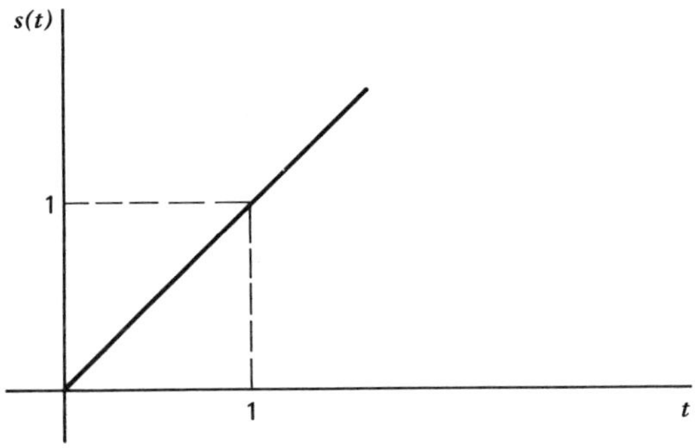

Fig. 1.48 Ramp function with zero time shift.

characterized by the equation

$$s(t) = \sum_{k=0}^{2} u(t-kT) \tag{1.87}$$

The Unit Ramp Function

The step function is also extremely useful in representing the shifted or delayed version of any given signal. For example, consider the *unit ramp function*

$$\rho(t) = tu(t) \tag{1.88}$$

shown in Fig. 1.48. Suppose the ramp is delayed by an amount $t = a$, as shown in Fig. 1.49. How do we represent the delayed version of the ramp? First, let us replace the variable t by a new variable $t' = t - a$. Then

$$\rho(t') = t'u(t') \tag{1.89}$$

When $\rho(t')$ is plotted against t', the resulting curve is identical to the plot of $s(t)$ versus t in Fig. 1.48. If, however, we substitute $t - a = t'$ in $\rho(t')$, we then have

$$\rho(t') = (t-a)u(t-a) \tag{1.90}$$

When we plot $\rho(t')$ against t, we have the delayed version of $\rho(t)$ shown in Fig. 1.49.

From the preceding discussion, it is clear that if any signal $f(t)u(t)$ is delayed by a time T, the delayed or shifted signal is given by

$$f(t') = f(t-T)u(t-T) \tag{1.91}$$

For example, let us delay the function $(\sin \pi t/T)u(t)$ by a period T. Then

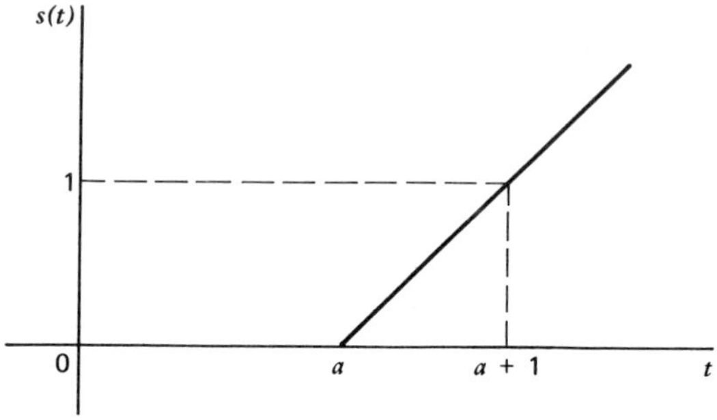

Fig. 1.49 Ramp function with time shift $-a$.

the delayed function $s(t')$, shown in Fig. 1.50, is

$$s(t') = \left[\sin \frac{\pi}{T}(t-T)\right] u(t-T) \tag{1.92}$$

As a final example, consider the waveform in Fig. 1.51, whose component parts are given in Fig. 1.52. For increasing t, the first nonzero component is the function $2(t-1)u(t-1)$, which represents the straight line of slope 2 at $t=1$. At $t=2$, the rise of the straight line is to be arrested, so we add to the first component a term equal to $-2(t-2)u(t-2)$ with a slope of -2. The sum is then a constant equal to 2. We then add a term $-2u(t-2)$ to bring the level down to zero. Thus,

$$s(t) = 2(t-1)u(t-1) - 2(t-2)u(t-2) - 2u(t-2) \tag{1.93}$$

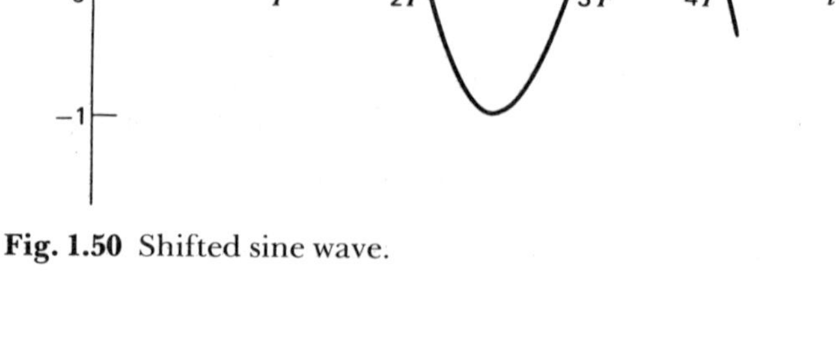

Fig. 1.50 Shifted sine wave.

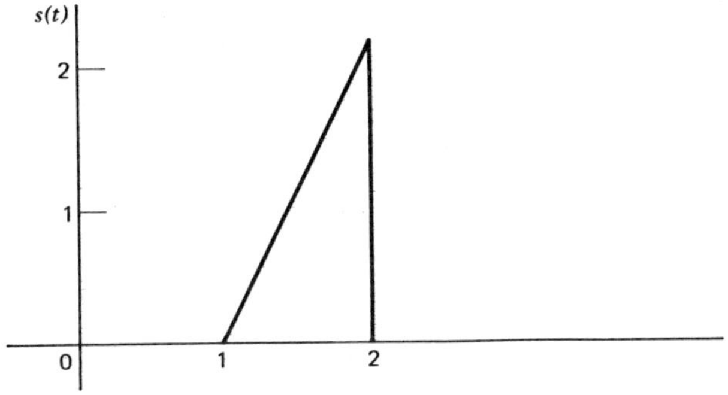

Fig. 1.51 Triangular pulse.

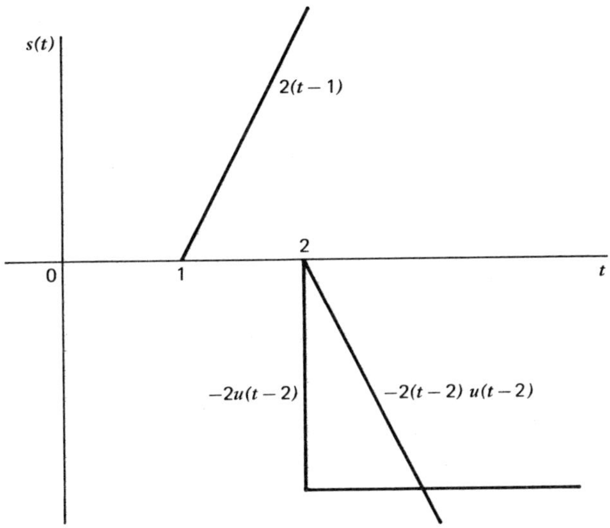

Fig. 1.52 Decomposition of the triangular pulse in Fig. 1.51.

The Unit Impulse Function

The *unit impulse,* or *delta function,* is a mathematical anomaly. P. A. M. Dirac first used it in his writings on quantum mechanics.* He defined the delta function $\delta(t)$ by the equations

$$\int_{-\infty}^{\infty} \delta(t)dt = 1 \tag{1.94}$$

$$\delta(t) = 0 \quad \text{for} \quad t \neq 0 \tag{1.95}$$

*P. A. M. Dirac, *The Principles of Quantum Mechanics,* Oxford University Press, 1930.

Its most important property is the *sifting property*, expressed by

$$\int_{-\infty}^{\infty} f(t)\delta(t)dt = f(0) \tag{1.96}$$

In this section we will examine the unit impulse from a nonrigorous approach. We will show that the unit impulse is the derivative of the unit step

$$\delta(t) = u'(t) \tag{1.97}$$

At first glance this statement is doubtful. After all, the derivative of the unit step is zero everywhere except at the jump discontinuity, and it does not even exist at that point! However, consider the function $g_\epsilon(t)$ in Fig. 1.53. It is clear that as ϵ goes to zero, $g_\epsilon(t)$ approaches a unit step, that is,

$$\lim_{\epsilon \to 0} g_\epsilon(t) = u(t) \tag{1.98}$$

Taking the derivative of $g_\epsilon(t)$, we obtain $g'_\epsilon(t)$, which is defined by the equations

$$\begin{aligned} g'_\epsilon(t) &= \frac{1}{\epsilon}; \qquad 0 \leqslant t \leqslant \epsilon \\ &= 0; \qquad t < 0, t > \epsilon \end{aligned} \tag{1.99}$$

as shown in Fig. 1.54. Now let ϵ go to zero. The unit impulse $\delta(t)$ is defined as

$$\delta(t) = \lim_{\epsilon \to 0} g'_\epsilon(t) \tag{1.100}$$

It should be stressed that this is not a rigorous definition, but merely

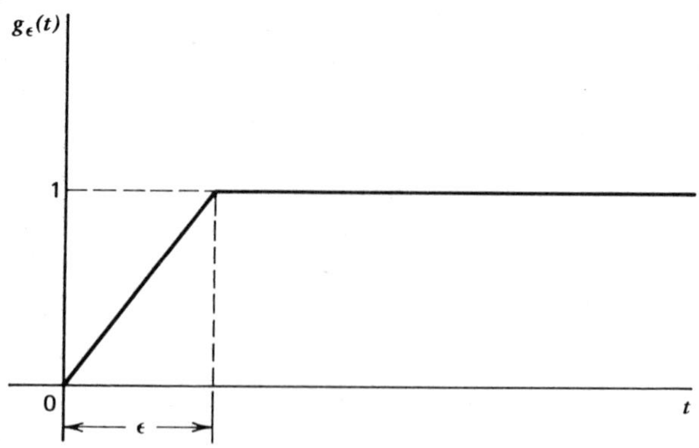

Fig. 1.53 Unit step when $\epsilon \to 0$.

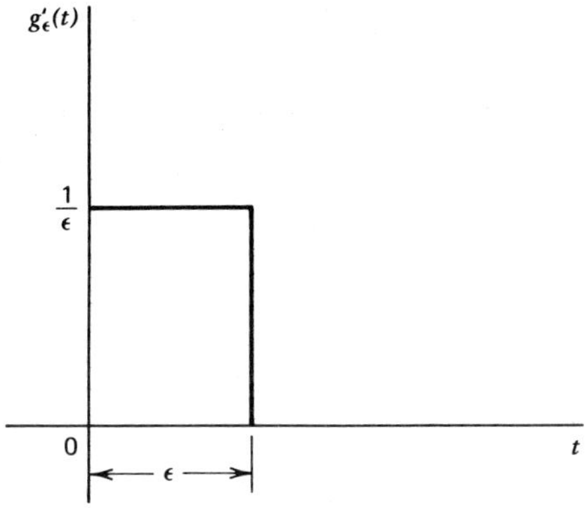

Fig. 1.54 Derivative of $g_\epsilon(t)$ in Fig. 1.53.

heuristic. From Fig. 1.54, we see that as $\epsilon \rightarrow 0$,

$$g_\epsilon(0) \underset{\epsilon \rightarrow 0}{\longrightarrow} \infty \tag{1.101}$$

so that we can think of the impulse or delta function as having the additional properties

$$\begin{aligned} \delta(0) &= \infty \\ \delta(t) &= 0 \quad \text{for} \quad t \neq 0 \end{aligned} \tag{1.102}$$

Continuing with this heuristic treatment, we say that the area "under" the impulse is unity, and, since the impulse is zero for $t \neq 0$, we have

$$\int_{-\infty}^{\infty} \delta(t)\mathrm{dt} = \int_{0-}^{0+} \delta(t)\,dt = 1 \tag{1.103}$$

Thus the entire area of the impulse is "concentrated" at $t = 0$. Consequently, any integral that does not integrate through $t = 0$ is zero, as seen by

$$\int_{-\infty}^{0-} \delta(t)\,dt = \int_{0+}^{+\infty} \delta(t)\,dt = 0 \tag{1.104}$$

The change of scale and time shift properties discussed earlier also apply for the impulse function. The derivative of a step function

$$s(t) = Au(t-a) \tag{1.105}$$

yields an impulse function

$$s'(t) = A\delta(t-a) \tag{1.106}$$

which is shown in Fig. 1.55. Graphically, we represent an impulse func-

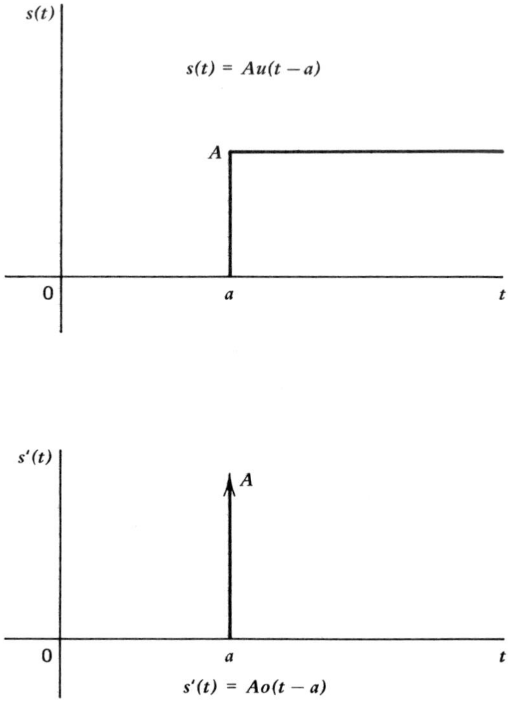

Fig. 1.55

tion by an arrowhead, with the constant multiplier A written next to the arrowhead. Note that A is the area under the impulse $A\delta(t-a)$.

Consider the implications of Eqs. 1.105 and 1.106. From these equations we see that the derivative of the step at the jump discontinuity of height A yields an impulse of area A at that same point $t=T$. Generalizing on this argument, consider any function $f(t)$ with a jump discontinuity at $t=T$. Then the derivative $f'(t)$ must have an impulse at $t=T$. As an example, consider $f(t)$ in Fig. 1.56. At $t=T$, $f(t)$ has a discontinuity of height A, which is given as*

$$A=f(T+)-f(T-) \tag{1.107}$$

Let us define $f_1(t)$ as being equal to $f(t)$ for $t<T$, and having the same shape as $f(t)$, but without the discontinuity for $t>T$, that is,

$$f_1(t)=f(t)-Au(t-T) \tag{1.108}$$

*$f(T+)=\lim_{\epsilon\to 0} f(T+\epsilon), f(T-)=\lim_{\epsilon\to 0} f(T-\epsilon)$.

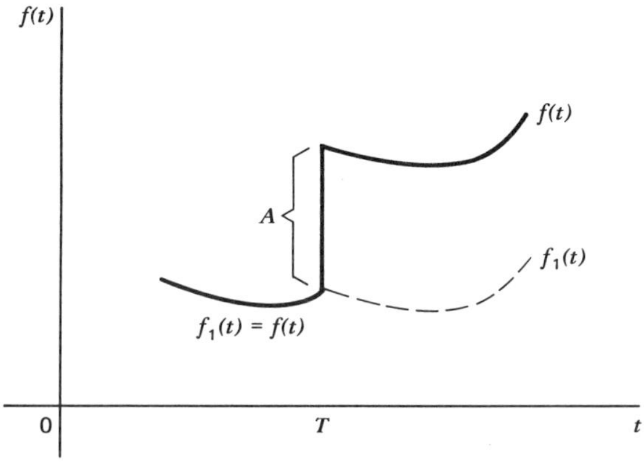

Fig. 1.56 Function with discontinuity at T.

The derivative $f'(t)$ is then

$$f'(t) = f_1'(t) + A\delta(t-T) \tag{1.109}$$

The following example illustrates this point more clearly. In Fig. 1.57*a*, the function $f(t)$ is

$$f(t) = Au(t-a) - Au(t-b) \tag{1.110}$$

Its derivative is

$$f'(t) = A\delta(t-a) - A\delta(t-b) \tag{1.111}$$

and is shown in Fig. 1.57*b*. Since $f(t)$ has two discontinuities, at $t = a$ and $t = b$, its derivative must have impulses at these points. The coefficient of the impulse at $t = b$ is negative because

$$f(b+) - f(b-) = -A \tag{1.112}$$

As a second example, consider the function $g(t)$ shown in Fig. 1.58*a*. We obtain $g'(t)$ by inspection, and note that the discontinuity at $t = 1$ produces the impulse in $g'(t)$ of area

$$g(1+) - g(1-) = 2 \tag{1.113}$$

as given in Fig. 1.58*b*.

Another interesting property of the impulse function is expressed by the integral

$$\int_{-\infty}^{+\infty} f(t)\delta(t-T)\,dt = f(T) \tag{1.114}$$

This integral is easily evaluated if we consider that $\delta(t-T) = 0$ for all

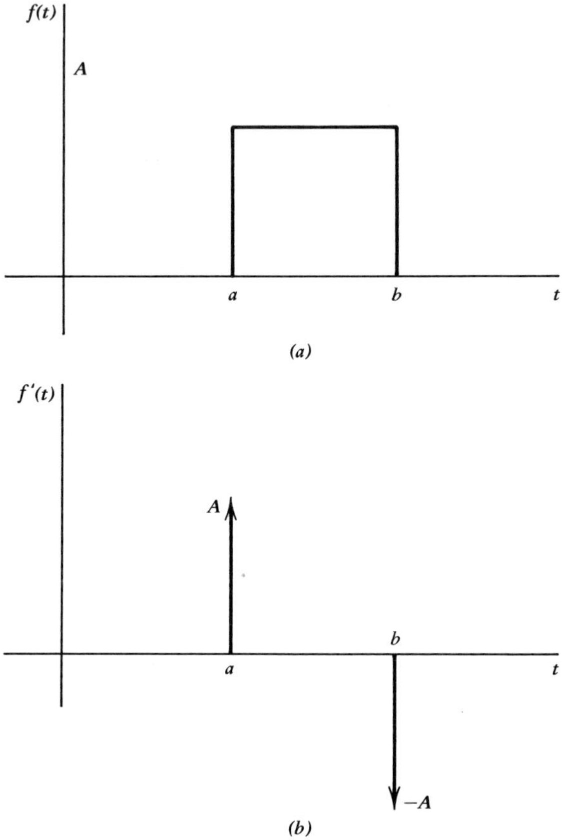

Fig. 1.57 (*a*) Square pulse. (*b*) Derivative of square pulse.

$t \neq T$. Therefore, the product

$$f(t)\delta(t-T) = 0 \qquad \text{all } t \neq T \tag{1.115}$$

If $f(t)$ is single-valued at $t = T$, $f(T)$ can be factored from the integral so that we obtain

$$f(T)\int_{-\infty}^{\infty} \delta(t-T)\,dt = f(T) \tag{1.116}$$

Figure 1.59 shows $f(t)$ and $\delta(t-T)$, where $f(t)$ is continuous at $t = T$. If $f(t)$ has a discontinuity at $t = T$, the integral

$$\int_{-\infty}^{+\infty} f(t)\delta(t-T)\,dt$$

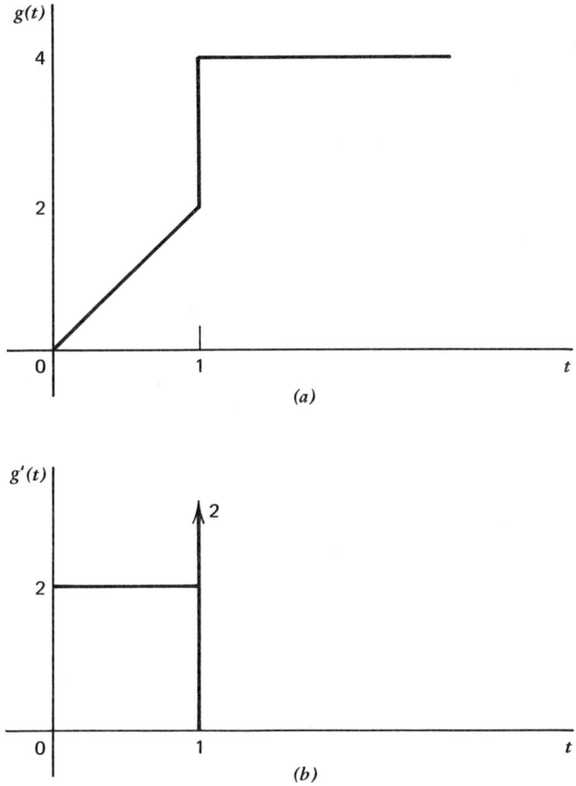

Fig. 1.58 (*a*) Signal. (*b*) Derivative.

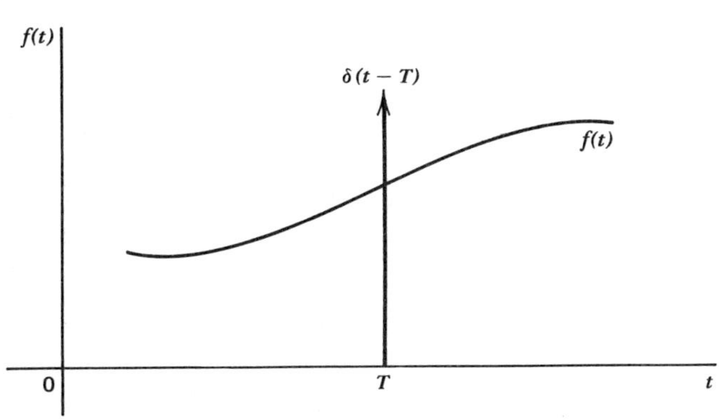

Fig. 1.59

is not defined because the value of $f(T)$ is not uniquely given. Conside the following examples.

Example A

$$f(t) = e^{j\omega t}$$

$$\int_{-\infty}^{+\infty} e^{j\omega t}\delta(t-T)dt = e^{j\omega T} \tag{1.117}$$

Example B

$$f(t) = \sin t$$

$$\int_{-\infty}^{+\infty} \sin t\,\delta\left(t-\frac{\pi}{4}\right)dt = \frac{1}{\sqrt{2}} \tag{1.118}$$

Consider next the case where $f(t)$ is continuous for $-\infty < t < \infty$. Le us direct our attention to the integral

$$\int_{-\infty}^{\infty} f(t)\delta(t-T)dt = f(T) \tag{1.119}$$

which holds for all t in this case. If T were varied from $-\infty$ to $+\infty$, ther $f(t)$ would be reproduced in its entirety. An operation of this sort cor responds to *scanning* the function $f(t)$ by moving a sheet of paper witl a thin slit across a plot of the function, as shown in Fig. 1.60.

Let us now examine higher order derivatives of the unit-step function Here we represent the unit impulse by the function $f_\epsilon(t)$ in Fig. 1.61 which, as $\epsilon \to 0$, becomes the unit impulse. The derivative of $f_\epsilon(t)$ is giver in Fig. 1.62. As ϵ approaches zero, $f_\epsilon'(t)$ approaches the derivative of the unit impulse $\delta'(t)$, which consists of a pair of impulses as seen in Fig. 1.63 The area under $\delta'(t)$, which is sometimes called a *doublet*, is equal to zero

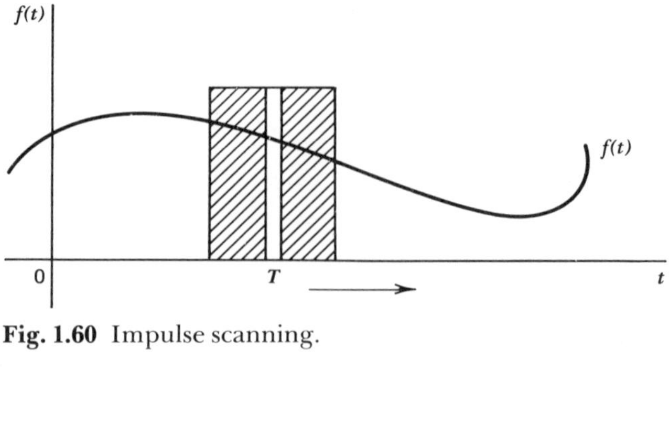

Fig. 1.60 Impulse scanning.

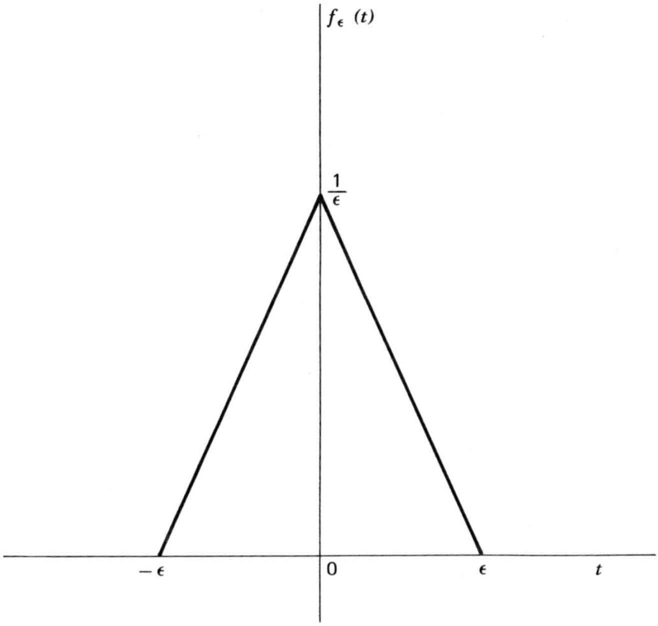

Fig. 1.61

Thus,

$$\int_{-\infty}^{\infty} \delta'(t)dt = 0 \tag{1.120}$$

The other significant property of the doublet is

$$\int_{-\infty}^{\infty} f(t)\delta'(t-T)dt = -f'(T) \tag{1.121}$$

where $f'(T)$ is the derivative of $f(t)$ evaluated at $t = T$ where, again, we

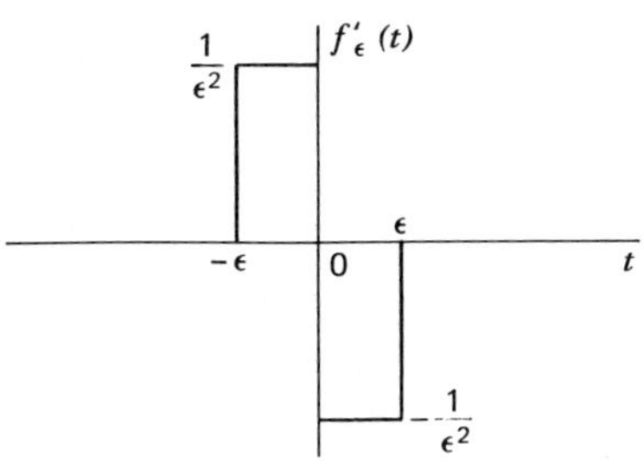

Fig. 1.62 Unit doublet as $\epsilon \to 0$.

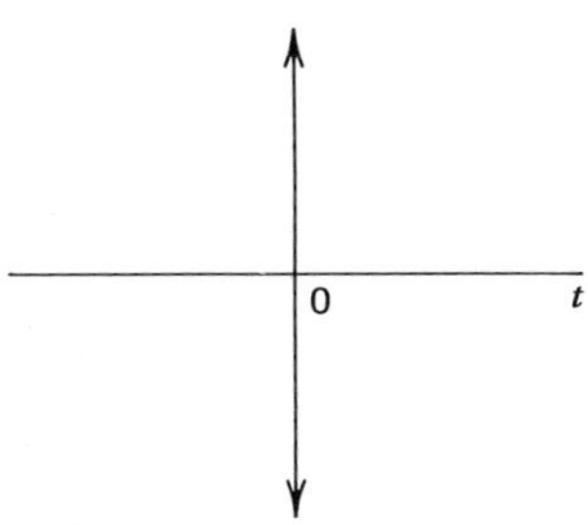

Fig. 1.63 The doublet $\delta''(t)$.

assume that $f(t)$ is continuous and differentiable. Equation 1.71 can be proved by integration by parts. Thus,

$$\int_{-\infty}^{\infty} f(t)\delta'(t-T)dt = f(t)\delta(t-T)\Big|_{-\infty}^{\infty} - \int_{-\infty}^{\infty} f'(t)\delta(t-T)dt$$
$$= -f'(T) \tag{1.122}$$

It can be shown in general that

$$\int_{-\infty}^{\infty} f(t)\delta^{(n)}(t-T)dt = (-1)^n f^{(n)}(T) \tag{1.123}$$

where $\delta^{(n)}$ and $f^{(n)}$ denote nth derivatives. The higher order derivatives of $\delta(t)$ can be evaluated in similar fashion.

Complex Exponentials

In Chapter 6 we will undertake a thorough discussion of complex numbers. We discuss here the use of complex numbers only in so far as they concern the signal descriptions. Both as excitations and as responses, signal waveforms arise that may be most conveniently described in terms of complex exponentials, that is, exponentials whose exponents are complex numbers.

A complex number is defined as

$$z = u + jv \tag{1.124}$$

where, u and v are real numbers, and

$$j^2 = -1 \tag{1.125}$$

or

$$j = \sqrt{-1} \tag{1.126}$$

Since the square root of a negative number is not real, j is called an *imaginary* number. Thus, jv is also an imaginary number. A complex number thus contains a real and an imaginary part or component. We will manipulate these quantities in the usual manner and substitute $j^2 = -1$ whenever necessary.

A complex exponential is an exponential with a complex exponent, that is,

$$e^z = e^{u+jv} \tag{1.127}$$

Complex numbers and exponentials must be consistent with their real counterparts. Therefore, since

$$e^{u_1+u_2} = e^{u_1} \cdot e^{u_2} \tag{1.128}$$

where u_1 and u_2 are real numbers, it follows that

$$e^z = e^{u+jv} = e^u \,.\, e^{jv} \tag{1.129}$$

Since e^u is real, its properties are known. We shall now consider the meaning of e^{jv}. The exponential of a real number can be expressed in a Taylor's series about the origin as

$$e^u = \sum_{k=0}^{\infty} \frac{u^k}{k!} \tag{1.130}$$

This infinite series representation is also true for complex numbers. So extending this to the complex exponential, and considering only the imaginary part, we have

$$e^{jv} = \sum_{k=0}^{\infty} \frac{(jv)^k}{k!} \tag{1.131}$$

Since $j = \sqrt{-1}$, $j^2 = -1$, we can separate the series given by Eq. 1.131 into even and odd powers of j. Then

$$e^{jv} = \underbrace{\sum_{m=0}^{\infty} \frac{(jv)^{2m}}{2m!}}_{\text{even part}} + \underbrace{\sum_{m=0}^{\infty} \frac{(jv)^{2m+1}}{(2m+1)!}}_{\text{odd part}} \tag{1.132}$$

Since $(j)^{2m} = (-1)^m$ and $(j)^{2m+1} = j(-1)^m$, Eq. 1.132 may be written as

$$e^{jv} = \underbrace{\sum_{m=0}^{\infty} \frac{(-1)^m v^{2m}}{2m!}}_{\text{real series}} + j \underbrace{\sum_{m=0}^{\infty} \frac{(-1)^m v^{2m+1}}{(2m+1)!}}_{\substack{\text{real series}\\ \text{multiplied}\\ \text{by } j}} \tag{1.33}$$

Taylor's series for sine and cosine are given by

$$\cos v = \sum_{m=0}^{\infty} \frac{(-1)^m v^{2m}}{2m!}$$

$$\sin v = \sum_{m=0}^{\infty} \frac{(-1)^m v^{2m+1}}{(2m+1)!} \tag{1.134}$$

Therefore, from Eqs. 1.133 and 1.134 we have the following

$$e^{jv} = \cos v + j \sin v \tag{1.135}$$

Replacing v by $-v$ we have

$$e^{-jv} = \cos(-v) + j\sin(-v)$$
$$= \cos v - j\sin v \tag{1.136}$$

From Eqs. 1.135 and 1.136 it may be shown that

$$\cos v = \tfrac{1}{2}(e^{jv} + e^{-jv})$$
$$\sin v = \frac{1}{2j}(e^{jv} - e^{-jv}) \tag{1.137}$$

From the above discussion it may be noted that signals which contain exponentials of real and complex are easily handled since the complex exponentials add or subtract in such a way that the ultimate answer turns out to be real trigonometric functions.

Some useful time functions and their complex exponential representations are given below.

$$i_1(t) = I_0 \cos(\omega t + \theta)$$
$$= \frac{I_0}{2}[e^{j(\omega t+\theta)} + e^{-j(\omega t+\theta)}]$$
$$= \frac{I_0 e^{j\theta}}{2} e^{j\omega t} + \frac{I_0 e^{-j\theta}}{2} e^{-j\omega t}$$
$$= I_1 e^{j\omega t} + I_2 e^{-j\omega t} \tag{1.138}$$

$$i_2(t) = I_0 e^{-at}\cos(\omega t + \theta)$$
$$= \frac{I_0 e^{j\theta}}{2} e^{(-a+j\omega)t} + \frac{I_0 e^{-j\theta}}{2} e^{(-a-j\omega)t}$$
$$= I_1 e^{p_1 t} + I_2 e^{p_2 t} \tag{1.139}$$

$$i_3(t) = I_0 \sin(\omega t + \theta)$$
$$= \frac{I_0}{2j}[e^{j(\omega t+\theta)} - e^{-j(\omega t+\theta)}]$$
$$= I_3 e^{j\omega t} + I_4 e^{-j\omega t} \tag{1.140}$$

Note that the constants I_1, I_2, I_3, I_4, p_1, and p_2 are complex numbers given by:

$$I_1 = \frac{I_0}{2} e^{j\theta} \qquad I_2 = \frac{I_0}{2} e^{-j\theta}$$

$$I_3 = \frac{I_0}{2j} e^{j\theta} \qquad I_4 = -\frac{I_0}{2j} e^{-j\theta}$$

$$p_1 = -a + j\omega \qquad p_2 = -a - j\omega$$

1.9 DIGITAL COMPUTATION I: PRELIMINARY

In describing the excitation-response relationships of circuits, we form differential equations or systems of linear algebraic equations. These equations may sometimes be solved analytically. However, for most practical cases we can only hope to get approximate numerical solutions for the problems at hand. Thus a digital computer is called upon to solve these problems numerically. We discuss in this section some basic concepts useful in digital computation.

Number System

To begin with, numbers must be stored in the computer memory before the actual computation takes place. Unlike the decimal system we use everyday, the *binary* number system is used to represent numbers in a computer. This is because modern computers read pulses sent by electrical components and the state of an electrical pulse is either *on* or *off* that could represent 1 or 0 respectively. Thus the number 20, for example, is expressed in decimal system as

$$20 = 2\times 10 + 0\times 10^0$$

where 10 is the *base* of the system. The same number 20 could also be expressed in binary system as

$$10100 = 1\times 2^4 + 0\times 2^3 + 1\times 2^2 + 0\times 2^1 + 0\times 2^0$$

and the base is 2.

In general a non-negative real number N will be represented in the binary system as

$$\begin{aligned} N &= a_n a_{n-1} \cdots a_1 a_0 \cdot a_{-1} a_{-2} \cdots a_{-m} \cdots \\ &= a_n 2^n + a_{n-1} 2^{n-1} + \cdots + a_1 2^1 + a_0 2^0 \\ &\quad + a_{-1} 2^{-1} + a_{-2} 2^{-2} + \cdots + a_{-m} 2^{-m} + \cdots \end{aligned}$$

where the coefficients a_k are either 0 or 1. The positively subscripted coefficients are the integer part of N and the negatively subscripted ones are used to represent the fraction of N. Let us now discuss separately the representations of the integer part and the fractional part of a number N in *binary digits* or *bits* a_k.

Suppose that N is expressed as

$$N = I + F$$

where I stands for the integer and F for the fraction of N. Now

$$I = a_n 2^n + a_{n-1} 2^{n-1} + \cdots + a_1 2^1 + a_0 2^0 \tag{1.141}$$

From 1.141 we see that the last bit a_0 is 0 if and only if I is even. In other words, a_0 is 0 if and only if the remainder of $I/2$ is zero. Similarly the second last bit a_1 is 0 if and only if the remainder of $(I-a_0)/2$ is zero. Thus we could set up a table of successive divisions to determine the bit representation of I as follows. Let the original I be denoted by I_0, then we have

$$
\begin{aligned}
2\,)\underline{I_0)}\,R_0 &= a_0 \\
2\,)\underline{I_1)}\,R_1 &= a_1 \\
&\vdots \\
2\,)\underline{I_{n-1})}\,R_{n-1} &= a_{n-1} \\
2\,)\underline{I_n)}\,R_n &= a_n \\
I_{n+1} &= 0
\end{aligned}
\qquad (1.142)
$$

where I_{k+1} is the quotient obtained when I_k is divided by 2 and R_k is the remainder of this division. Or analytically we have

$$2I_{k+1} = I_k - R_k \qquad k = 0, 1, 2, \cdots$$

and the table terminates when $I_{n+1} = 0$. In this case we need $n+1$ bits to represent I. For an example the table for the number 21 is

$$
\begin{aligned}
2\,)\underline{I_0 = 21)}\,R_0 &= a_0 = 1 \\
2\,)\underline{I_1 = 10)}\,R_1 &= a_1 = 0 \\
2\,)\underline{I_2 = 5)}\,R_2 &= a_2 = 1 \\
2\,)\underline{I_3 = 2)}\,R_3 &= a_3 = 0 \\
2\,)\underline{I_4 = 1)}\,R_4 &= a_4 = 1 \\
I_5 &= 0
\end{aligned}
$$

and the table terminates when $I_5 = 0$. Hence we need 5 bits to represent 21 as

$$1\ 0\ 1\ 0\ 1 = a_4 a_3 a_2 a_1 a_0$$

For the fractional part of N we have

$$F = a_{-1}2^{-1} + a_{-2}2^{-2} + \cdots + a_{-m}2^{-m} + \cdots \qquad (1.143)$$

where a_{-k} is either 0 or 1. Note that we might have unending bit-representation for F. To obtain a_{-1}, we simply multiply Eq. 1.143 by 2. The result is

$$2F = a_{-1} + a_{-2}2^{-1} + \cdots + a_{-m}2^{-m+1} + \cdots \qquad (1.144)$$

We see from Eq. 1.144 that a_{-1} is 0 if and only if $2F$ is less than 1. Similarly a_{-2} is 0 if and only if $2(2F - a_{-1})$ is less than 1. Therefore we could set up a table of successive multiplications to determine the bit-representation of F as follows. First we denote F by F_0. Then we calculate

$$F_k = 2F_{k-1} - S_k \qquad k = 1, 2, \cdots$$

where

$$S_k = \begin{cases} 0 & \text{if} \quad 2F_{k-1} < 1 \\ 1 & \text{if} \quad 2F_{k-1} \geqslant 1 \end{cases}$$

And we have

$$a_{-k} = S_k \qquad k = 1, 2, \cdots \tag{1.145}$$

For example, if $F = F_0 = 0{\cdot}33$, then we have the following results for $m = 6$:

k	F_k	S_k
1	0·66	0
2	0·32	1
3	0·64	0
4	0·28	1
5	0·56	0
6	0·12	1

Hence the binary representation of 0·33 is

$$.010101 \cdots = .a_{-1}a_{-2}a_{-3}a_{-4}a_{-5}a_{-6} \cdots$$

Note that the binary representations for integers are finite whereas the ones for the fractions might have infinite strings of bits. Thus the binary equivalent of 21.33 is

$$1\ 0\ 1\ 0\ 1\ .\ 0\ 1\ 0\ 1\ 0\ 1\ \cdots \tag{1.146}$$

Since the number of bits in a machine word is finite, only approximations of the numbers are represented by the machine numbers. We will use x^* to denote the machine approximation of a certain number x. If m bits are available in a machine word, then in general we have two alternatives to approximate x by x^*. The first one is the *chopped* approximation x_c^* which is formed by the m leading bits of x. Thus for the number $x = 21.33$ with $m = 8$, the chopped approximation is $x_c^* = 10101.010$. Another approximation is the *rounded* approximation x_r^* whose representation depends on the $(m+1)$st bit of x in the following way: if the $(m+1)$st bit is 0, then we have $x_r^* = x_c^*$; otherwise 1 is added to the mth bit of x and the first m bits of the *added* result together with the sign are retained to form x_r^*. Thus the rounded approximation of 21.33 using 8 bits is 10101.011 as the 9th bit of x is 1.

Fixed and Floating-Point Representations

For convenience of manipulation the machine numbers are formed in one of the following two forms: namely, the fixed or floating-point format. Let us assume that we have 10 bits in a machine word. Then the fixed-point approximation x^* will have 10 bits for its representation with the decimal point *fixed* by the binary representation of x. Thus for the number $x = 21.33$, the fixed-point approximation in chopped form is

$$x_c^* = 1\ 0\ 1\ 0\ 1\ .\ 0\ 1\ 0\ 1\ 0 \tag{1.147}$$

Note that the decimal point of x_c^* is fixed between the 5th and 6th bit. On the other hand, just as in the decimal system the number 21.33 can be expressed as 0.2133×10^2 or $.2133(+2)$ where the number inside the parenthesis represents the powers of 10, the approximation x_c^* of 21.33 can also be expressed as

$$x_c^* = .\ 1\ 0\ 1\ 0\ 1\ 0\ 1(1\ 0\ 1) \tag{1.148}$$

if 7 bits are allowed to represent the fractional part, or mantissa, of x_c^* and three bits are for the powers of 2. This is the so-called *normalized floating-point format* as the decimal point is floated to left or right depending on the magnitude of the number represented.

Since all machine words have finite bit-lengths, the range of numbers that are representable in floating-point format is far greater than that using fixed-point format. For example, using the format of Eq. 1.147, numbers ranging from 00000.00000 to 11111.11111, or from 0 to approximately 32, could be represented. On the other hand, numbers ranging from .0000000(000) to .1111111(111), or from 0 to approximately 128, could be represented using the format of Eq. 1.148. Hence floating-point format is used extensively in scientific computations. We will hereafter confine ourselves to the discussion of floating-point representations of numbers and the associated operations.

In general for finite-bit words the *normalized* floating-point machine number has the form

$$x^* = a(b) \tag{1.149}$$

where

$$\begin{aligned} a &= \pm .\, a_{-1}a_{-2} \cdots a_{-t} \qquad \frac{1}{2} \leqslant a < 1 \\ b &= \pm b_1 b_2 \cdots b_p. \end{aligned} \tag{1.150}$$

Thus we need t bits and p bits to represent the fraction a and the powers of two b, respectively. An extra bit is needed for the sign of a or b. The bits a_{-k} and b_k are either 1 or 0 with $a_{-1} = 1$ for nonzero x.

The binary fraction a is called the *mantissa* or *fractional part* of x^* and b

is the *exponent.* The number of bits from the first nonzero bit to the last correct bit in a is called the number of *significant figures* of x^*. Thus using a 10-bit word with a 7-bit mantissa, $x = 21.33$ can be approximated by

$$x_c^* = a(b) = +.1\ 0\ 1\ 0\ 1\ 0\ 1\ (+1\ 0\ 1) \tag{1.151}$$

and the number of significant figures is 7. Note the exponent b is negative if the magnitude of x is less than 1/2.

The error incurred in using x^* with finite bits to represent a number x that might have an infinite string of bits is called the *round-off* error. Thus for the number 21.33, we have the binary expression given in Eq. 1.146 and the round-off error is

$$x - x_c^* = +.0\ 0\ 0\ 0\ 0\ 0\ 0\ \underline{0\ 1\ 0\cdots}(+1\ 0\ 1) \tag{1.152}$$

where the underlined bits are missing in x_c^*. The magnitude of the error is proportional to the magnitude of the original number x.

In general if we have t bits for the mantissa a of a machine word and if the number x is representable using this word with the associated exponent b, then

$$x^* = x(1+\delta),\ |\delta| \leqslant \begin{cases} 2^{-t+1} & \text{(chopped format)} \\ 2^{-t} & \text{(rounded format)} \end{cases} \tag{1.153}$$

The derivation of Eq. 1.153 for chopped format is as follows. First assume that x is formed as an unending normalized floating number

$$x = a(b) \qquad a = \pm .\,a_{-1}a_{-2}\cdots a_{-t}\cdots \tag{1.154}$$

where $a_{-1} = 1$. Then the machine representation x_c^* is equal to

$$x_c^* = a^*(b) \qquad a^* = \pm .\,a_{-1}a_{-2}\cdots a_{-t} \tag{1.155}$$

From Eqs. 1.154 and 1.155, we observe that

$$|x_c^*| \leqslant |x|,\ |a| \geqslant \frac{1}{2} \tag{1.156}$$

Thus we have

$$|x - x_c^*| = .00\cdots 0a_{-t-1}a_{-t-2}\cdots(b) = a_e(b) \tag{1.157}$$

Observe that the magnitude of the unnormalized mantissa a_e is always less than 2^{-t} even if every bit after the tth bit is 1. Now dividing Eq. 1.157 by $|x|$ and using the fact that $a_e \leqslant 2^{-t}$ and $|a| \geqslant 1/2$, we have

$$\frac{|x - x_c^*|}{|x|} = \frac{a_e(b)}{|a(b)|} = \frac{a_e}{|a|} \leqslant 2^{-t+1} \tag{1.158}$$

Let $(x - x_c^*)/x$ be denoted by $-\delta$, then we have from Eq. 1.158 that

$$x_c^* = x(1+\delta) \qquad |\delta| \leqslant 2^{-t+1} \tag{1.159}$$

For an example, using a 7-bit mantissa and 3-bit exponent word we are able to represent numbers with magnitudes ranging from 0 to approximately 128; hence, 21.33 is representable using this word format. The round-off error using chopped approximation is given in Eq. 1.152. It is easily seen that

$$|\delta| = \frac{|x - x_c^*|}{|x|} = \frac{.\,0000000010 \cdots (000)}{.\,1010101010 \cdots (000)}$$

$$\leqslant \frac{.\,0000000100\,(000)}{.\,1000000000\,(000)} = 2^{-7}$$

which is certainly less than 2^{-t+1} or 2^{-6}.

Equation 1.159 is the basic equation used for finding the errors incurred when arithmetic operations are performed on machine numbers. We can similarly derive Eq. 1.153 for rounded approximations.

Floating-Point Operations

Once the given numbers are represented in the machine as machine numbers, then arithmetic operations can be performed to get the desired result. For easy illustration, we will only discuss the various operations and the associated errors incurred in normalized chopped floating operations. Consequently the subscript c is also dropped from all x_c^*.

Additive Operations. To add or subtract two binary floating-point machine numbers, the exponent of the number with a smaller magnitude is first raised such that it is equal to that of the number with a larger magnitude. This implies the adding of leading zeroes and the shifting of significant figures to the right in the mantissa of the smaller number. Once the exponent parts of the two numbers are equal, then addition or subtraction is done simply by adding or subtracting the corresponding mantissa of these two numbers. The final result is then normalized (so that $a_{-1} = 1$) and chopped to a t-bit mantissa for subsequent use. Note that more than t bits are needed to store the mantissa of the shifted number or the final unnormalized result. For example, let $t = 5$ and $x^* =$ .11101(101), $y^* =$ .11101(100). To obtain $(x^* + y^*)$ and $(x^* - y^*)$, the exponent of y^* must be raised to 101 and consequently the bits in the mantissa of y^* must be shifted one bit to the right. Thus we have

$$\begin{aligned} x^* &= .11101(101) \\ y^* &= .011101(101) \end{aligned} \tag{1.160}$$

The exact results are then obtained as

$$
\begin{aligned}
x^* + y^* &= .1010111(110) \\
x^* - y^* &= .11101(100)
\end{aligned}
\tag{1.161}
$$

and the chopped results are given by

$$
\begin{aligned}
(x^* + y^*)^* &= .10101(110) \\
(x^* - y^*)^* &= .11101(100)
\end{aligned}
\tag{1.162}
$$

Note in this example the final chopped result of subtraction is exact. Applying Eq. 1.153 to the exact results $(x^* \pm y^*)$ and the computed results $(x^* \pm y^*)^*$, we have in general that

$$
(x^* \pm y^*)^* = (x^* \pm y^*)(1+\delta) \qquad |\delta| \leqslant 2^{-t+1} \tag{1.163}
$$

Thus the round-off error is small if the magnitude of $(x^* \pm y^*)$ is small. However, if the two numbers are nearly equal, the significant figures of the result of subtraction might be reduced drastically. For example if $x^* = .11101(001)$ and $y^* = .11100(001)$ with five significant figures, then

$$
(x^* - y^*)^* = x^* - y^* = .10000(-011) \tag{1.164}
$$

While the round-off error is zero, the final result has only one significant figure.

The effect of loss of significant figures upon the accuracy of the computation can be seen best through the following problem. Suppose it is desired to find the smaller root of a quadratic equation

$$
ax^2 + bx + c = 0 \tag{1.165}
$$

where $b > 0$. The solution is easily found as

$$
x = \frac{-b + (b^2 - 4ac)^{\frac{1}{2}}}{2a} \tag{1.166}
$$

Let us assume that $b^2 \gg 4ac$. Then the numerator in Eq. 1.166 is the result of subtraction of two nearly equal numbers and thus some loss of significant figures is expected in the process of computing x. Specifically let $a = 1$, $b = 1.001$, and $c = 0.001$ be decimal numbers; then the exact answer is $x = 0.001 = .100000(-2)$. Using decimal floating-point operations with a 6-digit mantissa assigned for each number, the *computed* x is $.100050(-2)$ with *four* significant figures.

On the other hand, if we multiply both the numerator and the denominator of Eq. 1.169 by $(-b-\sqrt{b^2-4ac})$, then

$$x=\frac{2c}{b+(b^2-4ac)^{\frac{1}{2}}} \tag{1.167}$$

Using Eq. 1.167, the solution is found as .100000(−2) that has *six* significant figures. Thus it is important to avoid the loss of significant figures if it is possible.

Multiplicative Operations. For multiplication or division, the exponents of the two numbers are added or subtracted first, then the mantissa parts are multiplied or divided accordingly. Finally the result is normalized so that the first bit of the mantissa is 1. Consider two numbers $x^*=.11101(101)$ and $y^*=.11100(100)$ with $t=5$; we have

$$\begin{aligned} y^*x^* &= .1100101100(1001) \\ y^*/x^* &= .111101\cdots(-001) \end{aligned} \tag{1.168}$$

And the final results are

$$\begin{aligned} (y^*x^*)^* &= .11001(1001) \\ (y^*/x^*)^* &= .11110(-001) \end{aligned} \tag{1.169}$$

It can easily be shown that Eqs. 1.168 and 1.169 satisfy the general equations

$$\begin{aligned} (y^*x^*)^* &= y^*x^*(1+\delta_1) \\ (y^*/x^*)^* &= (y^*/x^*)(1+\delta_2) \end{aligned} \qquad |\delta_1|, |\delta_2| \leqslant 2^{-t+1} \tag{1.170}$$

Thus in general the round-off errors will be larger if two large numbers are multiplied or a large number is divided by a small number. These observations are useful when we discuss the method for solving systems of linear algebraic equations in later chapters.

Signal Representation

We have seen how numbers are represented and manipulated in a digital computer. Since a digital computer can only store and manipulate finite digit numbers, continuous functions can be handled by the computer only as sets of numbers. Usually these are discrete sets of numerical values (samples) the functions assume at equally spaced time instants. Thus if the function is given, the approximation samples are easily obtained. On the other hand, if the numerical samples are obtained as the intermediate output of the computer and it is desired to find approxi-

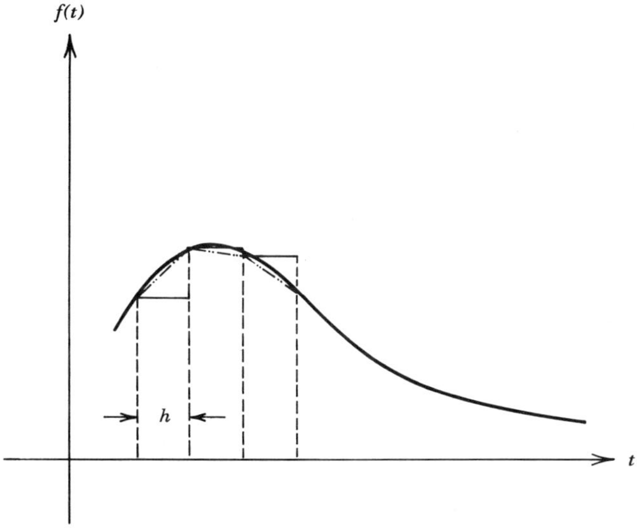

Fig. 1.64 Approximation of functions.

mations for the function values in-between the sampling instants, then we may either assume constant values from one time instant to another or a straight line can be drawn connecting each adjacent pair of samples as is shown in Fig. 1.64. We see that in general better approximations can be obtained using the latter approach. More elaborate methods for finding better approximations are also available. For any method it is generally true that the more frequent is the sampling, the closer is the approximation we can have.

The computing methods or algorithms used in obtaining the circuit response evaluate the successive samples of the response one at a time. The accuracy of the computation depends again on the sampling interval h. In general the error in evaluating the response is small if h is small. There is, of course, no simple way to determine a reasonable value of h. It depends on the particular problem at hand. A simple check on the results is obtained by recomputing with a new $h' = h/2$. If the results are almost the same, then they are considered of acceptable accuracy.

PROBLEMS

1-1. For a certain voltage source we have

$$v(t) = \sin 2t$$

(i) Find the voltage readings of the voltmeter V at

$$t = 0, \pi/4, \pi/2, 3\pi/4, \pi.$$

(ii) Interchange the terminals of V and repeat (i).

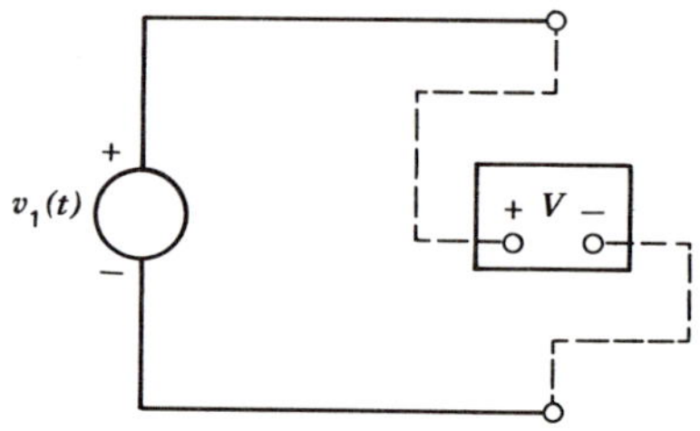

Fig. P1-1

1-2. Find the voltmeter readings at $t = 0, \pi/4, \pi/2, 3\pi/4$, and π of the following circuit with

$$v_1(t) = \sin 2t \qquad i_2(t) = 2\cos 2t$$

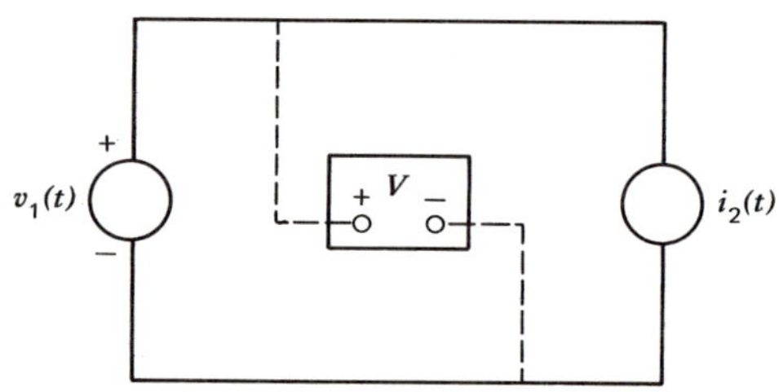

Fig. P1-2

1-3. In Fig. P1-3, we have

$$v_1(t) = 2v \qquad v_2(t) = 4v \qquad i(t) = 10a$$

(i) Find the power delivered to the circuit elements between pairs of nodes ab, ac, cd, and bd at $t = 0, \pi/4, \pi/2, 3\pi/4, \pi$.
(ii) Repeat (i) for $v_1(t) = \sin 2t$.
(iii) Find the energy delivered to those elements in (i) and (ii) for the time period $t = 0$ to $t = \pi$.
(iv) Repeat (i) if ab are short-circuited.

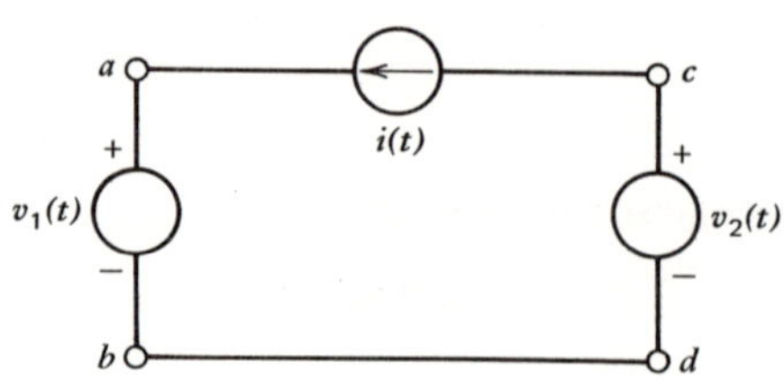

Fig. P1-3

1-4. For the circuit in Fig. P1-4, we have $i(t) = \sin 2t$.
(i) Find the voltage $v_L(t)$ across the inductor L.
(ii) Repeat (i) if the direction of $i(t)$ is reversed.

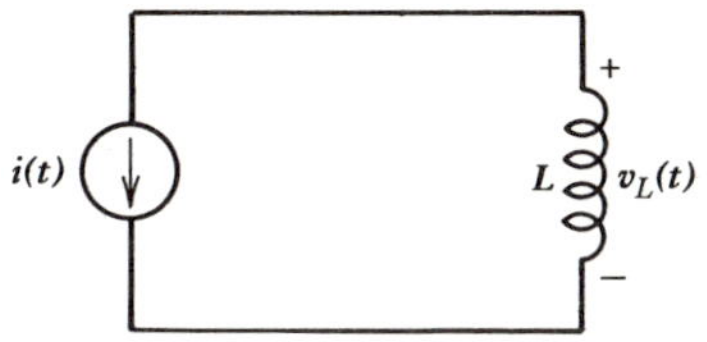

Fig. P1-4

1-4*a*. The practical voltage source can be represented as an ideal voltage source $v_s(t)$ in series with an internal resistance R_s as shown in Fig. P1-4*a*. Write the KVL equation and solve for $i(t)$. Plot $v_r(t)$ versus $i(t)$ for $R = 0\Omega$, 10Ω, 100Ω, $1\text{k}\Omega$.

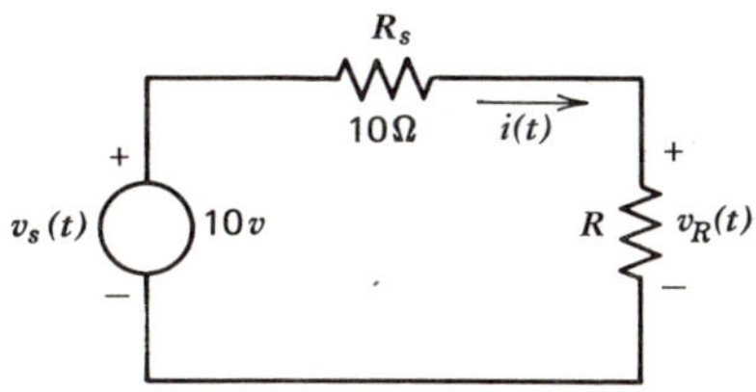

Fig. P1-4*a*

1-4*b*. For the following circuit, the current source $i(t)$ is given as $i(t) = (1 - e^{-t})$ amp.
(i) Find the voltage $v_L(t)$ across the inductor L.
(ii) Find the power delivered by the current source $i(t)$.
(iii) Find the power dissipated in the circuit.
(iv) Are the results of (ii) and (iii) equal? If not, explain.

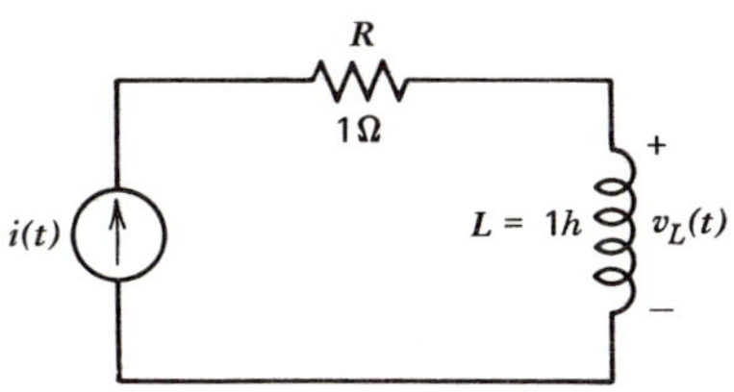

Fig. P1-4*b*

1-4*c*. For the circuit of Fig. P1-4, the excitation $i(t)$ is recorded as a set of numbers versus a discrete set of time instants. Find the voltage $v_L(t)$ approximately by using the approximation

$$v_L(t_0) \cong L \cdot \left[\frac{i(t_0+h) - i(t_0)}{t_0+h-t_0}\right]$$

for the following recorded table:

t(sec.)	0.0	0.1	0.2	0.3	0.4	0.5	0.6	0.7	0.8	0.9	1.0
$i(t)$(amp)	0.0	0.5	0.5	0.7	0.7	0.9	0.8	0.5	0.3	0.2	0.1

1-4*d*. Repeat 1-4c for the following set of data:

t(sec.)	0	0.1	0.25	0.4	0.5	0.65	0.75	0.85	0.95	1.0
$i(t)$(amp)	0	0.5	0.7	0.9	0.8	0.95	0.90	1.25	1.00	0.8

1-5. Find the current through the capacitor in Fig. P1-5 if we have

$$v(t) = e^{-t}u(t)$$

where $u(t)$ is a unit-step function. Is the capacitor short-circuited at $t = 0$? How large is the current at $t = 0$?

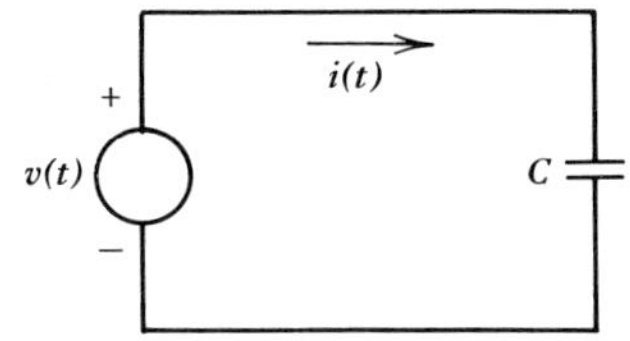

Fig. P1-5

1-5a. For the circuit of Fig. P1-5*a*, we have $v(t) = e^{-t}v$, $R = 1\Omega$, and $C = 1f$.
(i) Find $i_R(t)$ and $i_C(t)$.
(ii) Find the power delivered to the circuit by the source.
(iii) Find the power dissipated in the circuit.
(iv) Explain any difference between the results of (ii) and (iii).

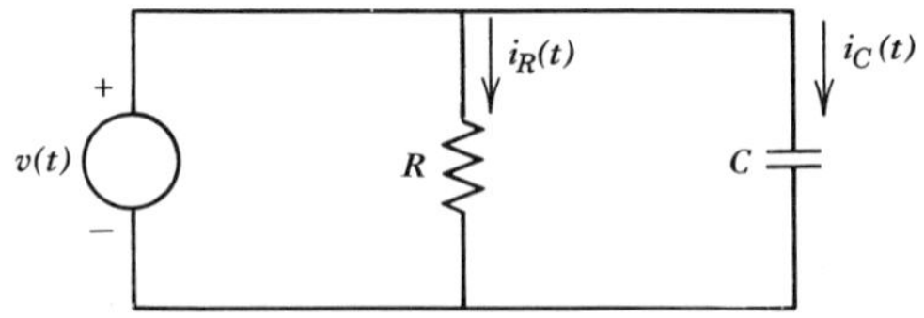

Fig. P1-5*a*

1-6. For the circuit shown in Fig. P1-6,
(i) Determine the number of independent KVL and KCL equations.
(ii) Write the KVL and KCL equations for the circuit.

(iii) If you are required to find the current through the voltage source, which set of equations would you choose? Why?

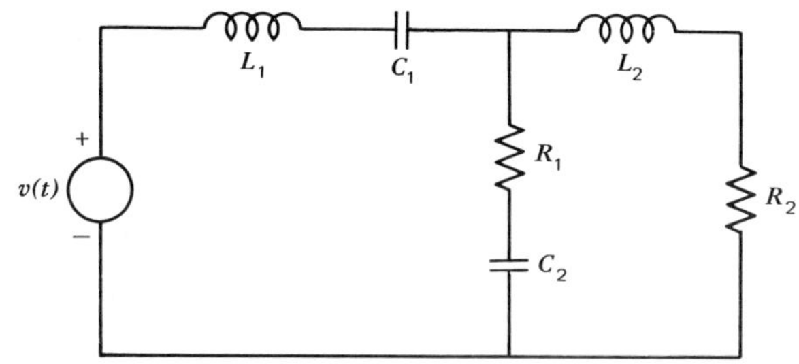

Fig. P1-6

1-6a. Repeat 1-6(i) and (ii) for the circuit shown in Fig. P1-6*a*.

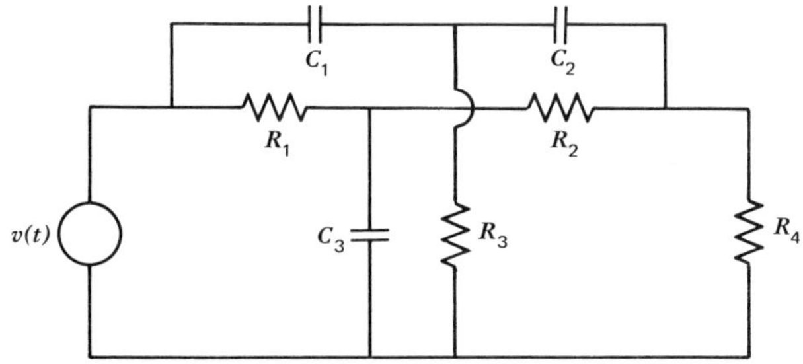

Fig. P1-6*a*

1-6b. Repeat 1-6(i) and (ii) for the circuit shown in Fig. P1-6*b*.

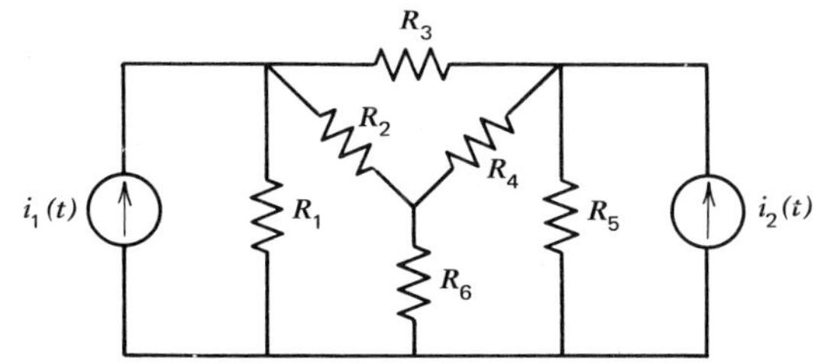

Fig. P1-6*b*

1-6c. Repeat 1-6(i) and (ii) for the circuit shown in Fig. P1-6*c*.

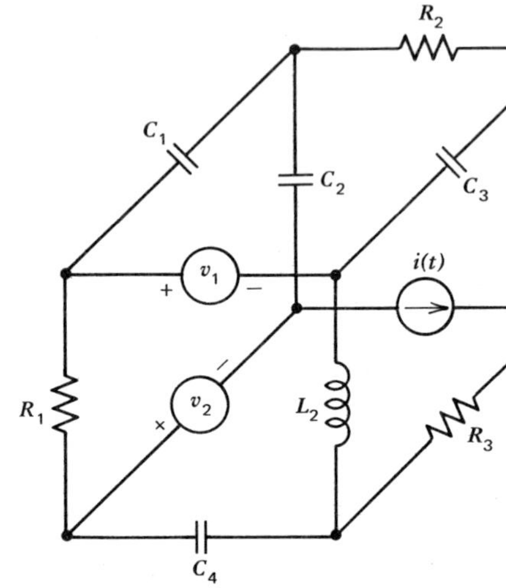

Fig. P1-6c

1-7. In order to solve 1-6, we need to know the initial states of the reactive elements. What is the vector representing the state of the circuit? The initial state?

1-7a. Repeat 1-7 for the circuit of Fig. P1-6*a*.

1-7b. Repeat 1-7 for the circuit shown in Fig. P1-6*c*.

1-7c. For the circuits shown in Fig. P1-7*c*, find the voltages $v(t)$ across the current sources $i(t)$. Are they equal?

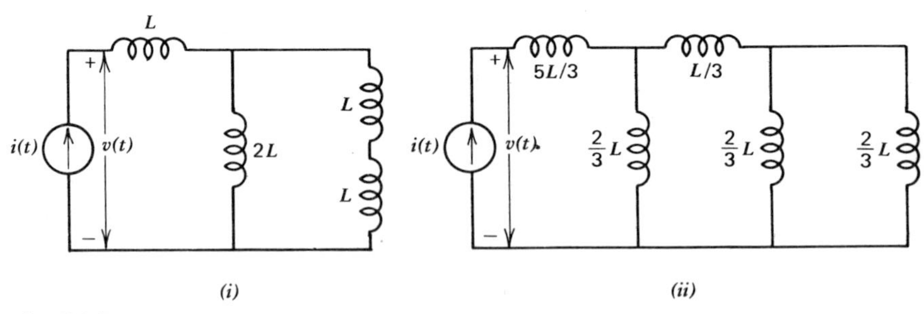

Fig. P1-7c

1-7d. The circuit shown in Fig. P1-7*d*(i) is composed of coupled inductors. Find an equivalent inductor L_{eq} such that the voltage across the same current source $i(t)$ of Fig. P1-7*d*(ii) is the same as that of Fig. P1-7*d*(i).

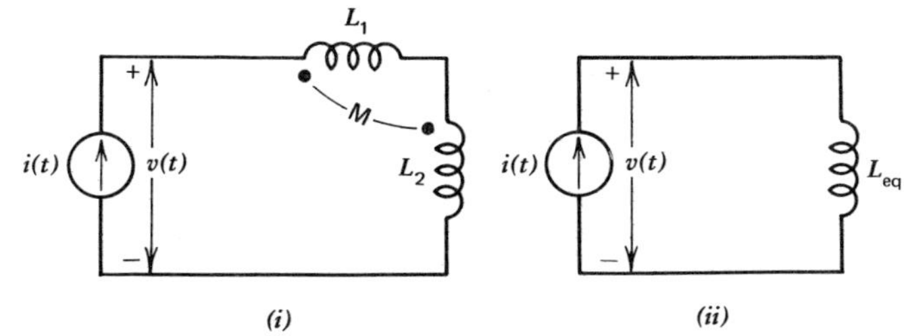

Fig. P1-7*d*

1-7e. Repeat 1-7d for the circuits shown in Fig. P1-7*e*.

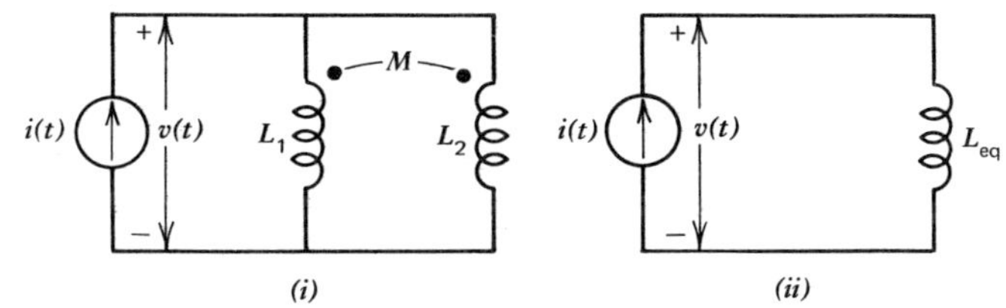

Fig. P1-7*e*

1-8. Specify for each of the following systems whether it is linear, causal, or whether insufficient information is given. The response $r(t)$ to the excitation $e(t)$ is
(i) $r(t) = \max(e(t), e(t-1))$.
(ii) $r(t) = 1 + e(t) \sin wt$.
(iii) $r(t) = \sin\{[1+e(t)]wt\}$.
(iv) $r(t) = 3e(t) + e'(t) + e(t^2)$.

1-9. For the network in Fig. P1-9, the voltage-current relationship is given as

$$v(t) = -Ki(t)$$

Let the excitation current be given as

$$i(t) = \sin(t/\pi)$$

Find the energy delivered to the network from $t = 0$ to $t = \pi$. Is this network passive?

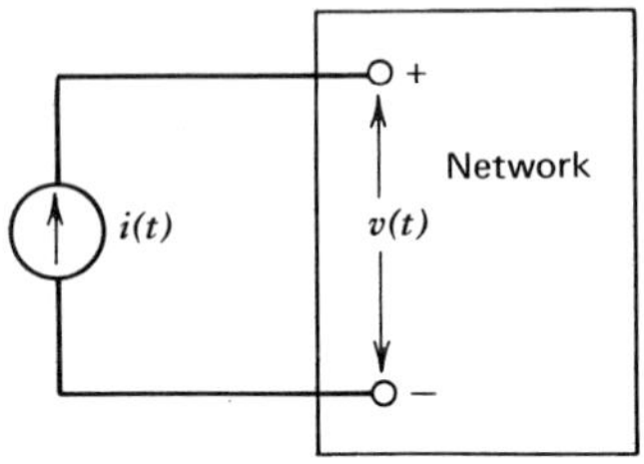

Fig. P1-9

1-9a. For the unit-coupled circuit in Fig. 1-9a we have $L_1 = 1h$, $L_2 = 4h$, and $M = 2h$. Find $i_R(t)$ for the following excitations:

(i) $v(t) = u(t)$
(ii) $v(t) = e^{-t}u(t)$
(iii) $v(t) = \sin t$

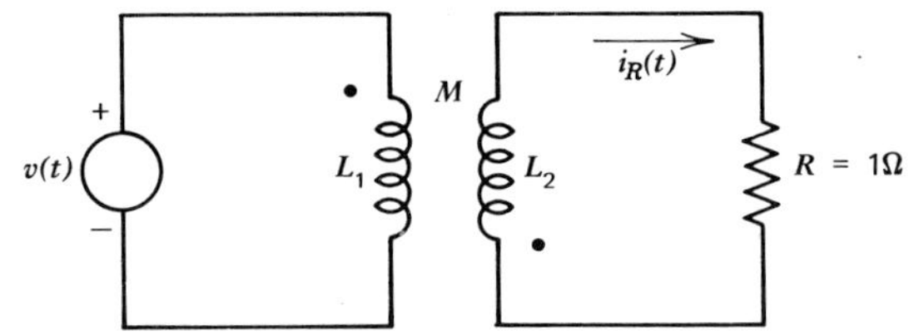

Fig. P1-9*a*

1-9b. Find the voltage response of the coupled circuit shown in Fig. P1-9*b* under the following excitations:

(i) $i_1(t) = i_2(t) = 10a$.
(ii) $i_1(t) = 10 \sin t$, $i_2(t) = 10 \cos t$.
(iii) $i_1(t) = 10e^{-t}$, $i_2(t) = 10e^{-t} \cos t$.

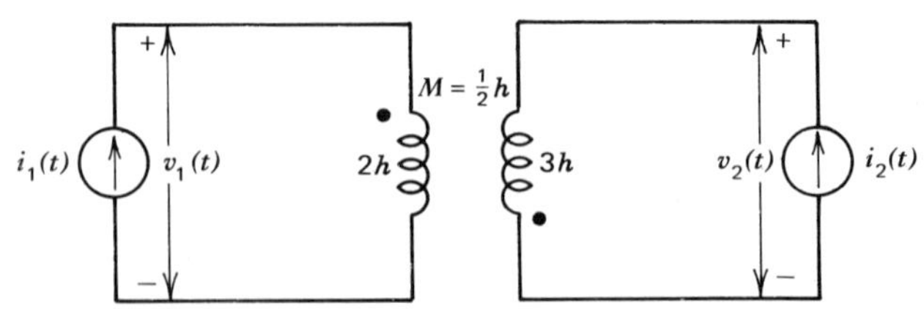

Fig. P1-9*b*

1-9c. For the ideal transformer shown in Fig. P1-9*c*;

(i) Find $i_R(t)$ for $v_1(t) = 10 \sin t$ and $R = 1\Omega$.
(ii) Find $v_1(t)$ if $i_R(t) = \sin t$.
(iii) Find $i_1(t)$ for both (ii) and (i).

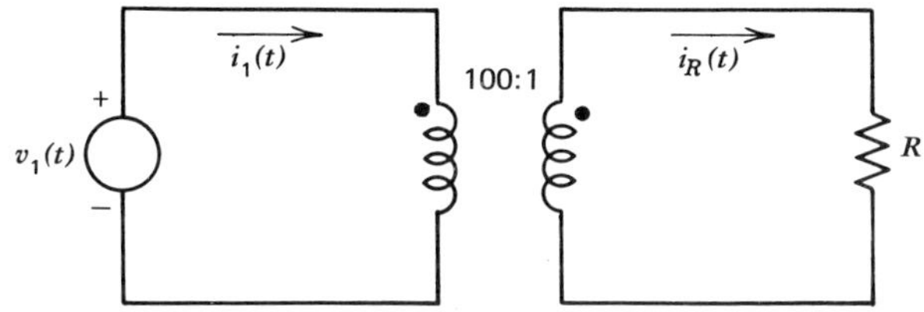

Fig. P1-9*c*

1-9d. The circuit shown in Fig. P1-9*d* is a voltage-controlled current source. It is given that $v_1(t) = 4$ volts and $K_1 = 10$ amp/volt. Find $i_R(t)$ and $v_R(t)$ for (i) $R = 0\Omega$, (ii) $R = 10\Omega$, (iii) $R = \infty\Omega$.

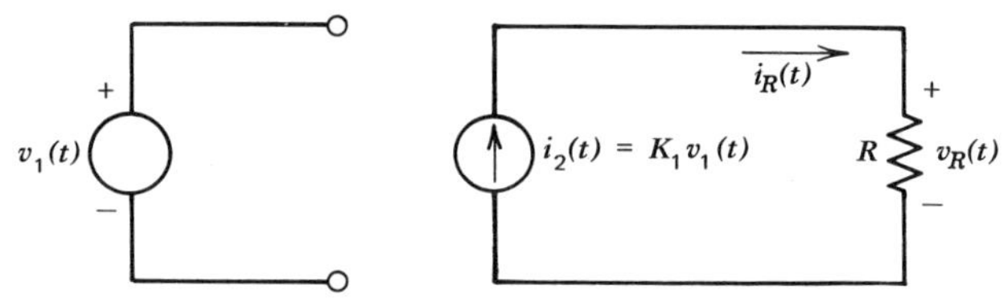

Fig. P1-9*d*

1-9e. Repeat 1-9d for the current-controlled voltage source shown in Fig. P1-9*e* with $K = 10$ volts/amp and $i_1(t) = 4$ amps.

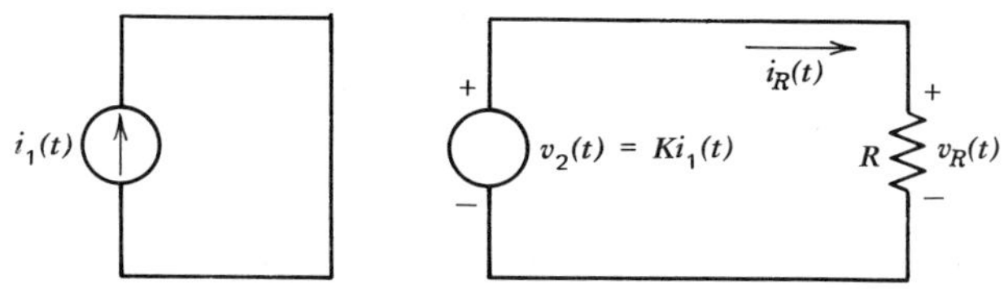

Fig. P1-9*e*

1-10. Write the equation for the waveforms of Fig. P1-10 using shifted step functions.

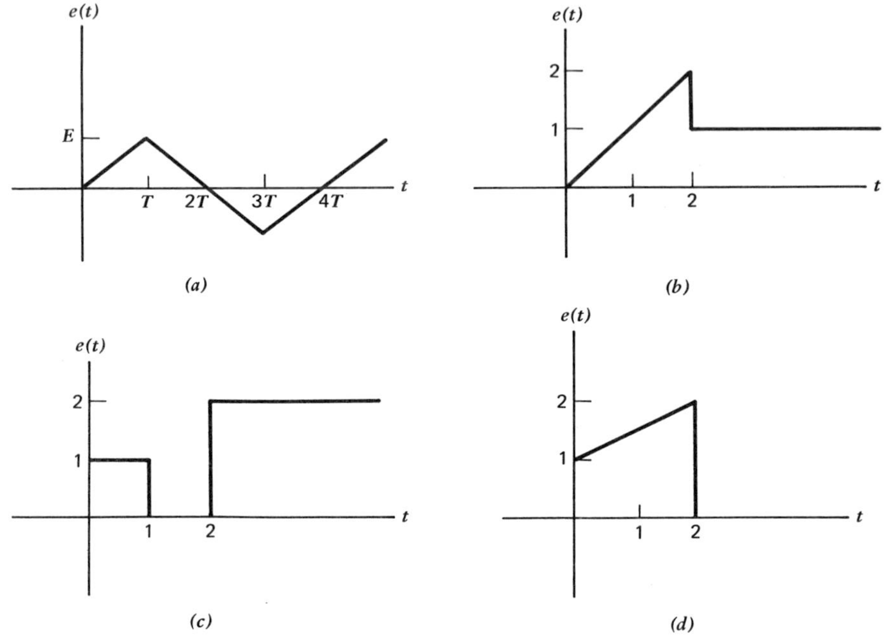

Fig. P1-10

1-11. Plot the derivatives of the waveforms of 1-10 and write the equations of the derivatives using shifted step and/or impulse functions.

1-12. Plot

(i) $e(t) = u(\cos t)$.

(ii) $e(t) = \delta(t - \sin t)$.

1-13. Evaluate the following integrals:

(i) $I = \int_{-\infty}^{\infty} \delta(t - T_1) u(t - T_2)\, dt,\ T_2 < T_1.$

(ii) $I = \int_{-\infty}^{\infty} \delta(w - w_0) \cos wt\, dw.$

(iii) $I = \int_{-\infty}^{\infty} [\delta(t) - A\delta(t - T_1) + 2A\delta(t - T_2)] e^{-jnwt}\, dt.$

1-14. For the waveforms shown in Fig. P1-14,

(i) Determine K such that $I = \int_{-\infty}^{\infty} f(t)\, dt = 0.$

(ii) Determine K such that $I = \int_{0+}^{\infty} f(t)\, dt = 0.$

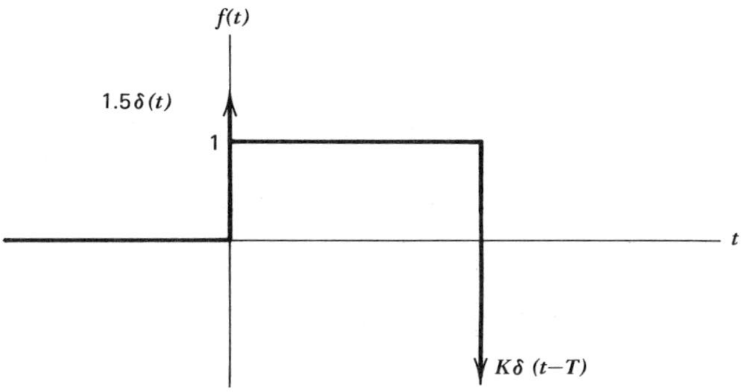

Fig. P1-14

1-15. If the response of a time-invariant system due to an impulse excitation $e(t) = \delta(t)$ is $r(t) = e^{-t}u(t)$, determine the response due to an excitation

$$e(t) = 2\delta(t-1) - 2\delta(t-2)$$

1-16. Find the binary equivalent of the following decimal numbers: (a) 3.14159 (b) 2.71828 (c) 142.59 (d) 407.391 (e) 4029 (f) 0.003345.

1-17. Find the chopped and rounded approximations to each of the binary numbers found in 1-16, using 10 significant bits.

1-18. Convert the results of 1-17 to a normalized floating-point format with a 10-bit mantissa. How many bits are necessary to represent all of the numbers thus found?

1-19. Convert the binary numbers found in 1-17 back to decimal numbers. Compare the results with the original numbers given in 1-16. How large are the absolute errors?

1-20. Verify Eq. 1.153 using the results of 1-18 and 1-16.

1-21. Given two machine numbers (binary numbers)

$$x^* = 0.110110(101)$$
$$y^* = 0.111101(010)$$

verify Eqs. 1.163 and 1.170 by performing the following normalized floating-point operations in chopped format:

(a) $x^* \times y^*$ (b) x^*/y^* (c) $x^* + y^*$ (d) $x^* - y^*$

1-21a. Repeat 1-21 for the following set of x^* and y^*:

(i) $x^* = 0.100010(001), y^* = 0.110111(011)$
(ii) $x^* = 0.110011(101), y^* = 0.111011(101)$
(iii) $x^* = 0.110001(000), y^* = 0.101010(111)$

1-22. Write equations similar to Eqs. 1.163 and 1.170 for rounded operations. Repeat 1-21 to verify your new equations.

1-22a. Repeat 1-22 for the numbers given in 1-21a.

1-22b. We have seen from Eq. 1.153 that if the number x is representable using a machine word with t-bit mantissa a, and the associated exponent b, then the machine approximation x^* of x satisfies

$$x^* = x(1+\delta) \begin{cases} |\delta| \leqq 2^{-t+1} & \text{for chopped format} \\ |\delta| \leqq 2^{-t} & \text{for rounded format} \end{cases}$$

Following similar arguments in deriving Eq. 1.159, show that the chopped machine approximation x^* of x satisfies also the equation

$$x^*(1+\Delta) = x \qquad |\Delta| \leqq 2^{-t+1}$$

1-22c. From 1-22b the following computational equations for chopped floating-point arithmetic can easily be proved:

$$\begin{aligned} (x^* \pm y^*)^*(1+\Delta) &= x^* \pm y^* \\ (x^* y^*)^*(1+\Delta') &= x^* y^* \qquad\qquad |\Delta|, |\Delta'|, |\Delta''| \leqq 2^{-t+1} \\ (x^*/y^*)^*(1+\Delta'') &= x^*/y^* \end{aligned}$$

Verify the above equations using the results of 1-21.

1-22d. Repeat 1-22c for the set of x^* and y^* given in 1-21a.

1-22e. Results similar to that of 1-22c can also be obtained for chopped decimal floating-point operations. In fact, we have

$$\begin{aligned} (x^* \pm y^*)^*(1+\theta) &= x^* \pm y^* \\ (x^* y^*)^*(1+\theta') &= x^* y^* \qquad\qquad |\theta|, |\theta'|, |\theta''| \leqq 10^{-t+1} \\ (x^*/y^*)^*(1+\theta'') &= x^*/y^* \end{aligned}$$

where t is the number of digits used for the mantissa. Given $x^* = 0.92345(5)$ and $y^* = 0.56789(-2)$, verify the above equations using 5-digit chopped floating-point operations.

1-22f. Repeat 1-22e for the following set of x^* and y^*:

(i) $x^* = 0.23479(1), y^* = 0.12345(10)$.
(ii) $x^* = 0.10009(12), y^* = 0.79865(12)$.
(iii) $x^* = 0.27358(6), y^* = 0.99799(12)$.

CHAPTER 2
Transient Analysis I: First-Order Systems

The necessary elementary concepts describing the signals, the circuit elements, the equilibrium laws of circuits, and the state of the circuit are discussed in Chapter 1. We will now discuss the analysis problem. The circuit parameters (i.e., the element values), the initial state of the circuit, and the specific excitation are given, and the circuit response is determined.

A voltage or current response signal anywhere in the circuit can be determined either from all of the node voltages with respect to the appropriate reference node(s) or from all of the independent loop currents. Our analysis objectives, therefore, will be to determine such node voltages or loop currents.

We begin by a consideration of some simple circuits that contain only one reactive element (capacitor or inductor). Specifically, the circuits considered in this chapter consist of a single resistor in series or in parallel with a capacitor or an inductor. The sources that excite these circuits are either constant excitations or exponential excitations.

Circuit equations for circuits with one reactive element give rise to first order differential equations. Both the explicit solutions and the numerical methods for solutions of such first order systems are considered.

2.1 A PARALLEL RC CIRCUIT UNDER CONSTANT EXCITATION

Consider a very simple circuit with a resistor and a capacitor in parallel driven by a current source as shown in Fig. 2.1, and assume the source current to be a constant $i(t) \equiv I$. Initially, the current is going through

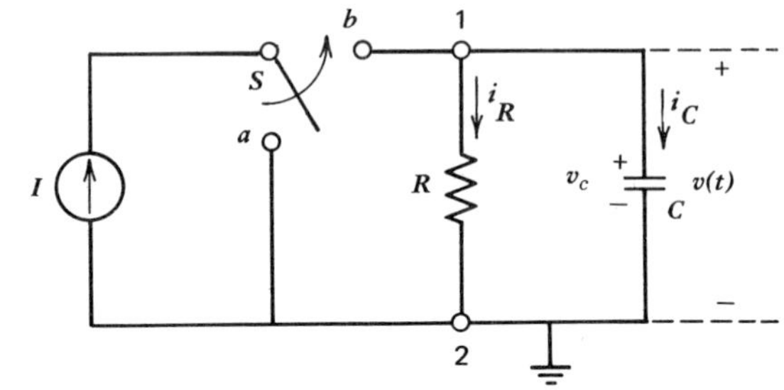

Fig. 2.1 Parallel *RC* circuit driven by a current source.

the short circuit, but at time $t = t_0$, the switch S is moved from position **a** to position **b**, and the current starts flowing into the circuit; that is, the entire circuit is connected together at $t = t_0$ and remains so for all $t \geqslant t_0$.

The initial state of the circuit is the initial voltage across the capacitor specified at $t = t_0$ an instant before the circuit connection to the source. We have observed in Chapter 1 that the state cannot change instantaneously, and it must have the same value at t_{0+}. We, therefore, need specify the initial state merely at t_0 and refer to t_{0-} or t_{0+} only when it is essential to distinguish between the instant before and after circuit connection.

For the circuit of Fig. 2.1, we thus specify the initial state or the initial voltage across the capacitor at $t = t_0$, that is, $v_C(t_0) = v_{C0}$. The voltage of node 1, with node 2 specified as the reference, is given by $v(t)$. The current in the resistor is i_R and that in the capacitor is i_C.

We are interested in determining the circuit behavior, in terms of $v(t)$, for all time t after the instant of connection t_0 (i.e., for all $t \geqslant t_0$).

Before we proceed with a mathematical analysis, let us examine the circuit behavior in an intuitive manner. The capacitor voltage at $t = t_{0-}$ is v_{C0} and cannot change instantly. Just before the switch is moved, the current in the resistor R is given by the voltage across it divided by R, that is, $i_R(t_{0-}) = v_{C0}/R$. Since the same current flows through all elements of a closed circuit, the current through the capacitor at $t = t_{0-}$ is $i_C(t_{0-}) = -i_R(t_{0-})$. Now at $t = t_{0+}$, that is, just after the current source is connected, the voltage v_{C0} across the capacitor remains the same. Consequently, the current through the resistor remains the same, or $i_R(t_{0+}) = v_{C0}/R$. The current through the capacitor becomes, according to KCL, $i_C(t_{0+}) = (I - v_{C0}/R)$. As this current starts flowing through the capacitor, the latter is charged and the voltage across it rises. As $v(t)$ rises, so does the current through the resistor. In the limit the current through the resistor, i_R, reaches the value I that through the capacitor must reach zero by KCL. With i_C reaching zero, there is no further rise in $v(t)$ which has reached IR.

Observe that i_R starts at v_{C0}/R and reaches I. The current i_C starts at $(IR - v_{C0}/R)$ and reaches zero. The voltage $v(t)$ starts at v_{C0} and reaches IR. The final values are maintained as long as the circuit is unchanged. These final values are also called steady-state values since the state of the circuit has reached some uniform or steady conditions.

A final note on the initial behavior of the capacitor. If $v_{C0} = 0$, then the initial current through the capacitor is the entire source current I, with $i_R = 0$. Initially, therefore, the capacitor behaves like a short-circuit drawing all the current through it. Such an interpretation is a common aid to intuitive thinking. Of course, appropriate adjustments must be made if the capacitor is initially charged.

Intuitive thinking is of the utmost importance in a real understanding of circuits. While we will engage for the most part in mathematical analysis, the reader is encouraged to develop his intuitive thinking in the above manner.

We now proceed with the mathematical analysis of the circuit for all $t \geqslant t_0$. The current $i(t)$ exciting the circuit is zero for $t < t_0$ and equal to I for $t \geqslant t_0$. Therefore,

$$i(t) = Iu(t - t_0) \tag{2.1}$$

where $u(t - t_0)$ is the unit step function with the step at $t = t_0$. Our interest, however, is in the circuit behavior for $t \geqslant t_0$, and the excitation current may be stated as I for $t \geqslant t_0$.

For the sake of consistency, we will express our equilibrium equations such that the known quantities are on the right-side side of the equations with the unknowns on the left-hand sides. As a rule, we will also assign positive directions to currents flowing out of a node.

For our circuit, the KCL states that the sum of currents going out of the node 1, i_R plus i_C, must equal current coming into the node that is the current i due to the generator or source.

$$\text{KCL} \qquad i_R + i_C = i \tag{2.2}$$

Now the current going through the capacitor is (by Coulomb's law)

$$i_C = C\frac{dv_C}{dt} = C\frac{dv}{dt} \tag{2.3}$$

By Ohm's Law, the current through the resistor is

$$i_R = \frac{v_R}{R} = \frac{v}{R} \tag{2.4}$$

The generator current $i = Iu(t - t_0)$, and the *circuit equilibrium* equation is

$$\text{CE} \qquad \frac{v}{R} + C\frac{dv}{dt} = Iu(t - t_0) \tag{2.5}$$

This equation is a *differential equation,* so named because the variable v is expressed in terms of the derivative dv/dt and v itself. The known function on the right-hand side of a differential equation is called the *driving* or *forcing function.* The solution we would like to obtain is the voltage $v(t)$. Our interest is in the behavior for $t \geqslant t_0$, and the differential equation becomes

$$\text{DE} \qquad \frac{v}{R}+C\frac{dv}{dt}=I \qquad t \geqslant t_0 \tag{2.6}$$

Driven Response

Since the driving function for $t \geqslant t_0$ is a constant, I, we may be able to find a solution v (of the differential equation) which is also a constant, that is, $v(t) = V$, a constant. Let us substitute this constant value, $v(t) = V$, in Eq. 2.6 to obtain

$$\frac{V}{R}+0=I$$

or

$$V = RI \tag{2.7}$$

Thus, the constant $V = RI$ is a solution of Eq. 2.6. This solution depends directly on the driving function I on the right-hand side of Eq. 2.6, and it is zero if the driving function is zero. This solution is, therefore, called the *driven solution* or *driven response* or *forced response*.

Observe that something is not quite right about this solution. From our intuitive thinking, we saw that $v(t)$ should have the value v_{C0} at $t = t_0$; and when t becomes large enough, the final value of $v(t)$ settles to IR. The actual solution for $v(t)$ must have, therefore, an additional term that provides a correct initial value and that vanishes for large t to leave the final value of the driven response. We determine such an additional term below.

Homogeneous Equation

Is the driven response the only possible solution or is it possible to have another solution to the differential equation? On the basis of the above discussion we will assume that there is another solution v, such that

$$\begin{aligned} v &= v_1 + v_2 \\ &= V + v_2 \end{aligned} \tag{2.8}$$

where v_2 = unknown, and $v_1 = V = RI$. Substituting this into the differential equation, Eq. 2.6, we find the condition that v_2 must satisfy, so that the new solution v is valid. The differential equation becomes:

$$\frac{V}{R}+\frac{V_2}{R}+\frac{dv_2}{dt}=I$$

and since $V/R = I$, we have

$$\frac{v_2}{R}+C\frac{dv_2}{dt}=0 \tag{2.9}$$

Observe that the new differential equation is the same as the original differential equation (Eq. 2.6) with the forcing function set to zero. A differential equation with the right-hand side set to zero has a special name, the *homogeneous equation*.

Thus, the solution to the homogeneous equation added to the driven response is a solution to the differential equation.

To solve the homogeneous equation we separate the variables in Eq. 2.9

$$\frac{dv_2}{v_2}=-\frac{1}{RC}dt$$

and integrate both sides to give

$$\ln v_2=-\frac{1}{RC}t+\text{constant}$$

or

$$v_2=Ae^{-t/RC} \tag{2.10}$$

where A is an arbitrary constant yet to be determined.

Total Solution

The general solution v is the driven response plus the solution to the homogeneous equation (Eq. 2.8):

$$v(t)=RI+Ae^{-t/RC} \tag{2.11}$$

Notice in Eq. 2.11 that the second term of the solution decreases with increasing time and approaches zero in the limit as t approaches infinity. The entire second term thus ultimately vanishes leaving only the driven solution. The term that vanishes with increasing time is called, therefore, the *transient* response or solution. The driven response is what remains as the response after the transient has vanished. The driven response is sometimes also called the *steady-state* response. (In rare cases, transient responses do not vanish, and steady-state responses are zero.)

The total solution for $v(t)$ has an arbitrary constant A that must be determined before the solution is complete. If the value of the variable $v(t)$ were precisely known at *any* instant of time, it is clear from Eq. 2.11

that the value of the constant A would also be determined. Initially at $t = t_0$, before the circuit is connected together, the initial state of the circuit is known and it cannot change in zero time. Hence, the initial voltage at $t = t_{0+}$ is given by $v(t_{0+}) = v_C(t_{0+}) = v_C(t_{0-}) = v_{C0}$. Observe that the value of any desired voltage or current variable at the initial instant can be determined from the initial state of the circuit. These initial values of the variables necessary to solve the differential equations are generally known as *initial conditions.* The desired initial conditions can be determined from the initial state, but the two are not always the same. One refers to the variables in the differential equations; the other refers to the energy variables for reactive elements in the circuit.

Remember that it is the initial state that is continuous at t_0 (i.e., it has the same value at t_{0-} and at t_{0+}). The initial condition may or may not be continuous at t_0. For problems defined for $t \geqslant t_0$, the initial conditions are desired at $t = t_{0+}$. These are determined from the initial states at $t = t_{0+}$, which in turn are determined from their values at $t = t_{0-}$.

The instants t_{0-} and t_{0+} will be used only when we wish to draw special attention to them. Otherwise, t_0 will be used when there is no possibility of confusion. Usually, t_0 will refer to t_{0+} whenever the circuit behavior is of interest for $t \geqslant t_0$.

The initial condition for our case is the value of $v(t)$ at $t = t_0$,

$$v(t_0) = v_C(t_0) = v_{C0} \tag{2.12}$$

In this case it so happens that the initial condition is the same as the initial state of the circuit. Substituting the value of $v(t_0)$ above in Eq. 2.11, we obtain

$$\begin{aligned} v(t_0) &= RI + Ae^{-t_0/RC} \\ &= v_{C0} \end{aligned}$$

or

$$A = (v_{C0} - RI)e^{t_0/RC} \tag{2.13}$$

This value of A is now substituted in Eq. 2.11 to give

$$v(t) = RI + (v_{C0} - RI)e^{-(t-t_0)/RC} \tag{2.14}$$

defined for all time $t \geqslant t_0$.

The mathematical solution of the problem is now complete. We now examine the solution to interpret the circuit behavior. Observe that the first term in Eq. 2.14 is the driven or forced response, also known as steady-state response. The second term represents the transient response.

$$v_1 = RI \rightarrow \text{driven, forced or steady-state response}$$
$$v_2 = (v_{C0} - RI)e^{-(t-t_0)/RC} \rightarrow \text{transient response}$$

Note that the transient solution waveform depends entirely on the character of the circuit while its magnitude is dependent on the difference between the initial and the final state. The transient is nonzero only when the initial state differs from the final state. In order words, the transient part bridges the gap between the initial state and the final state of the circuit.

The transient response is also called the natural response since its waveform depends entirely on the characteristics of the circuit elements. The transient part is a solution to the homogeneous equation free of any forcing or driving source function, and thus it is also called the free response.

The transient term v_2 is a decreasing exponential with the time constant $T = RC$. As we shall see, all RC circuits are governed by their time constants given by $T = RC$.

Plot of the Solution

The plot of the solution of the differential equation as a function of time can be decomposed into two functions, as in Eq. 2.14; one is a constant with time and the other is an exponential with time constant $T = RC$. The exponential term has the coefficient either minus or plus, depending upon whether v_{C0} is greater or less than RI. For our discussion, let us assume that the initial value v_{C0} is less than the value of the final state. Thus, our exponential term becomes negative; the exponential goes from some negative value, $v_{C0} - RI$, to zero, as $t \to \infty$.

Now adding $v_1 + v_2$, we obtain the total solution, as shown in Fig. 2.2, where at t_0, $v(t_0) = v_{C0}$. Thus, at $t = t_0$, the initial voltage is v_{C0}, and as t increases, the voltage exponentially increases to the steady-state or final value RI.

For convenience, let us set the initial capacitor voltage v_{C0} to zero. The voltage $v(t)$ for all $t \geqslant t_0$ is given by

$$v(t) = RI(1 - e^{-(t-t_0)/RC}) \tag{2.15}$$

and shown in Fig. 2.3. In the figure, the time axis is in terms of $(t - t_0)$ with the origin representing $t = t_0$.

The currents i_R and i_C can be obtained from the value of $v(t)$ in Eq. 2.15. From the circuit we observe that $i_R = v/R$; therefore, the resistor current curve is similar to $v(t)$ in Fig. 2.3 except for a constant $1/R$.

$$i_R(t) = \frac{1}{R}[RI(1 - e^{-(t-t_0)/RC})] = I(1 - e^{-(t-t_0)}) \qquad t \geqslant t_0 \tag{2.16a}$$

The KCL states that $I = i_R(t) + i_C(t)$. Thus, the current through the

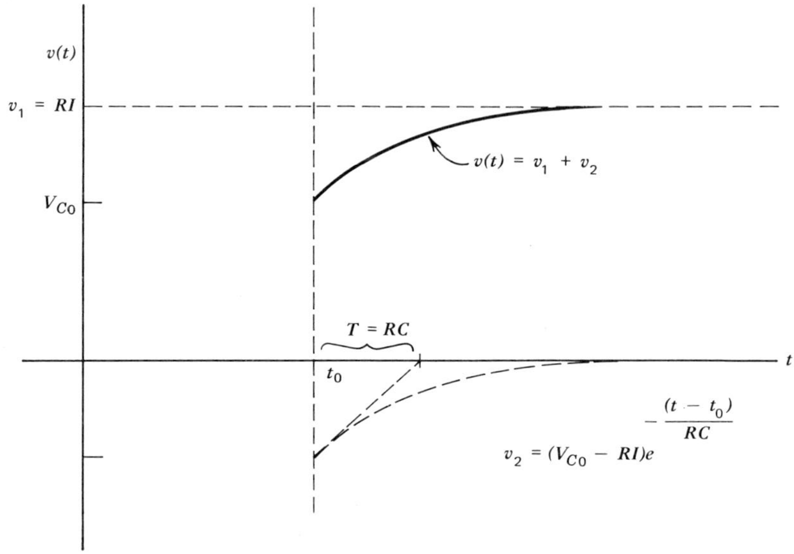

Fig. 2.2 Graph of the solution to the differential equation.

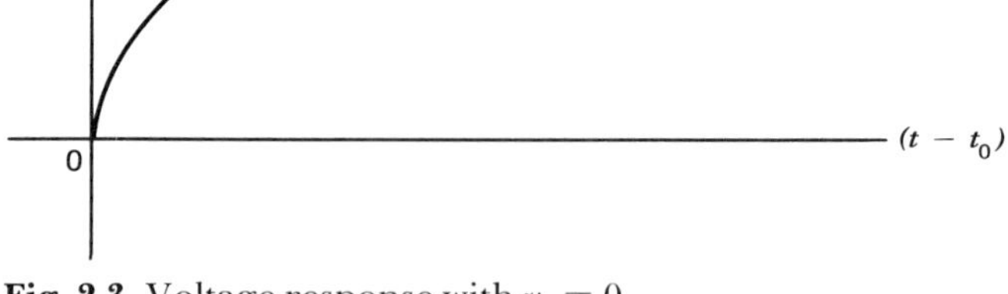

Fig. 2.3 Voltage response with $v_0 = 0$.

capacitor follows the curve $i_C(t) = I - v(t)/R$, or

$$i_C(t) = Ie^{-(t-t_0)/RC} \qquad t \geqslant t_0 \tag{2.16b}$$

Fig. 2.4 shows the plots of i_R and i_C.

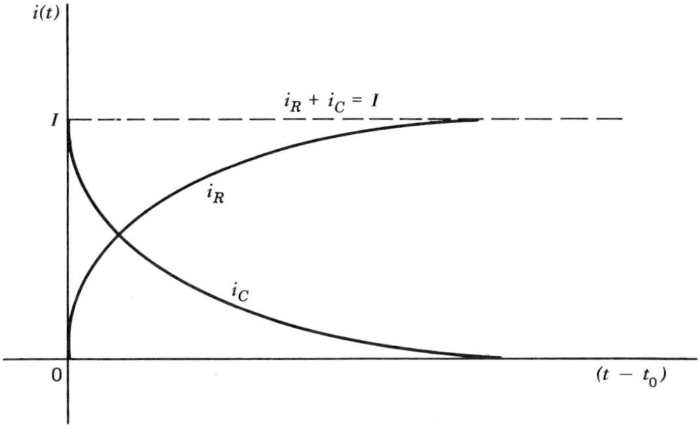

Fig. 2.4 Current responses in the circuit.

Summary of Procedure

In this section, we have considered a simple circuit and developed certain general procedures for solving the circuit equation. The KCL equilibrium equation for our circuit turned out to be a differential equation. The solution of the circuit differential equation needed the following steps:

1. First, we determined the driven response (forced response or steady-state response).
2. Next, we determined the transient response (free response, natural response). The transient is the solution to the homogeneous differential equation, obtained by putting the right-hand side of the circuit differential equation to zero. This response is exponential in nature.
3. Then we added together these two solutions, the steady-state and the transient responses. The transient response has an unspecified magnitude.
4. Then we used the initial condition to determine the magnitude of the transient response. The initial condition is obtained from the initial state. The complete solution is thus obtained.

It so happens that the circuit equilibrium equations are not always differential equations. They may contain integral expressions as well as differential expressions. Such equations are known as integro-differential equations. However, it is possible to obtain differential equations by differentiating the equilibrium equations.

In order to fix the magnitude of the transient solution of such a differential equation, we must use the initial value of the voltage (or current) variable. Quite often this value is not available directly, but must be obtained from the initial state of the circuit and the circuit equilibrium equation.

We discuss such a circuit in the next section.

2.2 A SERIES RC CIRCUIT UNDER CONSTANT EXCITATION

Consider a simple circuit shown in Fig. 2.5 that consists of a resistor and a capacitor in series, excited by a voltage source. The voltage source is a constant source (or, battery) of voltage V. The switch S is open for all $t < t_0$, closed at $t = t_0$ and remains closed thereafter. The circuit is excited for all $t \geqslant t_0$ by the closing of the switch S. The initial state is given by the initial capacitor voltage $v_C(t_0) = v_{C0}$. We are interested in determining the circuit behavior for all time $t \geqslant t_0$.

The circuit behavior is determined at the initial instant by the capacitor voltage v_{C0} which cannot change instantaneously. The voltage drop across the resistor is initially $V - v_{C0}$, and the initial current through the resistor is $i_R(t_0) = (V - v_{C0})/R$. As the current flows, the capacitor voltage increases gradually until it reaches V. The current i_R decreases as v_C increases and reaches zero when v_C approaches V. The final state is thus reached when the capacitor is charged to the source voltage.

A glance at the circuit shows that there is only a single loop of the circuit requiring only one KVL equilibrium equation. On the other hand, two KCL equations are indicated by the presence of three nodes. It seems more convenient then to write the KVL equation for the circuit. Let v_S be the source voltage with the given polarities. Then, KVL states that the sum of the voltage drops v_R, v_C and $(-v_S)$ must equal zero. Or,

$$\text{KVL} \qquad v_R + v_C = v_S \tag{2.17}$$

where v_R and v_C represent the voltage drops across the resistor and capacitor, respectively, with the polarities as shown in the figure.

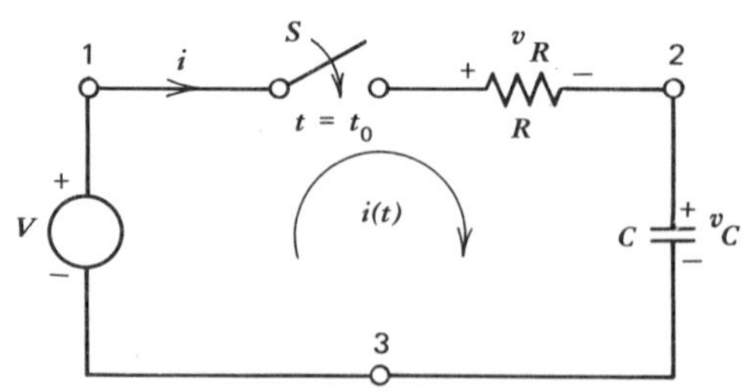

IS: $v_C(t_0) = v_{CO}$

Fig. 2.5 *RC* series circuits.

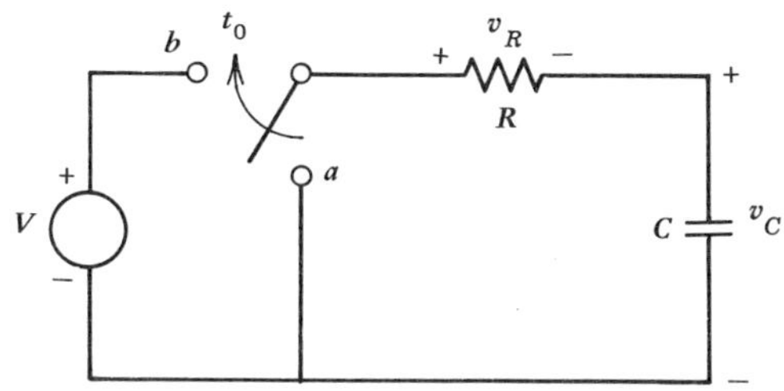

Fig. 2.6 KVL: $v_R + v_C = Vu(t-t_0)$.

The Eq. 2.17 is not defined for time $t < t_0$ because it is not possible to write it prior to $t = t_0$. Until the switch is closed at t_0, there is, in fact, no closed loop and KVL cannot be applied. Thus, v_S is not equal to $Vu(t-t_0)$; it is not existent for $t < t_0$. This is an important point and the difference is illustrated in Fig. 2.6. The switch is assumed in position **a** for all $t < t_0$ and moved to b for all $t \geqslant t_0$. Before t_0, there is a closed loop excited by zero voltage; after t_0, there is a closed loop excited by voltage V. In this case, the excitation is thus $Vu(t-t_0)$. The distinction is quite significant and must not be confused.

Let us now return to the circuit of Fig. 2.5 and the circuit equilibrium equation, Eq. 2.17, defined for $t \geqslant t_0$. Now assume that an unknown loop current $i(t)$ flows in the loop in the clockwise direction. We will solve for this function $i(t)$ and its sign, of course, will determine the actual direction of current flow as a function of time. It should be apparent that the entire circuit behavior is determined by $i(t)$.

Substituting the branch constraints of Eqs. 1.4 and 1.10 into Eq. 2.17, we obtain the circuit equilibrium equation:

$$Ri(t) + \frac{1}{C}\int_{t_0}^{t} i(\tau)d\tau + v_{C0} = V \tag{2.18}$$

for all $t \geqslant t_0$. Observe that Eq. 2.18 is an integro-differential equation and must be converted to a differential equation. Differentiating both sides of Eq. 2.18,

$$R\frac{di}{dt} + \frac{1}{C}i = 0 \qquad t \geqslant t_0 \tag{2.19}$$

Observe that the driving function on the right-hand side of Eq. 2.19 is zero. Hence, the steady-state solution for this equation is zero, and so the procedure described in the previous section requires us merely to find the transient response and then determine its magnitude.

By a comparison of Eq. 2.19 with Eq. 2.9, we know that the transient solution is an exponential function. Instead of separating variables and integrating, we can substitute an unknown exponential into Eq. 2.19 and determine the exponent as follows. Let the transient solution be

$$i(t) = Ae^{st} \tag{2.20}$$

where s is an unknown exponent to be determined. The arbitrary constant A will be determined by the initial value of $i(t)$ as in the last section. Substitute this value of $i(t)$ into Eq. 2.19 to obtain

$$\left(Rs + \frac{1}{C}\right) Ae^{st} = 0 \tag{2.21}$$

The exponential function is never zero for finite time and A is arbitrary, not necessarily zero. Eq. 2.21 thus implies

$$Rs + \frac{1}{C} = 0$$

or

$$s = -\frac{1}{RC} \tag{2.22}$$

Substituting into Eq. 2.20, the transient solution is

$$i_t(t) = Ae^{-t/RC} \tag{2.23}$$

According to our procedures of Sec. 2.1, the total solution is the sum of the steady-state solution and the transient solution. For the problem of Eq. 2.19, the steady-state solution $i_{ss} = 0$; so, the total solution (for $t \geqslant t_0$) is,

$$\begin{aligned} i(t) &= i_{ss} + i_t \\ &= Ae^{-t/RC} \end{aligned} \tag{2.24}$$

The constant A is determined by the value of $i(t_0)$. The initial state is given by $v_C(t_0) = v_{C0}$. To obtain $i(t_0)$, let us examine the circuit equilibrium equation, Eq. 2.18, at $t = t_0$. Eq. 2.18 is

$$Ri(t) + \frac{1}{C}\int_{t_0}^{t} i(\tau)\,d\tau + v_{C0} = V$$

At $t = t_0$, this becomes

$$Ri(t_0) + 0 + v_{C0} = V$$

or,

$$i(t_0) = \frac{V - v_{C0}}{R} \tag{2.25}$$

Substituting into Eq. 2.24

$$\begin{aligned} i(t_0) &= Ae^{-t_0/RC} \\ &= \frac{V - v_{C0}}{R} \end{aligned}$$

and
$$A = \left(\frac{V - v_{C0}}{R}\right) e^{t_0/RC} \tag{2.26}$$

Finally, this value of A and Eq. 2.24 gives, for all $t \geqslant t_0$

$$i(t) = \left(\frac{V - v_{C0}}{R}\right) e^{-(t-t_0)/RC} \tag{2.27}$$

The solution is complete and checks with our intuitive reasoning. Observe that the current through the capacitor decays to zero exponentially with the same time constant, $T = RC$, as in the previous section. Fig. 2.7 shows the current $i(t)$ as a function of $(t - t_0)$. Observe that the current $i(t)$ has a discontinuity at $t = t_0$. At t_{0-} the current is zero, whereas at $t = t_{0+}$ it is $(V - v_{C0})/R$. Thus, the initial condition is discontinuous at t_0, but the initial state is continuous as always for a bounded excitation.

We now examine the variation of voltage on the capacitor with time. The charge on the capacitor builds up and the voltage ultimately becomes equal to the source voltage. The expression for the voltage across the capacitor is (for $t \geqslant t_0$),

$$\begin{aligned} v_C &= \frac{1}{C}\int_{t_0}^{t} i\,d\tau + v_{C0} \\ &= \frac{1}{C}\int_{t_0}^{t} \frac{(V - v_{C0})}{R} e^{-(\tau - t_0)/RC}\,d\tau + v_{C0} \\ &= v_{C0} + (V - v_{C0})(1 - e^{-(t-t_0)/RC}) \end{aligned} \tag{2.28}$$

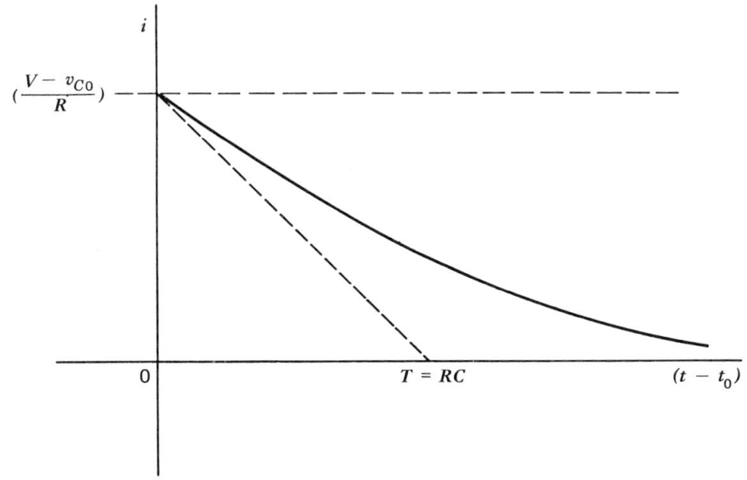

Fig. 2.7 Loop current $i(t)$.

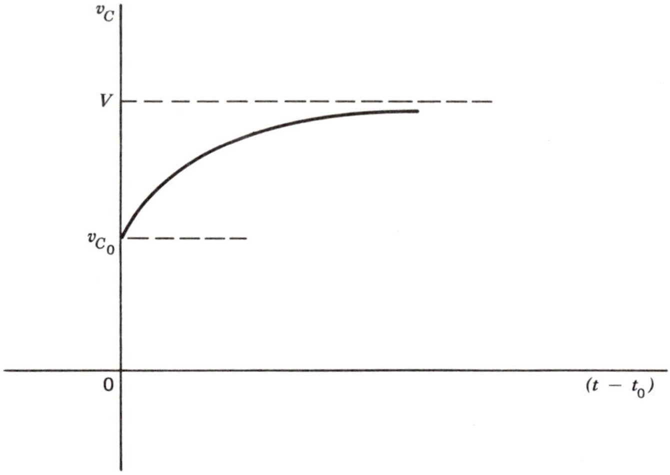

Fig. 2.8 Voltage across the capacitor.

This indicates that initially, at $t = t_0$,

$$v_C(t_0) = v_{C0} \tag{2.29}$$

When t becomes very large, the exponential factor, $(1 - e^{-(t-t_0)/RC})$, approaches unity. Eq. 2.28, in the limit as $t \to \infty$, reduces to

$$v_C(t \to \infty) = v_{C0} + V - v_{C0} = V \tag{2.30}$$

Starting with the initial value v_{C0} the capacitor voltage builds up exponentially to the value V, the source voltage. The capacitor voltage as a function of time is shown in Fig. 2.8.

Summary of Procedure

Let us summarize the procedure we have developed when the circuit equation is not a differential equation.

1. Write the integro-differential circuit equilibrium equation.
2. Differentiate the circuit equation to obtain a differential equation.
3. The initial value of the differential equation variable is obtained from the initial state and the equilibrium equation.
4. The differential equation is solved using the procedure of Sec. 2.1.

Example 2.2.1 The circuit of Fig. 2.5 can also be analyzed by using the KCL description. There are two independent nodes and we let node 3 be the reference. The KCL at node 1 states that the source current i_S

(unknown) equals the current in the resistor, that is,

$$\text{KCL-1:} \qquad i = i_R = i_S$$

or,

$$\frac{v_1 - v_2}{R} = i_S$$

The KCL at node 2 is,

$$\text{KCL-2:} \qquad -i_R + i_C = 0$$

or,

$$-\frac{v_1 - v_2}{R} + C\frac{dv_2}{dt} = 0$$

Thus, we have two KCL equations in two unknown voltages v_1 and v_2. But we already know $v_1 = V$. No equation at node 1 is thus needed and the only KCL that is relevant is given by:

$$-\frac{V - v_2}{R} + C\frac{dv_2}{dt} = 0$$

or,

$$C\frac{dv_2}{dt} + \frac{v_2}{R} = \frac{V}{R}$$

The initial condition is $v_2(t_0) = v_C(t_0) = v_{C0}$. The differential equation is then solved to give,

$$v_2(t) = V + (v_{C0} - V)e^{-(t-t_0)/RC}$$

This is identical to the solution for v_C obtained in Eq. 2.28.

The circuit can thus be solved by the KCL description as well.

2.3 SERIES RL CIRCUIT WITH EXPONENTIAL EXCITATION

In the first two sections of this chapter we considered circuits that are excited by constant voltage or current sources. We now discuss circuits excited by exponentially varying signal sources. Exponential signal sources are not encountered in practice. Their importance stems from the fact that many frequently encountered sources may be represented as sums of exponential sources. For example, the most frequently encountered source signal, the sinusoid, can be expressed as a sum of two exponential signals. For this reason, exponential signal sources are very important to our discussion.

Consider a series resistor, inductor (*RL*) circuit of Fig. 2.9 excited by the exponential source. The circuit is assumed connected together by the closing of the switch at $t = t_0$. The initial state of the circuit is defined by the initial current through the inductor,

$$i_L(t_0) = i_{L0} \tag{2.31}$$

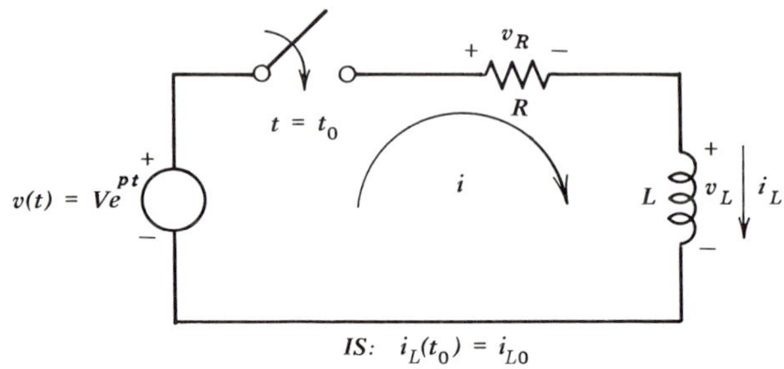

Fig. 2.9 Series *RL* circuit with exponential source.

Notice that no current can flow through the inductor prior to $t = t_0$ since the circuit in Fig. 2.9 is not complete when the switch is open. What is implied by the figure is in a sense symbolic. It merely signifies the instant of interconnection of the entire circuit and does not necessarily attempt to prescribe the past history of the circuit elements. It is entirely conceivable that the circuit elements were part of some other circuit prior to t_0 that is not shown in the figure. Furthermore, the value of i_{L0} may well be zero. We include the initial state description in a general way to develop as complete a theory as possible.

Now the KVL for the circuit for all $t \geqslant t_0$ is

$$\text{KVL:} \qquad v_L + v_R = v_S \tag{2.32}$$

where v_S is the source voltage. By using the branch constraints, we obtain the circuit equilibrium equation in terms of the loop current $i(t)$ shown in Fig. 2.9:

$$\text{DE} \qquad L\frac{di}{dt} + Ri = Ve^{pt} \qquad t \geqslant t_0 \tag{2.33}$$

where V and p are some prescribed constants, real or complex. Notice that the circuit equation is a differential equation and we use the procedure of Sec. 2.1 to solve it.

We first determine the initial value of the differential equation variable, $i(t_0)$, (initial condition) from the initial state in Eq. 2.31. Since the current $i(t)$ is the same as the current $i_L(t)$, we have

$$\text{IC} \qquad i(t_0) = i_L(t_0) = i_{L0} \tag{2.34}$$

Now we proceed to determine the driven response (or, steady state solution) due to the driving function Ve^{pt}. We attempt a steady-state solution of the same time-variation as the driving function, say

$$i_{ss} = Ie^{pt} \tag{2.35}$$

with the exponent p, but an unknown value of I. This is consistent with a constant value of driven response obtained in Sec. 2.1 for a constant excitation. Substitute Eq. 2.35 in Eq. 2.33 to obtain

$$(pL+R)Ie^{pt} = Ve^{pt} \tag{2.36}$$

or, provided

$$pL+R \neq 0 \tag{2.37}$$

$$I = \frac{V}{pL+R} \tag{2.38}$$

and by Eq. 2.35,

$$i_{ss} = \left(\frac{V}{pL+R}\right) e^{pt} \tag{2.39}$$

We have assumed in the above that $(pL+R) \neq 0$, The solution of Eq. 2.39 is not valid if $(pL+R)=0$. We develop later in this section a modified steady-state solution in the pathological case when $(pL+R)=0$. For now, we will assume $(pL+R) \neq 0$ in which case the above solution is obtained.

We next proceed to obtain the transient solution from the homogeneous equation:

$$\text{HE} \qquad L\frac{di}{dt}+Ri=0 \tag{2.40}$$

As in Sec. 2.2, we merely substitute an exponential transient solution with an unknown exponent. That is, we substitute

$$i_t = Ae^{st} \tag{2.41}$$

with the unknown exponent s and arbitrary A into Eq. 2.40 to obtain,

$$(sL+R)Ae^{st} = 0$$

or

$$s = -\frac{R}{L} \tag{2.42}$$

since $e^{st} \neq 0$ and A is arbitrary. The transient solution, from Eqs. 2.41 and 2.42, is

$$i_t = Ae^{-(R/L)t} \tag{2.43}$$

The total solution is the sum of the steady-state and the transient responses of Eqs. 2.39 and 2.43:

$$\begin{aligned} i(t) &= i_{ss}+i_t \\ &= \left(\frac{V}{pL+R}\right) e^{pt}+Ae^{-(R/L)t} \end{aligned} \tag{2.44}$$

where the arbitrary constant is yet to be determined.

To solve for the constant A we apply the initial condition of Eq. 2.34 to Eq. 2.44 and obtain

$$i(t_0) = i_{L0} = \left(\frac{V}{pL+R}\right) e^{pt_0} + Ae^{-(R/L)t_0} \tag{2.45}$$

Solving Eq. 2.45 for A we have

$$A = \left[i_{L0} - \left(\frac{V}{pL+R}\right) e^{pt_0}\right] e^{(R/L)t_0} \tag{2.46}$$

Hence, using Eqs. 2.44 and 2.46, the final solution is expressed as:

$$i(t) = \left(\frac{V}{pL+R}\right) e^{pt} + \left[i_{L0} - \left(\frac{V}{pL+R}\right) e^{pt_0}\right] e^{-(R/L)(t-t_0)} \tag{2.47}$$

Upon simplification, Eq. 2.47 reduces to

$$i(t) = \frac{Ve^{pt_0}}{pL+R}\left[e^{p(t-t_0)} - e^{(R/L)(t-t_0)}\right] + i_{L0}e^{-(R/L)(t-t_0)} \tag{2.48}$$

valid for all $t \geqslant t_0$ since our circuit equation, Eq. 2.33, is defined only for $t \geqslant t_0$. The function $i(t)$ can now be plotted against time. There are several important properties of all linear, time-invariant circuits that considerably simplify the descriptions of circuit behavior. We shall identify these properties before proceeding with the plot of the solution.

Time-Invariance

In Eq. 2.48, observe that the exponential time-variation is in terms of $(t-t_0)$ just as it was in Secs. 2.1 and 2.2. The time-variation is thus dependent on the time elapsed after the switch is closed and not upon the absolute time. Let

$$t - t_0 = t' \tag{2.49}$$

and,

$$Ve^{pt_0} = V_0 \tag{2.50}$$

The source voltage of Fig. 2.9 now becomes

$$v_S(t) = V_0 e^{pt'} \tag{2.51}$$

where V_0 is the initial source voltage. The solution in Eq. 2.48 becomes

$$i(t) = \left(\frac{V_0}{pL+R}\right)\left[e^{pt'} - e^{-(R/L)t'}\right] + i_{L0}e^{-(R/L)t'} \tag{2.52}$$

Observe that the response of the circuit is defined completely in terms of the elapsed time since the start of the excitation. No reference need be made to the precise moment of circuit connection. The circuit behavior is independent of the absolute time value of the initial instant t_0. It is merely a function of the relative or elapsed time t'. Circuits of time-invariant elements are all governed by this time-invariant property of the response.

The circuit behavior can thus be described just as well in terms of the elapsed time $(t-t_0)$, with t_0 equivalent to zero on the new scale. Equivalently, we can just assume that the switch is closed at time $t=0$ and that time is measured from that origin. We will assume this henceforth and omit the primes from Eq. 2.52.

Superposition Principle I

The total solution $i(t)$ is defined as the sum of the steady-state and the transient solutions. Eq. 2.48, or rather 2.52, defines the solution $i(t)$ in terms of other quantities equally significant. In Eq. 2.52, observe that the first term on the right-hand side is the solution if the initial state $i_{L0}=0$. For this reason, the first term is called the *zero-state solution*, that is, the solution obtained when the initial state is set equal to zero:

$$\begin{aligned}\text{Zero-state solution} &= i_{ZS}\\ &= \left(\frac{V_0}{pL+R}\right)[e^{pt}-e^{-(R/L)t}]\end{aligned} \tag{2.53}$$

where we have omitted the primes.

The second term on the right of Eq. 2.52 is the solution for the loop current if the source voltage were set to zero or $V_0=0$. The second term is called, therefore, the *zero-input solution*, that is, the solution obtained when the source signal or the input is set equal to zero:

$$\begin{aligned}\text{Zero-input solution} &= i_{ZI}\\ &= i_{L0}e^{-(R/L)t}\end{aligned} \tag{2.54}$$

where the primes are omitted again.

The total solution is thus the sum of the zero-state solution and the zero-input solution:

$$\text{Total solution} = \text{Zero-state solution} + \text{Zero-input solution} \tag{2.55}$$

It can be shown that this property is valid for all linear circuits. An examination of solutions of previous sections verifies the property for our simple circuits. Frequent use is made of this property in general discussions of linear circuits or systems. It allows us to develop a general theory for zero-state circuits. The zero-state solutions are then modified by the additions of zero-input solutions to obtain the actual solutions for circuits with non-zero initial states. The property in Eq. 2.55 will be called the Superposition Principle I because it allows us to superimpose the responses due to the signal sources and those due to the initially stored energy. The next section will discuss the Superposition Principle II.

Zero-Signal Sources

Let us examine the implications of setting the source signals equal to zero. Fig. 2.10*a* represents a zero signal voltage source. The voltage across the source terminals is constrained to be zero regardless of the current through the terminals and the source. Zero voltage across a source element regardless of the current through is equivalent to a short-circuit between the terminals. A zero signal voltage source thus represents a short-circuit as in Fig. 2.10*b*.

Similarly, a zero-signal current source requires zero current through the source element regardless of the voltage across it. This is equivalent to an open-circuit between the terminals. A zero-signal current source and its equivalent are shown in Fig. 2.11.

To summarize, when the source signals are set to zero for any purpose, the voltage sources are replaced by short-circuits and the current sources are replaced by open-circuits.

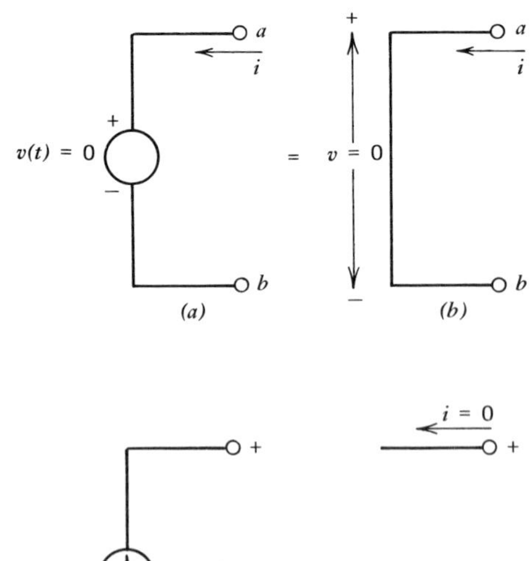

Fig. 2.10

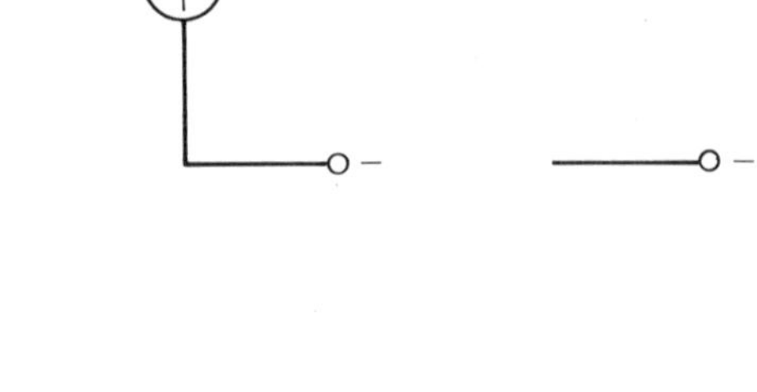

Fig. 2.11

Solution

The time-invariant property allows us to recast our original problem of Fig. 2.9 to the new circuit of Fig. 2.12. The switch is closed at $t = 0$.

The Superposition Principle I allows us to solve the circuit in two steps as shown in Fig. 2.13. The solution then is given by

$$i = i_{ZS} + i_{ZI} \tag{2.56}$$

which we have already stated in Eqs. 2.52–2.54—

$$i(t) = \left(\frac{V_0}{pL+R}\right)[e^{pt} - e^{-(R/L)t}] + i_{L0}e^{-(R/L)t} \tag{2.57}$$

valid for all $t \geqslant 0$.

The zero-state solution is shown in Fig. 2.14. It is the solution if $i_{L0} = 0$; otherwise, the zero-input solution must be added to obtain the solution. In Fig. 2.14, it is assumed that V_0 is real and positive, and that p is negative real, and $R/L > |p|$.

Finally, observe that the response to a constant voltage source, V_0, is obtained by letting $p = 0$ in Eq. 2.57.

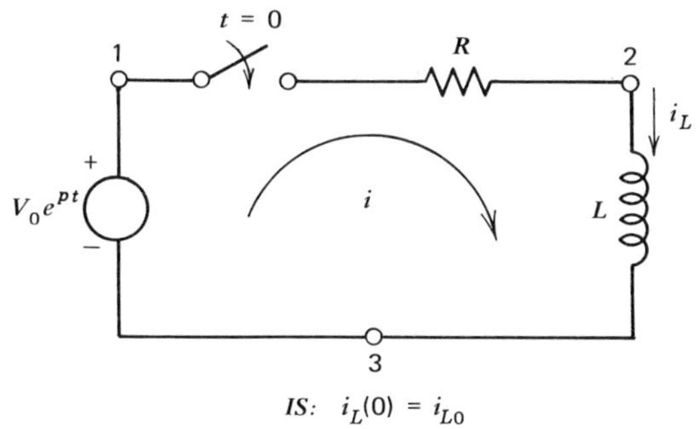

Fig. 2.12

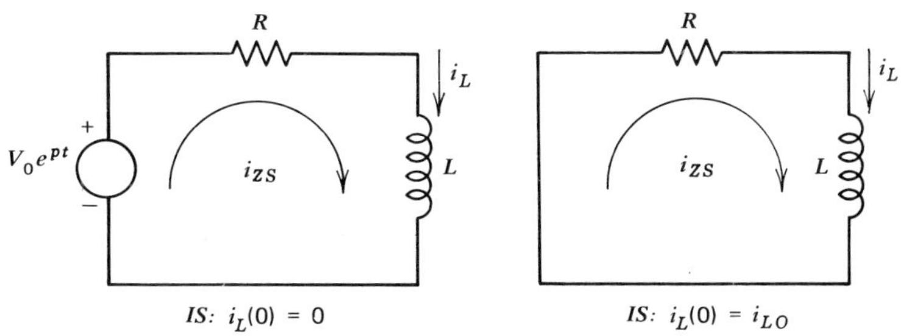

Fig. 2.13

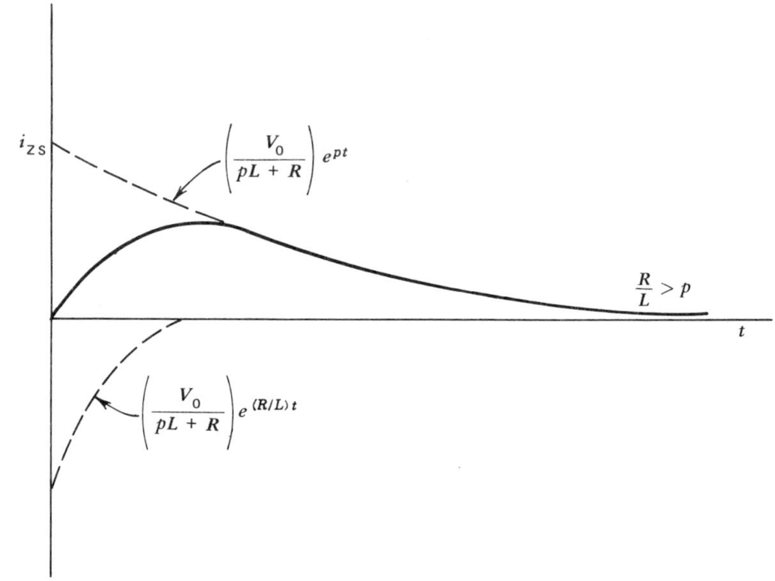

Fig. 2.14 Zero-state solution assuming $p =$ real, negative.

Case When $pL + R = 0$

If $pL + R = 0$ or $p = -R/L$, then we can see from Eq. 2.38 that the magnitude of the i_{ss} cannot be determined. Hence, in this case, we *cannot* assume that the steady-state solution is of the form

$$i_{ss} = Ie^{pt} \tag{2.58}$$

However, from the theory of differential equations, when $pL + R = 0$, a steady-state solution of the form

$$i_{ss} = I_1 te^{pt} \tag{2.59}$$

exists, and I_1 must be determined by substitution in the differential equation.

Substituting Eq. 2.59 into 2.33, we have

$$LI_1(pte^{pt} + e^{pt}) + RI_1 te^{pt} = V_0 e^{pt} \tag{2.60}$$

Now, since $pL + R = 0$, Eq. 2.60 reduces to

$$LI_1 e^{pt} = V_0 e^{pt}$$

or,

$$I_1 = \frac{V_0}{L} \tag{2.61}$$

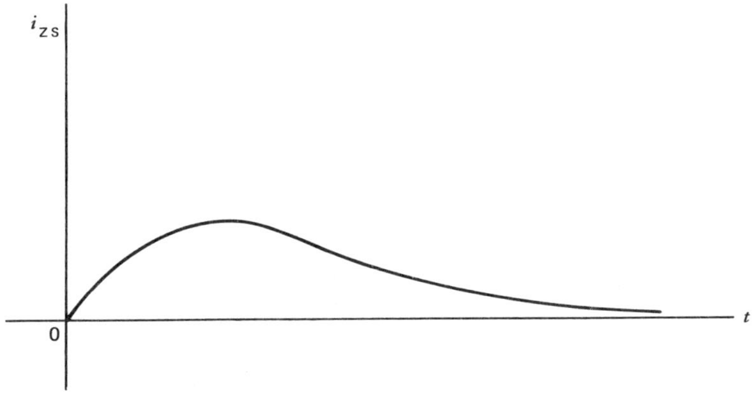

Fig. 2.15

From Eqs. 2.59 and 2.61,

$$i_{ss} = \frac{V_0}{L} te^{pt} \tag{2.62}$$

The total solution is the sum of the steady state solution i_{ss} and the transient solution i_t in Eq. 2.43. We have

$$i(t) = \frac{V_0}{L} te^{pt} + Ae^{-(R/L)t} \tag{2.63}$$

In terms of the circuit defined in Fig. 2.12, the initial state is defined at $t = 0$,

$$i(0) = i_L(0) = i_{L0} \tag{2.64}$$

Evaluating Eq. 2.63 at $t = 0$ and using Eq. 2.64, we have

$$A = i_{L0},$$

and,

$$i(t) = \frac{V_0}{L} te^{pt} + i_{L0}e^{-(R/L)t} \tag{2.65}$$

The first term on the right side of Eq. 2.65 is the zero-state solution and the second term gives the zero-input solution. The zero-state solution is shown in Fig. 2.15 for p real and negative.

Summary

1. Time-Invariance Principle:
   The response is a function of the elapsed time from the instant of excitation. A time translation of excitation results in a corresponding time translation of the response.

2. Superposition Principle I:
   The total response is the sum of the zero-state response and the zero-input response.
3. The steady-state solution for an exponential driving function of the form V_0e^{pt} is either Ie^{pt} or Ite^{pt} with the value of I determined by the circuit parameters,

Example 2.3.1 Let us consider the KCL description for the circuit in Fig. 2.12. Let node 3 be the reference. No KCL at node 1 is needed, since we know that $v_1 = v_0e^{pt}$. At node 2, the KCL gives,

$$\frac{1}{L}\int_0^t v_2(\tau)\,d\tau + i_{L0} - \frac{(v_1 - v_2)}{R} = 0$$

or,

$$\frac{v_2}{R} + \frac{1}{L}\int_0^t v_2(\tau)\,d\tau = \frac{V_0e^{pt}}{R} - i_{L0}$$

Then DE becomes

$$\frac{dv_2}{dt} + \frac{R}{L}v_2 = pV_0e^{pt}$$

and the IC is

$$v_2(0) = V_0 - i_{L0}R$$

Assuming that $p \neq -R/L$, we can develop the solution directly. The steady-state solution is

$$\frac{pV_0}{p + \dfrac{R}{L}}e^{pt}$$

and the transient solution is

$$Ae^{-(R/L)t}$$

Applying the IC to the total solution we determine A and obtain,

$$v_2(t) = \left[(V_0 - i_{L0}R) - \frac{pV_0}{p + \dfrac{R}{L}}\right]e^{-(R/L)t} + \frac{pV_0}{p + \dfrac{R}{L}}e^{pt}$$

The solution can be compared with that obtained in Eq. 2.57 since $L(di/dt) = v_2$.

2.4 MULTIPLE EXCITATIONS

Let us now consider the case of multiple excitations. We have seen in the last section that the total response consists of zero-state response and zero-input response. Therefore, only the zero-state response is considered in this section. The zero-input response is determined as before by setting all the source signals to zero. Presence of multiple sources affect only the determination of zero-state responses. We shall use exponential sources for our discussion but the derivations and conclusions will be valid for arbitrary sources.

Let us consider the same circuit as in Sec. 2.3, a series RL circuit, excited by two exponential sources shown in Fig. 2.16. The circuit is connected at $t = 0$. Only the zero-state solution is of interest and so we assume $i_L(0) = 0$. The differential equation and the initial condition for it are:

$$\text{DE:} \quad L\frac{di}{dt} + Ri = V_1 e^{p_1 t} + V_2 e^{p_2 t} \qquad t \geqslant 0$$
$$\text{IC:} \quad i_L(0) = i(0) = 0 \tag{2.66}$$

We will develop general procedures that apply to any circuit with multiple excitations.

We have developed in Sec. 2.3 the solution when one exponential source is present, that is, the other source signal is set to zero. Let us use this knowledge to develop the solution to Eq. 2.66. Let i_1 be the solution when $V_2 = 0$ in Eq. 2.66, or

$$\text{DE:} \quad L\frac{di_1}{dt} + Ri_1 = V_1 e^{p_1 t} \qquad t \geqslant 0$$
$$\text{IC:} \quad i_1(0) = 0 \qquad V_2 = 0 \tag{2.67}$$

Observe that i_1 is the zero-state solution to the differential equation for the circuit of Fig. 2.16. Similarly, let i_2 be the solution when $V_1 = 0$ in Eq. 2.66, or

$$\text{DE:} \quad L\frac{di_2}{dt} + Ri_2 = V_2 e^{p_2 t} \qquad t \geqslant 0$$
$$\text{IC:} \quad i_2(0) = 0 \qquad V_1 = 0 \tag{2.68}$$

where i_2 is the zero-state solution to Eq. 2.66 for $V_1 = 0$.

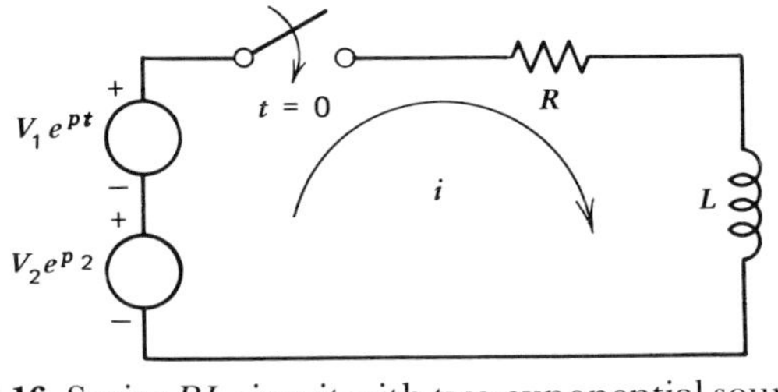

Fig. 2.16 Series RL circuit with two exponential sources.

The right-hand side of Eq. 2.66 is the sum of the right-hand sides of Eqs. 2.67 and 2.68. Let us then add the Eqs. 2.67 and 2.68:

$$\text{DE:} \quad L\frac{d}{dt}(i_1+i_2)+R(i_1+i_2)=V_1e^{p_1t}+V_2e^{p_2t}, \qquad t \geqslant 0$$

$$\text{IC:} \quad i_1(0)+i_2(0)=i_1+i_2|_{t=0}=0 \tag{2.69}$$

Compare Eq. 2.69 with the original circuit differential equation (Eq. 2.66) and observe that the solution

$$i=i_1+i_2 \tag{2.70}$$

is the zero-state response of the circuit in Fig. 2.16 with both sources present, i_1 is the zero-state response of the circuit with one of the source signals set to zero, and i_2 is the zero-state response with the second of the source signals set to zero.

The zero-state response of the circuit of Fig. 2.16 can thus be written using Eqs. 2.66–2.70 and Eq. 2.53:

For $t \geqslant 0$,

$$i(t)=\frac{V_1}{p_1L+R}[e^{p_1t}-e^{-(R/L)t}]+\frac{V_2}{p_2L+R}[e^{p_2t}-e^{-(R/L)t}] \tag{2.71}$$

where it is assumed that

$$\begin{aligned} p_1L+R &\neq 0 \\ p_2L+R &\neq 0 \end{aligned} \tag{2.72}$$

A modified zero-state response must be obtained from Eq. 2.65 if one of the above assumptions is not valid.

We conclude that the zero-state response of a circuit with two excitation sources is given by the sum of the zero-state responses of the circuit to each one of the sources with the other source set to zero. Similarly for a circuit excited by multiple sources, *the zero-state response is obtained by a sum of zero-state responses corresponding to individual sources taken one at a time with all other sources set equal to zero.* This property of linear circuits known as the Superposition principle will be called *Superposition Principle II.*

Figure 2.17 illustrates Superposition (Principle) II. Let a zero-state circuit be excited by excitations $e_1, e_2, \ldots, e_n$. Let r_1 be the response of the circuit to e_1 with $e_2, e_3, \ldots, e_n$ set to zero; r_2 the response to e_2 with $e_1, e_3, e_4, \ldots, e_n$ set to zero; and r_k for any k be the response to e_k with all excitations except e_k set to zero. Then the total response is the sum of all the responses $r_1, r_2, \ldots, r_n$.

Superposition II states that the locations of the excitations are arbitrary

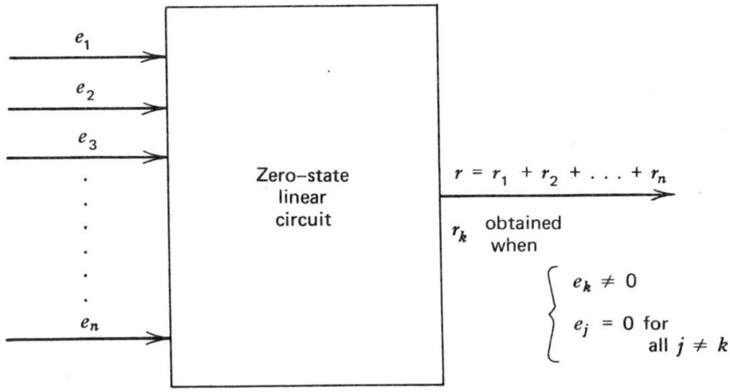

Fig. 2.17

as shown in Fig. 2.17. They are *not* required to be at the same location in the circuit as in our demonstration.

Example 2.4.1 For the circuit of Fig. 2.18, we wish to determine the current through the resistor in the direction shown for $t \geqslant 0$. By Superposition I:

$$i_R = i_{\text{ZI}} + i_{\text{ZS}}$$

and i_{ZI} is the current through R with source signals set to zero. The current source is open-circuited and the voltage source short-circuited. The situation of Fig. 2.19 occurs. In Eq. 2.27 we set $V = 0$, $t_0 = 0$ and reverse the sign to obtain the current in the direction shown. We then have

$$\begin{aligned} i_{\text{ZI}} &= \frac{v_C}{R} e^{-t/RC} \\ &= e^{-t} \text{ amps} \qquad t \geqslant 0 \end{aligned}$$

The current i_{ZS} must be obtained by using Superposition II. First, let the voltage source signal be set to zero. The current $i_{\text{ZS}}{}^1$ is obtained from Fig. 2.20.

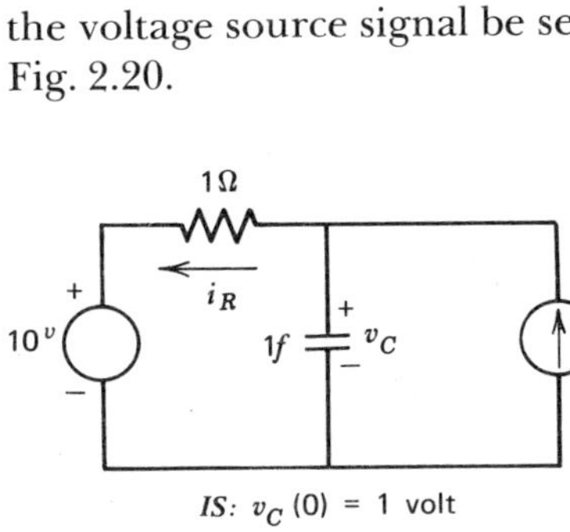

Fig. 2.18

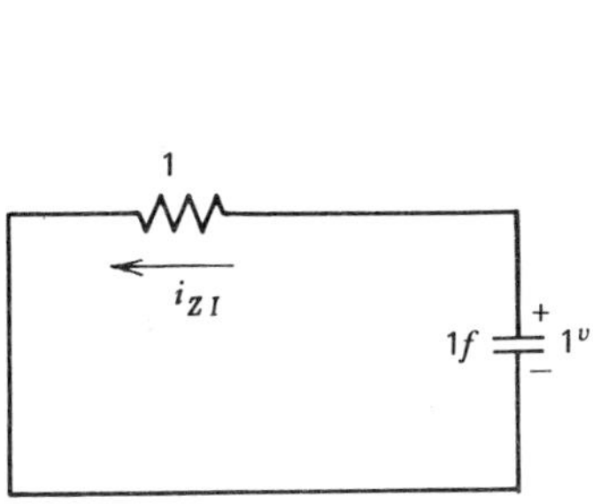

Fig. 2.19

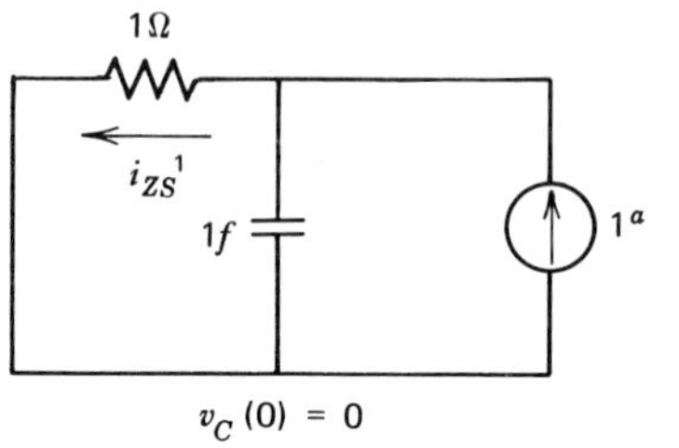

Fig. 2.20

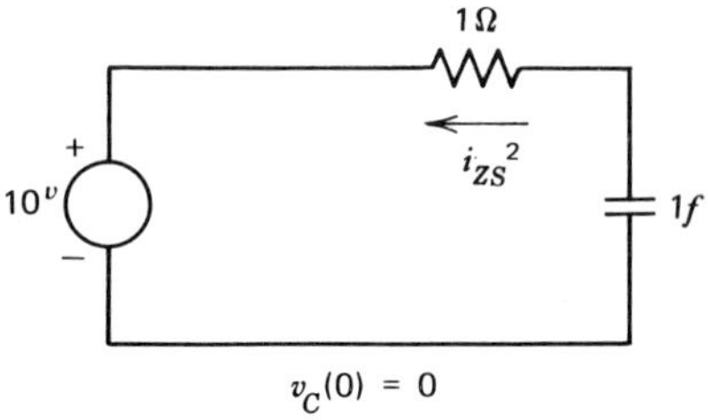

Fig. 2.21

From Eq. 2.16a

$$
\begin{aligned}
i_{ZS}^1 &= I\,(1-e^{-t/RC}) \\
&= (1-e^{-t})\text{ amps} \qquad t \geqslant 0
\end{aligned}
$$

Now, let the current source signal be set to zero to obtain i_{ZS}^2 as in Fig. 2.21. Note the direction of i_{ZS}^2 is opposite to that of the current i of Fig. 2.5 whose value is given in Eq. 2.27. Reversing the sign in Eq. 2.27 and setting $v_{C0} = 0$, $t_0 = 0$, we have

$$
\begin{aligned}
i_{ZS}^2 &= -\frac{(V - v_{C0})}{R}\, e^{-t/RC} \\
&= -(10)\, e^{-t}.
\end{aligned}
$$

The zero-state current is

$$
\begin{aligned}
i_{ZS} &= i_{ZS}^1 + i_{ZS}^2 \\
&= (1-e^{-t}) - 10\, e^{-t} \\
&= 1 - 11\, e^{-t} \qquad t \geqslant 0
\end{aligned}
$$

Finally, the total current is

$$
\begin{aligned}
i_R &= i_{ZI} + i_{ZS} \\
&= e^{-t} + 1 - 11\, e^{-t} \\
&= 1 - 10\, e^{-t}\text{ amps} \qquad t \geqslant 0
\end{aligned}
$$

Thus, the Superposition Principle II allows us to treat each source separately. Zero-state response is developed for each source with the remaining sources set to zero. The sum of such zero-state responses is the zero-state response of the circuit with all the sources present.

Note that the zero-input response is the transient that depends on the initial state. The transient in the zero-state response is the part that depends on the input signal. However, the entire steady-state response must be present only in the zero-state response. This allows us to apply Superposition II to the steady-state response. When the transients vanish, the zero-state response represents the steady-state response. Therefore, the total steady-state response for any circuit with multiple sources is the sum of the individual steady-state response due to each source with all others set to zero. Superposition II is thus very useful for handling multiple sources and is usually referred to as the Superposition Principle.

2.5 SINUSOIDAL EXCITATION

In this section we consider a first order *RC* circuit excited by a sinusoid. Two different methods are used to determine the driven solution. The direct method is treated first. While it appears simple, the second method has many advantages for circuit analysis and is used exclusively in the rest of the text. The second method is based on the decomposition of sinusoids into sums of exponentials with complex exponents described in Chapter 1. We are concerned here only with complex exponentials. An extensive discussion of complex numbers occur in Chapter 6.

Consider the circuit of Fig. 2.22 valid for $t \geqslant 0$ excited by a sinusoidal current source. The circuit differential equation with the initial condition is

$$\begin{aligned} \text{DE:} \quad & C\frac{dv}{dt}+\frac{v}{R}=I\cos(\omega t+\phi) \\ \text{IC:} \quad & v(0)=v_{CO} \end{aligned} \tag{2.73}$$

The driven solution v_{ss} for the differential equation is obtained by assuming it to be the same function with unknown amplitude and phase.

$$\begin{aligned} v_{ss} &= V\cos(\omega t+\theta) \\ &= V_1\cos\omega t+V_2\sin\omega t \end{aligned} \tag{2.74}$$

where $V_1 = V\cos\theta$ and $V_2 = -V\sin\theta$. Substituting this value of v_{ss} into Eq. 2.73,

$$C(-V_1\omega\sin\omega t+V_2\omega\cos\omega t)+\frac{V_1}{R}\cos\omega t+\frac{V_2}{R}\sin\omega t=I\cos(\omega t+\phi) \tag{2.75}$$

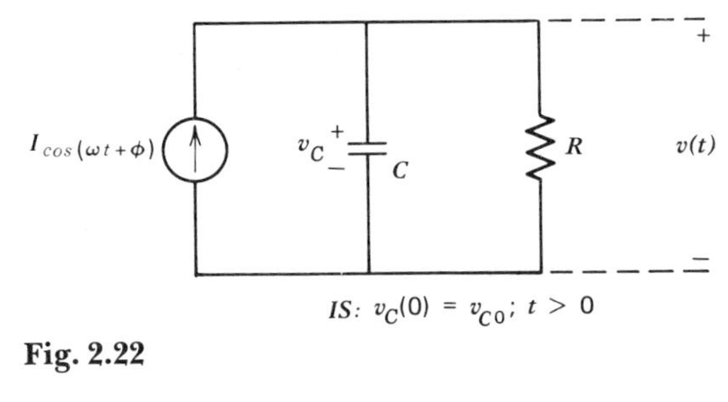

Fig. 2.22

For simplicity, and without loss of generality, let us choose $\phi = 0$. Then, by comparing the cosine and the sine terms on the two sides of Eq. 2.75, we have

$$\begin{aligned} GV_1 + \omega CV_2 &= I \\ -\omega CV_1 + GV_2 &= 0 \end{aligned} \tag{2.76}$$

where $G = 1/R$. Eq. 2.76 is now solved for V_1 and V_2 and the values substituted into Eq. 2.74.

$$\begin{aligned} V_1 &= \frac{G}{G^2 + \omega^2 C^2} I \\ V_2 &= \frac{\omega C}{G^2 + \omega^2 C^2} I \end{aligned} \tag{2.77}$$

and

$$v_{ss} = \frac{G}{G^2 + \omega^2 C^2} I \cos \omega t + \frac{\omega C}{G^2 + \omega^2 C^2} I \sin \omega t \tag{2.78}$$

The transient solution from the homogeneous equation is by Eq. 2.10,

$$v_t = Ae^{-t/RC} \tag{2.79}$$

and the total solution is

$$\begin{aligned} v(t) &= v_{ss} + v_t \\ &= Ae^{-t/RC} + \frac{G}{G^2 + \omega^2 C^2} I \cos \omega t + \frac{\omega G}{G^2 + \omega^2 C^2} I \sin \omega t \end{aligned} \tag{2.80}$$

We solve for A from the IC in Eq. 2.73,

$$v_{C0} = v(0) = A + \frac{G}{G^2 + \omega^2 C^2} I$$

or,

$$A = v_{C0} - \frac{G}{G^2 + \omega^2 C^2} I \tag{2.81}$$

Substituting the value of A into Eq. 2.80, we have

$$\begin{aligned} v(t) = v_{C0} e^{-t/RC} &- \frac{G}{G^2 + \omega^2 C^2} I e^{-t/RC} + \frac{G}{G^2 + \omega^2 C^2} I \cos \omega t \\ &+ \frac{\omega C}{G^2 + \omega^2 C^2} I \sin \omega t \end{aligned} \tag{2.82}$$

The first term in the solution is the zero-input solution ($I = 0$). The rest represents the zero-state solution.

Next, let us develop the solution by representing the driving function as a sum of complex exponentials and using the results of Secs. 2.3 and

2.4. The zero-input response obtained by setting all source signals to zero is of course the same and given by $v_{C0}e^{-t/RC}$. Only the zero-state response is affected by a change in representation of the source signal, but it must also be unchanged. We shall consider therefore the zero-state response with the sinusoid represented as a sum of exponentials and show that it is the same as before.

It is shown in Chapter 1 that

$$\cos x = \tfrac{1}{2}(e^{jx} + e^{-jx})$$

and,

$$\sin x = \frac{1}{2j}(e^{jx} - e^{-jx}) \tag{2.83}$$

where

$$j = \sqrt{-1} \tag{2.84}$$

is an imaginary number.

The circuit differential equation of Eq. 2.73 with $\phi = 0$ may be written using Eq. 2.83 as

$$C\frac{dv}{dt} + Gv = \frac{I}{2}(e^{j\omega t} + e^{-j\omega t}) \tag{2.85}$$

We wish to determine the zero-state solution of Eq. 2.85, that is, for $v(0) = v_{C0} = 0$.

Equation 2.85 is identical to Eq. 2.66 except for different coefficients and parameters. We can obtain therefore the zero-state solution by Superposition II as before. We first obtain the zero-state solution with one of the sources present, say $(I/2)e^{j\omega t}$, and the other source set to zero. The equation becomes

$$C\frac{dv}{dt} + Gv = \frac{I}{2}e^{j\omega t} \qquad v(0) = 0 \tag{2.85a}$$

The solution is given by

$$v_{ZS}{}^1 = \frac{I/2}{G + j\omega C}[e^{j\omega t} - e^{-t/RC}] \tag{2.86a}$$

Next, we must obtain the zero-state response to the source $(I/2)e^{-j\omega t}$ with the first source set to zero. That is, we must determine the solution to

$$C\frac{dv}{dt} + Gv = \frac{I}{2}e^{-j\omega t} \qquad v(0) = 0 \tag{2.85b}$$

The solution to the above equation is

$$v_{ZS}{}^2 = \frac{I/2}{G - j\omega C}[e^{-j\omega t} - e^{-t/RC}] \tag{2.86b}$$

The zero-state solution of Eq. 2.85 is now given by the sum of $v_{ZS}{}^1$ and $v_{ZS}{}^2$, that is,

$$v_{ZS}(t) = v_{ZS}{}^1 + v_{ZS}{}^2 = \frac{I/2}{j\omega C + G}\,[e^{j\omega t} - e^{-(G/C)t}] + \frac{I/2}{-j\omega C + G}\,[e^{-j\omega t} - e^{-(G/C)t}] \tag{2.86}$$

To this response must be added the zero-input response which in our case is $v_{C0}e^{-t/RC}$ to obtain the total response. Let us examine if our zero-state response in Eq. 2.86 is the same as the one obtained by letting $v_{C0} = 0$ in Eq. 2.82. Equation 2.86 involves complex numbers and, therefore, we must try an appropriate grouping of terms and use Eq. 2.83 to obtain a result in terms of sinusoids.

Let us group the driven and the transient response terms of the two zero-state responses in Eq. 2.86. The two transient terms give

$$\begin{aligned} v_{ZS}{}^t &= -\frac{I}{2}\,e^{-(G/C)t}\left[\frac{1}{j\omega C + G} + \frac{1}{-j\omega C + G}\right] \\ &= -\frac{IG}{G^2 + \omega^2 C^2}\,e^{-(G/C)t}, \quad (\text{since } j^2 = -1) \end{aligned} \tag{2.87}$$

The two steady-state terms in Eq. 2.86 must sum to the steady-state response in Eq. 2.78 by the superposition principle. Now, remembering $j^2 = -1$, we have from Eq. 2.86,

$$\begin{aligned} v_{ZS}{}^{ss} &= \frac{I}{2}\left[\frac{e^{j\omega t}}{G + j\omega C} + \frac{e^{-j\omega t}}{G - j\omega C}\right] \\ &= \frac{I}{2}\left[\frac{(G - j\omega C)e^{j\omega t} + (G + j\omega C)e^{-j\omega t}}{G^2 + \omega^2 C^2}\right] \\ &= \frac{I}{2(G^2 + \omega^2 C^2)}\,[G(e^{j\omega t} + e^{-j\omega t}) - j\omega C(e^{j\omega t} - e^{-j\omega t})] \\ &= \frac{GI}{G^2 + \omega^2 C^2}\cos\omega t + \frac{\omega C I}{G^2 + \omega^2 C^2}\sin\omega t \end{aligned} \tag{2.88}$$

since, from Eqs. 2.83

$$\frac{e^{j\omega t} + e^{-j\omega t}}{2} = \cos\omega t \tag{2.89}$$

and

$$\begin{aligned} \frac{j}{2}\,(e^{-j\omega t} - e^{j\omega t}) &= \frac{1}{2j}\,(e^{j\omega t} - e^{-j\omega t}) \\ &= \sin\omega t \end{aligned} \tag{2.90}$$

The result in Eq. 2.88 shows that $v_{ZS}{}^{ss} = v_{ss}$, the steady-state solution obtained in Eq. 2.78. The sum of the responses in Eqs. 2.87 and 2.88 give the desired zero-state response.

We have thus obtained the response to a sinusoidal excitation by an appropriate decomposition into sums of exponentials and the use of the Superposition principle.

Let us now consider the plot of the response $v(t)$ in Eq. 2.82 with $v_{C0} = 0$. That is, we consider the zero-state response:

$$v_{ZS} = -\frac{G}{G^2 + \omega^2 C^2} I e^{-t/RC} + \frac{I}{G^2 + \omega^2 C^2} [G \cos \omega t + \omega C \sin \omega t] \tag{2.91}$$

We use the trigonometric identity

$$A \cos x + B \sin x = C \cos (x + \theta) \tag{2.92}$$

where

$$C = \sqrt{A^2 + B^2} \tag{2.93}$$

and

$$\theta = \tan^{-1} \left(\frac{B}{A}\right)$$

Then

$$G \cos \omega t + \omega C \sin \omega t = \sqrt{G^2 + \omega^2 C^2} \cos (\omega t + \theta) \tag{2.94}$$

where

$$\theta = \tan^{-1} \left(\frac{\omega C}{G}\right) \tag{2.95}$$

Using identities of Eqs. 2.94 and 2.95, the solution in Eq. 2.91 becomes

$$\begin{aligned} v_{ZS} &= -\frac{G}{G^2 + \omega^2 C^2} I e^{-t/RC} + \frac{I}{\sqrt{G^2 + \omega^2 C^2}} \cos (\omega t + \theta) \\ &= \frac{I}{\sqrt{G^2 + \omega^2 C^2}} [\cos (\omega t + \theta) - \cos \theta e^{-t/RC}] \end{aligned} \tag{2.96}$$

Observe that $v_{ZS}(0) = 0$. The transient and steady-state responses are plotted in Fig. 2.23. The steady-state response is a cosine wave with phase angle θ and the transient is the decreasing exponential. The total zero-state response is shown in Fig. 2.24.

If the driving force or function in Eq. 2.73 is a cosine function with a phase angle, the procedure is unchanged in either approach. Let the driving function be

$$I \cos (\omega t + \phi) = I_1 \cos \omega t + I_2 \sin \omega t \tag{2.97}$$

where $I_1 = I \cos \phi$ and $I_2 = -I \sin \phi$. The driven response is still assumed as in Eq. 2.74 with unknown coefficients V_1 and V_2. Upon substitution of the value of v_{SS} of Eq. 2.74 into Eq. 2.73, the resulting simultaneous equations for V_1 and V_2 (Eq. 2.76) are slightly modified. The right-hand sides

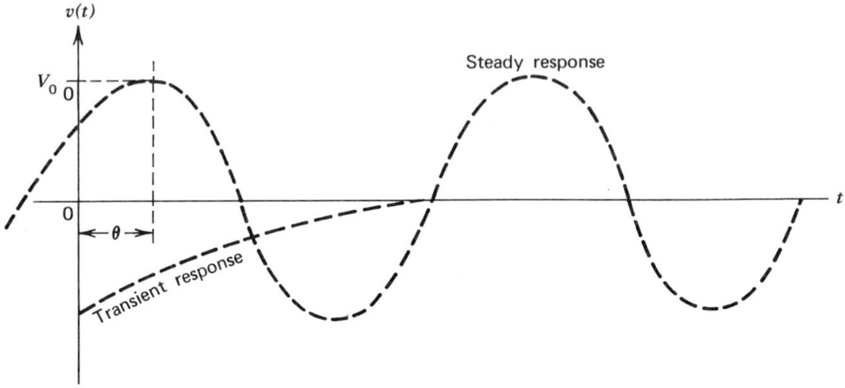

Fig. 2.23 Steady state and transient responses ($V_0 = I/\sqrt{G^2+\omega^2C^2}$).

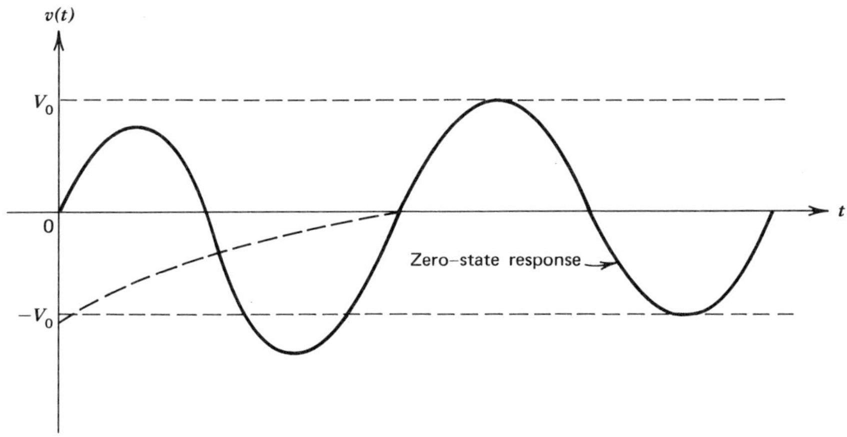

Fig. 2.24 Zero-state response ($V_0 = I/\sqrt{G^2+\omega^2C^2}$).

of the two equations in Eq. 2.76 will now be I_1 and I_2, respectively. The rest of the procedure is then carried out as before to obtain the complete solution.

In the procedure involving the decomposition of the sinusoid, there is only a minor change. From Eq. 2.83

$$
\begin{aligned}
I\cos(\omega t+\phi) &= \frac{I}{2}\left(e^{j(\omega t+\phi)} + e^{j(\omega t+\phi)}\right) \\
&= \frac{I}{2}e^{j\phi}e^{j\omega t} + \frac{I}{2}e^{-j\phi}e^{-j\omega t} \\
&= I_1 e^{j\omega t} + I_2 e^{-j\omega t} \qquad (2.98)
\end{aligned}
$$

The only difference is that the coefficients of the two exponential driving functions in Eq. 2.85 are now I_1 and I_2 instead of $I/2$. The rest of the procedure is unchanged.

The procedures are identical if a sine function were involved instead of a cosine function. Eq. 2.74 remains the same except for the value of θ which is not really relevant to the procedure. In the decomposition procedure, the second of the two equations in Eq. 2.83 is used to represent a sine function as a sum of exponentials. We thus see that the above procedures may be applied to the analysis of circuits with arbitrary sinusoidal excitations.

A final observation concerns the frequency of the steady-state response. Recall that our steady-state solutions are sinusoids with the same angular frequency ω as that of the driving function. Thus, linear, time-invariant circuits merely change the amplitude and phase of the excitation, but they never alter the frequency. This is an important property of linear, time-invariant circuits, and we shall have occasion to refer to it repeatedly in our discussion of steady-state behavior of circuits.

Summary

1. This section has introduced the decomposition of sinusoids into sums of exponentials namely, $\cos x = 1/2\ (e^{jx} + e^{-jx})$ and $\sin x = 1/2j\ (e^{jx} - e^{-jx})$, where $j^2 = -1$. The zero-state solution is then obtained using Superposition II.
2. Superposition principle (Superposition II) can also be applied for obtaining the steady state response alone (Eq. 2.88). The steady-state terms are unaffected by any assumptions of initial state.
3. The steady-state response of linear, time-invariant circuits under sinusoidal excitation exhibits the same frequency as the excitation.

2.6 NUMERICAL SOLUTION OF FIRST ORDER DIFFERENTIAL EQUATIONS

In the previous sections we have seen that each of the *RL* or *RC* circuits could be characterized by a first order differential equation. Thus, for the *RL* series circuit shown in Fig. 2.25, the relation between the excitation voltage $v(t)$ and the response current $i(t)$ is characterized by the following first order differential equation:

$$v(t) = Ri(t) + L\frac{di(t)}{dt} \tag{2.99}$$

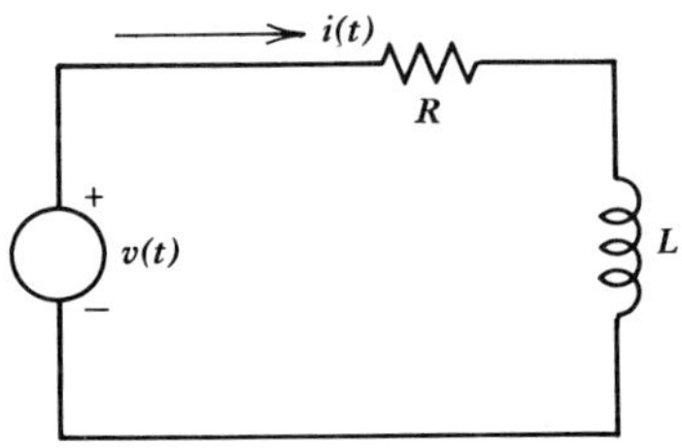

Fig. 2.25 An *RL* series circuit.

In other cases, assume that a first order differential equation is obtained after differentiation of the circuit equation.

We have seen that the general solution of Eq. 2.99 will contain an arbitrary constant in the transient part giving rise to a family of solutions for all possible values of the arbitrary constant. Since there is one arbitrary constant, this is called a one-parameter family. The unique solution is obtained from a knowledge of the initial value of the differential equation variable. The mathematical problem is thus termed an initial-value problem:

$$L\frac{di(t)}{dt} + Ri(t) = v(t) \qquad i(t_0) = i_0 \tag{2.100}$$

and the solution $i(t)$ is unique.

We have shown that the solution for Eq. 2.100 can be evaluated analytically when the excitation function $v(t)$ is expressible in terms of simple elementary time-functions, such as sinusoids, exponentials, constants, or unit steps. However, in many cases of practical interest, the excitation can only be specified as a series of observed values taken over a period of time. Examples are electrocardiograms, human speech, and so on. In such cases, the solution can be obtained only through numerical methods of computation on a digital computer.

Numerical methods are also necessary even when the excitation is simply specified. The circuit model may be valid over a small range of values of certain variables. Many nonlinear circuits are thus specified by different linear models over small ranges of variables (i.e., piecewise linear models). Numerical computational methods are the only way to obtain solutions in such cases.

Finally, for linear circuits of higher order (several reactive elements), numerical methods are essential even for simple excitations. For such circuits, the form of the solution is known but the values cannot be determined analytically. Numerical methods for higher order circuits are based upon those for first order circuits and they are easier to understand for first order situations.

Numerical methods employ the idea that a function can be represented by Taylor's Series as the infinite sum of terms containing successively higher order derivatives (starting with zeroth order). The coefficients of these derivatives decrease greatly in value as the order of the derivative increases. In all of the numerical methods we are going to consider those where only zeroth and first order derivatives are used to approximate the response.

We shall consider two elementary methods for solving general first order initial-value problems in the following form:

$$r' = f(t,r) \qquad r(t_0) = r_0 \tag{2.101}$$

where r denotes the response of the circuit and $f(t,r)$ is a function of both excitation and response expressed as a function of time and response explicitly. Thus, we could write Eq. 2.100 as

$$\frac{di(t)}{dt} = \frac{1}{L}[v(t) - Ri(t)] \qquad i(t_0) = i_0 \tag{2.102}$$

and hence

$$r(t) = i(t) \qquad f(t,r) = \frac{1}{L}[v(t) - Ri(t)]$$

Numerical methods solve the differential equations completely with both the initial conditions and the excitations included at the same time in the process of solution. Therefore, the numerical solution includes both the transient and the steady-state parts and satisfies the initial condition. The complete solution is thus obtained in a direct manner. Furthermore, the numerical procedures described are also valid for time-varying and nonlinear cases.

Euler's Method

If $r(t)$ is the exact solution of Eq. 2.101, then we can expand $r(t)$ into a Taylor's series about the point t near t_0:

$$r(t) = r(t_0) + (t - t_0)r'(t_0) + \frac{(t-t_0)^2}{2!}r''(t_0) + \cdots + \frac{(t-t_0)^k}{k!}r^{(k)}(t_0) + \cdots \tag{2.103}$$

where $r^{(k)}(t_0)$ is the value of the kth derivative of $r(t)$ evaluated at t_0. If we know $r(t_0)$, $r'(t_0), \ldots, r^{(k)}(t_0), \ldots$, then the exact value of $r(t)$ could be obtained by adding the infinite sequence of values in Eq. 2.103. However, if we take $t - t_0 = h$ to be sufficiently small, then a finite sum of the first few terms is enough to approximate the true value $r(t)$. If only the first two terms are used, then we do not even need the second and

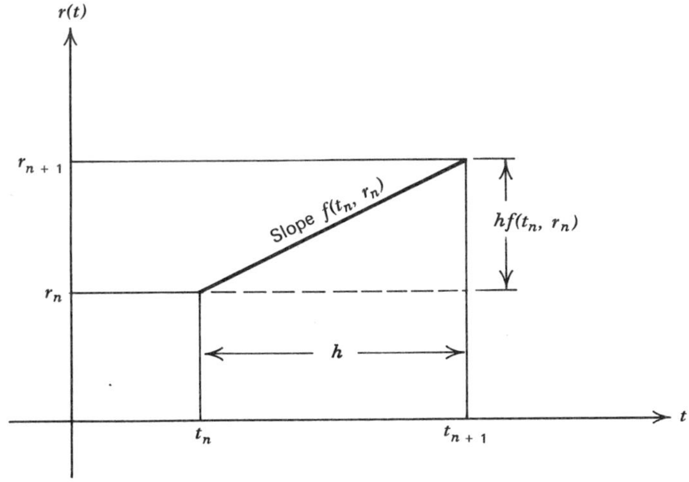

Fig. 2.26 Geometry of Euler's method.

higher derivatives of the function $r(t)$ and the method is called Euler's Method. Hence we have

$$r(t) \cong r(t_0) + (t - t_0)\, r'(t_0) \tag{2.104}$$

For $t - t_0 = h$, we evaluate r at $t_0 + h$ from Eq. 2.104 and then a similar approximation is used again for the next step. For evenly spaced base points with increment h, we have $t_n = t_0 + nh$ $(n = 0, 1, \cdots)$, and a discrete set of values r_n could be obtained that are the approximations to the true solution at the points $t_n = t_0 + nh$ $(n = 0, 1, \cdots)$. From now on we shall use $r(t_n)$ to denote the exact solution at a point t_n and r_n as the approximation for $r(t_n)$. Therefore, we have the following algorithm:

$$r_{n+1} = r_n + hf(t_n, r_n) \tag{2.105}$$

Figure 2.26 gives us a geometrical interpretation of Eq. 2.105. Thus for given t_n and r_n, the slope $f(t_n, r_n)$ is evaluated first and then the product $hf(t_n, r_n)$ is added to the given r_n to form the subsequent value r_{n+1}.

Example 2.6.1 For the circuit shown in Fig. 2.25, let $R = 1\Omega$, and $L = 1$ henry. Find the response current $i(t)$ at $t = 0.04$ if $v(t) = 0$ and $i(0) = 1$ by Euler's method.

Solution. We have for our circuit the following equation:

$$\frac{di(t)}{dt} = -i(t) \qquad i(0) = 1$$

Using Eq. 2.105 we have, therefore,

$$i_{n+1} = i_n - h i_n \tag{2.106}$$

If the step size h is chosen as 0.01, then we have the following results by retaining five decimal places:

$$\begin{aligned}
i(0.01) &\approx i_1 = 1 - 0.01(1) = 0.99000 \\
i(0.02) &\approx i_2 = 0.99000 - 0.01(0.99000) = 0.98010 \\
i(0.03) &\approx i_3 = 0.98010 - 0.01(0.98010) = 0.97030 \\
i(0.04) &\approx i_4 = 0.97030 - 0.01(0.97030) = 0.96060
\end{aligned}$$

Since the exact solution for this example is $i(t) = e^{-t}$, the exact value of $i(t)$ correct to five decimal places at $t = 0.04$ is 0.96079. Therefore, the error at $t = 0.04$ for $i(t)$ is of the magnitude 2×10^{-4}. If the step size is chosen as $h = 0.005$, then the approximations for $i(t)$ are:

$$\begin{aligned}
i(0.005) &\approx i_1 = 0.99500 \\
i(0.010) &\approx i_2 = 0.99002 \\
i(0.015) &\approx i_3 = 0.98507 \\
i(0.020) &\approx i_4 = 0.98014 \\
i(0.025) &\approx i_5 = 0.97524 \\
i(0.030) &\approx i_6 = 0.97036 \\
i(0.035) &\approx i_7 = 0.96551 \\
i(0.040) &\approx i_8 = 0.96068
\end{aligned}$$

And the error at $t = 0.04$ for $i(t)$ is of the magnitude 1×10^{-4}. Thus we see that in order to get fairly accurate approximations, the step size should be sufficiently small. Naturally the computational effort would increase as the step size gets smaller and smaller. This is clearly seen in Example 2.6.1 where we need a total of either four or eight derivative evaluations to get the approximation for $i(0.04)$ for a step size of 0.01 or 0.005 respectively.

We could certainly get more accurate results without decreasing the step size h by including more terms of the Taylor's expansion of the function $r(t)$ in our approximation. However, the calculation of the higher derivatives of $r(t)$ becomes increasingly complex unless $f(t, r(t))$ is a very simple function. Hence, this approach is not practical.

Fortunately, there are methods that do not need the evaluation of higher derivatives; they nevertheless achieve the same accuracy as would be achieved if the higher derivatives were included in the Taylor's series expansion of $r(t)$. These are the so-called Runge-Kutta methods.

Second Order Runge-Kutta Method

Euler's method tries to approximate the exact value of $r(t)$ by a sum of the first *two* terms of its Taylor's series expansion. A second order Runge-Kutta method tries to approximate $r(t)$ by the following formula:

$$r_{n+1} = r_n + ak_1 + bk_2 \tag{2.107}$$

where

$$k_1 = hf(t_n, r_n)$$

$$k_2 = hf(t_n + \alpha h,\, r_n + \beta k_1)$$

and a, b, α, β are constants to be determined so that Eq. 2.107 will agree with the sum of the first *three* terms of the Taylor's series expansion of $r(t)$. Therefore, we expect Eq. 2.107 to give us better results than Euler's method for the same step size h.

On expanding $r(t_{n+1})$ around t_n through terms of h^3 we have

$$r(t_{n+1}) = r(t_n) + hr'(t_n) + \frac{h^2}{2} r''(t_n) + \frac{h^3}{6} r'''(t_n) + \cdots \tag{2.108}$$

Since $r'(t) = f(t, r(t))$ is a function of t and $r(t)$, the following identities are useful in evaluating r'' and r''':

$$r'' = f'(t,r) = \frac{\partial f}{\partial t} + \frac{\partial f}{\partial r}\frac{\partial r}{\partial t} = \frac{\partial f}{\partial t} + \frac{\partial f}{\partial r} f$$

$$r''' = f''(t,r) = \frac{\partial f'}{\partial t} + \frac{\partial f'}{\partial r}\frac{\partial r}{\partial t} = \frac{\partial}{\partial t}\left[\frac{\partial f}{\partial t} + \frac{\partial f}{\partial r} f\right] + \frac{\partial}{\partial r}\left[\frac{\partial f}{\partial t} + \frac{\partial f}{\partial r} f\right]\frac{\partial r}{\partial t}$$

$$= \frac{\partial^2 f}{\partial t^2} + 2\frac{\partial^2 f}{\partial t \partial r} f + \frac{\partial f}{\partial r}\frac{\partial f}{\partial t} + \left(\frac{\partial f}{\partial r}\right)^2 f + \frac{\partial^2 f}{\partial r^2} f^2$$

Now let $\frac{\partial(\cdot)}{\partial r} \triangleq (\cdot)_r$ and $\frac{\partial(\cdot)}{\partial t} \triangleq (\cdot)_t$ where the argument inside the parentheses could be any function. Using these notations for r'' and r''', we have

$$r'' = f_t + f_r f$$

$$r''' = f_{tt} + 2f_{tr} f + f_r f_t + f_r^2 f + f_{rr} f^2 \tag{2.109}$$

Substituting Eq. 2.109 into Eq. 2.108, we have

$$r(t_{n+1}) = r(t_n) + hf(t_n, r_n) + \frac{h^2}{2}[f_t + f_r f]_n$$

$$+ \frac{h^3}{6}[f_{tt} + 2f_{tr} f + f_{rr} f^2 + f_t f_r + f_r^2 f]_n + \cdots \tag{2.110}$$

where the subscript n means that all functions involved are to be evaluated at (t_n, r_n).

Now the coefficient k_2 in Eq. 2.107 is a function of two variables. Using Taylor's series expansion for a function of two variables, k_2 could be expressed as:

$$k_2 = hf(t_n + \alpha h, r_n + \beta k_1)$$
$$= h\left\{f(t_n, r_n) + [\alpha h f_t + \beta k_1 f_r]_n + \left[\frac{\alpha^2 h^2}{2} f_{tt} + \alpha h \beta k_1 f_{tr} + \frac{\beta^2 {k_1}^2}{2} f_{rr}\right]_n + \cdots\right\}$$

Substituting this expression for k_2 into Eq. 2.107 and grouping terms in powers of h, we have

$$r_{n+1} = r_n + (a+b)h[f]_n + bh^2[\alpha f_t + \beta f_r f]_n + bh^3\left[\frac{\alpha^2}{2} f_{tt} + \alpha\beta f_{tr} f + \frac{\beta^2}{2} f_{rr} f^2\right]_n + \cdots \qquad (2.111)$$

Now Eq. 2.111 is used to approximate Eq. 2.110 such that the sum of the first three terms of Eq. 2.111 is equal to the sum of the first three terms of Eq. 2.110. Hence, it is seen that we must have

$$a + b = 1$$
$$b\alpha = b\beta = \tfrac{1}{2} \qquad (2.112)$$

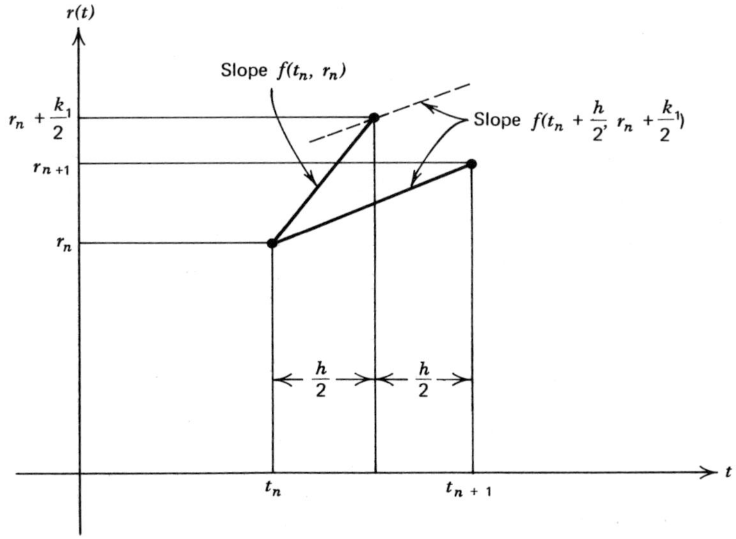

Fig. 2.27 Geometry of the second order Runge-Kutta method.

One of the choices to satisfy Eq. 2.112 is to set $a = 0$, $b = 1$, $\alpha = \beta = 1/2$, and we have the following algorithm:

$$r_{n+1} = r_n + hf\left(t_n + \frac{h}{2}, r_n + \frac{k_1}{2}\right)$$
$$k_1 = hf(t_n, r_n) \tag{2.113}$$

Other choices of a, b, α, and β may also be made to satisfy Eq. 2.112. The above choice of the constants is to be preferred since it gives a comparatively simple algorithm of Eq. 2.113. A geometrical interpretation of the above algorithm is shown in Fig. 2.27. We see that Euler's method is employed twice in sequence: first, the approximation $r_n + k_1/2$ is evaluated at the halfway point $t_n + h/2$ by using the given $f(t_n, r_n)$; second, the derivative $f(t,r)$ is evaluated for $t = t_n + h/2$, $r = r_n + k_1/2$, and then it is used as the average derivative for proceeding over the whole interval from t_n to t_{n+1}.

Example 2.6.2 Solve the problem in Example 2.6.1 by using the second order Runge-Kutta method.

Solution. The equation in Example 2.6.1 is

$$\frac{di(t)}{dt} = -i(t) \qquad i(0) = 1$$

By using the algorithm in Eq. 2.113 we have

$$i_{n+1} = i_n - h\left(i_n + \frac{k_1}{2}\right)$$

and

$$k_1 = -h(+i_n)$$

For $h = 0.01$ we have the following results by retaining five decimal places

$t = 0$, $k_1 = -0.01(1) = -0.0100$
$i(0.01) \approx i_1 = 1 - 0.01[1 - \frac{1}{2}(0.01000)] = 0.99005$

$t = 0.01$, $k_1 = -0.01(0.99005) = -0.00990$
$i(0.02) \approx i_2 = 0.99005 - 0.01[0.99005 - \frac{1}{2}(0.00990)] = 0.98020$

$t = 0.02$, $k_1 = -0.01(0.98020) = -0.00980$
$i(0.03) \approx i_3 = 0.98020 - 0.01[0.98020 - \frac{1}{2}(0.00980)] = 0.97045$

$t = 0.03$, $k_1 = -0.01(0.97045) = -0.00970$
$i(0.04) \approx i_4 = 0.97045 - 0.01[0.97045 - \frac{1}{2}(0.00970)] = 0.96079$

We see that the result of i_4 is correct to five decimal places as compared with the true value of $i(0.04)$. Therefore, we could say there is effectively no error at all for this particular problem!

*Errors

We have described two methods in solving a first order initial value problem numerically. These methods approximate the true values of $r(t)$ at a set of discrete points $t_0, t_1, \cdots, t_n, \cdots$. The error introduced in approximating $r(t)$ by any particular method is termed the *truncation error* of that method. In judging numerical methods used for solving first order initial value problems, we compare the truncation errors introduced at each step, say from t_n to t_{n+1}, by using different methods with the same step size. In this case the errors are called the *local truncation errors*. Let us assume now for convenience that we have an exact solution for $r(t)$ at the base point t_n; to approximate $r(t_{n+1})$ we have to use a specific numerical method to get an approximation r_{n+1}.

Using Euler's method, we have

$$r_{n+1} = r_n + hf(t_n, r_n) = r(t_n) + hr'(t_n)$$

Now the Taylor's expansion for $r(t_{n+1})$ with a remainder could be written as

$$r(t_{n+1}) = r(t_n) + hr'(t_n) + \frac{h^2}{2!} r''(\xi_n) \qquad t_n \leqslant \xi_n \leqslant t_{n+1}$$

Hence the local truncation error is

$$e_{n+1} \overset{\Delta}{=} r(t_{n+1}) - r_{n+1} = \frac{h^2}{2} r''(\xi_n) \tag{2.114}$$

$r(t)$ is generally unknown, so is $r''(\xi_n)$. All we can say is that the local truncation error e_{n+1} is proportional to the square of the step size h, or h^2.

Now for the second order R-K method we have, after subtracting Eq. 2.111 from Eq. 2.110,

$$e_{n+1} \overset{\Delta}{=} r(t_{n+1}) - r_{n+1} = h^3 k_2 + 0(h^4) \tag{2.115}$$

where k_2 depends (in a complicated way usually) upon $f(t,r)$ and its higher order partial derivatives and $0(h^4)$ stands for "terms of order h^4 and higher." If one assumes that h is sufficiently small, so that the error is dominated by the first term, then the local truncation error for the second order R-K method is directly proportional to h^3. This is certainly a better algorithm than Euler's method if the same step size h is used in

*This may be omitted without loss of continuity.

both methods. The price we pay for better results is that we must evaluate the derivative $f(t,r)$ twice instead of once.

If we desire even higher accuracy, then formulas of the R-K type by including higher order terms in the Taylor's expansion of $r(t)$ could be derived. However, the derivations become exceedingly complicated. We will only give the results of a commonly used Runge-Kutta method which attempts to include terms up to the *fourth order* derivatives of the Taylor's expansion of $r(t)$ in the approximation.

Fourth Order Runge-Kutta Method

For the first order initial value problem

$$r'(t) = f(t,r(t)) \qquad r(t_0) = r_0 \tag{2.116}$$

The fourth order Runge-Kutta method finds the approximation to $r(t)$ using the following algorithm:

$$r_{n+1} = r_n + \tfrac{1}{6}(k_1 + 2k_2 + 2k_3 + k_4) \tag{2.117}$$

where

$$\begin{aligned} k_1 &= hf(t_n,r_n) \\ k_2 &= hf\left(t_n + \frac{h}{2}, r_n + \frac{k_1}{2}\right) \\ k_3 &= hf\left(t_n + \frac{h}{2}, r_n + \frac{k_2}{2}\right) \\ k_4 &= hf(t_n + h, r_n + k_3) \end{aligned} \tag{2.118}$$

Example 2.6.3 To compare the algorithm Eq. 2.117 with the second order R-K method, let us again solve the problem in Example 2.6.2 using fourth order R-K method. We have

$$\frac{di(t)}{dt} = -i(t) \qquad i(0) = 1$$

Using a step size $h = 0.04$ and retaining five decimal places, we obtain

$$\begin{aligned} k_1 &= 0.04(-1) = -0.04000 \\ k_2 &= 0.04\left[-\left(1 - \frac{0.04}{2}\right)\right] = -0.03920 \\ k_3 &= 0.04\left[-\left(1 - \frac{0.03920}{2}\right)\right] = -0.03922 \\ k_4 &= 0.04[-(1 - 0.03922)] = -0.03843 \end{aligned}$$

Hence, we have

$$\begin{aligned} i(0.04) \approx i_1 &= 1 + \tfrac{1}{6}[-0.04000 - 0.07840 - 0.07844 - 0.03943] \\ &= 0.96079 \end{aligned}$$

which is correct to five decimal places. Comparing with the result by using the second order R-K method, we see that we have achieved the same accuracy by using a larger step size. The local truncation error for the fourth order R-K method can be shown as

$$e_{n+1} \stackrel{\Delta}{=} r(t_{n+1}) - r_{n+1} = h^5 k_4 + 0(h^6) \tag{2.119}$$

where k_4 depends upon $f(t,r)$ and its higher order partial derivatives. Hence, for a sufficiently small h, the errors are directly proportional to h^5. This high accuracy is achieved at the expense of requiring four derivative evaluations for each step.

Computer Program

To implement the fourth order Runge-Kutta method of the algorithm Eq. 2.117 on a computer, we need the following information:
(a) The function $r'(t) = f(t, r(t))$.
(b) The initial values t_0 and $r(t_0) = r_0$.
(c) The step size h.
(d) The final terminating time t_f.

Once (a), (b), (c), (d) are given, then we could write a FORTRAN program based on the flowchart in Fig. 2.28. The comments in this flowchart correspond to the following:

1. The function $f(t,r)$ could be defined in a FORTRAN program in either a direct Statement Function or a Function Subprogram. For example, if we have $r'(t) = t - r^2 = f(t,r)$, then we could either use

   $$F(T,R) = T - R * R$$

   in the beginning of the main program or use a separate Function Subprogram as follows,

   ```
   FUNCTION F(T,R)
   F = T - R * R
   RETURN
   END
   ```

2. t_0, r_0, and t_f are usually given; however, a suitable choice for h is not easy. In order to have a fairly accurate result, we should choose a sufficiently small step size h. On the other hand, too small a step size h would result in excessive computation time as four derivative evaluations are needed for each step. A rule of thumb to compromise these needs are the following:

   I. Set $h = (t_f - t_0)/20$ and run the program once.

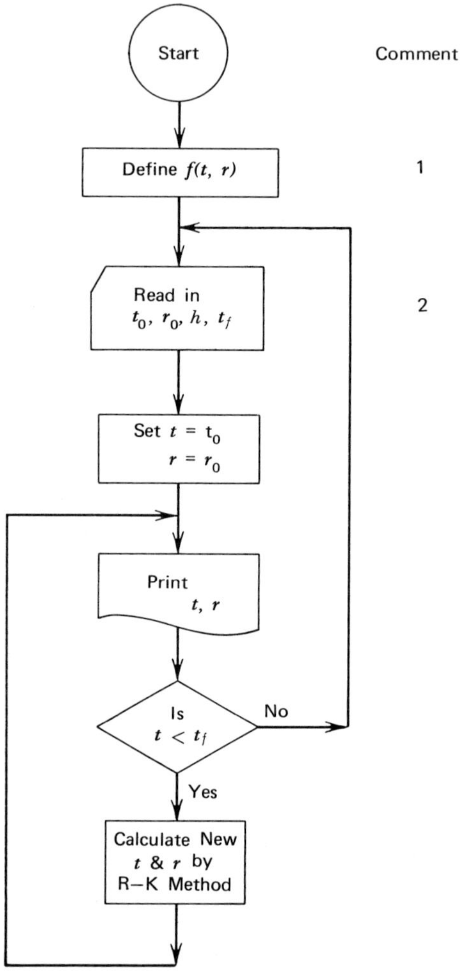

Fig. 2.28 Runge-Kutta method flow chart.

II. Set h to half of the step size in I and run the program once more.
III. If the results from I and II are pretty close, then they are acceptable; otherwise, reduce h and repeat I, II, and III.

Example 2.6.4 Using the fourth order Runge-Kutta method, find the solution of the following differential equation from $t = 0$ to $t = 1$.

$$\frac{di(t)}{dt} = -i(t) \qquad i(0) = 1$$

Solution. The FORTRAN program, input data listing, and a sample computer output are illustrated in Figs. 2.29 to 2.31. The parameters are the

```
C --- FOURTH-ORDER RUNGE-KUTTA METHOD
      F(T,R)=-R
    1 READ 2,H,RO,TO,TF
    2 FORMAT(4F16.8)
      PRINT 3
    3 FORMAT(16X,18HRUNGE-KUTTA METHOD/)
      PRINT 4
    4 FORMAT(15X,1HT,19X1HR/)
      T=TO
      R=RO
    5 PRINT 7,T,R
    7 FORMAT(2F20.8)
      IF(T-TF) 6,1,1
    6 XK1=H*F(T,R)
      XK2=H*F(T+H/2.,R+XK1/2.)
      XK3=H*F(T+H/2.,R+XK2/2.)
      XK4=H*F(T+H,R+XK3)
      RK=(XK1+2.*XK2+2.*XK3+XK4)/6.
      T=T+H
      R=R+RK
      GO TO 5
      END
```

Fig. 2.29 FORTRAN program for Runge-Kutta method.

H	RO	TO	TF
0.01	1.00	0.00	1.00

Fig. 2.30 Input data listing for Example 2.6.4.

RUNGE-KUTTA METHOD

T	R
0.0	1.00000000
0.01000000	0.99004984
0.02000000	0.98019868
0.02999999	0.97044551
0.03999999	0.96078938
0.04999999	0.95122933
0.05999999	0.94176441
0.06999993	0.93239367
0.07999992	0.92311615
0.08999991	0.91393095
0.09999990	0.90483719
0.10999990	0.89583391
0.11999989	0.88692021
0.12999988	0.87809521
0.13999987	0.86935800
0.14999986	0.86070770
0.15999985	0.85214347
0.16999984	0.84366447
0.17999983	0.83526987
0.18999982	0.82695878
0.19999981	0.81873035
0.20999980	0.81058383
0.21999979	0.80251837
0.22999978	0.79453313
0.23999977	0.78662735
0.24999976	0.77880025

Fig. 2.31 Sample computer output.

following:

T	Time variable
R	Response current
$F(T,R)$	Derivative function
TO	Initial T
RO	Initial response
TF	Terminating T
H	Step size
$XK1$, $XK2$, $XK3$, $XK4$	The four R-K equations
RK	The R-K formula

Example 2.6.5 Using the fourth order Runge-Kutta method, find the solution of the circuit represented by the following equation:

$$\frac{di(t)}{dt} + i(t) = e^{-t} \qquad i(0) = 1$$

We can implement a computer program for such differential equations. In certain computers, exponential is a built-in function. Otherwise a Function Subprogram is needed.

The FORTRAN program and a sample computer output are listed in Figs. 2.32 and 2.33. The parameters and input data listing are the same as in Example 2.6.4.

```
C --- FOURTH-ORDER RUNGE-KUTTA METHOD
      F(T,R)=-R + EXP(-T)
    1 READ 2,H,RO,TO,TF
    2 FORMAT(4F16.8)
      PRINT 3
    3 FORMAT(16X,18HRUNGE-KUTTA METHOD/)
      PRINT 4
    4 FORMAT(15X,1FT,19X1HR/)
      T=TO
      R=RO
    5 PRINT 7,T,R
    7 FORMAT(2F20.8)
      IF(T-TF) 6,1,1
    6 XK1=H*F(T,R)
      XK2=H*F(T+H/2.,R+XK1/2.)
      XK3=H*F(T+H/2.,R+XK2/2.)
      XK4=H*F(T+H,R+XK3)
      RK=(XK1+2.*XK2+2.*XK3+XK4)/6.
      T=T+H
      R=R+RK
      GO TO 5
      END
```

Fig. 2.32 FORTRAN program for Runge-Kutta method.

RUNGE-KUTTA METHOD

T	R
0.0	1.00000000
0.01000000	0.99995035→
0.02000000	0.99980265
0.02999999	0.99955887
0.03999999	0.99922097
0.04999999	0.99879080
0.05999999	0.99827027
0.06999993	0.99766123
0.07999992	0.99696547
0.08999991	0.99618477
0.09999990	0.99532092
0.10999990	0.99437565

Fig. 2.33 Sample computer output.

PROBLEMS

2-0a. Find the general solution of each of the following equations:

(i) $3x'(t)+2x(t)=0$

(ii) $5x'(t)+6x(t)=0$

(iii) $4x'(t)+x(t)=0$

2-0b. Find the total solution of each of the following equations:

(i) $3x'(t)+2x(t)=2u(t), x(0+)=0$

(ii) $5x'(t)+6x(t)=10e^{-6t/5}u(t), x(0+)=5$

(iii) $4x'(t)+x(t)=2u(t)+\sin 2t, x(0+)=1$

2-0c. Repeat 2-0b for the following equation:

$$x'(t)-x(t)=1.5e^{-0.5t}u(t), x(0+)=1$$

Is the solution bounded?

2-1. For the circuit of Fig. P2-1, we have

$$i(t)=10u(t), v(0-)=1.$$

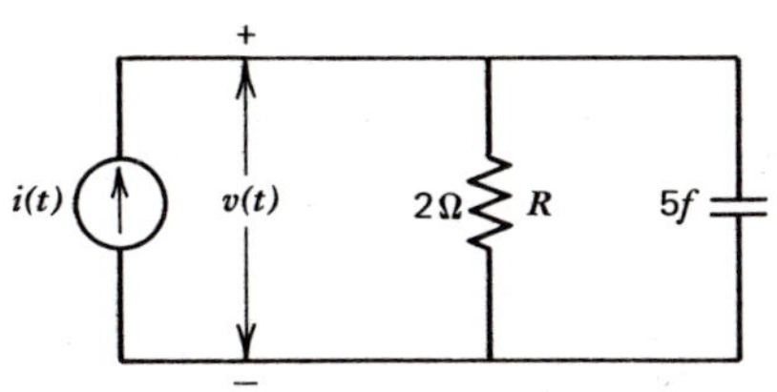

Fig. P2-1

(i) Find $v_{ss}(t)$ (the steady-state response).

(ii) Find $v_{tr}(t)$ (the transient response).

(iii) Find $v_{ZS}(t)$ (the zero-state response).

(iv) Find $v_{ZI}(t)$ (the zero-input response).

2-1a. Using the results of 2-1, find the transient, the steady-state, the zero-state, and the zero-input responses of the current $i_C(t)$ through the capacitor C of the circuit of Fig. P2-1.

2-1b. Repeat 2-1a for the current $i_R(t)$ through the resistor R. Since that state has no meaning for resistors, how do you interpret the zero-state and zero-input responses of $i_R(t)$?

2-1c. The relationship between the capacitor current $i_C(t)$ and the terminal voltage $v_C(t)$ is given by Eq. 1-10 as

$$v_C(t) = \frac{1}{C}\int_{t_0}^{t} i_C(\tau)d\tau + v_C(t_0-)$$

Thus we conclude that $v_C(t_0+) = v_C(t_0-)$ if $i_C(t)$ is bounded. For the circuit shown in Fig. P2-1*c* the switch S is closed at $t - t_0$. Discuss whether $i_C(t)$ is bounded for the following cases:

(i) $v(t) > v_C(t_0-)$
(ii) $v(t) = v_C(t_0-)$
(iii) $v(t) < v_C(t_0-)$

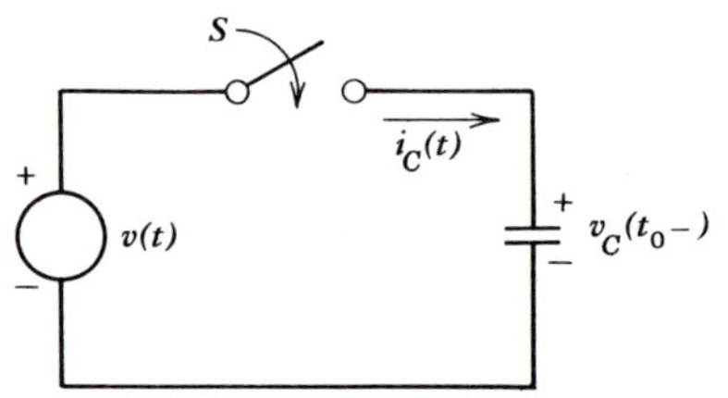

Fig. P2-1*c*

2-1d. The switch S of the circuit shown in Fig. P2-1*d* is opened at $t = 0$. Determine whether $v_L(t)$ is bounded for the following cases:

(i) $i(t) > i_L(0-)$
(ii) $i(t) = i_L(0-)$
(iii) $i(t) < i_L(0-)$

Will $i_L(0+)$ be equal to $i_L(0-)$ for all the cases?

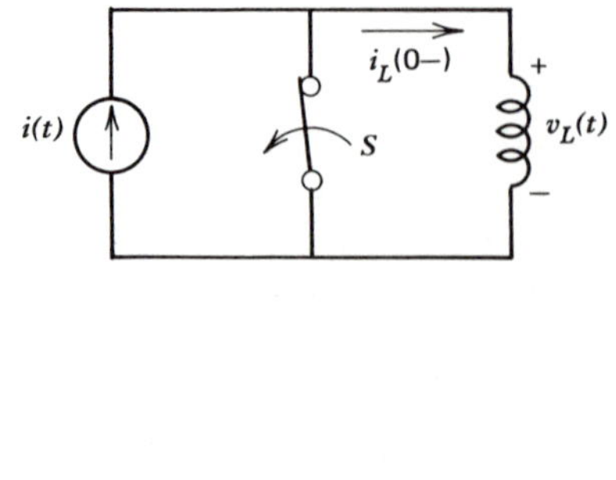

Fig. P2-1*d*

2-1e. From the results of 2-1c and 2-1d we see that the state of a capacitor or an inductor cannot be changed instantaneously provided the initial current through the capacitor or the initial voltage across the inductor, respectively, is bounded. One way to test whether this criterion is satisfied is as follows. Replace each capacitor with a short circuit in series with an ideal voltage source equal to its initial voltage. For an inductor an infinite resistance in parallel with an ideal current source equal to its initial current is replaced. The KVL and KCL equations are then obtained. The resultant current through the short circuit and the voltage across the infinite resistor are then equal to the initial current through the capacitor and the initial voltage across the inductor, respectively.

(i) Apply the above procedure to verify the results of 2-c.

(ii) Repeat (i) for the results of 2-1d.

2-1f. Apply the procedure of 2-1e to the circuit of Fig. P2-1 to see if the initial state can be changed instantaneously.

2-1g. Repeat 2-1f for the following circuits:

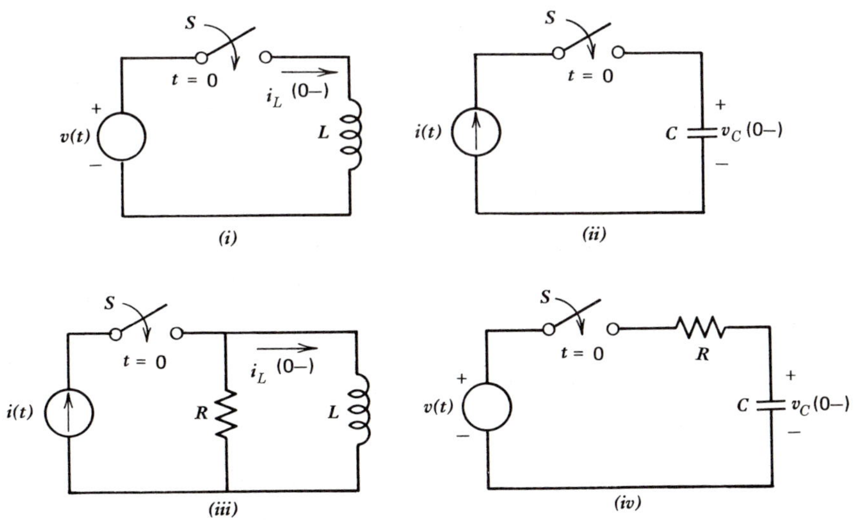

Fig. P2-1g

2-1h. Repeat 2-1f for the following circuits:

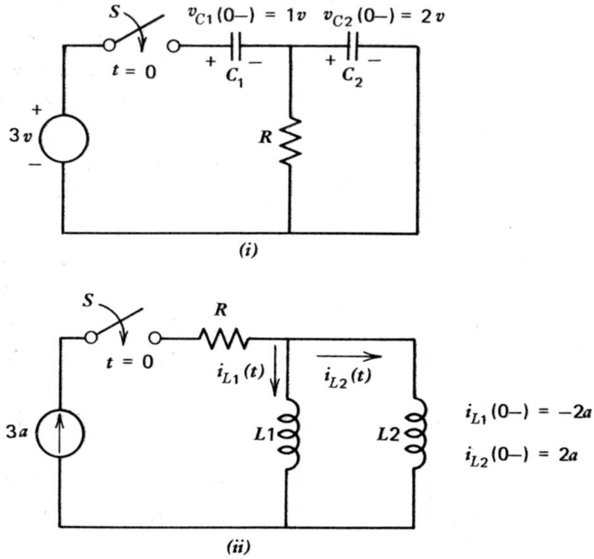

Fig. P2-1*h*

2-2. Repeat 2-1 for the following excitations:
  (i) $i(t) = 10e^{-t}u(t)$.
  (ii) $i(t) = 10(\sin 2t)u(t)$.
  (iii) $i(t) = 10e^{-t/RC}u(t)$.

2-3. For the circuit of Fig. P2-3, $v(t) = 10v$, $i(0-) = 1a$.
  (i) Find $i_{ss}(t)$ (the steady-state response).
  (ii) Find $i_{tr}(t)$ (the transient response).

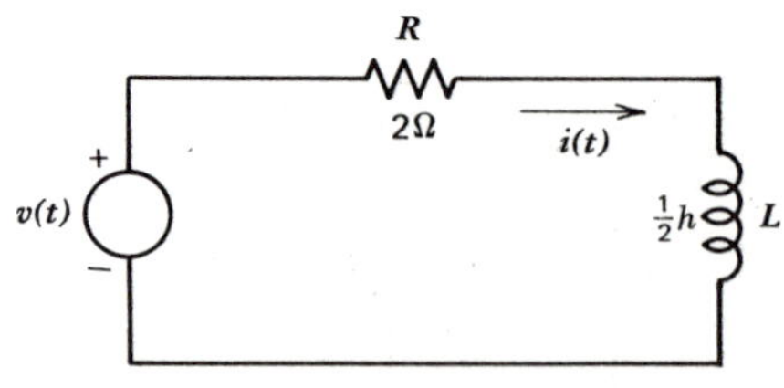

Fig. P2-3

2-4. Repeat 2-3 for the following conditions:
  (i) $v(t) = 0, i(0-) = 1$.
  (ii) $v(t) = 10u(t), i(0-) = 0$.
  (iii) $v(t) = 10(1-e^{-2t})u(t), i(0-) = 1$.

2-5. For the circuit of Fig. P2-5, we have

$$i_1(t) = e^{j\omega t}, i_2(t) = e^{-j\omega t}, i_L(0-) = -1$$

The switch S is closed at $t = 0$. Find $i_L(t)$ for $t \geqslant 0$. Put your answer in the simplest form.

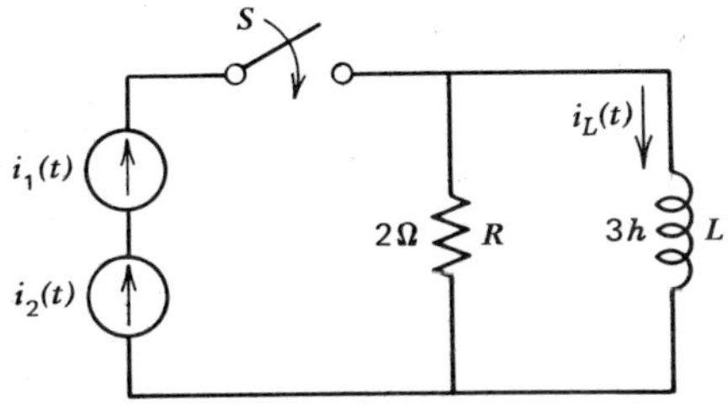

Fig. P2-5

2-6. From the results of the previous problems, is it true that initially the capacitor behaves like a short-circuited path? What about an inductor? Steady-state capacitors and inductors?

2-7. Using the results of 2-1 and the linearity principle, find $v(t)$ of the circuit in Fig. P2-1 subjected to the excitations shown in Fig. P2-7 below: $(v(0-) = 0)$.

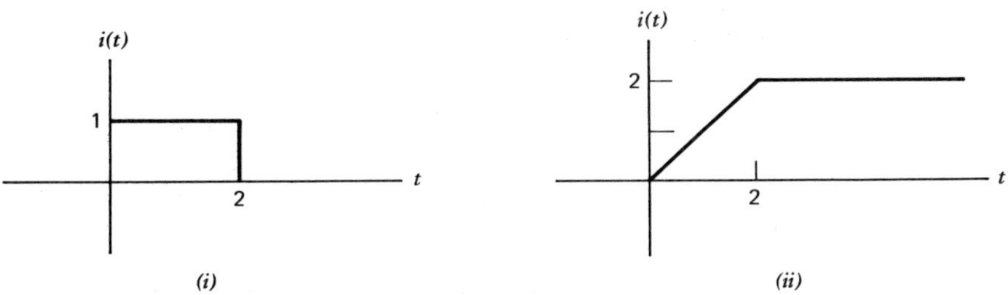

Fig. P2-7

2-8. For the circuit of Fig. P2-8, switch S is opened at $t = 0$ after the circuit had been in steady-state. Find and sketch $i(t)$ for $t \geqslant 0$.

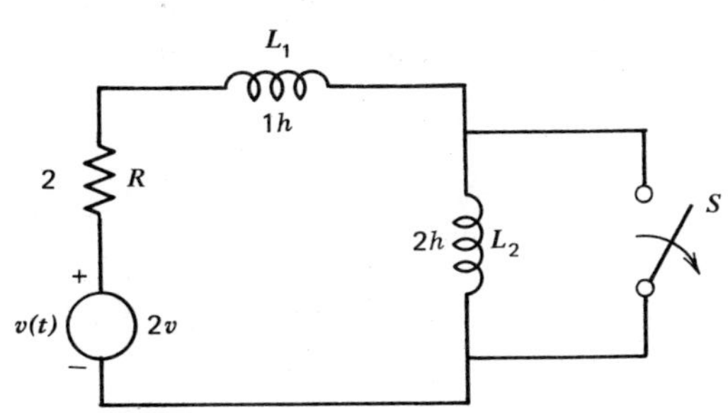

Fig. P2-8

2-9. For the circuit of Fig. P2-9, $i(t) = 10\ mA$, $R = 10\ k\Omega$, $L = 1\ mH$. If the circuit has been connected for a long time and switch S is moved from a to b at $t = 0$, find and sketch the responses $v(t)$, $i_R(t)$, and $i_L(t)$ for $t \geqslant 0$.

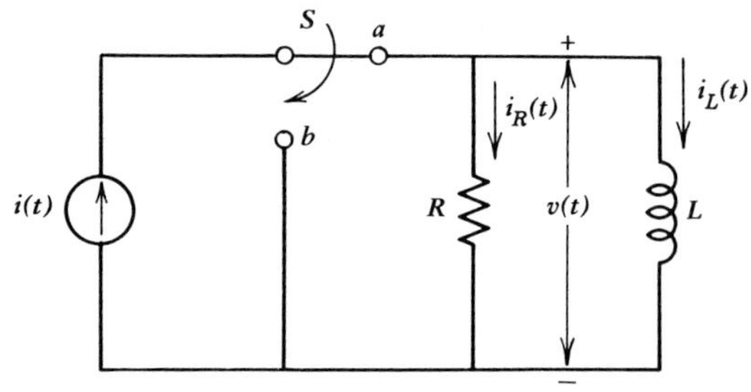

Fig. P2-9

2-10. For the circuit shown in Fig. P2-10, $v_2(0-) = 2$. Find $v_2(t)$ for $t \geqslant 0$ with the following excitations:

(i) $v_1(t) = 0$.
(ii) $v_1(t) = 2u(t)$.

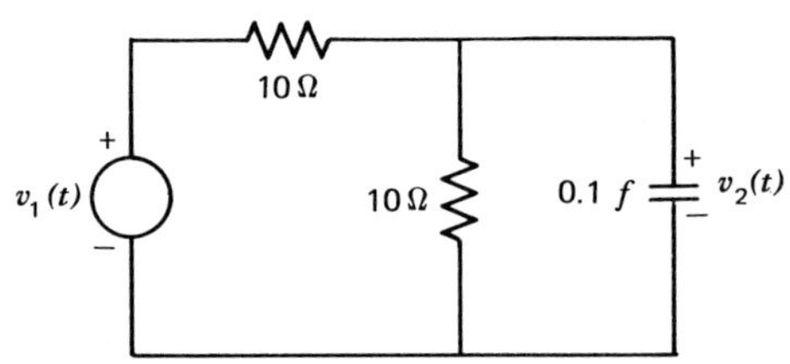

Fig. P2-10

2-11. In the circuit shown in Fig. P2-11, $v(t) = 2u(t)$ and $i_L(0-) = 2$.

(i) Find and sketch $i_2(t)$ for $t \geqslant 0$.
(ii) Find the energy delivered to R_1, R_2, and L respectively between $t = 0$ sec. to $t = 1$ sec.

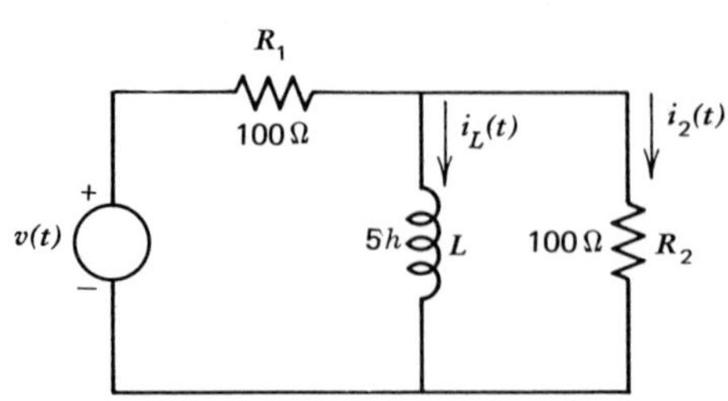

Fig. P2-11

2-12. For the circuit of Fig. P2-12, find $i(t)$, the current through the 2Ω resistor. Simplify your answer. (Hint: use superposition.)

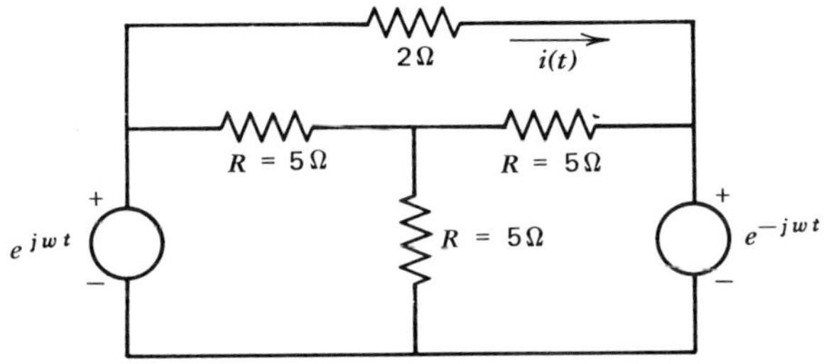

Fig. P2-12

2-13. Repeat 2-12 with $R = 100\Omega$. Is the result independent of the value of R? Why?

2-14. For the circuit shown in Fig. P2-14, find $i_C(t)$ for $t \geqslant 0$, assuming $v_C(0-) = 0$.

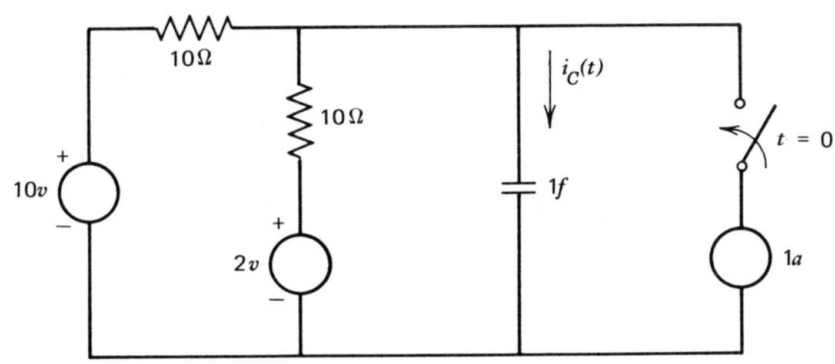

Fig. P2-14

2-15. The network of Fig. P2-15 consists of two charged capacitors in series with a resistor. Before the switch is closed, the voltmeter

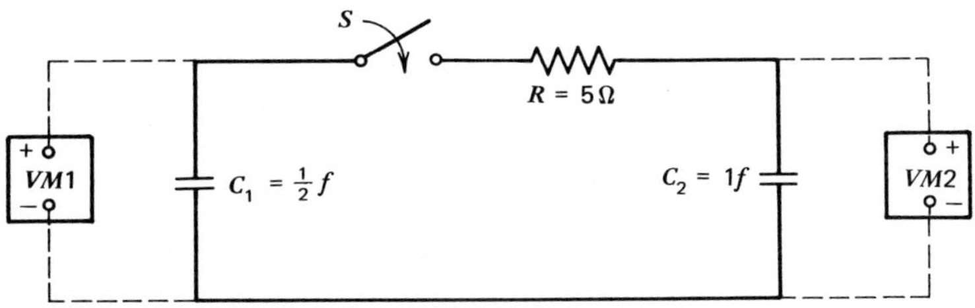

Fig. P2-15

marked $VM1$ reads $+10V$ and $VM2$ reads $+4V$. The switch is closed at time $t = 0$. Find the voltmeter readings when the system has reached steady state.

2-16. Repeat 2-15 if $R = 0\Omega$.

2-17. In the circuit shown in Fig. P2-17, $v(t) = A$. The switch S has been in position 1 for a long time. At $t = 0$, S is moved from 1 to 2. It then leaves 2 back to 1 at $t = T$.

(i) What is the value of $v_C(0-)$?

(ii) Find $i(t)$ for $0 \leqslant t \leqslant T$.

(iii) Find $i(t)$ for $T \leqslant t < \infty$.

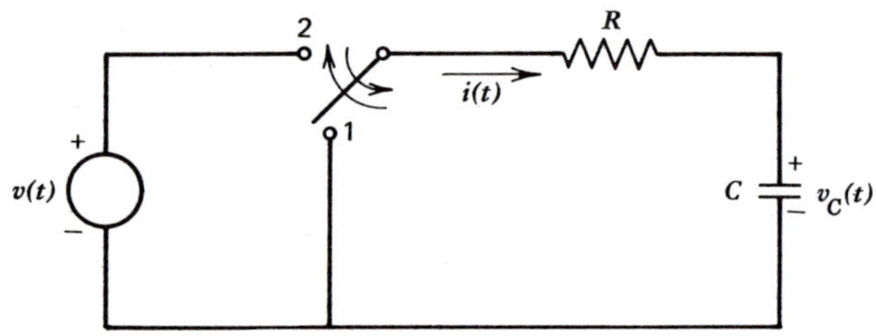

Fig. P2-17

2-17a. Find the zero-state and zero-input responses of the current $i_C(t)$ through the capacitor C in Fig. P2-17a.

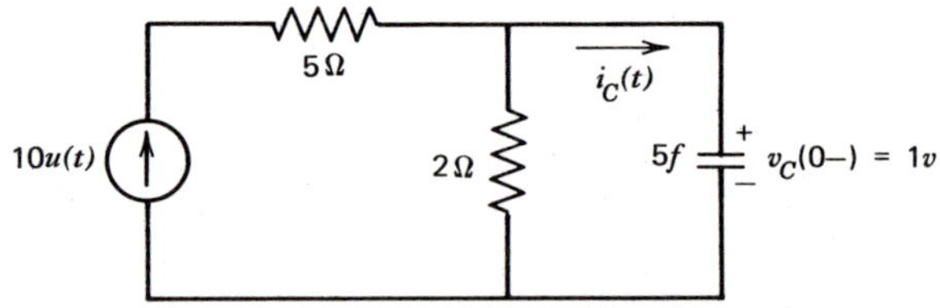

Fig. P2-17*a*

2-17b. Find the zero-state and zero-input responses of the current $i_L(t)$ through the inductor L shown in the circuit of Fig. P2-17b with $i_L(0-) = 1$.

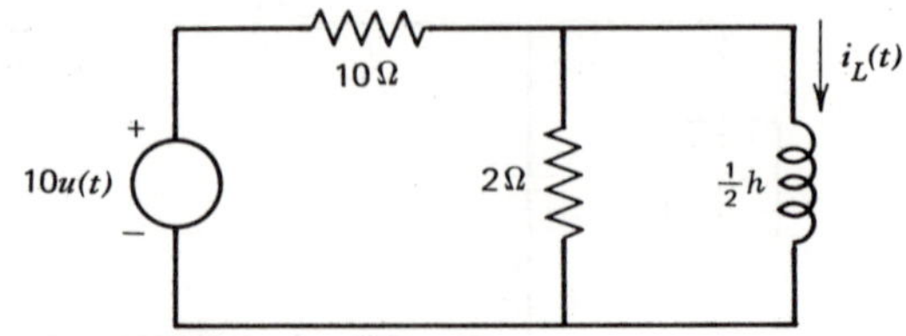

Fig. P2-17*b*

2-17c. Find the current response $i(t)$ of the circuit of Fig. P2-17c with $i_{L1}(0-) = i_{L2}(0-) = 0$.

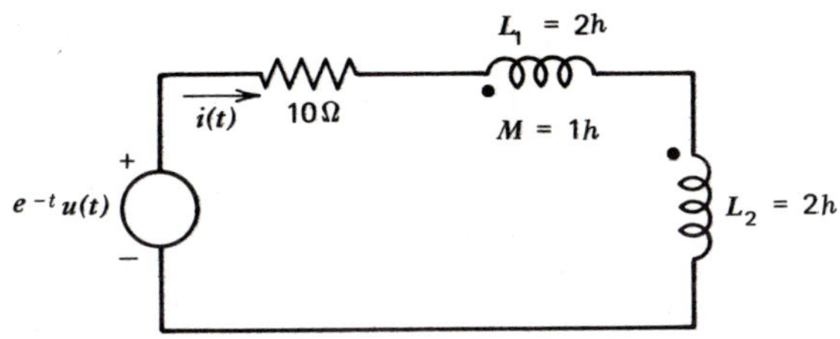

Fig. P2-17*c*

2-17d. Repeat 2-17c for the circuit shown in Fig. P2-17d with $i_{L1}(0-) = i_{L2}(0-) = 0$.

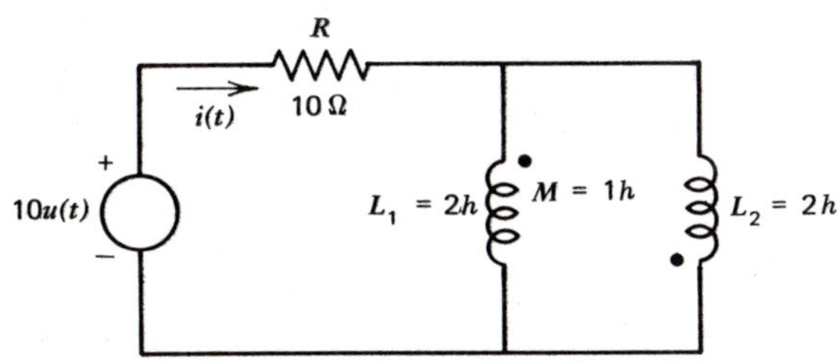

Fig. P2-17*d*

2-17e. For the circuit shown in Fig. P2-17e, $v_{C1}(0-) = 5$ and $v_{C2}(0-) = 10$. Find the voltage $v(t)$ across the current source $i(t)$.

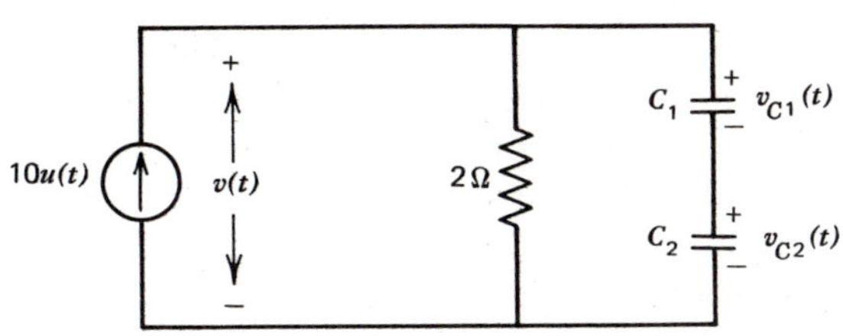

Fig. P2-17*e*

2-17f. For the circuit shown in Fig. P2-17f, $v(t) = 10u(t)$, $i_{L1}(0-) = 1$ and $i_{L2}(0-) = -2$. Find $i(t)$.

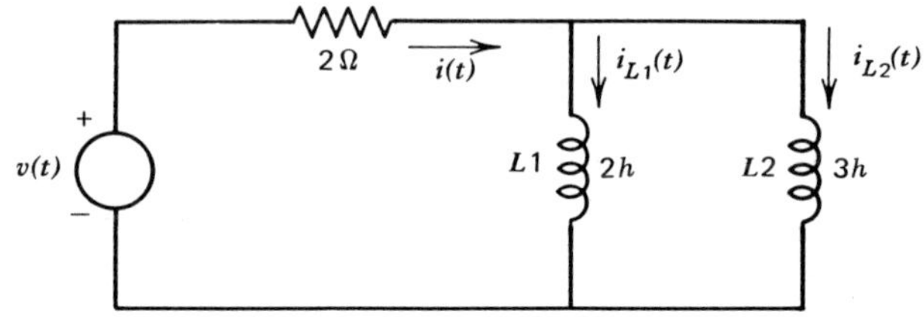

Fig. P2-17f

2-17g. Using Euler's method with 3-digit chopped floating-point arithmetic, find the solutions of the following equations: [The notation $t = a(h)b$ signifies the values of t ranging from $t = a$ to $t = b$ at intervals of h units; thus $t = 0(0.1)0.3$ stands for $t = 0.0, 0.1, 0.2, 0.3$.]

(i) $x'(t) = t + x(t), x(0+) = 0, t = 0(0.05)0.5$.

(ii) $x'(t) = [x(t)]^2 + 1, x(1+) = 3, t = 1(0.1)2$.

(iii) $x'(t) = tx(t), x(1+) = 1, t = 1(0.3)3$.

Answers: (i) $x(t) = e^t - t - 1$; (ii) $\tan^{-1} x(t) = t - 1 + \tan^{-1} 3$; (iii) $x(t) = e^{1/2t}$.

2-17b. Repeat 2-17g to solve 2-0b for $t = 0(0.1)2$. Check your answers with the exact ones.

2-18. Supply the details in deriving Eq. 2.111 of the second order Runge-Kutta method.

2-18a. Solve the equations of 2-17g using the second order R-K method given by Eq. 2.113. Are the results better?

2-18b. Repeat 2-18a to solve equations given in 2-0b for $t = 0(0.1)2$.

2-19. Derive a new set of second order R-K formulas by choosing $a = b = 1/2$ in Eq. 2.112. Solve Example 2.6.1 by the new formulas using a step-size $h = 0.01$ with 5-digit mantissa chopped floating-point operations. Compare the result at $t = 0.04$ with the exact solution.

2-20. Write a program to solve 2-7 by the fourth order Runge-Kutta algorithm of Eq. 2.117 with step sizes $h = 0.01, 0.05, 0.2$. Terminate the computations at $t = 1$. Compare the numerical results with the exact ones.

2-21. Repeat 2-20 to solve 2-14.

2-21a. Write a program to solve 2-17g using the fourth order R-K method of Eq. 2.117. Compare the results with that of 2-18a.

2-21b. Repeat 2-21a to solve equations given in 2-0b for $t = 0(0.1)2$. Compare the results with that of 2-18b.

2-22. For the circuit in Fig. P2-22, $v_a(t)$ is shown as a triangular pulse. The rectifier is used such that whenever $v_a \geqslant v_b$, the current i_{ab}

from a to b obeys the relation

$$i_{ab} = k(v_a - v_b)^{3/2}$$

where k is a constant. Otherwise $i_{ab} = 0$. The circuit parameters are: $V_m = 100\,V$, $T = 0.1$, 0.01, 0.001 sec., $C = 10\,\mu f$, $R = 500\Omega$, $k = 0.003$. Using the fourth order R-K method with appropriate step size h, write a program to find $v_b(t)$. Terminate the computations whenever $t \geqslant 20T$.

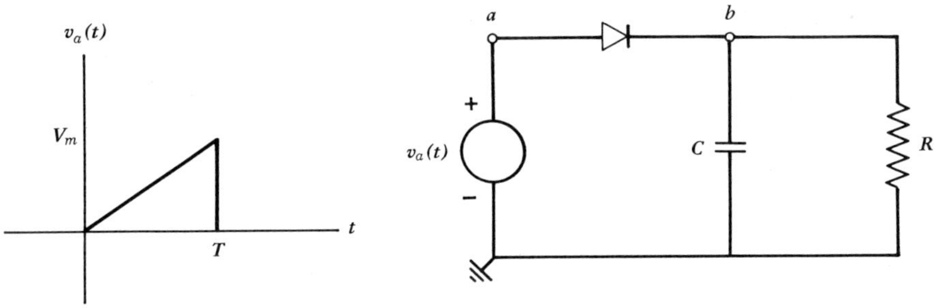

Fig. P2-22

CHAPTER 3

Transient Analysis II: Second and Higher Order Systems

We discussed in Chapter 2 the transient behavior of circuits containing only one energy-storage element (an inductor or a capacitor). The resulting differential equations are first order differential equations. In this chapter we discuss the behavior of circuits containing two or more energy storage elements. The resulting differential equations are of second and higher orders.

We begin with a simple second order circuit containing a resistor, a capacitor and an inductor in series or in parallel. Either node equations (KCL) or loop equations (KVL) may be written for such circuits with relative ease. For higher order circuits, however, we shall concentrate on node equations since they are most conveniently determined, as discussed in Chapter 1.

State-variable description of circuits is considered briefly and in an intuitive way. A rigorous derivation of state equations requires a graph-theoretic discussion of circuits that is beyond the scope of the text. Furthermore, state equations bring no special advantage to circuit analysis other than providing a neat and consistent formalism. Our discussion brings out this formalism and also motivates the computational format for all circuit equations.

3.1 SERIES RLC CIRCUIT DRIVEN BY A CONSTANT VOLTAGE SOURCE

Consider a series RLC circuit driven by a constant voltage source as shown in Fig. 3.1.

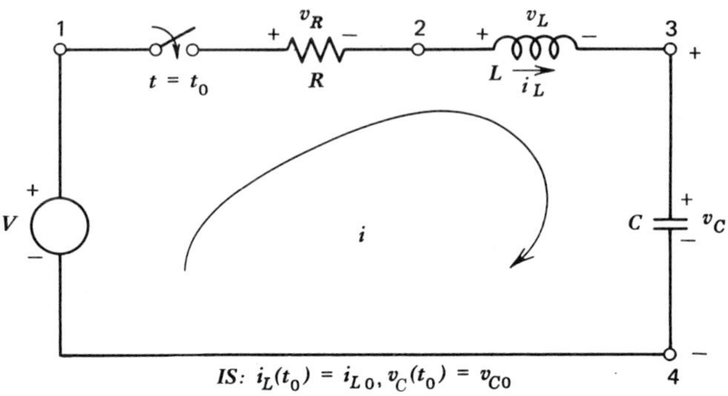

Fig. 3.1 Series *RLC* circuit.

From physical considerations, we can somewhat anticipate the behavior of the circuit. Let us assume zero initial state. The initial current in the circuit is then zero since the inductor will not allow a sudden change in the current. The capacitor is uncharged; there is no current through the resistor and the entire voltage V appears initially across the inductor. The flux starts building up slowly as the current through the inductor rises. As the current rises, there is a voltage drop across the resistor and the capacitor is charged. As the capacitor and resistor voltage rise, there is less voltage across the inductor with the current diminishing in due time. In the limit the capacitor is charged to the source voltage V and the current reaches zero under steady-state conditions.

The precise manner in which these events occur is a function of the circuit parameters and requires a mathematical analysis as follows.

Assume that the circuit is connected at $t = t_0$, and an arbitrary initial state is specified. This provides us with complete generality. Later, for simplicity, we will consider the initial state zero and let $t_0 = 0$. Recall that such assumptions do not compromise the generality of our discussion. A time translation of the excitation corresponds to a similar translation of the response, and addition of the zero-input response to the zero-state response corresponds to the response for a nonzero initial state. The initial state is given by the values of the initial current flowing in the inductor, i_L, and the initial voltage across the capacitor, v_C.

$$\text{IS:} \qquad \begin{aligned} i_L(t_0) &= i_{L0} \\ v_C(t_0) &= v_{C0} \end{aligned} \tag{3.1}$$

There is only one closed loop in the circuit, and the use of KVL would give us a single equilibrium equation:

$$\text{KVL:} \qquad v_R + v_L + v_C = v_S \tag{3.2}$$

where the source voltage $v_S = V$.

Next we express the voltage drops across each element of the circuit in terms of the unknown loop current, i, flowing in the direction shown:

$$\begin{aligned} v_R &= Ri \\ v_L &= L\frac{di}{dt} \\ v_C &= v_0 + \frac{1}{C}\int_{t_0}^{t} i(\tau)\,d\tau \end{aligned} \tag{3.3}$$

Substituting these branch constraints into Eq. 3.2 gives us the circuit equilibrium equation for $t \geqslant t_0$,

$$\text{CE:} \qquad Ri + L\frac{di}{dt} + v_0 + \frac{1}{C}\int_{t_0}^{t} i(\tau)\,d\tau = V \tag{3.4}$$

We use the procedure described in Chapter 2 first to obtain the differential equation by differentiating both sides of Eq. 3.4. Next, recall that the initial conditions for the differential equation variable are obtained from the initial state of the circuit and the circuit equilibrium equation, Eq. 3.4. The differential equation is,

$$\text{DE:} \qquad L\frac{d^2i}{dt^2} + R\frac{di}{dt} + \frac{1}{C}\,i = 0 \qquad (t \geqslant t_0) \tag{3.5}$$

The initial state is specified in Eq. 3.1. The initial current $i_L(t_0) = i_{L0}$ states that

$$i(t_0) = i_L(t_0) = i_{L0} \tag{3.6}$$

Evaluating Eq. 3.4 at $t = t_0$ and substituting the above value of $i(t_0)$ and the initial capacitor voltage $v_C(t_0) = v_{C0}$, we have

$$Ri_{L0} + L\left.\frac{di}{dt}\right|_{t=t_0} + v_{C0} = V$$

since the integral in Eq. 3.4 vanishes for $t = t_0$. Therefore,

$$\left.\frac{di}{dt}\right|_{t=t_0} = i'(t_0) = \frac{V - v_{C0} - Ri_{L0}}{L} \tag{3.7}$$

where the prime denotes the derivative with respect to time. The initial state in Eq. 3.1 thus gives the initial values of both $i(t)$ and $i'(t)$. As we shall see, we need both these initial values to solve Eq. 3.5, the circuit differential equation. Eqs. 3.6 and 3.7 represent the desired initial conditions for the solution of Eq. 3.5.

We solve the differential equation by the same procedure as in Sec. 2.1:

1. Find the steady-state response.
2. Find the transient response.

3. Add the steady-state response and the transient response whose magnitude is unspecified.
4. Calculate the magnitude of the transient response to satisfy the initial conditions.

The differential equation of the circuit, Eq. 3.5, has the driving function or the right-hand side equal to zero. Since the right-hand side is zero, the steady-state solution is zero, that is,

$$i_{ss} = 0 \tag{3.8}$$

As in Chapter 2, we determine the transient response by assuming it is of the form e^{st} with an arbitrary magnitude. We now substitute this exponential into the homogeneous differential equation, Eq. 3.5, to determine the value of s. We have

$$Ls^2e^{st} + Rse^{st} + \frac{1}{C}e^{st} = 0 \tag{3.9}$$

Since the exponential is never zero, we can divide both sides of Eq. 3.9 by e^{st}. The result is the following quadratic equation in s:

$$Ls^2 + Rs + \frac{1}{C} = 0 \tag{3.10}$$

This is called the *characteristic equation*. The roots of this equation are

$$s = -\frac{R}{2L} \pm \sqrt{\left(\frac{R}{2L}\right)^2 - \frac{1}{LC}}$$

or

$$\begin{aligned} s_1 &= -\frac{R}{2L} + \sqrt{\left(\frac{R}{2L}\right)^2 - \frac{1}{LC}} \\ s_2 &= -\frac{R}{2L} - \sqrt{\left(\frac{R}{2L}\right)^2 - \frac{1}{LC}} \end{aligned} \tag{3.11}$$

Equation 3.11 shows that there are two possible values of s: s_1, and s_2, for which the exponential is a transient solution of the differential equation, Eq. 3.5. That is, Ae^{s_1t} is a solution with an arbitrary A and so is Be^{s_2t} with an arbitrary B. The most general transient solution is, of course, an arbitrary combination of the two exponentials:

$$i_t = Ae^{s_1t} + Be^{s_2t} \tag{3.12}$$

Finally, since the steady-state solution of Eq. 3.5 is zero, the total solution is,

$$i(t) = i_{ss} + i_t = Ae^{s_1t} + Be^{s_2t} \qquad (t \geqslant t_0) \tag{3.13}$$

where the constants A and B must be determined from the initial conditions for $i(t)$. Note that there are two arbitrary constants in this case and, therefore, both the two initial conditions of Eqs. 3.6 and 3.7 are needed. Each initial condition applied to Eq. 3.13 gives an equation in the unknowns A and B. With two equations it is thus possible to solve for A and B.

The actual solution in Eq. 3.13 naturally depends on the exponents s_1 and s_2 given by Eq. 3.11. An examination of Eq. 3.11 shows that s_1 and s_2 can be real or complex depending upon the sign of the quantity $[(R/2L)^2 - 1/LC]$ under the square-root sign. A third case occurs when the quantity under the square-root is zero. We now examine these cases separately.

Case I: $$\left[\left(\frac{R}{2L}\right)^2 - \frac{1}{LC}\right] > 0$$

First, we consider Case I when $R/2L > 1/\sqrt{LC}$ in Eq. 3.11. In this case we have two distinct real roots s_1 and s_2. The solution in Eq. 3.13 is thus a linear combination of two real exponentials with unknown coefficients A and B. The values of A and B are determined by the initial conditions of Eqs. 3.6 and 3.7 for the current $i(t)$ and its derivative $i'(t)$.

We have from Eq. 3.13, for $t \geqslant t_0$,

$$i(t) = Ae^{s_1 t} + Be^{s_2 t}$$

and

$$i'(t) = s_1 Ae^{s_1 t} + s_2 Be^{s_2 t} \tag{3.14}$$

We now evaluate Eq. 3.14 at $t = t_0$ and substitute the initial conditions from Eqs. 3.6 and 3.7:

$$i(t_0) = i_{L0} = Ae^{s_1 t_0} + Be^{s_2 t_0}$$

and

$$i'(t_0) = \frac{V - v_{C0} - Ri_{C0}}{L} = s_1 Ae^{s_1 t_0} + s_2 Be^{s_2 t_0} \tag{3.15}$$

The two unknown A and B can be now determined from the above two equations. For convenience, let $t_0 = 0$. The two constants from Eq. 3.15 are given by

$$A = \frac{i'(0) - s_2 i(0)}{s_1 - s_2}$$

and

$$B = -\frac{i'(0) - s_1 i(0)}{s_1 - s_2} \tag{3.16}$$

where the values of $i(0)$ and $i'(0)$ are as given in Eq. 3.15 for $t_0 = 0$.

Substituting Eq. 3.16 into Eq. 3.13 or 3.14, we have for $t \geqslant 0$.

$$i(t) = \frac{V}{L(s_1 - s_2)} [e^{s_1 t} - e^{s_2 t}] - \frac{(v_{C0} + Ri_{L0})}{L(s_1 - s_2)} [e^{s_1 t} - e^{s_2 t}]$$

$$+ \frac{i_{L0}}{(s_1 - s_2)} [s_1 e^{s_2 t} - s_2 e^{s_1 t}] \tag{3.17}$$

The first term on the right-hand side of Eq. 3.17 is the zero-state response and the remaining terms represent the zero-input response, that is,

$$i_{ZS} = \frac{V}{L(s_1 - s_2)} [e^{s_1 t} - e^{s_2 t}] u(t) \tag{3.18}$$

$$i_{ZI} = -\frac{(v_{C0} + Ri_{L0})}{L(s_1 - s_2)} [e^{s_1 t} - e^{s_2 t}] u(t) + \frac{i_{L0}}{(s_1 - s_2)} [s_1 e^{s_2 t} - s_2 e^{s_1 t}] u(t) \tag{3.19}$$

where $u(t)$ is the unit step function signifying that the current is present only for $t \geqslant 0$.

A sketch of the zero-state current, Eq. 3.18, is shown in Fig. 3.2. To draw the sketch, we first determine the time constants related to s_1 and s_2. Let T_1 be the time constant corresponding to s_1, and T_2 be the time constant corresponding to s_2, that is, $(1/T_1) = s_1$ and $(1/T_2) = s_2$. Then

$$s_1 = -\frac{1}{T_1} = -\frac{R}{2L} + \sqrt{\left(\frac{R}{2L}\right)^2 - \frac{1}{LC}}$$

$$s_2 = -\frac{1}{T_2} = -\frac{R}{2L} - \sqrt{\left(\frac{R}{2L}\right)^2 - \frac{1}{LC}} \tag{3.20}$$

Subtracting s_1 and s_2, we have

$$s_1 - s_2 = 2\sqrt{\left(\frac{R}{2L}\right)^2 - \frac{1}{LC}} > 0 \tag{3.21}$$

Furthermore, the magnitude of s_1 is less than the magnitude of s_2, that is, $1/T_1 < 1/T_2$ or $T_2 < T_1$. The sketch of Fig. 3.2 follows. This behavior of the current as a function of time is called an *overdamped response* because the response decays fairly rapidly.

Now, consider the effect of changing the value of the resistance in our series circuit. With L and C constant, as the value of R is reduced the values of the time constants T_1 and T_2 in Eq. 3.20 increase. Hence the current in Fig. 3.2 decays slower. We conclude that reducing the resistance R reduces the damping effect.

Suppose that the value of R is reduced so that $R/2L = 1/\sqrt{LC}$. Observe that this is the limiting case before s_1 and s_2 become complex. We now

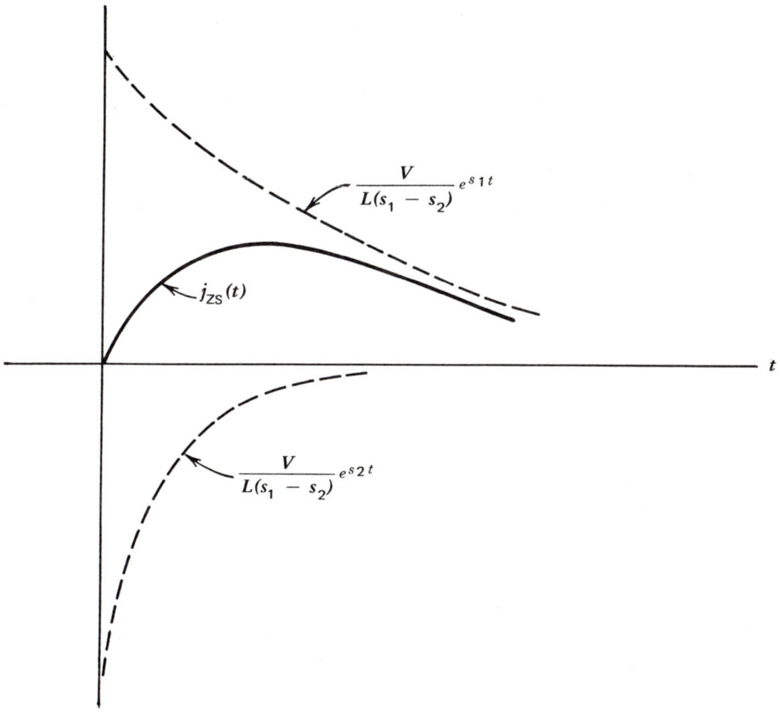

Fig. 3.2 Zero-state response of *RLC* series circuit: overdamped case.

examine the limiting case.

$$\text{Case II: } \left[\left(\frac{R}{2L}\right)^2 - \frac{1}{LC}\right] = 0$$

Recall from Eq. 3.11 that

$$s_1 = -\frac{R}{2L} + \sqrt{\left(\frac{R}{2L}\right)^2 - \frac{1}{LC}}$$

and

$$s_2 = -\frac{R}{2L} - \sqrt{\left(\frac{R}{2L}\right)^2 - \frac{1}{LC}}$$

The expressions for s_1 and s_2 can be conveniently expressed in terms of two new parameters, s_0 and h. These parameters are defined as follows:

$$s_0 = -\frac{R}{2L}$$
$$h = \sqrt{\left(\frac{R}{2L}\right)^2 - \frac{1}{LC}} \tag{3.22}$$

Then s_1 and s_2 can be rewritten as

$$s_1 = s_0 + h$$

and

$$s_2 = s_0 - h \tag{3.23}$$

As $h \to 0$, $s_1 = s_2 = s_0$; that is, the limiting case occurs when the characteristic equation, Eq. 3.10, has a double root at $s = s_0$.

Case II occurs in the limit as h approaches zero in Eq. 3.23, that is, as $[(R/2L)^2 - 1/LC]^{1/2} = h \to 0$. We will examine the behavior of our solution for this case by considering the limit as $h \to 0$ of the solution of Case I in Eqs. 3.17–3.19.

Let us first consider the zero-state solution for $t \geq 0$ in Eq. 3.18:

$$i_{ZS} = \frac{V}{L}\left[\frac{e^{s_1 t} - e^{s_2 t}}{s_1 - s_2}\right]$$

and substitute Eq. 3.23 for s_1 and s_2. Then

$$\begin{aligned} i_{ZS} &= \frac{V}{L}\left[\frac{e^{(s_0+h)t} - e^{(s_0-h)t}}{2h}\right] \\ &= \frac{Ve^{-ht}}{L}\left[\frac{e^{(s_0+2h)t} - e^{s_0 t}}{2h}\right] \end{aligned} \tag{3.24}$$

The limit of this expression as $h \to 0$ is given by

$$\begin{aligned} \lim_{h\to 0} i_{ZS} &= \frac{V}{L}\left[\lim_{h\to 0} \frac{e^{(s_0+2h)t} - e^{s_0 t}}{2h}\right] \\ &= \frac{V}{L}\left[\frac{d}{ds} e^{st}\Big|_{s=s_0}\right] \\ &= \frac{V}{L} te^{s_0 t} \qquad (t \geq 0) \end{aligned} \tag{3.25}$$

Next, let us examine the limit as $h \to 0$ of the zero-input solution of Eq. 3.19 for $t \geq 0$:

$$\lim_{h\to 0} i_{ZI} = -\lim_{h\to 0}\left(\frac{v_{C0} + Ri_{L0}}{L}\right)\left(\frac{e^{s_1 t} - e^{s_2 t}}{s_1 - s_2}\right) + \lim_{h\to 0} i_{L0}\left(\frac{s_1 e^{s_2 t} - s_2 e^{s_1 t}}{s_1 - s_2}\right) \tag{3.26}$$

Let us consider the second term since the limit of the first term is known from Eq. 3.24 and 3.25. The second term of Eq. 3.26 gives:

$$\lim_{h\to 0} i_{L0}\left[-s_0\left(\frac{e^{s_1 t} - e^{s_2 t}}{s_1 - s_2}\right) + h\left(\frac{e^{(s_0-h)t} + e^{(s_0+h)t}}{2h}\right)\right] = i_{L0}[-s_0 te^{s_0 t} + e^{s_0 t}] \tag{3.27}$$

From Eqs. 3.25 and 3.27, we see that Eq. 3.26 as $h \to 0$ becomes

$$i_{ZI} = -\left(\frac{v_{C0} + Ri_{L0}}{L}\right) te^{s_0 t} - i_{L0} s_0 te^{s_0 t} + i_{L0} e^{s_0 t}, \quad (t \geq 0) \tag{3.28}$$

The total current is the sum of the zero-state and the zero-input responses in Eqs. 3.25 and 3.28. Recall that the circuit differential equation, Eq. 3.5, is a homogeneous equation with the steady-state solution zero. The transient solution represents the total solution. Eq. 3.28 shows that when the characteristic equation has a double root at $s = s_0$, the transient solution contains terms of the form $e^{s_0 t}$ and $te^{s_0 t}$. The solution could be obtained then by assuming an arbitrary linear combination of the two terms. Thus,

$$i(t) = i_t = Ae^{s_0 t} + Bte^{s_0 t} \tag{3.29}$$

where $s_0 = -R/2L$. The constants A and B are then obtained from the initial conditions.

Let us use Eq. 3.29 to obtain the zero-state solution, that is, $i_{L0} = v_{C0} = 0$. By Eq. 3.6, for $t_0 = 0$, Eq. 3.29 yields

$$i(0) = 0 = A$$

By Eq. 3.7, the derivative of Eq. 3.29 (with $A = 0$) yields

$$i'(0) = \frac{V}{L} = B$$

Substituting in Eq. 3.29,

$$i_{ZS}(t) = \frac{V}{L} te^{s_0 t} \qquad (t \geq 0) \tag{3.30}$$

This checks with the zero-state solution obtained before in Eq. 3.25. The zero-state solution is shown in Fig. 3.3.

Notice that as t increases, the magnitude of the exponential $e^{s_0 t}$ decreases more rapidly than t increases. Therefore, $i(t)$ builds up with time until $e^{s_0 t}$ becomes dominant. Then $i(t)$ decays to zero. Clearly, the plot of this solution, Fig. 3.3, does not differ much from the plot of the solution in the overdamped case (Case I) except that it decays slower. The damping in Case II corresponds to $te^{s_0 t}$, whereas in Case I it corresponds to $e^{s_1 t}$.

The peak in the plot in Fig. 3.3 occurs when the derivative of the current is zero. The time instant at which the peak occurs is determined below:

$$\frac{di_{ZS}}{dt} = \frac{d}{dt}\left[\frac{V}{L} te^{s_0 t}\right] = 0$$

$$\frac{V}{L}\left[ts_0 e^{s_0 t} + e^{s_0 t}\right] = 0$$

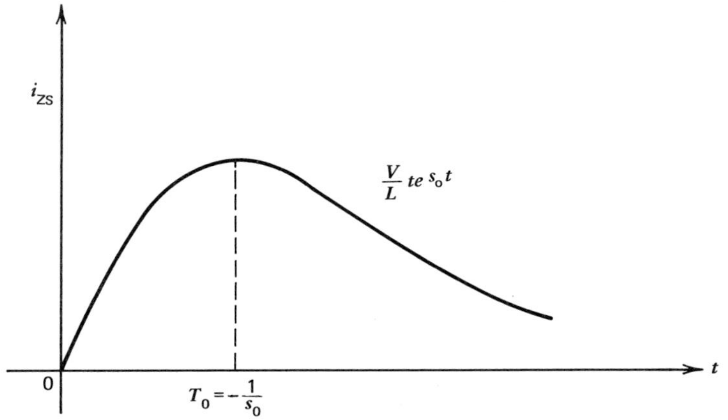

Fig. 3.3 Zero-state response of series *RLC* circuit when $R/2L = 1\sqrt{LC}$.

Dividing by $V/Le^{s_0 t}$, which is nonzero for all t, the relation becomes

$$ts_0 + 1 = 0$$

Obviously, the relation is satisfied when

$$t = -\frac{1}{s_0} = \frac{1}{R/2L} = T_0$$

Notice that $-1/s_0 = T_0$ is the time constant of the exponential $e^{s_0 t}$.

The current reaches a maximum within one time constant of the circuit, that is, the currents build up very rapidly before decaying to zero. This implies that this Case II may be the borderline between the overdamped case and the case where we have oscillations due to complex exponentials. Therefore, we call Case II *critically damped.*

If R is reduced further in the expression for the roots, Eq. 3.11, the roots become complex giving rise to Case III.

$$\text{Case III:} \qquad \left[\left(\frac{R}{2L}\right)^2 - \frac{1}{LC}\right] < 0$$

If $R/2L < 1/\sqrt{LC}$, the quantity under the square root of Eq. 3.11 is negative, and the roots s_1 and s_2 become complex. Thus,

$$\begin{aligned} s_1 &= -\frac{R}{2L} + j\sqrt{\frac{1}{LC} - \left(\frac{R}{2L}\right)^2} = -\alpha + j\beta \\ s_2 &= -\frac{R}{2L} - j\sqrt{\frac{1}{LC} - \left(\frac{R}{2L}\right)^2} = -\alpha - j\beta \end{aligned} \tag{3.31}$$

where $j = \sqrt{-1}$. Recall that the steady-state response is zero. So, the response current in this case will be given by

$$\begin{aligned} i(t) &= Ae^{s_1 t} + Be^{s_2 t} \\ &= Ae^{(-\alpha + j\beta)t} + Be^{(-\alpha - j\beta)t} \end{aligned} \tag{3.32}$$

where A and B must be determined by using Eqs. 3.6 and 3.7. The roots s_1 and s_2 are distinct, and the procedure is identical to that followed for Case I giving us the same expression as in Eq. 3.17 with s_1 and s_2 defined by Eq. 3.31.

Let us examine the zero-state response given in Eq. 3.18 with s_1 and s_2 substituted from Eq. 3.31:

$$\begin{aligned} i_{ZS} &= \frac{V}{L}\left[\frac{e^{s_1 t} - e^{s_2 t}}{s_1 - s_2}\right] u(t) \\ &= \frac{Ve^{-\alpha t}}{L}\left[\frac{e^{j\beta t} - e^{-j\beta t}}{2j\beta}\right] u(t) \\ &= \frac{V}{L\beta} e^{-\alpha t} \sin \beta t \, u(t) \end{aligned} \tag{3.33}$$

where we have used the expressions for complex exponentials defined in Eq. 1.137.

The zero-input current is obtained similarly by substituting Eq. 3.31 into Eq. 3.19:

$$\begin{aligned} i_{ZI} &= -\frac{(v_{C0} + Ri_{L0})}{L}\left[\frac{e^{s_1 t} - e^{s_2 t}}{s_1 - s_2}\right] u(t) + i_{L0}\left[\frac{s_1 e^{s_2 t} - s_2 e^{s_1 t}}{s_1 - s_2}\right] u(t) \\ &= -\frac{(v_{C0} + Ri_{L0})}{L\beta} e^{-\alpha t} \sin \beta t \, u(t) \\ &\quad + \frac{i_{L0}}{2j\beta}\left[(-\alpha + j\beta) e^{(-\alpha - j\beta)t} - (-\alpha - j\beta) e^{(-\alpha + j\beta)t}\right] u(t) \\ &= -\frac{(v_{C0} + Ri_{L0})}{L\beta} e^{-\alpha t} \sin \beta t \, u(t) + \frac{i_{L0}}{\beta} e^{-\alpha t}\left[\alpha \sin \beta t + \beta \cos \beta t\right] u(t) \end{aligned} \tag{3.34}$$

The total response is given by the sum of i_{ZS} and i_{ZI} in Eqs. 3.33 and 3.34. For simplicity, let us assume that the initial state is zero, that is, $v_{C0} = i_{L0} = 0$. Then $i_{ZI} = 0$ and the response is given by the zero-state response of Eq. 3.33. A plot of this zero-state response is shown in Fig. 3.4. The function is a damped sinusoid with its envelope given by $V/\beta L \, e^{-\alpha t} u(t)$. The current oscillates between positive and negative values until it ultimately decays to zero.

The zero-state capacitor voltage behavior is obtained by appropriate integration of the zero-state current. Figure 3.5 shows the zero-state response for the capacitor voltage. Observe that the voltage builds up very

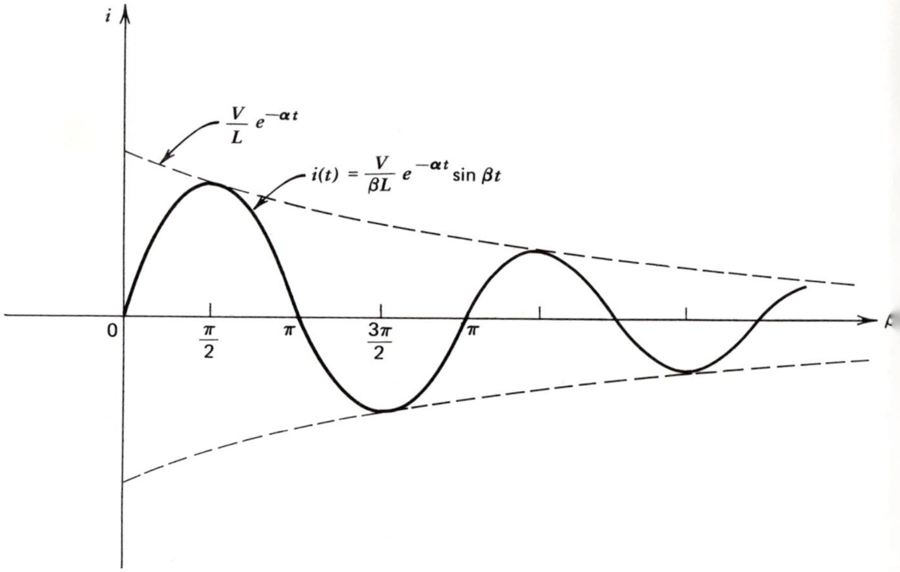

Fig. 3.4 Zero-state current response for Case III.

fast and goes higher than the steady-state value equal to the source voltage V. It then oscillates about this value V until it finally settles down to the steady-state value. Sudden applications of excitations to such circuits, therefore, give rise to capacitor voltages in excess of the source voltages. This can be a serious problem in some situations.

A final observation concerns the response when the resistance R is negligible or approaches zero. Equation 3.33 shows that the response is a

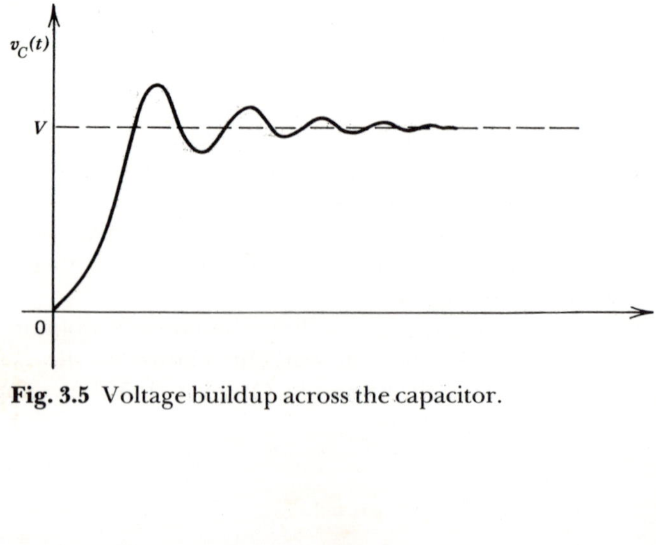

Fig. 3.5 Voltage buildup across the capacitor.

pure sinusoid of frequency $\omega_0 = 1/\sqrt{LC}$ rads/sec. Thus, LC circuits, ($R = 0$), are commonly used to generate sinusoidal excitations. Of course, resistance in practical circuits can be made very small but never zero. Some energy loss in the resistance always occurs and this creates a very small damping of the sinusoid. Active elements are then used to counteract such damping and loss of energy giving rise to circuits called *oscillators*. Oscillators are the major excitation sources in many electronic circuits.

Example 3.1.1 Let us consider again the circuit of Fig. 3.1, and write the node or KCL equations for it.

Let node 4 be the reference and let $t_0 = 0$. The voltage v_1 at node 1 is known, that is, $v_1 = V$, and no node equation is needed for node 1. We must write the KCL node equations for nodes 2 and 3 in terms of their node voltages v_2 and v_3.

$$\text{Node 2:} \qquad \frac{1}{L}\int_0^t (v_2 - v_3)d\tau + i_{L0} - \frac{(V - v_2)}{R} = 0$$

$$\text{Node 3:} \qquad C\frac{dv_3}{dt} - \frac{1}{L}\int_0^t (v_2 - v_3)d\tau - i_{L0} = 0$$

Let $G = 1/R$ and $\Gamma = 1/L$. Then by differentiating and rearranging terms, we have

$$\left(G\frac{dv_2}{dt} + \Gamma v_2\right) - \Gamma v_3 = G\frac{d}{dt}V$$

$$-\Gamma v_2 + \left(\Gamma v_3 + C\frac{d^2 v_3}{dt^2}\right) = 0$$

We thus have two differential equations with two unknowns v_2 and v_3. They can be combined to give a single differential equation in one of the unknowns as shown in Sec. 3.3. For example, the differential equation in v_3 turns out to be

$$GC\frac{d^2 v_3}{dt^2} + \Gamma C\frac{dv_3}{dt} + G\Gamma v_3 = G\Gamma V$$

The initial conditions are $v_3(0) = v_{C0}$ and from node 3 equation $v_3'(0) = (1/C)i_{L0}$. The second order differential equation can now be solved as before.

In sections 3.3 and 3.4 we discuss how the simultaneous differential equations are combined into a single one. All circuits with more than one KCL or KVL equation lead to similar situations as we shall see.

3.2 SERIES RLC: EXPONENTIAL EXCITATION

The *RLC* circuit has been thoroughly examined in Sec. 3.1 for a constant excitation. The steady-state response was zero and the transient response determined the entire solution. We now examine the same series RLC circuit with exponential excitation. Since the transient response for this circuit has been already discussed at length in the last section, we concentrate for the most part on the steady-state response. Moreover, we consider only the zero-state circuit.

Consider the zero-state circuit of Fig. 3.6 excited by the voltage Ve^{pt} for $t \geqslant 0$. The value of p in general is complex. Then, the KVL circuit equation is given by

$$L\frac{di}{dt}+Ri+\frac{1}{C}\int_0^t i(\tau)d\tau+v_{C0}=Ve^{pt} \qquad t\geqslant 0 \tag{3.35}$$

where $v_{C0}=0$ and $i(0)=i_L(0)$. The unit step function $u(t)$ in the figure specifies that the excitation is Ve^{pt} for $t \geqslant 0$, and it is zero for $t<0$. The zero-state circuit exhibits no response for $t<0$ since both the excitation and the initial state are zero. We thus consider the response for $t \geqslant 0$ as in Eq. 3.35.

The circuit differential equation is obtained by differentiating Eq. 3.35,

$$L\frac{d^2i}{dt^2}+R\frac{di}{dt}+\frac{1}{C}i=pVe^{pt} \qquad t\geqslant 0 \tag{3.36}$$

and the initial conditions, for $v_{C0}=i_{L0}=i(0)=0$, are

$$\begin{aligned} i(0)&=0 \\ i'(0)&=\frac{V}{L} \end{aligned} \tag{3.37}$$

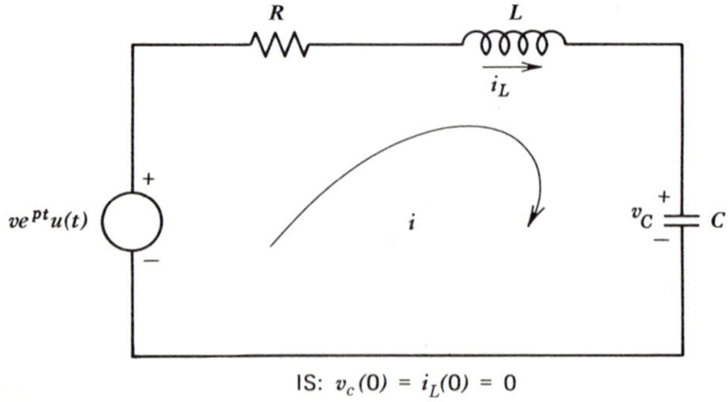

Fig. 3.6 An *RLC* series circuit under exponential excitation.

The steady-state solution for Eq. 3.36 is obtained as in Sec. 2.3 by assuming

$$i_{ss} = Ie^{pt} \tag{3.38}$$

where I is to be determined. Substituting Eq. 3.38 into Eq. 3.36, we have

$$I\left(Lp^2+Rp+\frac{1}{C}\right)e^{pt} = pVe^{pt} \tag{3.39}$$

Since $e^{pt} \neq 0$ for all t, we could simplify Eq. 3.39 and obtain

$$I\left(Lp^2+Rp+\frac{1}{C}\right) = pV$$

or,

$$I = \frac{pV}{Lp^2+Rp+\frac{1}{C}} \tag{3.40}$$

The steady-state solution is now obtained from Eqs. 3.38 and 3.40:

$$i_{ss} = Ie^{pt} = \left(\frac{pV}{Lp^2+Rp+\frac{1}{C}}\right)e^{pt} \tag{3.41}$$

where

$$\left(Lp^2+Rp+\frac{1}{C}\right) \neq 0 \tag{3.42}$$

If Eq. 3.42 is not valid, the steady-state solution must be modified as in Sec. 2.3. We discuss this situation later.

The transient solution for Eq. 3.36 is the solution to the homogeneous equation obtained by setting the right-hand side of Eq. 3.36 to zero. We have already obtained this solution in Sec. 3.1 and write it as

$$i_t = Ae^{s_1t}+Be^{s_2t} \tag{3.43}$$

where A and B are arbitrary constants, and s_1 and s_2 are the roots of the characteristic equation

$$Ls^2+Rs+\frac{1}{C} = 0 \tag{3.44}$$

We saw in Sec. 3.1 that three different cases describe the actual behavior of Eq. 3.43. However, Eq. 3.43 describes the general transient solution adequately for our purposes.

The total solution is given by the sum of Eqs. 3.41 and 3.43:

$$i(t) = Ae^{s_1t}+Be^{s_2t}+Ie^{pt} \qquad t \geqslant 0 \tag{3.45}$$

where the value of I is given in Eq. 3.40. The initial conditions of Eq. 3.37

substituted in Eq. 3.45 give us

$$i(0) = A + B + I = 0 \tag{3.46}$$

$$i'(0) = s_1 A + s_2 B + pI = \frac{V}{L}$$

We solve these equations to obtain

$$A = \frac{V - (p - s_2)LI}{L(s_1 - s_2)}$$

$$B = \frac{V - (p - s_1)LI}{L(s_1 - s_2)} \tag{3.47}$$

The solution is thus completely defined by Eqs. 3.45, 3.47, and 3.40.

In finding the steady-state solution, we have assumed that $Lp^2 + Rp + 1/C \neq 0$. What happens if $Lp^2 + Rp + 1/C = 0$? In that case the value of I in Eq. 3.40 will be indeterminate. Hence, Ie^{pt} is not a right candidate for the steady-state solution. As in Sec. 2.3 we try $I_1 te^{pt}$ as a steady-state solution for this case. Substituting $i_{ss} = I_1 te^{pt}$ into Eq. 3.36 and simplifying, we have

$$I_1 \left[\left(Lp^2 + Rp + \frac{1}{C} \right) te^{pt} + (2Lp + R) e^{pt} \right] = pVe^{pt}$$

Since $Lp^2 + Rp + 1/C = 0$, we have

$$I_1 = \frac{pV}{2Lp + R} \tag{3.48}$$

provided $2Lp + R \neq 0$.

In case $2Lp + R = 0$, then we should try $I_2 t^2 e^{pt}$. The results are

$$I_2 \left[\left(Lp^2 + Rp + \frac{1}{C} \right) t^2 e^{pt} + 2(2Lp + R) te^{pt} + 2Le^{pt} \right] = pVe^{pt}$$

Hence we have

$$I_2 = \frac{pV}{2L}. \tag{3.49}$$

Notice that when $Lp^2 + Rp + 1/C = 0$, then p is a root of the characteristic equation (Eq. 3.44),

$$Ls^2 + Rs + \frac{1}{C} = 0$$

If, furthermore, we have $2Lp + R = 0$, then p is a double root of the characteristic equation. Hence, as a rule in solving for the circuit response, we solve for the transient solution first. This will enable us to know the nature of the roots of the characteristic equation, Eq. 3.44. If p

happens to be none of the roots of Eq. 3.44, then Ie^{pt} is used as a steady-state solution. If p is equal to one of the roots of Eq. 3.44, then I_1te^{pt} is used. Finally, if Eq. 3.44 has double roots and p is equal to it, then $I_2t^2e^{pt}$ should be used as the steady-state solution.

The total solution is then obtained as the sum of the transient solution and the steady-state solution, and the arbitrary constants are determined from the initial conditions.

Response of the circuit to sinusoidal excitation is obtained by expressing the sinusoid as a sum of complex exponentials and then using the superposition principle to obtain the zero-state response. Of course, superposition may be used to first obtain the steady-state response alone, The transient solution is then added and the arbitrary constants determined from the initial conditions.

3.3 HIGHER ORDER CIRCUITS

In Chapter 1 we have presented the procedures for writing the complete set of linearly independent but mathematically coupled equilibrium equations. The KVL or loop equations for complicated circuits require great care in ensuring that the complete set is obtained. The KCL or node equations, on the other hand, can be written in a routine and straightforward fashion. One of the nodes is chosen as the reference node and a KCL is written for each of the remaining nodes to obtain the desired complete set of node equations.

The solutions of the equilibrium equations for simple circuits of Chapters 2 and 3 have shown that the same methods are used for both node and loop equations. The equilibrium equations either are converted to differential equations whose initial conditions are determined from the specified initial state variables of the circuits. The procedures for solving the differential equations are then applied to obtain the desired circuit responses. Thus, the node equations are preferable only in so far as they are easy to describe for arbitrary circuits. The solution methods are the same for both node and loop equations.

The simple circuits discussed so far contain only one or two reactive elements. They give rise to first or second order differential equations, respectively. Furthermore, each of the circuits has been described by a single equilibrium equation by an appropriate choice of loop or node equation description. The circuits considered have either two nodes or one loop, and they give rise to a single node equation (KCL) or a single loop equation (KVL), respectively. The single loop circuits were also described by node equations, usually two or more. The single loop equa-

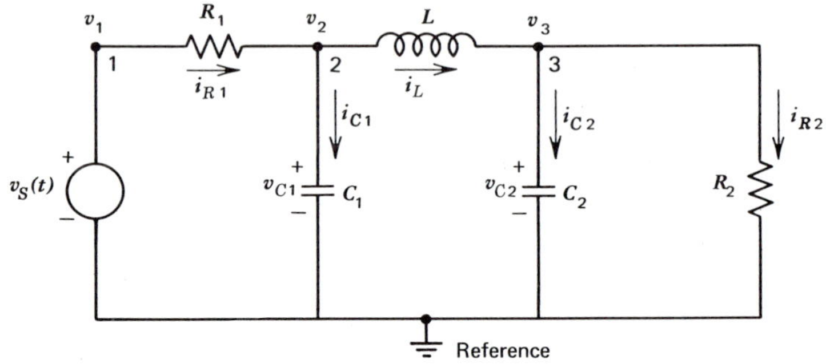

Fig. 3.7 IS: $i_L(0) = i_{L0}$, $v_{C1}(0) = v_{20}$, $v_{C2}(0) = v_{30}$.

tion provided simplicity over node equations and it could be written by inspection.

The choice of node or loop equations for simple circuits is thus governed by convenience and simplicity of writing the circuit equation. For arbitrarily complex circuits, on the other hand, systematic procedures for writing the circuit equations are paramount. The node equations with their straightforward procedure, thus, are to be preferred for general circuit description. In this section, we examine the method of solution with reference to the node equations for circuits of arbitrary order. Of course, the solution method is equally valid for the loop equations as we shall see.

Consider the circuit of Fig. 3.7 with four nodes and three reactive elements driven by a voltage source. There are four nodes with one of the nodes specified as the reference as shown. The remaining three nodes would normally yield the three node (KCL) equations. However, the voltage at node 1 is specified by the voltage source, and a node equation at this node is not necessary. The node equations are written at each node in terms of the unknown voltage of that node with respect to the reference node. The equations are then solved to obtain the node voltages. If a node voltage is already known, then obviously we do not need to solve for it. We thus need write node equations only at nodes 2 and 3:

$$\begin{aligned} &\text{Node 2:} \quad i_{R1} - i_L - i_{C1} = 0 \\ &\text{Node 3:} \quad i_L - i_{C2} - i_{R2} = 0 \end{aligned} \tag{3.50}$$

or

$$\begin{aligned} &\text{Node 2:} \quad \frac{v_S(t) - v_2}{R_1} - \frac{1}{L}\int_0^t (v_2 - v_3)\,d\tau - i_{L0} - C_1\frac{dv_2}{dt} = 0 \\ &\text{Node 3:} \quad \frac{1}{L}\int_0^t (v_2 - v_3)\,d\tau + i_{L0} - C_2\frac{dv_3}{dt} - \frac{v_3}{R_2} = 0 \end{aligned} \tag{3.51}$$

for all $t \geqslant 0$, where the initial state is given by i_{L0}, v_{20}, and v_{30}. The node voltages v_2 and v_3 are also the capacitance voltages v_{C1} and v_{C2}, respectively. The node voltage $v_1 = v_S$ is the source voltage.

Let us rearrange Eq. 3.51 with the known quantities on the right-hand side:

$$C_1\frac{dv_2}{dt}+\frac{v_2}{R_1}+\frac{1}{L}\int_0^t v_2 d\tau-\frac{1}{L}\int_0^t v_3 d\tau=\frac{v_S(t)}{R_1}-i_{L0}$$

and,
$$-\frac{1}{L}\int_0^t v_2 d\tau+C_2\frac{dv_3}{dt}+\frac{v_3}{R_2}+\frac{1}{L}\int_0^t v_3 d\tau=i_{L0} \tag{3.52}$$

for $t \geqslant 0$, and with the initial state values

$$\begin{aligned} i_L(0) &= i_{L0} \\ v_2(0) &= v_{20} \\ v_3(0) &= v_{30} \end{aligned} \tag{3.53}$$

Eq. 3.52 represents two simultaneous integro-differential equations in the unknowns v_2 and v_3.

The procedure that we follow is first to obtain a differential equation in the desired unknown variable by appropriate differentiation and combination of the two equations in Eq. 3.52. We then obtain the initial conditions for the unknown variable from Eqs. 3.53 and 3.52. The differential equation is then solved to obtain the unknown variable uniquely. The approach is quite general and applicable to circuits of arbitrary order.

For ease in manipulation, let us define the derivative operation by an operator

$$D=\frac{d}{dt}$$

then
$$D^{-1}=\int_0^t d\tau \tag{3.54}$$

since $DD^{-1} = D^{-1}D = 1$ except for some constants of integration. The manipulations must be carried out carefully so that the information on the right-hand sides of Eq. 3.52 does not vanish prematurely. In fact, it is always easier to complete all manipulations first before letting any terms vanish under differentiation.

Then Eq. 3.52 becomes

$$\left(C_1D+\frac{1}{R_1}+\frac{1}{L}D^{-1}\right)v_2-\left(\frac{1}{L}D^{-1}\right)v_3=\frac{v_S}{R_1}-i_{L0}$$

$$-\left(\frac{1}{L}D^{-1}\right)v_2+\left(C_2D+\frac{1}{R_2}+\frac{1}{L}D^{-1}\right)v_3=i_{L0} \tag{3.55}$$

or,

$$
\begin{aligned}
D^{-1}(C_1D^2+G_1D+\Gamma)v_2-D^{-1}(\Gamma)v_3 &= G_1v_S-i_{L0} \\
-D^{-1}(\Gamma)v_2+D^{-1}(C_2D^2+G_2D+\Gamma)v_3 &= i_{L0}
\end{aligned}
\tag{3.56}
$$

where we have put $G_1 = 1/R_1$, $G_2 = 1/R_2$, $\Gamma = 1/L$. Observe also that $D^2 = D \cdot D = (d/dt)^2$.

The two equations in Eq. 3.56 can now be solved formally for v_2 and v_3 as if they were two ordinary simultaneous equations. Let us first examine the process of solving simultaneous equations. Consider two equations in two unknowns x_1 and x_2:

$$
\begin{aligned}
a_{11}x_1+a_{12}x_2 &= b_1 \\
a_{21}x_1+a_{22}x_2 &= b_2
\end{aligned}
\tag{3.57}
$$

where a_{ik}'s and b_i's are known. Suppose we wish to solve for one of the unknowns, say x_2. We obtain then, a new equation by a linear combination of the two given equations such that the coefficient of x_1 is zero in the new equation. For example, if we multiply the first equation in Eq. 3.57 by $(-a_{21})$, the second by (a_{11}) and add the two, we get

$$(-a_{21}a_{11}+a_{11}a_{21})x_1+(-a_{21}a_{12}+a_{11}a_{22})x_2 = -a_{21}b_1+a_{11}b_2$$

or

$$(a_{11}a_{22}-a_{12}a_{21})x_2 = a_{11}b_2-a_{21}b_1 \tag{3.58}$$

which is then solved for x_2. The value of x_2 is then substituted into Eq. 3.57 to obtain the value of x_1 or an equation similar to Eq. 3.58 is obtained for x_1:

$$(a_{11}a_{22}-a_{12}a_{21})x_1 = a_{22}b_1-a_{12}b_2 \tag{3.59}$$

We can now solve Eq. 3.56 for v_2 and v_3 by direct comparison with Eqs. 3.57–3.59. Let

$$
\begin{aligned}
a_{11} &= D^{-1}(C_1D^2+G_1D+\Gamma), \qquad a_{12} = -D^{-1}\Gamma \\
a_{21} &= -D^{-1}\Gamma, \qquad a_{22} = D^{-1}(C_2D^2+G_2D+\Gamma)
\end{aligned}
\tag{3.60}
$$

and

$$
\begin{aligned}
x_1 &= v_2, \qquad b_1 = G_1v_S-i_{L0} \\
x_2 &= v_3, \qquad b_2 = i_{L0}
\end{aligned}
\tag{3.61}
$$

Then from Eq. 3.58, for v_3 we have

$$
\begin{aligned}
&D^{-2}[(C_1D^2+G_1D+\Gamma)(C_2D^2+G_2D+\Gamma)-\Gamma^2]v_3 \\
&\qquad = D^{-1}\Gamma(G_1v_S-i_{L0})+D^{-1}(C_1D^2+G_1D+\Gamma)i_{L0}
\end{aligned}
\tag{3.62}
$$

or

$$
\begin{aligned}
&D^{-2}[C_1C_2D^4+(G_1C_2+G_2C_1)D^3+(\Gamma C_1+\Gamma C_2+G_1G_2)D^2+(G_1\Gamma+G_2\Gamma)D]v_3 \\
&\qquad = \Gamma G_1D^{-1}v_S+D^{-1}(C_1D^2+G_1D)i_{L0}
\end{aligned}
\tag{3.63}
$$

Finally the differential equation in v_3 is obtained by multiplying both sides by D,

$$[C_1C_2D^3 + (G_1C_2 + G_2C_1)D^2 + (\Gamma C_1 + \Gamma C_2 + G_1G_2)D + (G_1\Gamma + G_2\Gamma)]v_3 = \Gamma G_1 v_S \qquad (3.64)$$

in this last step we have put $D^2 i_{L0} = D i_{L0} = 0$.

Similarly, a differential equation in v_2 can be obtained by using Eqs. 3.59–3.61:

$$[C_1C_2D^3 + (G_1C_2 + G_2C_1)D^2 + (\Gamma C_1 + \Gamma C_2 + G_1G_2)D + (G_1\Gamma + G_2\Gamma)]v_2 = (C_2D^2 + G_2D + \Gamma)G_1 v_S \qquad (3.65)$$

Observe that Eqs. 3.64 and 3.65 represent the same differential equation with different driving functions. The transient response will be of the same form for both v_2 and v_3 since it is governed only by the homogeneous equation. The driven response on the other hand will be different for both cases, and the total response will be determined by the initial conditions. Observe also that, in general, the driving function contains derivatives of the known source signal. Further, the circuit differential equation must always be independent of the initial state of the circuit. Only the solution of the differential equation is affected (through the initial conditions) by the initial state. Thus, the same differential equation would be obtained if the derivation ignored the initial state values. Of course, the initial state values must be included in the equilibrium equations, Eq. 3.52.

Let us use a general third order differential equation to represent either Eq. 3.64 or 3.65:

$$(a_3D^3 + a_2D^2 + a_1D + a_0)x = (b_2D^2 + b_1D + b_0)y \qquad (3.66)$$

where the coefficients a_k, b_k and the source signal y are known. The signal x is to be determined.

To solve Eq. 3.66 we first determine the transient response. The transient response is given by considering the homogeneous equation and assuming $x_t \sim e^{st}$; s is obtained by substituting e^{st} for x in the homogeneous part of Eq. 3.66. The characteristic equation is given by

$$(a_3s^3 + a_2s^2 + a_1s + a_0) = 0 \qquad (3.67)$$

whose roots determine the values of s. A third degree polynomial equation of Eq. 3.67 has three roots, say s_1, s_2, and s_3. These must be determined numerically since there are no general expressions available beyond the quadratic equation. The transient solution is

$$x_t = A_1e^{s_1t} + A_2e^{s_2t} + A_3e^{s_3t} \qquad (3.68)$$

where A_1, A_2, and A_3 are arbitrary constants, provided s_1, s_2, and s_3 are distinct. If there is a double root, say $s_1 = s_2$, then a term of the form $A_2te^{s_1t}$ must be used. If there is a three-fold root at s_1, then in addition to $A_1e^{s_1t}$ and $A_2te^{s_1t}$ we must use a term of the form $A_3t^2e^{s_1t}$ in our transient solution. We shall use the transient solution of Eq. 3.68 for our discussion.

Next, we determine the driven or steady-state solution. For a constant, an exponential, or a sinusoidal signal, y may be treated as exponential:

$$y = Ye^{pt} \tag{3.69}$$

where $p = 0$, real, or a complex exponent. The complex exponential is one of the terms of the sinusoid represented as a sum of exponentials. The steady state response is of the same form as y provided p is not equal to one of the roots of the characteristic equation. That is,

$$x_{ss} = Xe^{pt} \qquad (p \neq s_1, s_2, s_3) \tag{3.70}$$

By substitution of Eqs. 3.69 and 3.70 into Eq. 3.66, we have

$$X = \frac{b_2p^2 + b_1p + b_0}{a_3p^3 + a_2p^2 + a_1p + a_0} \tag{3.71}$$

and the steady-state response is given by Eq. 3.70 with the value of X substituted.

If p is equal to a single root of the characteristic equation Eq. 3.67, then the steady-state solution is of the form Xte^{pt}. If p is equal to a double or a triple root of Eq. 3.67, x_{ss} is of the form Xt^2e^{pt} or Xt^3e^{pt}, respectively. We assume Eq. 3.70 to represent the steady-state solution.

The total solution for all $t \geqslant 0$ is

$$x(t) = x_{ss} + x_t = A_1e^{s_1t} + A_2e^{s_2t} + A_3e^{s_3t} + Xe^{pt} \tag{3.72}$$

The constants A_i in Eq. 3.72 are determined from the initial conditions for x. Since there are three constants to be determined, we need three initial conditions—$x(0)$, $x'(0)$, $x''(0)$. These three conditions on Eq. 3.72 will give three equations:

$$\begin{aligned} x(0) &= A_1 + A_2 + A_3 + X \\ x'(0) &= s_1A_1 + s_2A_2 + s_3A_3 + pX \\ x''(0) &= s_1{}^2A_1 + s_2{}^2A_2 + s_3{}^2A_3 + p^2X \end{aligned} \tag{3.73}$$

These equations are then solved for A_1, A_2, and A_3 and the values substituted in Eq. 3.72 to obtain the solution.

The initial conditions are obtained from the initial state, Eq. 3.53, and the circuit equilibrium equations, Eq. 3.52. Let us obtain the initial conditions when $x = v_2$. From Eq. 3.53,

$$v_2(0) = v_{20} \tag{3.74}$$

From the first equation of Eq. 3.52,

$$C_1 \left.\frac{dv_2}{dt}\right|_{t=0} + G_1 v_{20} = G_1 v_S(0) - i_{L0}$$

or,

$$v_2'(0) = \frac{G_1 v_S(0) - G_1 v_{20} - i_{L0}}{C_1} \tag{3.75}$$

Finally, differentiating Eq. 3.52,

$$C_1 v_2''(0) + G_1 v_2'(0) + \Gamma v_2(0) - \Gamma v_3(0) = G_1 v_S'(0)$$

$$v_2''(0) = \frac{G_1 v_S'(0) - G_1 v_2'(0) - \Gamma v_{20} + \Gamma v_{30}}{C_1} \tag{3.76}$$

The desired initial conditions are thus obtained for v_2. The initial conditions for v_3 can be obtained in a similar manner. The solution is thus completely defined and the same general procedure is applicable to all higher order multinode circuits.

The presence of voltage sources creates no difficulty in setting down the node equations. When one of the voltage source terminals is connected to the reference node, we know the node voltage of the other terminal, and no node equation is necessary for that node. When a voltage source constrains two nodes, neither of which is the reference, the situation is equally easy to handle. We illustrate the procedure by a simple example.

Example 3.3.1 Consider the circuit of Fig. 3.8. There are two voltage sources present constraining the node voltages as follows:

$$v_1 = v_{s1}$$

$$v_3 - v_2 = v_{s2}$$

The procedure for writing the KCL node equations merely assumes unknown source currents supplied by the voltage sources. We then write a node equation at each of the nodes except the reference. Thus, let node 4 be the reference and let i_1 and i_2 be the unknown source currents as shown. Then the node equations are:

Node 1: $$G_1(v_1 - v_2) = i_1$$

Node 2: $$C_1 \frac{dv_2}{dt} - G_1(v_1 - v_2) + G_2(v_2 - v_3) = -i_2$$

Node 3: $$C_2 \frac{dv_3}{dt} - G_2(v_2 - v_3) = i_2$$

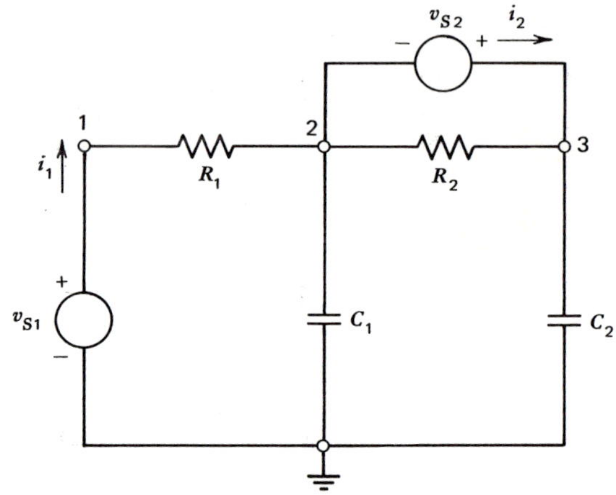

Fig. 3.8

Eliminate the node equation at node 1 since we know the node voltage $v_1 = v_{s1}$. Thus, if the unknown source current appears in only one of the equations, we merely delete the equation and replace the corresponding node voltage by the voltage source constraint, that is, $v_1 = v_{s1}$.

Next, nodes 2 and 3 are constrained by $v_{s2} = v_3 - v_2$. Only one of the nodes can have an independent voltage. Therefore, there should be only one equation to describe the behavior of the independent node voltage. Observe that the unknown source current must appear in both the equations corresponding to the constrained nodes. This current is eliminated by combining the two equations into a single one. We thus have a single equation, and the unknown source current is eliminated. Finally, one of the constrained node voltages is eliminated by using the constraint $v_{s2} = v_3 - v_2$ in the single combined equation.

Suppose node 2 is assumed to be the independent one. Then we first sum the equations for nodes 2 and 3 to obtain

$$C_1 \frac{dv_2}{dt} - G_1(v_{s1} - v_2) + C_2 \frac{dv_3}{dt} = 0$$

where we have put $v_1 = v_{s1}$. Next we set $v_3 = v_{s2} + v_2$ to obtain the equation in v_2:

$$(C_2 + C_1)\frac{dv_2}{dt} + G_1 v_2 = G_1 v_{s1} - C_2 \frac{dv_{s2}}{dt}$$

The circuit description is defined by this single equation in v_2 that may be solved in the standard way. Then, if needed, we can determine $v_3 = v_2 + v_{s2}$.

To summarize, if an unknown source current appears in only one of the node equations the equation is simply deleted and the corresponding node voltage replaced in all equations by the constrained value. If the unknown source current appears in two node equations, these are combined into one and the unknown current eliminated. One of the corresponding nodes is selected as the independent one and the voltage constraint is used to eliminate the constrained voltage variable.

3.4 GENERAL PROCEDURE

The above procedure is easily generalized for arbitrary circuits. For convenience, we assume that the circuit is electrically connected into a single part and requires a single reference node. If there are several electrical parts of the circuit, the same procedure described here is applied to each one of the electrical parts.

Let us consider a circuit with $(N+1)$ nodes. Then one of the nodes is specified as the reference, say the $(N+1)$th node. The KCL node equations are then written for the remaining N nodes. Of course, the presence of a voltage source reduces the number of node equations by one.

Let

$$a_{ij} = \pm (C_{ij}D + G_{ij} + \Gamma_{ij}D^{-1}) \tag{3.77}$$

where C_{ij}, G_{ij} and Γ_{ij} represent the elements that connect nodes i and j. If they are all zero, then the nodes are not connected directly by single circuit elements. When $i=j$, all the elements that connect node i to all other nodes including the reference node must be included. The signs are negative for $i \neq j$ and positive for $i=j$. These consistent signs are obtained by assigning positive signs to the currents flowing out of a node.

When two-port elements are present in the circuit, the a_{ij}'s are not expressible in a simple way. They must be modified to include the constraints governing the two-port elements involved. However, only the a_{ij}'s are modified without any substantive change in the procedure below.

The node equations can then be written as

$$\begin{aligned}
a_{11}v_1 + a_{12}v_2 + \cdots + a_{1N}v_N &= i_1 \\
a_{21}v_1 + a_{22}v_2 + \cdots + a_{2N}v_N &= i_2 \\
\vdots \qquad \vdots \qquad\quad \vdots \quad & \\
a_{N1}v_1 + a_{N2}v_2 + \cdots + a_{NN}v_N &= i_N
\end{aligned} \tag{3.78}$$

where $v_1, \ldots, v_N$ are the unknown node voltages and $i_1, i_2, \ldots, i_N$ are the known values of possible source signals including any initial currents.

Let us assume that the initial state vector is of order n and specified by

$$x(0) = \begin{pmatrix} i_{L1}(0) \\ i_{L2}(0) \\ \cdot \\ \cdot \\ \cdot \\ v_{C1}(0) \\ v_{C2}(0) \\ \cdot \\ \cdot \\ \cdot \end{pmatrix} \tag{3.79}$$

where the total number of independent initial values is n.

The Eq. 3.78 is then solved in the desired variable v_k to obtain the desired differential equation. If there is only one source, say i_1, then the differential equation is given by

$$(a_nD^n + a_{n-1}D^{n-1} + \cdots + a_1D + a_0)v_k = (b_mD^m + b_{m-1}D^{m-1} + \cdots + b_1D + b_0)i_1 \tag{3.80}$$

where the coefficients a_ν, b_ν are obtained from the solution of Eq. 3.78 as discussed below. The initial conditions $v_k(0)$, $v_k'(0)$, $v_k''(0)$, $\cdots$, $v_k^{(n-1)}(0)$ are obtained from Eqs. 3.78 and 3.79.

The transient solution is given by

$$v_k{}^t = A_1e^{s_1t} + \cdots + A_ne^{s_nt} \tag{3.81}$$

where $s_1, \cdots, s_n$ are the roots of the characteristic equation

$$(a_ns^n + a_{n-1}s^{n-1} + \cdots + a_1s + a_0) = 0 \tag{3.82}$$

The steady-state or driven response is obtained by substituting $v_k{}^{ss} = V_ke^{pt}$ into Eq. 3.80. Then,

$$v_k{}^{ss} = \frac{b_mp^m + \cdots + b_1p + b_0}{a_np^n + \cdots + a_1p + a_0} I_1e^{pt} \tag{3.83}$$

where it is assumed that

$$i_1 = I_1e^{pt} \tag{3.84}$$

The total solution is then given by

$$v_k = v_k{}^{ss} + v_k{}^t \tag{3.85}$$

and the arbitrary constants $A_1, \cdots, A_n$ determined from the initial conditions $v_k(0)$, $v_k'(0)$, $\cdots$, $v_k{}^{(n-1)}(0)$.

The solution is modified if there are any degeneracies in the roots of the characteristic equation. If there is d-fold root of Eq. 3.82, say s_q, the transient solution has terms of the form $e^{s_q t}, te^{s_q t}, t^2 e^{s_q t}, \cdots, t^{d-1}e^{s_q t}$. The driven response is also modified if p is equal to any root of the characteristic equation, say, s_q of order d. Then $v_k{}^{ss} = V_k t^d e^{pt}$ where V_k is obtained by substituting the value of $v_k{}^{ss}$ into Eq. 3.80. Finally, the differential equation is modified if there are several sources. Then Eq. 3.80 will contain terms on the right-hand side corresponding to each source. The steady-state solution is obtained by superposition.

We conclude the discussion by expressing the solution for the simultaneous equations, Eq. 3.78, in terms of the determinant and cofactors of the coefficients as discussed in the Appendix. If the determinant of the coefficients a_{ij} is given by

$$|A| = \begin{vmatrix} a_{11} & \cdots\cdots & a_{1N} \\ a_{21} & \cdots\cdots & a_{2N} \\ \vdots & & \vdots \\ a_{N1} & \cdots\cdots & a_{NN} \end{vmatrix} \tag{3.86}$$

then its cofactor

$$A_{ij} = (-1)^{i+j} M_{ij} \tag{3.87}$$

where M_{ij} is the determinant of the coefficients with the ith row and jth column removed:

$$M_{ij} = \begin{vmatrix} a_{11} & \cdots & a_{1j} & \cdots & a_{1N} \\ & & a_{2j} & & \\ & & a_{3j} & & \\ & & & & \\ a_{i1} & \cdots & a_{ij} & \cdots & a_{iN} \\ a_{N1} & \cdots & a_{Nj} & \cdots & a_{NN} \end{vmatrix} \begin{matrix} \\ \\ \\ \\ i\text{th row} \\ \\ \end{matrix} \tag{3.88}$$

jth column

The solution of Eq. 3.78 is then given by

$$|A| v_k = A_{1k} i_1 + A_{2k} i_2 + \cdots + A_{Nk} i_N \tag{3.89}$$

for $k = 1, 2, \cdots, N$. The right-hand side of Eq. 3.89 contains terms corresponding to sources only. The initial current terms will vanish since the differential equation is independent of the initial state.

The general solution of Eq. 3.78 is given by Eqs. 3.81–3.85. There are many difficulties in the development of the solution. The determinant and the cofactors must be evaluated in symbolic form to derive Eq. 3.89. For very large circuits, this is not an easy task. Numerical computations of circuits avoid the use of general solutions as a rule. Numerical methods are based on the notion of numerical integration that is done over small intervals of time as in Chapter 2. The same general idea is used in Sec. 3.6 for numerical computation of response of arbitrary circuits. The formalism of Eq. 3.80 or 3.89 is not the most convenient for numerical integration, and a new formalism for this purpose is introduced in Sec. 3.5.

The general formalism of Eq. 3.80 is, however, very useful for an understanding of the circuit behavior. It is also of considerable importance in many computations, for example, the steady-state response. In many cases, the steady-state response is the only one of concern, as we shall see in later chapters. Then the computation of Eq. 3.83 can be performed directly for a known value of p. An examination of Eqs. 3.80 and 3.83 shows that the coefficients in Eq. 3.80 occur in the numerator and the denominator of Eq. 3.83. The computation of the steady-state response can be performed by a substitution of $v_k^{ss} = V_k e^{pt}$, $k = 1, 2, \cdots, N$, into Eq. 3.78. All the initial state currents must be set to zero since we are only concerned with the steady-state response.

The Eq. 3.78 become equations in the unknowns $V_1, V_2, \cdots, V_N$ upon substitution of $v_k^{ss} = V_k e^{pt}$. Let us assume there is only one source $i_1 = I_1 e^{pt}$. Then Eq. 3.78 becomes:

$$
\begin{aligned}
a_{11}V_1 + a_{12}V_2 + \cdots + a_{1N}V_N &= I_1 \\
a_{21}V_1 + a_{22}V_2 + \cdots + a_{2N}V_N &= 0 \\
&\vdots \\
a_{N1}V_1 + a_{N2}V_2 + \cdots + a_{NN}V_N &= 0
\end{aligned}
\tag{3.90}
$$

where

$$
a_{ij} = \pm \frac{(C_{ij}p^2 + G_{ij}p + \Gamma_{ij})}{p} \tag{3.91}
$$

The solution of Eq. 3.90 is obtained in the same manner as in Eq. 3.89:

$$
|A|V_k = A_{1k}I_1
$$

or

$$
V_k e^{pt} = \frac{b_m p^m + \cdots + b_1 p + b_0}{a_n p^n + \cdots + a_1 p + a_0} I_1 e^{pt} \tag{3.92}
$$

Eq. 3.83 and 3.92 represent the same steady-state response. For a given value of p, it can be computed directly from Eq. 3.78.

Loop Equations

It is possible to prescribe systematic procedures for writing loop equations for a class of circuits known as *planar* circuits. Planar circuits can be placed on a *plane* surface with no overlapping wires. For such circuits the independent loops are given by the individual *windows* or *pockets* in the circuit diagram.

Figure 3.9 shows a planar circuit with the boxes representing the circuit elements. The independent loops and loop currents are as shown. The KVL Loop equations for the indicated loops form a complete set. If the loop currents are all taken in the same direction, say clockwise, then the loop equations are given by

$$\begin{aligned} b_{11}i_1 + \cdots + b_{1n}i_n &= v_1 \\ b_{21}i_1 + \cdots + b_{2n}i_n &= v_2 \\ &\vdots \\ b_{n1}i_1 + \cdots + b_{nn}i_n &= v_n \end{aligned} \tag{3.93}$$

where the i_k's are the unknown loop currents, v_k's are the known sources and possible initial voltages, and

$$b_{ij} = \pm\left(L_{ij}D + R_{ij} + \frac{1}{C_{ij}}D^{-1}\right) \tag{3.94}$$

In Eq. 3.94, L_{ij}, R_{ij}, and C_{ij} are elements common to loops i and j. When $i = j$, all the elements in the loop are included. The sign is positive when

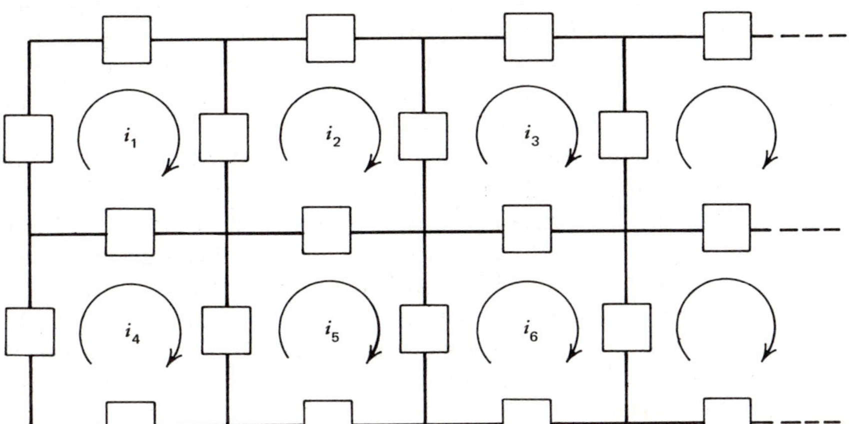

Fig. 3.9

$i=j$ and negative when $i \neq j$. Two-port elements will alter the b_{ij}'s to include the two-port constraints. If any loop currents are specified by current sources, the corresponding loop equations are not present in Eq. 3.93.

The solution procedure is the same as discussed before for the node equations.

3.5 STATE EQUATIONS

We defined the state of the circuit in Chapter 1 in terms of the capacitor voltages and the inductor currents. The state vector defined at any instant of time completely prescribes the currents and voltages anywhere in the circuit at that instant of time. We have used this fact to obtain the initial conditions for the differential equations from the initial state and the equilibrium equations. Therefore, if we were able to determine the behavior of the state vector as a function of time, then we would know the behavior of the entire circuit. Differential equations that represent the behavior of the state vector or the state variables are called *state-differential equations, state-variable equations,* or *state equations.*

A rigorous discussion of state equations requires a knowledge of graph theory that is beyond the scope of this text. Furthermore, the state equations are neither essential nor advantageous in most cases and, in fact, they quite often complicate the circuit description. Why then bother with the state equations? The answer is that they provide a very useful and elegant formalism; they directly relate to the energy variables; and, they are particularly convenient and sometimes the only possible description for non-linear, time-varying circuits. The state equations also provide the behavior of the entire circuit that may be useful at times even though it means greater complexity. Finally, the format of the state equations is convenient and consistent for numerical computations.

We will, therefore, develop the state-variable circuit description in an intuitive manner and without rigor. The discussion is intended primarily to illustrate the concepts. Consider the circuit of Fig. 3.7. The state variables are v_{C1}, i_L, and v_{C2}. We express the rate of change of these variables by using appropriate Kirchhoff's laws. For our description to be consistent, the rate of change of these variables (i.e., the derivatives), must be expressible entirely in terms of the same unknown variables. All other unknown variables that occur must be somehow expressed in terms of the state variables.

Let us start with the derivative of v_{C1}. The derivative of v_{C1} is proportional to the current i_{C1} and so the KCL at node *2* should be used:

$$C_1 \frac{dv_{C1}}{dt} = i_{R1} - i_L$$
$$= G_1(v_s - v_{C_1}) - i_L$$

or,

$$\frac{dv_{C1}}{dt} = -\frac{G_1}{C_1} v_{C1} - \frac{1}{C_1} i_L + \frac{G_1}{C_1} v_s \qquad (3.95)$$

Next, the derivative of i_L is proportional to the voltage drop across it; and, so the KVL is indicated for the loop consisting of C_1, L, and C_2. Thus,

$$L \frac{di_L}{dt} = v_{C1} - v_{C2}$$

or,

$$\frac{di_L}{dt} = \frac{1}{L} v_{C1} - \frac{1}{L} v_{C2} \qquad (3.96)$$

Finally, the derivative of v_{C2} is obtained form a KCL at node *3*:

$$C_2 \frac{dv_{C2}}{dt} = i_L - i_{R2}$$
$$= i_L - G_2 v_{C2}$$

or,

$$\frac{dv_{C2}}{dt} = \frac{1}{C_2} i_L - \frac{G_2}{C_2} v_{C2} \qquad (3.97)$$

The three equations, Eqs. 3.95–3.97, with the given initial state completely define the circuit behavior in terms of the state variables for all $t \geqslant 0$.

Observe that the state-differential equations involve a set of simultaneous equations. The derivative of each state variable is represented as the sum of a linear combination of the state variables and the source functions. The initial conditions for each of the state variables is specified directly by the initial state. The computational methods of numerical integration can be directly applied to these equations.

The procedure illustrated above can be applied quite generally to arbitrary circuits. Difficulties occur when some capacitor voltages or inductor currents are not independent variables but are constrained by other state variables. Such dependent variables create difficulties in writing the state equations. The elements that generate dependent variables are called *excess* elements. The difficulties are removed by putting dummy inductors in series with excess capacitors and dummy capacitors across excess inductors. This will allow us to write the state equations without any difficulty. Once the equations are written, the dummy elements are allowed to go to zero and the excess variables expressed in terms of the other state variables. We illustrate by an example.

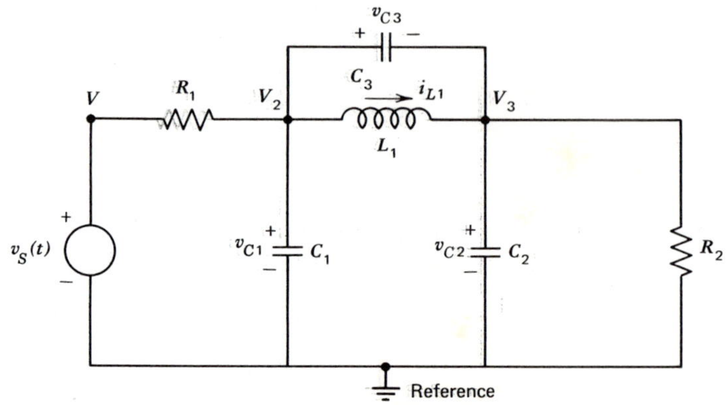

Fig. 3.10

Consider the circuit of Fig. 3.7 with a capacitor between nodes *2* and *3*. The circuit is shown in Fig. 3.10. The voltage across this capacitor is constrained by v_{C1} and v_{C2}. The difficulty occurs in writing the state equation for v_{C_1}:

$$C_1 \frac{dv_{C1}}{dt} = G_1(V - v_{C1}) - i_{L1} - i_{C3} \tag{3.98}$$

The equation contains $i_{C3} = C_3(dv_{C3}/dt)$ on the right-hand side. The derivative of v_{C1} is not expressed anymore as a linear combination of the state variables; the derivative of v_{C3} is also involved. The difficulty is removed by the introduction of a dummy inductor L_3 in series with the excess capacitor C_3 as shown in Fig. 3.11.

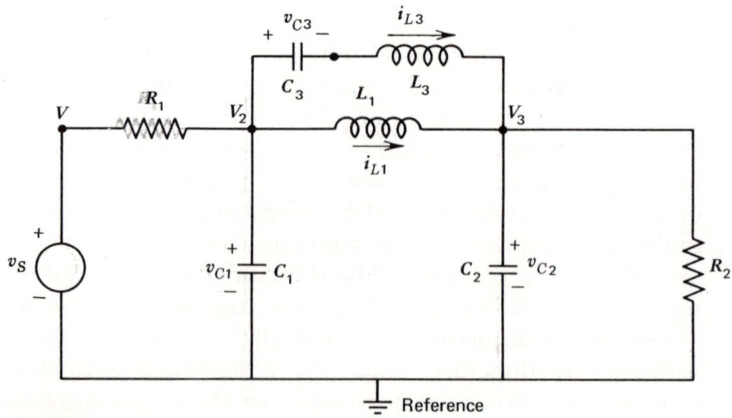

Fig. 3.11

Then the state equations are:

$$C_1 \frac{dv_{C1}}{dt} = (V - v_{C1})G_1 - i_{L1} - i_{L3}$$

$$L_1 \frac{di_{L1}}{dt} = v_{C1} - v_{C2}$$

$$C_2 \frac{dv_{C2}}{dt} = i_{L1} - v_{C2}G_2 + i_{L3} \qquad (3.99)$$

$$C_3 \frac{dv_{C3}}{dt} = i_{L3}$$

$$L_3 \frac{di_{L3}}{dt} = v_{C1} - v_{C2} - v_{C3}$$

The state equations are written without any difficulty whatever. The dummy inductor is now allowed to go to zero and the variables i_{L3} and v_{C3} eliminated as follows. The last two equations in Eq. 3.99, with $L_3 = 0$, give

$$v_{C3} = v_{C1} - v_{C2}$$

and

$$i_{L3} = C_3(v'_{C1} - v'_{C2}) \qquad (3.100)$$

where $v'_C = dv_C/dt$. Substituting Eq. 3.100 into the first three equations of Eq. 3.99, we have

$$\begin{aligned} C_1 v'_{C1} &= -G_1 v_{C1} - i_{L1} - C_3(v'_{C1} - v'_{C2}) + G_1 V \\ C_2 v'_{C2} &= -G_2 v_{C2} + i_{L1} + C_3(v'_{C1} - v'_{C2}) \\ L_1 i'_{L1} &= v_{C1} - v_{C2} \end{aligned} \qquad (3.101)$$

Collecting terms and solving for v'_{C1} and v'_{C2}, the state equations are now obtained:

$$\begin{aligned} v'_{C1} &= -\frac{G_1(C_2 + C_3)}{|C|} v_{C1} - \frac{G_1 C_3}{|C|} v_{C2} + \frac{G_1(C_2 + C_3)}{|C|} V \\ v'_{C2} &= -\frac{G_1 C_3}{|C|} v_{C1} - \frac{G_1(C_1 + C_3)}{|C|} v_{C2} + \frac{C_1}{|C|} i_{L1} + \frac{G_1 C_3}{|C|} V \\ i'_{L1} &= \frac{1}{L_1} v_{C1} - \frac{1}{L_1} v_{C2} \end{aligned} \qquad (3.102)$$

where $|C| = C_1C_2 + C_2C_3 + C_3C_1$.

The state equations in Eq. 3.102 are now in terms of the independent state variables with the excess element variables removed. The first two equations of Eq. 3.101 had to be solved simultaneously for v'_{C1} and v'_{C2}. Other situations also occur when some of the equations must be processed (or solved) to obtain the desired form of state equations. Such situations occur when several resistor currents or voltages must be expressed as linear combinations of the state variables. In any case, the procedure described always results in the state-variable equations of the form of Eq. 3.102.

The general form of the state-differential equations may be expressed as:

$$
\begin{aligned}
x'_1 &= a_{11}x_1 + a_{12}x_2 + \cdots + a_{1n}x_n + F_1 \\
x'_2 &= a_{21}x_1 + a_{22}x_2 + \cdots + a_{2n}x_n + F_2 \\
&\cdot \\
&\cdot \\
&\cdot \\
x'_n &= a_{n1}x_1 + a_{n2}x_2 + \cdots + a_{nn}x_n + F_n
\end{aligned}
\tag{3.103}
$$

where the state-variables $x_1, x_2, \cdots, x_n$ are the independent capacitor voltages and inductor currents. The $F_1, F_2, \cdots, F_n$ represent the driving functions and the coefficients a_{ij} are the constants determined by the circuit and the element values. The initial state specifies the initial values of the state variables and represents the initial conditions for the differential equations. If $\mathbf{x}$ is the column vector consisting of the state variables $x_1, x_2, \cdots, x_n$, and $\mathbf{F}$ is the column vector consisting of $F_1, F_2, \cdots, F_n$, then Eq. 3.103 may be expressed in matrix form

$$\mathbf{x}' = A\mathbf{x} + \mathbf{F} \tag{3.104}$$

where x' is the vector of derivatives $x'_1, x'_2, \cdots, x'_n$, and A is the matrix of coefficients a_{ij}:

$$
A = \begin{pmatrix}
a_{11} & a_{12} & \cdots\cdots & a_{1n} \\
a_{21} & a_{22} & \cdots\cdots & a_{2n} \\
\cdot & \cdot & & \cdot \\
\cdot & \cdot & & \cdot \\
\cdot & \cdot & & \cdot \\
a_{n1} & a_{n2} & & a_{nn}
\end{pmatrix}
\tag{3.105}
$$

The initial state is then given by $x(0)$.

A great deal of mathematical theory exists for equations of the form of Eq. 3.104 and, consequently, a good deal of the modern circuit and system analysis is based on the state-variable formalism. The form of Eq. 3.104 is also convenient for numerical integration as we shall see in the next section. For these reasons, the scalar differential equations of Eq. 3.80 are frequently expressed as vector or state differential equations of the form in Eq. 3.104. We show the relationships between the two by considering the scalar differential equation

$$(a_nD^n + a_{n-1}D^{n-1} + \cdots + a_1D + a_0)x = y(t) \tag{3.106}$$

Let the state variables be defined by

$$\begin{aligned} x_1 &= x \\ x_2 &= x_1' \\ x_3 &= x_2' \\ &\vdots \\ x_n &= x_{n-1}' \end{aligned} \tag{3.107}$$

the state variables are

$$\begin{aligned} x_1' &= x_2 \\ x_2' &= \quad x_3 \\ x_3' &= \qquad x_4 \\ &\vdots \\ x_n' &= -\frac{a_0}{a_n}x_1 - \frac{a_1}{a_n}x_2 - \cdots - \frac{a_{n-1}}{a_n}x_n + y \end{aligned} \tag{3.108}$$

or,

$$\mathbf{x}' = A\mathbf{x} + \mathbf{F} \tag{3.109}$$

where

$$A = \begin{pmatrix} 0 & 1 & 0 & \cdots & 0 \\ 0 & 0 & 1\ 0 & \cdots & 0 \\ \vdots & & & & \\ -\frac{a_0}{a_n} & -\frac{a_1}{a_n} & \cdots & \cdots & -\frac{a_{n-1}}{a_n} \end{pmatrix} \tag{3.110}$$

and

$$F = \begin{pmatrix} 0 \\ \vdots \\ 0 \\ y \end{pmatrix} \tag{3.111}$$

The initial state is given in terms of $x_k(0) = x^{k-1}(0)$, where x^{k-1} is the $(k-1)$th derivative of x.

We have thus seen that the state equation formalism is compact, consistent, and universal. It also provides a convenient numerical integration format. The numerical method discussed in the next section is, therefore, developed entirely for this format.

3.6 NUMERICAL SOLUTION OF HIGHER ORDER DIFFERENTIAL EQUATIONS

In Chapter 2, we have discussed three methods for solving first order initial value problems, namely, Euler's method, and the second and fourth order Runge-Kutta methods. These methods require information about the solution at a *single point* $t = t_m$, from which the methods proceed to obtain the solution $r(t)$ at the next point $t = t_{m+1}$. A close examination of the fourth order Runge-Kutta method shows that essentially it applies Euler's method four times to obtain four different estimates for the derivatives. These estimates are then averaged for proceeding over the whole interval h. The local truncation errors are directly proportional to h^2, h^3, and h^5 for the three methods, respectively, where h is the step size.

Euler's method requires one derivative evaluation for each step. However, the second and fourth order Runge-Kutta methods require two and four derivative evaluations, respectively, to achieve higher accuracy. The price for the favorable local truncation errors may be considerable computer time for those problems in which the function $f(t, r)$ is very complicated. Therefore, it is natural to ask if there is any other method that requires fewer derivative evaluations and at the same time achieves an accuracy comparable to, say, the fourth order Runge-Kutta method. The answer is yes. However, such methods require information about the solution at *several points*; hence, they are called *multistep methods*.

Multistep Methods

Multistep methods are highly accurate and easily adaptable to the numerical integration of higher order differential equations. We first discuss the multistep methods in the context of a first order differential equation. The approach is then extended to the second and higher order equations by expressing the differential equations as a set of first order differential equations. Such a set of first order differential equations may be specified as in the state-variable description or derived from the scalar differential equation description as discussed in Sec. 3.5.

Multistep Methods: First Order Equations

Let us consider the first order initial value problem as in Eq. 2.101:

$$r' = f(t, r), \qquad r(t_0) = r_0 \tag{3.112}$$

where r is the response of the circuit, r' is its time derivative, and $f(t,r)$ a function of both excitation and response is expressed as a function of time and the response. Let us assume that the actual $r'(t)$ or $f(t,r)$ is given as a continuous function of time t. This is shown in Fig. 3.12. Then

$$\int_{t_m}^{t_{m+1}} r'(t)\,dt = r(t_{m+1}) - r(t_m) = \int_{t_m}^{t_{m+1}} f(t,r)\,dt \tag{3.113}$$

where

$$\int_{t_m}^{t_{m+1}} f(t,r)\,dt$$

is the area under the curve $f(t,r)$ from t_m to t_{m+1}. Hence, if the true curve $f(t,r)$ vs. t is given and $r(t_m)$ is known, then there is no problem in finding an exact $r(t_{m+1})$ from Eq. 3.113. This is of course impossible if we have not solved for $r(t)$ explicitly as a function of t. Euler's method

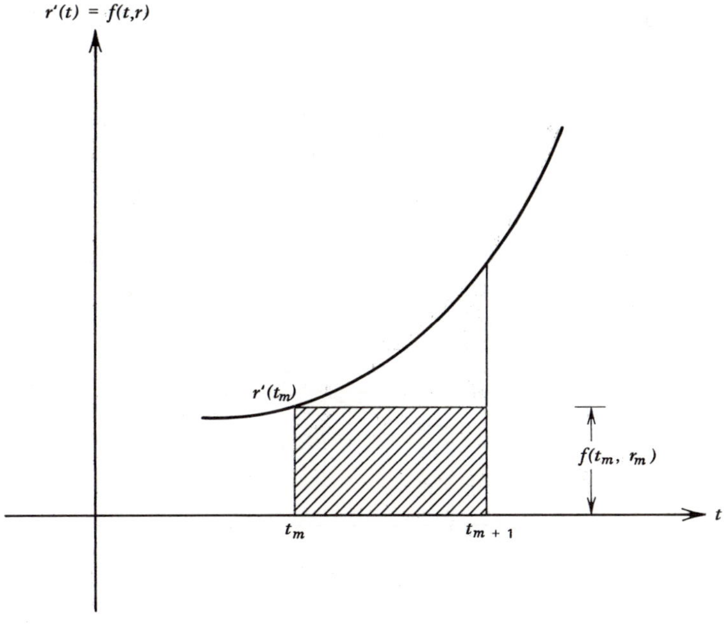

Fig. 3.12 Euler's method.

attempts to approximate $f(t,r)$ between t_m and t_{m+1} by a constant $f(t_m, r_m)$ in which case we have from Eq. 3.113,

$$r_{m+1} = r_m + hf(t_m, r_m) \tag{3.114}$$

Now suppose that the values of $r(t)$ and $f(t,r)$ are known at the points $t = t_{m-1}$ and $t = t_m$. Then we can draw a straight line through $(t_{m-1}, f(t_{m-1}, r_{m-1}))$ and $(t_m, f(t_m, r_m))$ and extend it over the interval from t_m to t_{m+1}. The situation is shown in Fig. 3.13. Let us denote $f(t_m, r_m)$ by $\bar{f}_m$, then the straight line through $(t_{m-1}, \bar{f}_{m-1})$ and $(t_m, \bar{f}_m)$ is a better approximation to the actual curve $f(t,r)$ than the constant used in Euler's method. In fact, from Fig. 3.13, we have the straight-line approximation to $f(t,r)$ given by

$$\phi(t) = \frac{(\bar{f}_m - \bar{f}_{m-1})}{(t_m - t_{m-1})}(t - t_m) + \bar{f}_m$$

or,

$$\phi(t) = \frac{(\bar{f}_m - \bar{f}_{m-1})}{h}(t - t_m) + \bar{f}_m \tag{3.115}$$

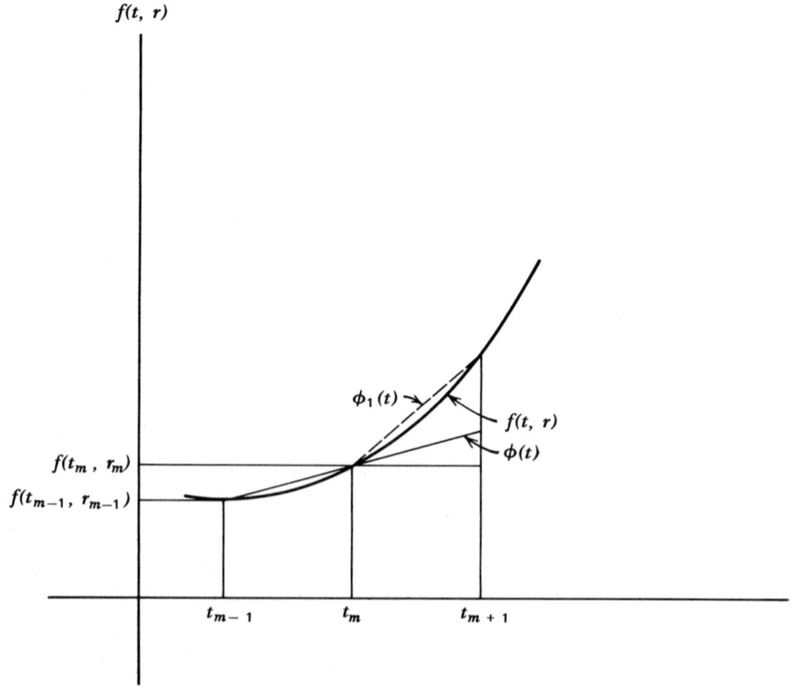

Fig. 3.13 Two different approximations for $f(t,r)$.

Note that $\phi(t_m) = \bar{f}_m$ and $\phi(t_{m-1}) = \bar{f}_{m-1}$ in Eq. 3.115. The straight-line approximation of Eq. 3.115 is better than the constant approximation used in Euler's method. Therefore, the use of Eq. 3.115 in Eq. 3.113 should give us a better increment to be added to r_m to form r_{m+1}. The algorithm for r_{m+1} given by Eqs. 3.113 and 3.115 is:

$$
\begin{aligned}
r_{m+1} &= r_m + \int_{t_m}^{t_{m+1}} \phi(t)\,dt \\
&= r_m + \frac{(\bar{f}_m - \bar{f}_{m-1})}{h} \cdot \frac{1}{2}(t - t_m)^2 \Big|_{t_m}^{t_{m+1}} + h\bar{f}_m \\
&= r_m + \frac{h}{2}(3\bar{f}_m - \bar{f}_{m-1})
\end{aligned} \tag{3.116}
$$

This algorithm of Eq. 3.116 is called the *predictor* algorithm since it is an attempt to predict the value of r_{m+1} by a straight-line approximation to the curve $f(t, r)$. We illustrate its use through an example.

Example 3.6.1 Solve the circuit in Example 2.6.1 of Chapter 2 using the algorithm of Eq. 3.116.

Solution. The equation is

$$\frac{di(t)}{dt} = -i(t) \qquad i_0 = i(0) = 1$$

With a step size $h = 0.01$, we first find $i(0.01) \approx i_1$ by using the second order RK method. Then with the given i_0 and $i_1 = 0.99005$ we apply Eq. 3.116 for $m = 1, 2, 3$. Retaining five decimal places, we have the following results:

$$
\begin{aligned}
i(0.00) &\approx i_0 = 1.00000 \\
i(0.01) &\approx i_1 = 0.99005 \\
i(0.02) &\approx i_2 = 0.99005 + \frac{0.01}{2}(-3(0.99005) + 1) = 0.98020 \\
i(0.03) &\approx i_3 = 0.98020 + \frac{0.01}{2}(-3(0.98020) + 0{\cdot}99005) = 0.97045 \\
i(0.04) &\approx i_4 = 0.97045 + \frac{0.01}{2}(-3(0.97045) + 0.98020) = 0.96079
\end{aligned}
$$

The result for $i(0.04)$ is correct to five decimal places. The local truncation error for Eq. 3.116 can be shown to be

$$e_{m+1} \triangleq r(t_{m+1}) - r_{m+1} = \frac{5}{12}h^3 r'''(\xi) \qquad t_{m-1} \leq \xi \leq t_m \tag{3.117}$$

which is comparable to the second order R-K method.

A glance at Fig. 3.13 shows that if we knew $f(t_{m+1}, r_{m+1})$, then the integration of a straight line passing through $\bar{f}_m$ and $\bar{f}_{m+1}$ from t_m to t_{m+1} might give us an even better approximation for the increment. In this case we have,

$$\phi_1(t) = \frac{(\bar{f}_{m+1} - \bar{f}_m)}{h}(t - t_{m+1}) + \bar{f}_{m+1} \tag{3.118}$$

With Eq. 3.118 as an approximation for $f(t, r)$, we have from Eq. 3.113:

$$\begin{aligned} r_{m+1} &= r_m + \int_{t_m}^{t_{m+1}} \phi_1(t)\,dt \\ &= r_m + \frac{h}{2}(\bar{f}_{m+1} + \bar{f}_m) \end{aligned} \tag{3.119}$$

The local truncation error for the algorithm of Eq. 3.119 can be found to be:

$$e_{m+1} \triangleq r(t_{m+1}) - r_{m+1} = -\frac{1}{12}h^3 r'''(\eta) \qquad t_m \leqslant \eta \leqslant t_{m+1} \tag{3.120}$$

Observe that the algorithm of Eq. 3.119 yields a smaller magnitude of the local truncation error given by Eq. 3.120 than that given by Eq. 3.117 for the algorithm of Eq. 3.116. Although the algorithm of Eq. 3.119 leads to a smaller truncation error, it cannot be used directly to solve for r_{m+1} since $\bar{f}_{m+1} = f(t_{m+1}, r_{m+1})$ is itself a function of r_{m+1} which is unknown. If a very good approximation to r_{m+1} or $\bar{f}_{m+1}$ were known, however, the algorithm of Eq. 3.119 can be used to obtain a better or corrected value of r_{m+1}. This is the manner in which it is commonly used and therefore Eq. 3.119 is called the *corrector* algorithm. The predictor algorithm of Eq. 3.116 is used to predict the values of r_{m+1} and $\bar{f}_{m+1}$; then the corrector algorithm of Eq. 3.119 is used to obtain a better value of r_{m+1}. The steps involved are:

(i) Use Eq. 3.116 to predict an approximation for r_{m+1}

$$r_{m+1,p} = r_m + \frac{h}{2}(3\bar{f}_m - \bar{f}_{m-1}) \tag{3.121}$$

(ii) Use this value of $r_{m+1,p}$ to find

$$\bar{f}_{m+1} \approx \bar{f}_{m+1,p} = f(t_{m+1}, r_{m+1,p})$$

(iii) Finally use Eq. 3.119 to correct the predicted value $r_{m+1,p}$

$$r_{m+1,c} = r_m + \frac{h}{2}(\bar{f}_{m+1,p} + \bar{f}_m) \tag{3.122}$$

Let us see what happens in actual computation. Again we use the problem in Example 2.6.1 of Chapter 2. The step size is chosen as $h = 0.01$, the starting value is calculated by using the second order R-K method with $h = 0.01$. Retaining five decimal places and applying Eqs. 3.121–3.122, we have

$$i(0.01) \approx i_1 = 0.99005 \qquad \text{(By second order R-K method)}$$

$$i_{2,p} = 0.99005 + \frac{0.01}{2}(-3(0.99005)+1) = 0.98020$$

$$i(0.02) \approx i_{2,c} = 0.99005 + \frac{h}{2}(-0.98020 - 0.99005) = 0.98020$$

$$i_{3,p} = 0.98020 + \frac{0.01}{2}(-3(0.98020)+0.99005) = 0.97045$$

$$i(0.03) \approx i_{3,c} = 0.98020 + \frac{h}{2}(-0.97045 - 0.98020) = 0.97045$$

$$i_{4,p} = 0.97045 + \frac{0.01}{2}(-3(0.97045)+0.98020) = 0.96079$$

$$i(0.04) \approx i_{4,c} = 0.97045 + \frac{0.01}{2}(-0.96079 - 0.97045) = 0.96079$$

The predicted value is already a very good approximation for r_{m+1} and therefore the improvement in accuracy by using Eq. 3.122 is negligible in our simple problem. As a rule, for nontrivial cases, the corrector algorithm is an important adjunct to the predictor to ensure a high degree of accuracy. In practice, the two algorithms are used together and called the *predictor-corrector* method.

We have illustrated above a two-step predictor-corrector method. The method involves a straight-line approximation to the curve $f(t,r)$ and is the simplest multistep method. The constant or zeroth degree polynomial approximation equals in value (i.e., matches) $f(t,r)$ at only one point and gives only fair results for the value of r_{m+1}. The straight-line or first degree polynomial approximation matches $f(t,r)$ at two points, allows predictor-corrector algorithms, and gives good results for r_{m+1}. Much better (i.e., highly accurate) results are obtained by using higher degree polynomials to approximate $f(t,r)$. A kth degree polynomial approximation matches $f(t,r)$ at $(k+1)$ points and, of course, allows the predictor-corrector approach.

A high degree of accuracy is obtained in most cases by the use of a third degree polynomial to approximate $f(t,r)$ in the interval from $t = t_m$ to $t = t_{m+1}$. The resulting predictor-corrector method is the most widely used and is called the Adams-Moulton (A-M) method. We give below the A-M algorithms and their local truncation errors:

A-M Predictor

$$\begin{aligned} r_{m+1} &= r_m + \frac{h}{24}\left(55\bar{f}_m - 59\bar{f}_{m-1} + 37\bar{f}_{m-2} - 9\bar{f}_{m-3}\right) \\ e_{m+1} &= \frac{251}{720} h^5 r^{(5)}(\xi) \qquad t_{m-3} \leq \xi \leq t_m \end{aligned} \tag{3.123}$$

A-M Corrector

$$\begin{aligned} r_{m+1} &= r_m + \frac{h}{24}\left(9\bar{f}_{m+1} + 19\bar{f}_m - 5\bar{f}_{m-1} + \bar{f}_{m-2}\right) \\ e_{m+1} &= -\frac{19}{720} h^5 r^{(5)}(\xi) \qquad t_{m-2} \leq \xi \leq t_{m+1} \end{aligned} \tag{3.124}$$

where $r^{(5)}$ is the fifth derivative of r. The local truncation error in Eqs. 3.123 or 3.124 is proportional to h^5, which is comparable to that of the fourth order Runge-Kutta method. Furthermore, using Eqs. 3.123 and 3.124 together we need two derivative evaluations for each step. This is a considerable advantage of this predictor-corrector method over the fourth order Runge-Kutta method that requires four derivative evaluations per step. Of course, we need at least the response at the first four points in time. The fourth order R-K method is commonly used for this purpose.

Second and Higher Order Systems

The fourth order R-K method and the A-M multistep method discussed for the first order systems are easily extended to higher order systems. The A-M method is the primary method used to obtain the response, but the R-K method is needed to obtain the response values r_1, r_2, and r_3 at $(t-t_0) = h$, $2h$, and $3h$, respectively, from the initial value $r_0 = r(t_0)$. Once the values of r are obtained at the first four points in time, the A-M method takes over the computation.

The extension of the procedure to the second and higher order systems is performed by expressing them into a set of first order systems. The extension for the nth order system is identical to that for the second order system. Therefore, we first examine the second order system and then state the algorithms for the nth order system.

Let the second order circuit equations be expressed in the state-variable format:

$$\begin{aligned} x_1' &= a_{11}x_1 + a_{12}x_2 + F_1 \qquad x_1(t_0) = x_{10} \\ x_2' &= a_{21}x_1 + a_{22}x_2 + F_2 \qquad x_2(t_0) = x_{20} \end{aligned} \tag{3.125}$$

These equations are either obtained for the state variables of the circuit or else the second order circuit differential equation is expressed in the above form as discussed in the last section. The response is determined by the variables $r_1 = x_1$, $r_2 = x_2$. We can express Eq. 3.125 in the computational format of the first order system:

$$\begin{aligned} r_1' &= f_1(t, r_1, r_2) \qquad r_1(t_0) = r_{10} = x_{10} \\ r_2' &= f_2(t, r_1, r_2) \qquad r_2(t_0) = r_{20} = x_{20} \end{aligned} \tag{3.126}$$

where f_1 and f_2 are the right-hand sides of Eq. 3.125.

R-K Method

Each of the equations in Eq. 3.126 is exactly in the same form as Eq. 3.112 or 2.116. The algorithm in Eq. 2.117 for the fourth order R-K method is, therefore, applied to each of the two equations in Eq. 3.126, that is,

$$\begin{aligned} r_{1,m+1} &= r_{1,m} + \tfrac{1}{6}(k_{11} + 2k_{12} + 2k_{13} + k_{14}) \\ r_{2,m+1} &= r_{2,m} + \tfrac{1}{6}(k_{21} + 2k_{22} + 2k_{23} + k_{24}) \end{aligned} \tag{3.127}$$

where $r_{1,m}$ and $r_{2,m}$ are the approximations to $r_1(t_m)$ and $r_2(t_m)$, respectively. The parameters $k_{\nu 1}$, $k_{\nu 2}$, $k_{\nu 3}$, and $k_{\nu 4}$ for $\nu = 1, 2$ are given by expressions analogous to Eq. 2.118:

$$\begin{aligned} k_{\nu 1} &= hf_\nu(t_m, r_{1,m}, r_{2,m}) \\ k_{\nu 2} &= hf_\nu\left(t_m + \frac{h}{2}, r_{1,m} + \frac{k_{11}}{2}, r_{2,m} + \frac{k_{21}}{2}\right) \\ k_{\nu 3} &= hf_\nu\left(t_m + \frac{h}{2}, r_{1,m} + \frac{k_{12}}{2}, r_{2,m} + \frac{k_{22}}{2}\right) \\ k_{\nu 4} &= hf_\nu(t_m + h, r_{1,m} + k_{13}, r_{2,m} + k_{23}) \end{aligned} \tag{3.128}$$

A-M Method

The R-K method is used to obtain the values of r_1 and r_2 at the first four points in time. With the initial values given at t_0, only three steps need be computed using the R-K method. After $(t - t_0) = 3h$, the computation is performed using the more advantageous A-M predictor-corrector method. The algorithms are written as in Eqs. 3.123 and 3.124 for each of the equations in Eq. 3.126:

Predictor

$$r_{1,m+1} = r_{1,m} + \frac{h}{24}\,[55\overline{f}_{1,m} - 59\overline{f}_{1,m-1} + 37\overline{f}_{1,m-2} - 9\overline{f}_{1,m-3}]$$
$$r_{2,m+1} = r_{2,m} + \frac{h}{24}\,[55\overline{f}_{2,m} - 59\overline{f}_{2,m-1} + 37\overline{f}_{2,m-2} - 9\overline{f}_{2,m-3}] \qquad (3.129)$$

Corrector

$$r_{1,m+1} = r_{1,m} + \frac{h}{24}\,[9\overline{f}_{1,m+1} + 19\overline{f}_{1,m} - 5\overline{f}_{1,m-1} + \overline{f}_{1,m-2}]$$
$$r_{2,m+1} = r_{2,m} + \frac{h}{24}\,[9\overline{f}_{2,m+1} + 19\overline{f}_{2,m} - 5\overline{f}_{2,m-1} + \overline{f}_{2,m-2}] \qquad (3.130)$$

where $\overline{f}_{\nu,m}$ denotes $f_\nu(t_m, r_{1,m}, r_{2,m})$ for $\nu = 1, 2$. Note that the predicted values of $r_{\nu,m+1}$ from Eq. 3.129 are used to evaluate $\overline{f}_{\nu,m+1}$ for use in Eq. 3.130.

Example 3.6.2 Consider the second order circuit of Fig. 3.14 whose explicit solution under simple excitation has been discussed in Sec. 3.1.

The circuit differential equation using KVL is

$$\frac{d^2 i}{dt^2} + 3\frac{di}{dt} + 2i = 0 \qquad (3.131)$$

and the initial conditions are

$$i(0) = 1$$
$$i'(0) = \frac{v(t) - Ri(0) - v_{C0}}{L} = 0 \qquad (3.132)$$

The explicit solution of Eq. 3.131 has been discussed in Sec. 3.1 and is

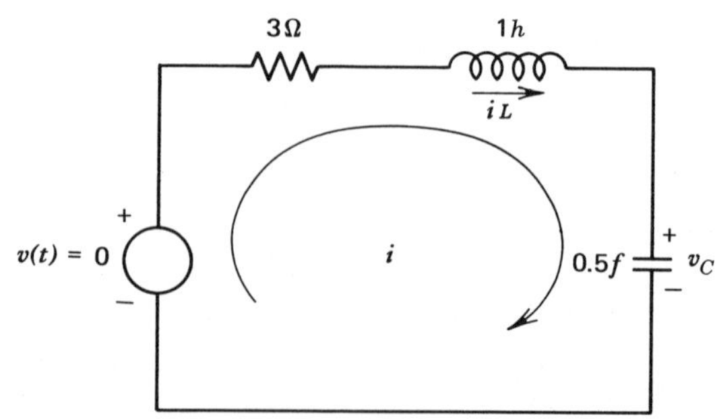

Fig. 3.14 IS: $i_1(0) = +1$ amp, $v_C(0) = -3$ volts.

given by

$$i(t) = -e^{-2t} + 2e^{-t} \qquad t \geqslant 0 \tag{3.133}$$

Now, let us solve this simple problem by the computational methods discussed above. To convert the differential equation to the state-variable format, let

$$r_1(t) = i(t)$$

$$r_2(t) = \frac{di}{dt} = r_1'$$

Then

$$\begin{aligned} r_1' &= r_2 \qquad r_1(0) = 1 \\ r_2' &= -2r_1 - 3r_2 \qquad r_2(0) = 0 \end{aligned} \tag{3.134}$$

Let us solve Eq. 3.134 by the R-K method by assuming a step size of $h = 0.04$ and retaining five decimal places. The parameters of Eq. 3.128 are given by

$$\begin{aligned}
k_{11} &= 0.04(0) = 0 \\
k_{21} &= 0.04[-2(1) - 3(0)] = -0.08000 \\
k_{12} &= 0.04[-\tfrac{1}{2}(0.08000)] = -0.00160 \\
k_{22} &= 0.04[-2(1) - 3(-\tfrac{1}{2})(0.08000)] = -0.07520 \\
k_{13} &= 0.04[0 - \tfrac{1}{2}(0.07520)] = -0.00150 \\
k_{23} &= 0.04[-2(1 - \tfrac{1}{2}(0.00160)) - 3(-\tfrac{1}{2}(0.07520))] = -0.07542 \\
k_{14} &= 0.04[-0.07542] = 0.00302 \\
k_{24} &= 0.04[-2(1 - 0.00150) - 3(-0.07542)] = -0.07083
\end{aligned}$$

Hence, we obtain

$$r_1(0.04) \approx r_{1,1} = 1 + \tfrac{1}{6}(k_{11} + 2k_{12} + 2k_{13} + k_{14}) = 0.99846$$

$$r_2(0.04) \approx r_{2,1} = 0 + \tfrac{1}{6}(k_{21} + 2k_{22} + 2k_{23} + k_{24}) = 0.07535$$

The parameters are then determined for the second step and the next-step solution is obtained, and so on.

Let us compare the solution at $t = 0.04$ obtained above with the exact solution given by Eq. 3.133:

$$r_1(t) = i(t) = -e^{-2t} + 2e^{-t}$$

$$r_2(t) = i'(t) = 2e^{-2t} - 2e^{-t}$$

At $t = 0.04$, the exact values given by the above expressions are $r_1(0.04) = 0.99846$ and $r_2(0.04) = 0.07534$ correct to five decimal places. Thus, the R-K method gave us $r_{1,1}$ correct to five decimal places and $r_{2,1}$ correct to four decimal places compared to the exact solutions given.

The multistep A-M method is commonly used to develop the whole solution after the first four points are determined by the R-K method. We develop the solution for the above example by the R-K/A-M method using the computer program discussed at the conclusion of this section. The solution is given in Fig. 3.21.

*N*th Order System

The extensions of the R-K and the A-M methods to nth order systems are obvious and straightforward. The circuit equations are assumed either available or put into the state-variable format. They may be expressed as

$$r_\nu' = f_\nu(t, r_1, r_2, \cdots, r_n), \qquad r_\nu(0) = r_{\nu 0} \tag{3.135}$$

for $\nu = 1, 2, \cdots, n$.

From Eqs. 3.127 and 3.128, the fourth order R-K method becomes for $\nu = 1, 2, \cdots, n$:

$$r_{\nu,m+1} = r_{\nu,m} + \tfrac{1}{6}(k_{\nu 1} + 2k_{\nu 2} + 2k_{\nu 3} + k_{\nu 4}) \tag{3.136}$$

where

$$\begin{aligned} k_{\nu 1} &= hf_\nu(t_m, r_{1,m}, r_{2,m}, \cdots, r_{n,m}) \\ k_{\nu 2} &= hf_\nu\left(t_m + \frac{h}{2}, r_{1,m} + \frac{k_{11}}{2}, \cdots, r_{n,m} + \frac{k_{n1}}{2}\right) \\ k_{\nu 3} &= hf_\nu\left(t_m + \frac{h}{2}, r_{1,m} + \frac{k_{12}}{2}, \cdots, r_{n,m} + \frac{k_{n2}}{2}\right) \\ k_{\nu 4} &= hf_\nu(t_m + h, r_{1,m} + k_{13}, \cdots, r_{n,m} + k_{n3}) \end{aligned} \tag{3.137}$$

From Eqs. 3.129 and 3.130, the A-M method for an nth order system can be expressed as follows for $\nu = 1, 2, \cdots, n$:

Predictor

$$r_{\nu,m+1} = r_{\nu,m} + \frac{h}{24}\left[55f_{\nu,m} - 59f_{\nu,m-1} + 37f_{\nu,m-2} - 9f_{\nu,m-3}\right] \tag{3.138a}$$

Corrector

$$r_{\nu,m+1} = r_{\nu,m} + \frac{h}{24}\left[9f_{\nu,m+1} + 19f_{\nu,m} - 5f_{\nu,m-1} + f_{\nu,m-2}\right] \tag{3.138b}$$

The computer implementation of the A-M method with the R-K method used for starting the problem is discussed next for nth order systems.

Computer Program

Let us now consider the implementation of the Adams-Moulton method for nth-order systems using the fourth order Runge-Kutta method for starting the computation. The given initial-value problem is of the form shown in Eq. 3.135. If we denote

$$\mathbf{r}(t) = \begin{bmatrix} r_1(t) \\ r_2(t) \\ \cdot \\ \cdot \\ \cdot \\ r_n(t) \end{bmatrix} \qquad \mathbf{f}(t, \mathbf{r}) = \begin{bmatrix} f_1(t, r_1, r_2, \cdots, r_n) \\ f_2(t, r_1, r_2, \cdots, r_n) \\ \cdot \\ \cdot \\ \cdot \\ f_n(t, r_1, r_2, \cdots, r_n) \end{bmatrix} \tag{3.139}$$

that is, $\mathbf{r}(\mathbf{t})$ and $\mathbf{f}(t, \mathbf{r})$ are vectors with n components. Then we can rewrite Eq. 3.135 as

$$\mathbf{r}'(t) = \mathbf{f}(t, r) \qquad \mathbf{r}(t_0) = \mathbf{r}_0 \tag{3.140}$$

where,

$$\mathbf{r}'(t) = \begin{bmatrix} r_1'(t) \\ r_2'(t) \\ \cdot \\ \cdot \\ \cdot \\ r_n'(t) \end{bmatrix} \qquad \mathbf{r}(t_0) = \mathbf{r}_0 = \begin{bmatrix} r_{10} \\ r_{20} \\ \cdot \\ \cdot \\ \cdot \\ r_{n0} \end{bmatrix}$$

We represent the initial value $\mathbf{r}_0$, the response $\mathbf{r}$ and the derivative function $\mathbf{f}$ as three vectors stored in three blocks of n storage registers in the computer named as *RO*, *R*, and *F* respectively. The Runge-Kutta and Adams-Moulton method could then be used to find new t, $\mathbf{r}$, and $\mathbf{f}$ once the initial time t_0, terminating time t_f, step size h and system order n are given.

A flowchart of a general-purpose routine for solving n first-order initial value problems is given in Fig. 3.15 with the following comments.

1. To compute new $\mathbf{r}$, we need to evaluate the values of $\mathbf{f}(t, \mathbf{r})$ at several points. The function $\mathbf{f}(t, \mathbf{r})$ can best be simulated in a FORTRAN SUBROUTINE. For example, if we have a system of two first order differential equations,

$$r_1'(t) = f_1(t, r_1, r_2) = r_1 + r_2$$

$$r_2'(t) = f_2(t, r_1, r_2) = 3r_1 - r_2$$

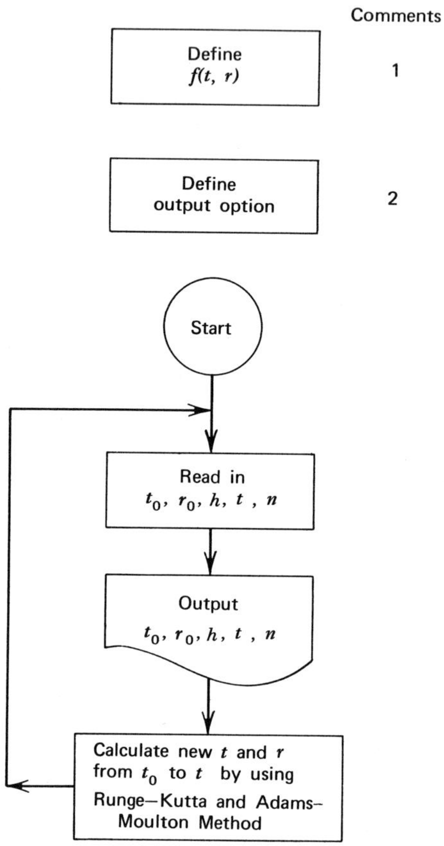

Fig. 3.15 General-purpose differential equation solver.

then we can write the following FORTRAN statement

```
SUBROUTINE FUN (T, R, F)
DIMENSION R(2), F(2)
F(1) = R(1) + R(2)
F(2) = 3.*R(1) - R(2)
RETURN
END
```

A call to subroutine FUN (T, R, F) with given T, $R(1)$ and $R(2)$ will return with the new function values $F(1)$ and $F(2)$.

2. After each step, the results of t_{m+1}, $\mathbf{r}_{m+1}$, and $\mathbf{f}_{m+1}$ are ready for output. We could either ask for printed output or have cards punched. If we

ask for a tabulated output after each step, we could then write a FORTRAN subroutine as follows,

```
SUBROUTINE OUT(T, R, F)
DIMENSION R(2), F(2)
FORMAT (E14.6)
PRINT 1, T, R(1), R(2)
RETURN
END
```

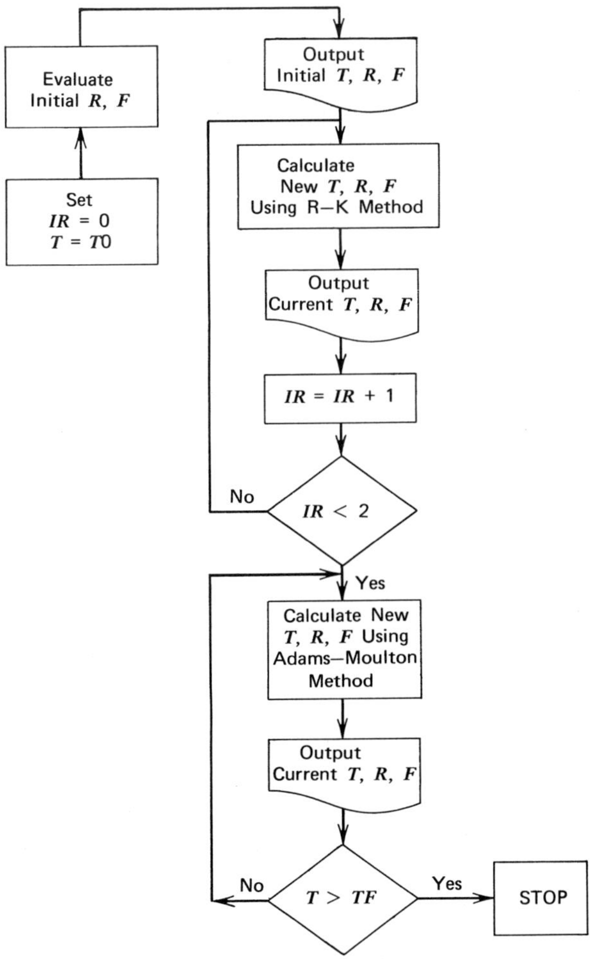

Fig. 3.16 THE RKAM subroutine.

A call to this OUT(T, R, F) with T, R, F given will print the values of T, $R(1)$, $R(2)$ in the FORMAT specified.

A FORTRAN subroutine RKAM is written to solve up to a system of 20 first order differential equations. The flowchart is shown in Fig. 3.16. In order to use RKAM, one must specify the following:

a. The parameters TO, TF, N, H, and RO (defined in Ex. 3.6.3).

b. The derivative subroutine FUN(T, R, F) which is the specific system of first-order differential equations.

c. The output subroutine OUT(T, R, F) for transferring the computed results.

In case we have a system with more than 20 first order differential equations, then the dimension of each vector has to be changed to the specific n.

Example 3.6.3. Solve the circuit in Fig. 3.14 by using RKAM. The given differential equations are

$$r' = r_2 \qquad r_1(0) = 1$$
$$r_2' = -2r_1 - 3r_2 \qquad r_2(0) = 0$$

Solution. The FORTRAN program, input data listing, and a sample computer output are illustrated in Figs. 3.17 to 3.22.

```
  DIMENSION RO(20)
1 READ 2,H,TO,TF,N
2 FORMAT(3F10.2,I2)
  PRINT 2,H,TO,TF,N
  READ 3,(RO(I), I=1,N)
3 FORMAT(2F10.2)
  PRINT 3, (RO(I), I=1,N)
  CALL RKAM(TO,RO,TF,H,N)
  GO TO 1
  END
```

Fig. 3.17 Main routine for the differential equation solver.

```
  SUBROUTINE FUN(T,R,F)
  DIMENSION R(20),F(20)
  F(1)=R(2)
  F(2)=-2.*R(1)-3.*R(2)
  RETURN
  END
```

Fig. 3.18 Derivative function subroutine.

```
  SUBROUTINE OUT(T,R,F)
  DIMENSION R(20),F(20)
1 FORMAT(3E16.5)
  PRINT 1,T,R(1),R(2)
  RETURN
  END
```

Fig. 3.19 Output subroutine.

The parameters are the following:

H Step size
TO Initial time

TF	Terminating time
N	Number of differential equations
RO	Initial values of the response (dimension N).
$RKAM$	Runge-Kutta & Adams-Moulton subroutine
FUN	Derivative function subroutine
OUT	Output subroutine
T	Time variable
R	System response (dimension N)
F	Derivative value (dimension N)
$XK1$, $XK2$, $XK3$, $XK4$	R-K equations (dimension N)
$F0$, $F1$, $F2$, $F3$	A-M parameters (dimension N)
RK	R-K formula
AM	A-M formula
IR	R-K counter (Number of starting points)

PROBLEMS

3-0a. Show that

$$x_1(t) = M_1 e^{-t} \sin 2t$$

and

$$x_2(t) = M_2 e^{-t} \cos 2t$$

are solutions for the equation

$$x''(t) + 2x'(t) + 5x(t) = 0$$

Show that $x_1(t) + x_2(t)$ is also a solution.

3-0b. Determine the roots of the characteristic equation of the following equations:

(i) $x''(t) + 4x'(t) + 3x(t) = 0$

(ii) $x''(t) + 8x'(t) + 5x(t) = 0$

(iii) $x''(t) + 5x'(t) = 0$

(iv) $x''(t) + 6x'(t) + 25x(t) = 0$

3-0c. Given the initial conditions $x(0+) = 1$, $x'(0+) = -1$, determine the solutions for the following homogeneous equations:

(i) $x''(t) + 6x'(t) + 25x(t) = 0$

(ii) $x''(t) + 8x'(t) + 16x(t) = 0$

(iii) $x''(t) + 4.81x'(t) + 5.76x(t) = 0$

```
      SUBROUTINE RKAM(TO,RO,TF,H,N)
      DIMENSION RO(20),R(20),F(20),F3(20),F2(20),F1(20)
      DIMENSION FO(20),XK1(20),XK2(20),XK3(20),XK4(20)
      IR=0
      T=TO
      DO 10 I=1,N
        R(I)=RO(I)
   10 CONTINUE
      CALL FUN(T,R,F)
      CALL OUT(T,R,F)
      DO 100 I=1,N
        F3(I)=F(I)
  100 CONTINUE
    9 DO 201 I=1,N
        XK1(I)=H*F(I)
  201 CONTINUE
      T=TO+0.5*H
      DO 202 I=1,N
        R(I)=RO(I)+0.5*XK1(I)
  202 CONTINUE
      CALL FUN(T,R,F)
      DO 203 I=1,N
        XK2(I)=H*F(I)
  203 CONTINUE
      T=TO+0.5*H
      DO 204 I=1,N
        R(I)=RO(I)+0.5*XK2(I)
  204 CONTINUE
      CALL FUN(T,R,F)
      DO 205 I=1,N
        XK3(I)=H*F(I)
  205 CONTINUE
      T=TO+H
      DO 206 I=1,N
        R(I)=RO(I)+XK3(I)
  206 CONTINUE
      CALL FUN(T,R,F)
      DO 207 I=1,N
        XK4(I)=H*F(I)
  207 CONTINUE
      DO 208 I=1,N
        RK=(XK1(I)+2.*XK2(I)+2.*XK3(I)+XK4(I))/6.
        R(I)=RO(I)+RK
        RO(I)=R(I)
  208 CONTINUE
      TO=T
      CALL FUN(T,R,F)
      CALL OUT(T,R,F)
      IR=IR+1
      IF(IR-2) 101,102,103
  101 DO 1 I=1,N
        F2(I)=F(I)
    1 CONTINUE
      GO TO 9
  102 DO 2 I=1,N
        F1(I)=F(I)
    2 CONTINUE
      GO TO 9
  103 DO 3 I=1,N
        FO(I)=F(I)
```

Fig. 3.20 Subroutine RKAM.

```
  3 CONTINUE
300 T=T+H
    DO 301 I=1,N
      AM=(55.*F0(I)-59.*F1(I)+37.*F2(I)-9.*F3(I))/24.
      R(I)=R0(I)+H*AM
301 CONTINUE
    CALL FUN(T,R,F)
    DO 302 I=1,N
      AM=(9.*F(I)+19.*F0(I)-5.*F1(I)+F2(I))/24.
      R(I)=R0(I)+H*AM
      R0(I)=R(I)
302 CONTINUE
    CALL FUN(T,R,F)
    CALL OUT(T,R,F)
    IF(T-TF) 401,401,501
401 DO 402 I=1,N
      F3(I)=F2(I)
      F2(I)=F1(I)
      F1(I)=F0(I)
      F0(I)=F(I)
402 CONTINUE
    GO TO 300
501 RETURN
    END
```

Fig. 3.20 *Continued*

```
0.01     0.0     1.00  2
1.00     0.0
```

Fig. 3.21 Input data listing.

T	R(1)	R(2)
0.0	0.10000E 01	0.0
0.10000E-01	0.99990E 00	-0.19702E-01
0.20000E-01	0.99961E 00	-0.38818E-01
0.30000E-01	0.99913E 00	-0.57362E-01
0.40000E-01	0.99846E 00	-0.75346E-01
0.50000E-01	0.99762E 00	-0.92784E-01
0.60000E-01	0.99661E 00	-0.10969E 00
0.70000E-01	0.99543E 00	-0.12607E 00
0.80000E-01	0.99409E 00	-0.14194E 00
0.90000E-01	0.99259E 00	-0.15732E 00
0.10000E 00	0.99094E 00	-0.17221E 00
0.11000E 00	0.98915E 00	-0.18663E 00
0.12000E 00	0.98721E 00	-0.20058E 00
0.13000E 00	0.98514E 00	-0.21409E 00
0.14000E 00	0.98293E 00	-0.22715E 00
0.15000E 00	0.98060E 00	-0.23978E 00
0.16000E 00	0.97814E 00	-0.25199E 00
0.17000E 00	0.97556E 00	-0.26379E 00
0.18000E 00	0.97286E 00	-0.27519E 00
0.19000E 00	0.97006E 00	-0.28619E 00
0.20000E 00	0.96714E 00	-0.29682E 00
0.21000E 00	0.96412E 00	-0.30707E 00
0.22000E 00	0.96100E 00	-0.31696E 00
0.23000E 00	0.95778E 00	-0.32650E 00
0.24000E 00	0.95447E 00	-0.33569E 00
0.25000E 00	0.95107E 00	-0.34454E 00

Fig. 3.22 Sample computer output.

3-0d. Given the initial conditions $x(0+) = 1$, $x'(0+) = 0$, determine the solutions for the following inhomogeneous equations:

(i) $x''(t) + 4x'(t) + 3x(t) = 5e^{-t} \sin 2t$

(ii) $x''(t) + 6x'(t) + 25x(t) = 2 \cos t$

(iii) $x''(t) + 8x'(t) + 16x(t) = 2$

3-0e. Given the initial condition $x(0-) = -2$, solve the following integro-differential equations:

(i) $x'(t) + 5x(t) + \int_0^t x(\tau)d\tau = 2 \sin t$

(ii) $x'(t) + 2x(t) + 2\int_0^t x(\tau)d\tau = \frac{3}{4}e^{-t}$

3-0f. Solve the following set of equations:

(i) $x'(t) + x(t) + y(t) = \sin t$
$x(t) + y'(t) + 2y(t) = 1$
with $x(0+) = y(0+) = 0$.

(ii) $2x'(t) + 2x(t) + y'(t) - y(t) = 2$
$x'(t) + x(t) + y'(t) + 2y(t) = 3e^{-t}$
with $x(0+) = 1$, $y(0+) = 0$.

3-0g. Solve the equation
$x'''(t) + 3x''(t) + 3x'(t) + x(t) = 2e^{-t}$
with $x'''(0+) = x''(0+) = x'(0+) = 0$.

3-1. The network shown has reached steady state before the switch S moves from a to b at $t = 0$. Determine the initial states of the network and also the final states (as $t \to \infty$) of the network for $v_1(t) = 10v$ and $v_2(t) = 5v$.

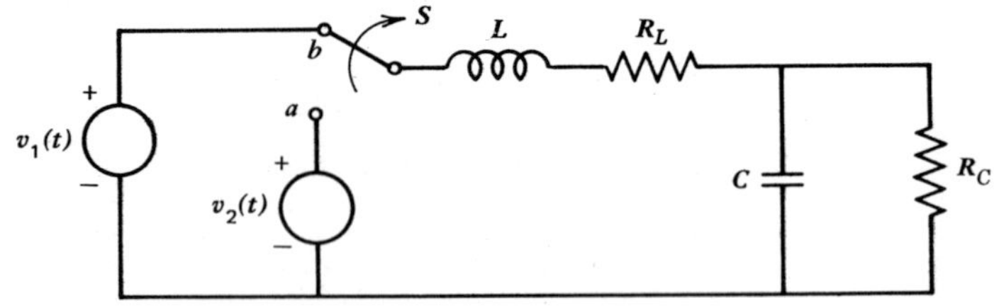

Fig. P3-1

3-2. The series connection of L and R_L of Fig. P3-1 can be considered as a nonideal inductor. Similarly the parallel connection of C and R_C is considered as a practical capacitor. Ideally $R_L = 0$ and $R_C = \infty$. So the network is equivalent to a practical series L-C circuit.

(i) Formulate a differential equation to find $v_C(t)$ for $t \geqslant 0$.

(ii) Find the range of R_C such that for fixed L, C, and R_L we have
 (a) underdamped $v_C(t)$,
 (b) critically damped $v_C(t)$,
 (c) overdamped $v_C(t)$.
(iii) Repeat (ii) for R_L with fixed L, C, and R_C.
(iv) From the results of (ii) and (iii), describe how the damping of this network is affected by R_L and R_C.

3-3. Another equivalent network for a practical series LC circuit is shown in Fig. P3-3. Repeat 3-2 for this circuit.

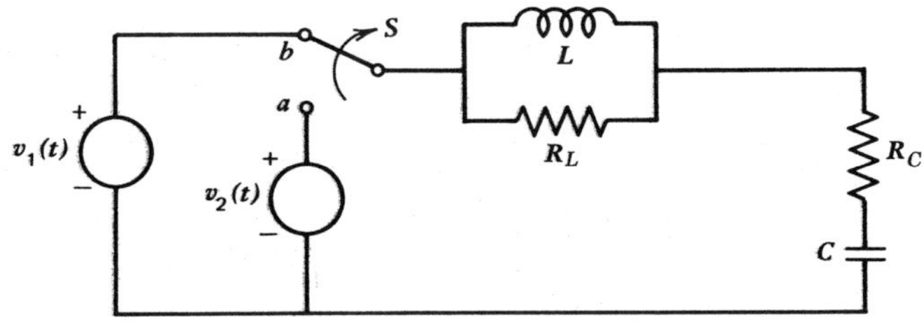

Fig. P3-3

3-4. Solve $v_C(t)$ of the network of Fig. P3-1 for $t \geqslant 0$ of the following parameters: $R_L = 0$, $L = 1H$. $R_C = 1/4\Omega$, $C = 1F$, $v_1(t) = 10V$ and $v_2(t) = 5V$. Separate the final solutions in terms of a sum of zero-state response and zero-input response.

3-5. For the network shown, before the switch moves from a to b, steady state conditions prevailed. Find the current $i(t)$. Any damping effect for $i(t)$? Explain why.

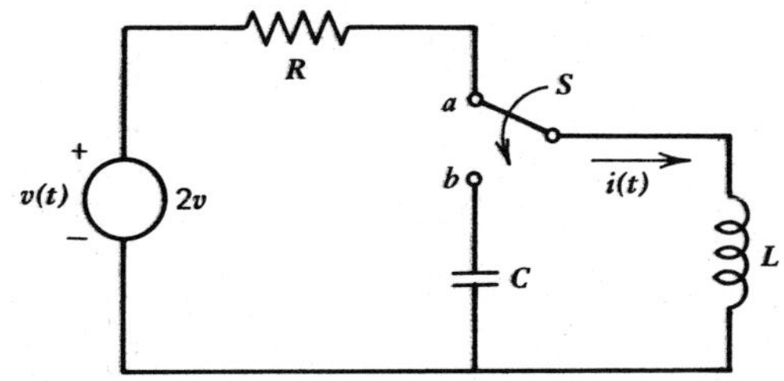

Fig. P3-5

3-5a. The circuit of Fig. P3-5a is in the steady state with the switch S open. At $t = 0$, S is closed. Find $i(t)$.

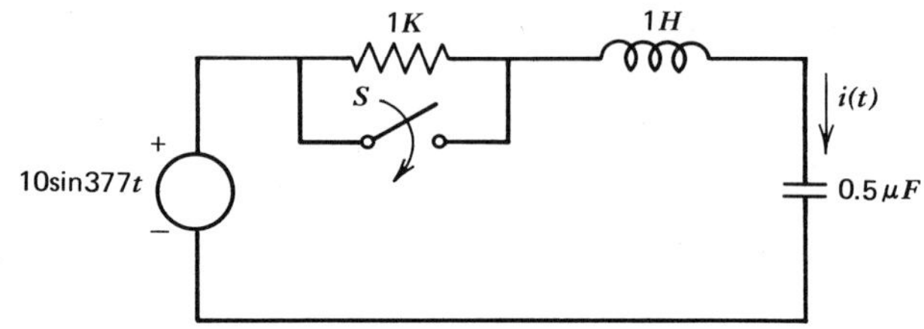

Fig. P3-5*a*

3-6. At $t = 0$, the switch S goes from 1 to 2. Find $i(t)$, given that $v(t) = e^{-t} \sin 2t$. Assume the circuit had been in steady state for $t < 0$.

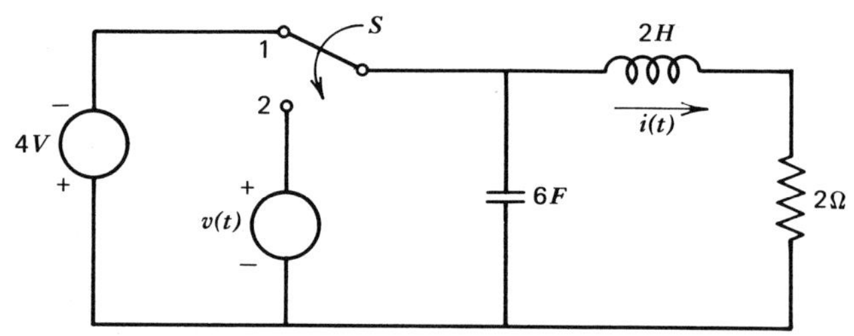

Fig. P3-6

3-7. For the circuit shown, $i(t) = 4e^{-2t}u(t)$. Assuming zero initial states, find $v(t)$ for $t \geqslant 0$.

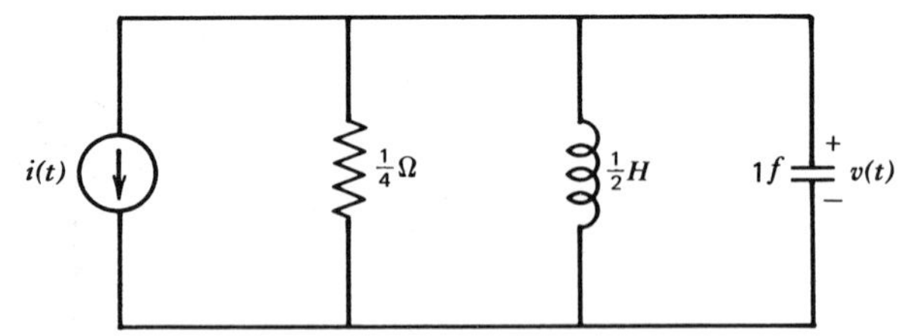

Fig. P3-7

3-7a. The circuit of Fig. P3-7*a* is in the steady state before the switch S is closed at $t = 0$. Find $v(t)$ for $t > 0$.

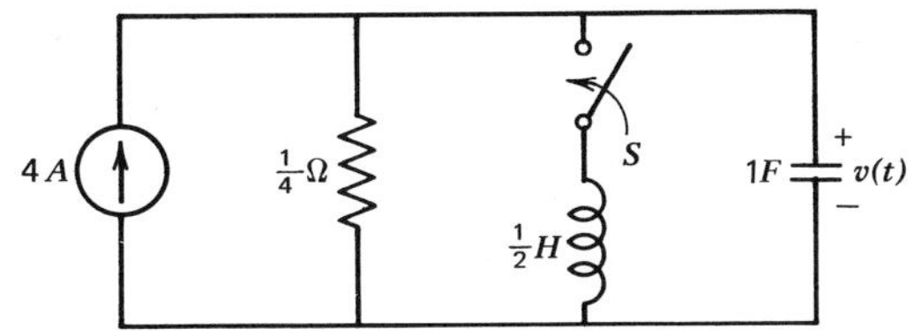

Fig. P3-7*a*

3-8. For the transformer circuit shown, the switch S closes at $t = 0$. Assume zero initial states.

(i) Formulate a system of differential equations for the unknowns $i_1(t)$ and $i_2(t)$.

(ii) Reduce the result of (i) to a differential equation of $i_1(t)$ by the procedure of Sec. 3.3.

(iii) Repeat (ii) for $i_2(t)$.

(iv) Solve (ii) for $t \geqslant 0$.

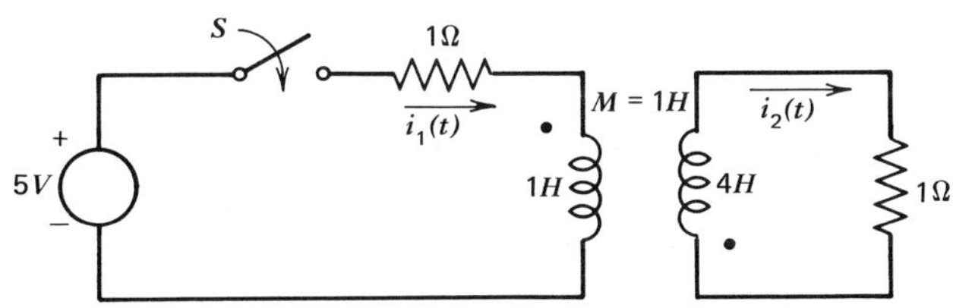

Fig. P3-8

3-9. For the network shown, v_a and v_d are constrained by the voltage sources $v_1(t)$ and $v_2(t)$. Assume zero initial states.

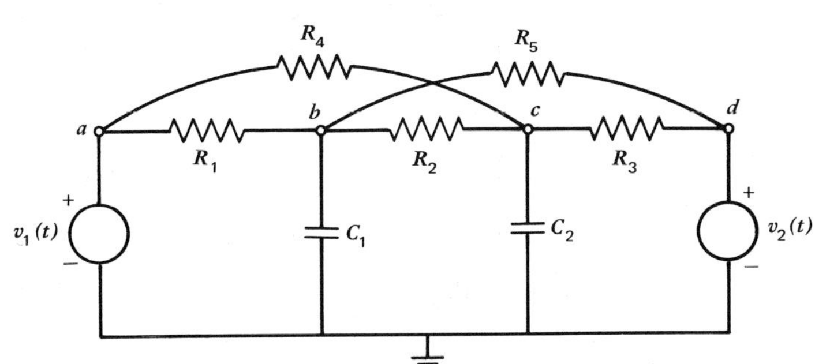

Fig. P3-9

(i) Formulate *a* system of differential equation for v_b and v_c.
(ii) From (i), find a differential equation for v_b.
(iii) Repeat (ii) for v_c.
(iv) Solve (ii) and (iii), and find all the currents through the resistors with $R_k = k\Omega$ $(k = 1, 2, \ldots, 5)$, $C_1 = 1F$, $C_2 = 2F$, and $v_1(t) = v_2(t) = 2V$, $v_{C1}(0-) = v_{C2}(0-) = 0$.

3-10. Form a set of independent node equations for the network shown in Fig. P3-10 using node 4 as the reference node.

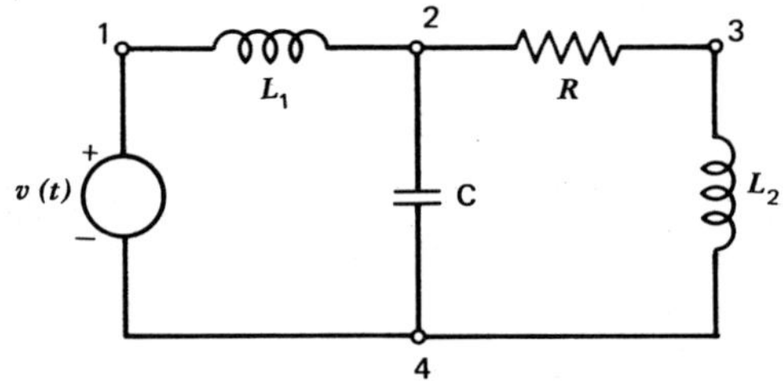

Fig. P3-10

3-11. Repeat 3-10 using node 2 as the reference node.
3-12. Reduce the results of 3-10 to a single differential equation of the unknown node voltage v_2.
3-13. Repeat 3-12 for the node voltage v_3 of Fig. P3-10. Compare the result with that of 3-12.
3-14. Write a set of state equations for the circuit of Fig. P3-10. Simplify the results to a single differential equation of the unknown state of capacitor. Is the resultant equation identical to that of the result of 3-12?
3-15. Write a set of state equations for the circuit of Fig. P3-15.

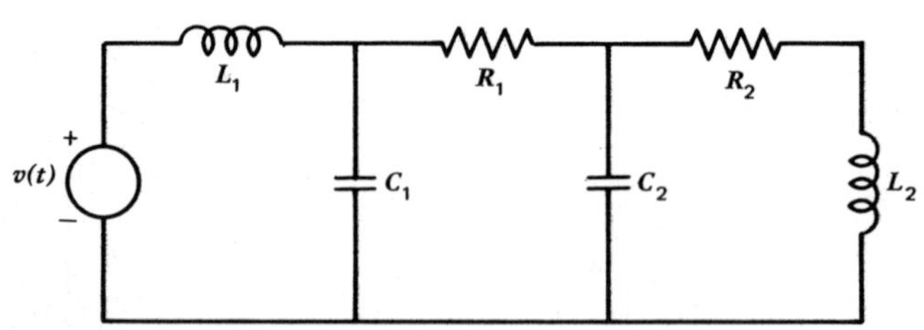

Fig. P3-15

3-16. Repeat 3-15 for $R_2 = 0$.
3-17. Write a set of state equations for the circuit of Fig. P3-17.

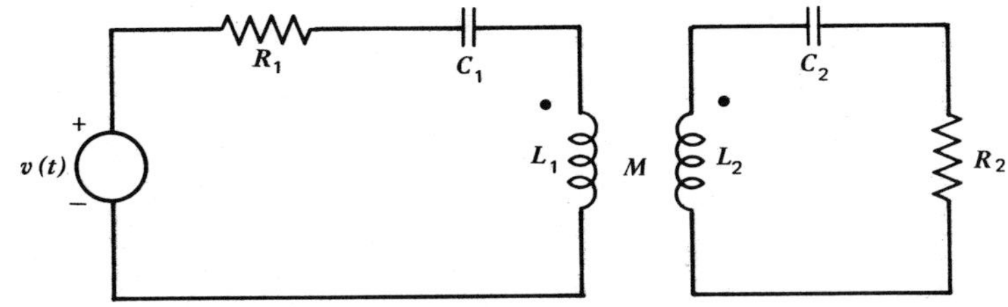

Fig. P3-17

3-18. Write a set of state equations for the network shown in Fig. P3-18.

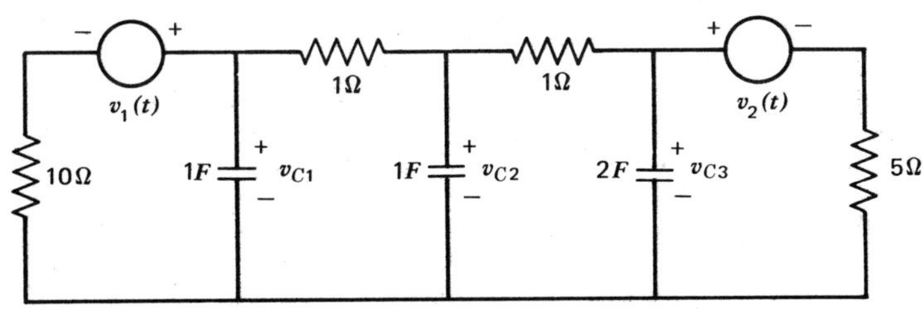

Fig. P3-18

3-18a. Write a set of state equations for the network shown in Fig. P1-6*a*.

3-18b. Repeat 3-18*a* for the circuit of Fig. P1-6*c*.

3-19. Put the result of 3-12 into a set of state differential equation of the form of Eq. 3.103.

3-20. Write equivalent set of state equations for the following differential equations:

(i) $v^{(4)}(t) + 2v''(t) + v(t) = i(t)$.

(ii) $(t-1)^2 v''(t) + t(t-1)v'(t) - v(t) = 1 - t^2$.

3-20a. Transform the equations of 3-0*c* into a system of first order differential equation with appropriate initial conditions.

3-20b. Repeat 3-20a for the equations of 3-0d.

3-20c. Repeat 3-20a for the equation of 3-0g.

3-20d. For the following circuit, steady-state conditions prevailed before the switch S is closed at $t = 0$.

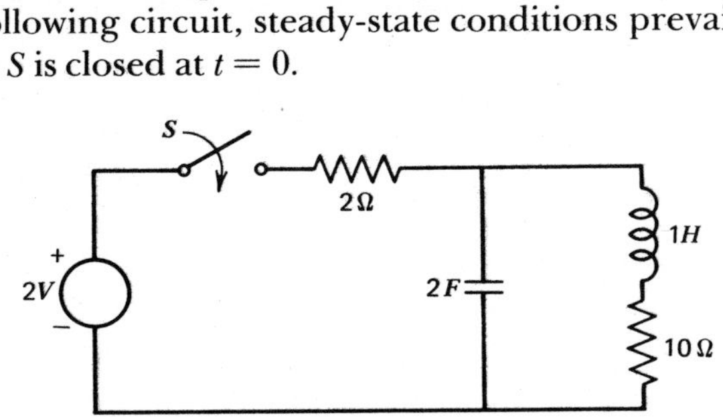

Fig. P3-20*d*

(i) Find the initial current through the capacitor and the initial voltage across the inductor. Are they bounded? (Check your answer using the method of 2-1e.)

(ii) Solve for the states of the reactive elements.

3-20e. Repeat 3-20d for the following circuits:

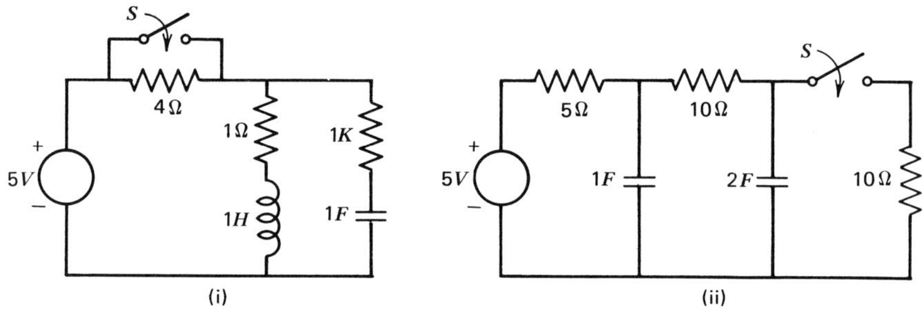

Fig. P3-20*e*

3-20f. Solve for $i(t)$ of the following zero-state circuit where $M = 2H$.

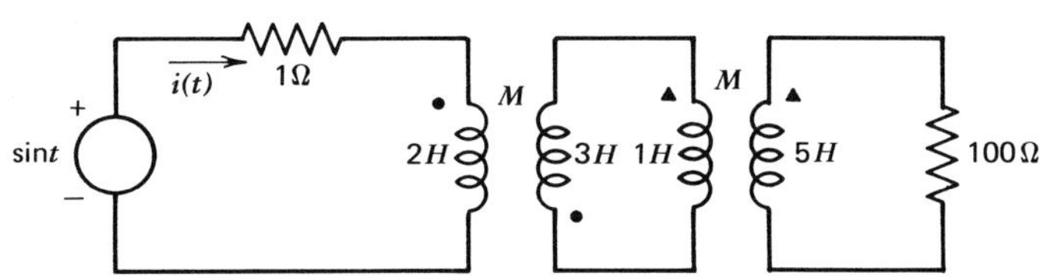

Fig. P3-20*f*

3-21. With the initial states found in 3-1, compute a few point-wise solutions for $v_c(t)$ of the circuit of Fig. P3-1 by first setting up two first order state equations using the result of 3-4 and then applying the fourth order R-K method of Eq. 3.127 for the solution. Choose a suitable step size and then compare the results with that of 3-4.

3-21a. Solve the following equations using the subroutine RKAM.

(i) $(t^2-1)v''(t)+t(t-1)v'(t)+v(t)=t^2$,
$v(2+)=2,\ v'(2+)=1,\ t=2(0.1)3.$

(ii) $\begin{cases} x'(t)-1/2y(t)=-1/2. \\ 3x(t)y'(t)=2t,\ x(1+)=2/3,\ y(1+)=3,\ t=1(0.1)2. \end{cases}$

Answers: (i) $v(t)=t$, (ii) $x(t)=2/3t^{3/2}$, $y(t)=2t^{1/2}+1$.

3-22. Write a program to solve 3-6 using the subroutine RKAM. Terminate the computation at $t=5$ sec. Program your output subroutine such that the results are printed for every tenth step. Use step sizes $h = 0.001$, 0.01, 0.1 sec. Compare the results with that found in 3-6.

3-22a. Solve the equations of 3-0d using the subroutine RKAM for $t = 0(0.1)1$.

3-22b. Repeat 3-21a for $t = 0(0.05)1$. Are the results better?

3-22c. Solve the equation of 3-0g using the subroutine RKAM for $t = 0(0.1)1$. Compare with the exact answer of 3-0g.

3-22d. Solve the resultant state equations of 3-18a using the subroutine RKAM for $t = 0(0.1)10$. The parameters are as follows: $v(t) = 10V$, $R_k = k\Omega$, $C_k = kF$ for $k = 1, 2, 3, 4$ with zero initial states.

3-22e. Solve the resultant state equations of 3-18b using RKAM for $t = 0(0.1)10$. The parameters are as follows:
$R_k = k\Omega$, $C_k = kF$, $L_k = kH$ for $k = 1, 2, 3, 4$.
$v_1 = v_2 = 1$ volt, $i(t) = 1$ amp with zero initial states.

3-23. Write a program using subroutine RKAM to solve 3-18 for the *zero-input* response with the following initial states:
(i) $v_{C1}(0-) = -0.2078V$, $v_{C2}(0-) = 0.5609V$, $v_{C3}(0-) = -0.8015V$.
(ii) $v_{C1}(0-) = 0.6064V$, $v_{C2}(0-) = 0.5986V$, $v_{C3}(0-) = 0.5235V$.
How small should the step size h be in order to get a fairly "accurate" answer for each case?

CHAPTER 4

Step and Impulse Response: Convolution

We have completed our discussion of the determination of the circuit response for linear, time-invariant circuits under various excitations. Explicit solutions have been attempted only for circuits excited by constant, exponential, and sinusoidal signals. These signals are describable by elementary mathematical functions and are, therefore, called *elementary* signals. The response to the elementary signals is represented by the response to the exponential signals, where the exponents are appropriately zero, real, or complex. Therefore, explicit solutions have been attempted only for circuits excited by elementary or exponential signals. Numerical computational methods must be used in all cases where the excitations are nonelementary functions.

Even for circuits under elementary or exponential excitations, explicit solutions were obtained only for simple first and second order circuits. For higher order circuits only the *form* of the solution can be stated explicitly. While the steady-state solutions can be obtained directly for circuits of any order under exponential excitations, the transient solutions cannot be obtained without extensive numerical computations. The transient solutions are determined by the roots of the characteristic equations. For characteristic equations of high order, the task is quite formidable even on a digital computer.

Numerical computational methods for solving the circuit differential equations are, therefore, essential to circuit analysis both in the case of higher order circuits and in the case of nonelementary excitations. Explicit complete solutions should usually be looked for only in the special case of first or second order circuits under elementary excitations.

An understanding of circuit behavior, however, can be obtained in large measure from a thorough appreciation of the general properties exhibited by the circuits. The forms of the solutions, superposition and time-invariance properties, all lead to a general theory of circuits that goes far beyond anything remotely possible through mere computation of circuit response. The rest of this text is devoted mostly to the development of this general theory of linear, time-invariant circuits together with the associated computational methods.

Recall that the response of a linear circuit whose initial state is not zero can be expressed as a sum of the zero-state response and the zero-input response (Superposition I). That is, the effect of arbitrary initial state is isolated and added on to the zero (initial) state response. Zero-state is viewed thus as a normal or basic situation with other terms added to the response to account for any departure from the base. Consequently, the general theory of circuits will be developed only under the zero initial state assumptions. We also assume henceforth that the initial instant of excitation is $t_0 = 0$, since for time-invariant circuits the response is dependent only on time elapsed since excitation.

In this chapter, we use the properties of linear, time-invariant circuits discussed in the previous chapters to develop a major result of importance not only for circuits but also for many problems in engineering and science. In particular, we develop the concept of *convolution*, which states that the response of a zero-state circuit or system to an arbitrary excitation is expressed in terms of its response to the elementary unit-step function or, rather, to its derivative. The derivative of the unit step is the unit impulse, and the responses to these two excitations are called the step and impulse response, respectively. In this context, it should be noted that the step and impulse response are defined for zero-state circuits. We discuss these responses before proceeding with the development of the convolution idea.

4.1 STEP AND IMPULSE FORCING FUNCTIONS

All circuits are henceforth assumed to be in zero-state unless otherwise stated. That is, at the instant prior to excitation, $t = 0-$, all the state variables are assumed to be zero. There is no excitation for $t < 0$ and the initial energy at $t = 0-$ is zero; therefore, the circuit exhibits zero response for $t < 0$. When the excitation is applied, the state at $t = 0+$ is still zero since the state does not change instantly for all physical situations. The voltage or current variable in the circuit differential equation and its derivatives at $t = 0-$ are all zero since the response is zero for negative time. Their values for $t = 0+$ (i.e., initial conditions) are determined from the zero initial state at $t = 0+$. These values may not be all zero anymore. Thus, while the state variables are continuous at $t = 0$, the initial conditions are not necessarily so.

The continuous nature of the state variables is only valid under excitations that are bounded signals. If these signals are infinitely large, then, of course, the state may change instantly and, in fact, does under impulse excitations. The impulse functions are not physically realistic signals but are important mathematical abstractions. Therefore, we must consider these impulse excitations with special care since the initial state is not continuous anymore. The initial conditions are thus available only at $t = 0-$, and their values at $t = 0+$ must be determined without reference to the continuity of the state variables.

We develop below the general procedures for obtaining the initial conditions at $t = 0+$ from their known values at $t = 0-$. The forcing functions for the circuit differential equations often involve derivatives of the excitation signals. For simplicity in discussion, we shall only consider the step and impulse forcing functions. As we shall see later, the impulse response is obtained simply from the step response, and the step response is obtained in the usual manner discussed in Chapters 2 and 3.

In this section we discuss solutions of differential equations with step or impulse forcing functions. We are concerned here with the mathematical problem of solving for the impulse and step response, given a linear differential equation with initial conditions at $t = 0-$.

From Chapter 1, recall that the definition of the unit-step function was

$$\begin{aligned} u(t) &= 1 \qquad t \geqslant 0 \\ &= 0 \qquad t < 0 \end{aligned}$$

and the unit impulse was shown to have the properties:

$$\begin{aligned} \delta(t) &= \infty \qquad t = 0 \\ &= 0 \qquad t \neq 0 \end{aligned}$$

and

$$\int_{0-}^{0+} \delta(t)dt = 1$$

In addition, we have the relationship

$$\delta(t) = \frac{du(t)}{dt}$$

As the definitions of $\delta(t)$ and $u(t)$ indicate, both functions have discontinuities at $t = 0$. In dealing with initial conditions for step and impulse drivers, we must then recognize that the solution $x(t)$ *and its derivatives*

$x'(t)$, $x''(t)$, etc., *may not be continuous at* $t = 0$. In other words, it may be that

$$
\begin{aligned}
x^{(n)}(0-) &\neq x^{(n)}(0+) \\
x^{(n-1)}(0-) &\neq x^{(n-1)}(0+) \\
&\cdots \\
x(0-) &\neq x(0+)
\end{aligned}
$$

In physical problems, the initial conditions, obtained from the initial state, are specified at $t = 0-$. However, to evaluate the unknown constants of the total solution, we must have the initial conditions at $t = 0+$. Our task, then, is to determine the conditions at $t = 0+$, given the initial conditions at $t = 0-$. The method discussed here is borrowed from electromagnetic theory and is often referred to as "integrating through a Green's function."*

Consider the differential equation with an impulse forcing function

$$a_n x^{(n)}(t) + a_{n-1} x^{(n-1)}(t) + \cdots + a_0 x(t) = A\delta(t) \tag{4.1}$$

To insure that the right-hand side of Eq. 4.1 will equal the left-hand side, one of the terms $x^{(n)}(t), x^{(n-1)}(t), \ldots, x(t)$ must contain an impulse. The question is, "Which term contains the impulse?" A close examination shows that the highest derivative term $x^{(n)}(t)$ *must* contain the impulse, because if $x^{(n-1)}(t)$ contained the impulse, $x^{(n)}(t)$ would contain a *doublet* $C\delta'(t)$. This argument holds, similarly, for all lower derivative terms of $x(t)$. If the term $x^{(n)}(t)$ contains the impulse, then $x^{(n-1)}(t)$ would contain a step and $x^{(n-2)}(t)$, a ramp. *We conclude therefore that, for an impulse forcing function, the two highest derivative terms are discontinuous at* $t = 0$. For a step function, only the highest derivative term is discontinuous at $t = 0$.

Since initial conditions are usually given at $t = 0-$, our task is to determine the values $x^{(n)}(0+)$ and $x^{(n-1)}(0+)$ for an impulse forcing function. Referring to Eq. 4.1, let us integrate the equation between $t = 0-$ and $t = 0+$, namely,

$$a_n \int_{0-}^{0+} x^{(n)}(t)\,dt + a_{n-1} \int_{0-}^{0+} x^{(n-1)}(t)\,dt + \cdots + a_0 \int_{0-}^{0+} x(t)\,dt = A \int_{0-}^{0+} \delta(t)\,dt \tag{4.2}$$

After integrating, we obtain

$$a_n[x^{(n-1)}(0+) - x^{(n-1)}(0-)] + a_{n-1}[x^{(n-2)}(0+) - x^{(n-2)}(0-)] + \cdots = A \tag{4.3}$$

We know that all derivative terms below $(n-1)$ are continuous at $t = 0$.

*The Green's function is another name for impulse response; for example, see Morse and Freshbach, *Methods of Theoretical Physics*, McGraw-Hill, New York, 1952, Chapter 7.

Consequently, Eq. 4.3 simplifies to

$$a_n[x^{(n-1)}(0+) - x^{(n-1)}(0-)] = A \tag{4.4}$$

so that

$$x^{(n-1)}(0+) = \frac{A}{a_n} + x^{(n-1)}(0-) \tag{4.5}$$

We must next determine $x^{(n)}(0+)$. At $t = 0+$, the differential equation in Eq. 4.1 is

$$a_n x^{(n)}(0+) + a_{n-1}x^{(n-1)}(0+) + \cdots + a_0 x(0+) = 0 \tag{4.6}$$

Since all derivative terms below $(n-1)$ are continuous, and since we have already solved for $x^{(n-1)}(0+)$, we find that

$$x^{(n)}(0+) = -\frac{1}{a_n}[a_{n-1}x^{(n-1)}(0+) + \cdots + a_1 x'(0+) + a_0 x(0+)] \tag{4.7}$$

For a step forcing function $Au(t)$, all derivative terms except $x^{(n)}(t)$, are continuous at $t = 0$. To determine $x^{(n)}(0+)$, we derive in a manner similar to Eq. 4.7, the expression

$$x^{(n)}(0+) = \frac{A}{a_n} - \frac{1}{a_n}[a_{n-1}x^{(n-1)}(0+) + \cdots + a_0 x(0+)] \tag{4.8}$$

In determining the unknown constants in the total solution of the differential equation, the initial conditions needed involve derivatives up to the order $(n-1)$. The nth derivative initial value is not needed. Thus, for a step forcing function all the initial conditions are continuous while for an impulse forcing function only the initial condition for the $(n-1)$th derivative is discontinuous. For a complete understanding, however, the above discussion includes the initial derivatives of the nth order as well.

The process of determining initial conditions when the forcing function is an impulse or one of its higher derivatives is simplified by the visual process shown in Eqs. 4.9 and 4.10. Above each derivative term we draw its associated highest order singularity. Note that we need only go as low as a step in this visual aid.

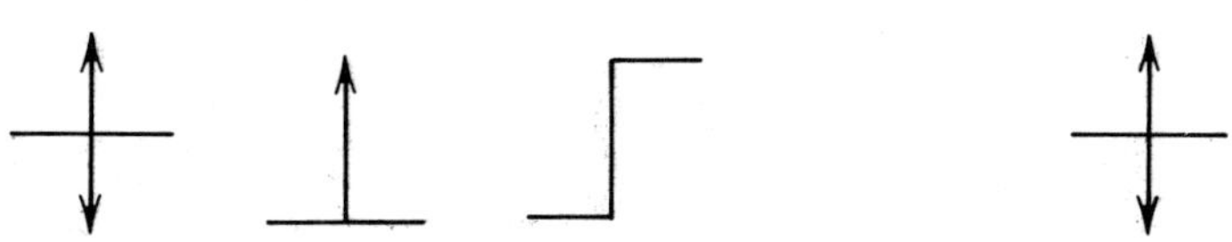

$$a_n x^{(n)}(t) + a_{n-1}x^{(n-1)}(t) + a_{n-2}x^{(n-2)}(t) + \cdots + a_0 x(t) = \delta'(t) \tag{4.9}$$

$$a_n x^{(n)}(t) + a_{n-1} x^{(n-1)}(t) + a_{n-2} x^{(n-2)}(t) + \cdots + a_0 x(t) = \delta(t) \qquad (4.10)$$

It should be noted that if a derivative term contains a certain singularity – for example, a doublet – it also contains all lower derivative terms. For example, in the equation

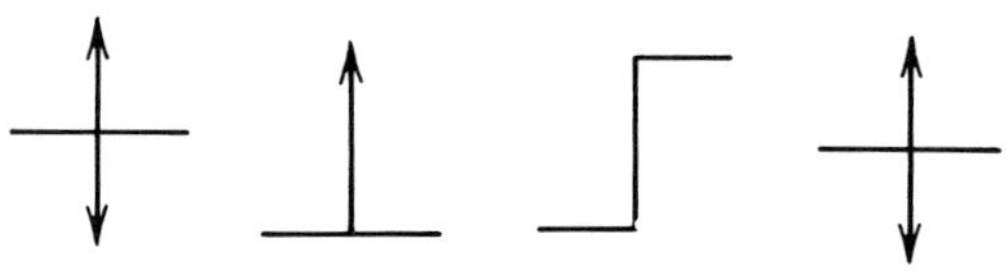

$$x''(t) \quad + \quad 3x'(t) \quad + \quad 2x(t) \quad = \quad 4\delta'(t) \qquad (4.11)$$

we assume the following forms for the derivative terms at $t = 0$:

$$\begin{aligned} x''(t) &= A\delta'(t) + B\delta(t) + Cu(t) \\ x'(t) &= A\delta(t) + Bu(t) \\ x(t) &= Au(t) \end{aligned} \qquad (4.12)$$

Substituting Eq. 4.12 into Eq. 4.11, we obtain

$$A\delta'(t) + B\delta(t) + Cu(t) + 3A\delta(t) + 3Bu(t) + 2Au(t) = 4\delta'(t) \qquad (4.13)$$

Or in a more convenient form, we have

$$A\delta'(t) + (B + 3A)\delta(t) + (C + 3B + 2A)u(t) = 4\delta'(t) \qquad (4.14)$$

Equating like coefficients on both sides of Eq. 4.14 gives

$$\begin{aligned} A &= 4 \\ B + 3A &= 0 \\ C + 3B + 2A &= 0 \end{aligned} \qquad (4.15)$$

from which we obtain $B = -12$ and $C = 28$. Therefore, at $t = 0$, it is true that

$$\begin{aligned} x''(t) &= 4\delta'(t) - 12\delta(t) + 28u(t) \\ x'(t) &= 4\delta(t) - 12u(t) \\ x(t) &= 4u(t) \end{aligned} \qquad (4.16)$$

The $u(t)$ terms in Eq. 4.16 gives rise to the discontinuities in the initial conditions at $t = 0$. We are given the initial conditions at $t = 0-$. Once we evaluate the A, B, and C coefficients in Eq. 4.12, we can obtain the initial conditions at $t = 0+$ by referring to the coefficients of the step terms. For example, if

$$\begin{aligned} x(0-) &= -2 \\ x'(0-) &= -1 \\ x''(0-) &= 7 \end{aligned}$$

Then from Eq. 4.16 we obtain

$$\begin{aligned} x(0+) &= -2+4 = 2 \\ x'(0+) &= -1-12 = -13 \\ x''(0+) &= 7+28 = 35 \end{aligned} \tag{4.17}$$

The total solution of Eq. 4.11 is obtained as though it were a homogeneous equation, since $\delta'(t) = 0$ for $t \neq 0$. The only influence of the doublet driver is to produce discontinuities in the initial conditions at $t = 0$. Having evaluated the initial conditions at $t = 0^+$, we obtain easily the total solution. Thus

$$x(t) = A_1 e^{-t} + A_2 e^{-2t} \tag{4.18}$$

From Eq. 4.17 we readily obtain

$$x(t) = -9e^{-t} + 11e^{-2t} \qquad t \geqslant 0+ \tag{4.19}$$

The total solution of a differential equation with a step or impulse forcing function is obtained in an equally straightforward manner. For a step forcing function, only the highest derivative term has a discontinuity at $t = 0$. Since we do not need the initial condition of the highest derivative term for our solution, we proceed as if we were solving a standard nonhomogeneous equation with a constant forcing function. For an impulse driver, once we determine the initial conditions at $t = 0+$, the equation is solved in the same manner as a homogeneous equation.

The solution for $t < 0$ is zero for zero-state circuits. Zero initial state variables at $t = 0-$ yield zero initial conditions at $t = 0-$ and the solution is therefore zero for $t < 0$. The solution that is zero for $t < 0$ and nonzero for $t \geqslant 0$ is commonly described by employing a multiplicative unit-step function. In the above example, we have purposely used nonzero initial conditions at $t = 0-$ and, hence, the solution in Eq. 4.19 is described for $t \geqslant 0$ without assuming zero solution for $t < 0$. Generally, our discussion will be for zero-state circuits, and we commonly employ the unit-step function to indicate zero response for negative time.

Example 4.1.1 Find the step and impulse response for the equation

$$x''(t) + 3x'(t) + 2x(t) = f(t)$$

where $f(t) = \delta(t)$ and $f(t) = u(t)$, respectively. The initial conditions at

$t = 0-$ are

$$x(0-) = x'(0-) = x''(0-) = 0$$

Solution. Let us first find the impulse response. Note that the x'' term contains an impulse and a step; the x' term contains a step; the x term contains a ramp, and is therefore continuous at $t = 0$. Thus $x(0+) = x(0-) = 0$. To obtain $x'(0+)$, we use Eq. 4.5

$$x'(0+) = \frac{K}{a_2} + x'(0-) = \frac{1}{2} + 0 = \frac{1}{2} \tag{4.20}$$

Note that we actually need only $x(0+)$ to evaluate the constants for the second order differential equation. The general solution is

$$x(t) = A_1e^{-t} + A_2e^{-2t} \tag{4.21}$$

Substituting the initial conditions at $t = 0+$, we obtain

$$\begin{aligned} x(0+) &= 0 = A_1 + A_2 \\ x'(0+) &= \tfrac{1}{2} = -A_1 - 2A_2 \end{aligned} \tag{4.22}$$

from which we find $A_1 = 1/2$, $A_2 = -1/2$. The initial conditions at $0-$ are all zero, and the solution is zero for $t < 0$. Hence, the impulse response is

$$x(t) = \tfrac{1}{2}(e^{-t} - e^{-2t})u(t) \tag{4.23}$$

For the step response, the general solution is

$$x(t) = \tfrac{1}{2} + A_1e^{-t} + A_2e^{-2t} \tag{4.24}$$

Since $x'(t)$ and $x(t)$ must be continuous for a step forcing function,

$$\begin{aligned} x(0+) &= x(0-) = 0 \\ x'(0+) &= x'(0-) = 0 \end{aligned} \tag{4.25}$$

Substituting these initial conditions into $x(t)$ and $x'(t)$, we find that $A_1 = -1$ and $A_2 = 1/2$. Therefore, the step response is

$$x(t) = \tfrac{1}{2}(1 - 2e^{-t} + e^{-2t})u(t) \tag{4.26}$$

4.2 PROPERTIES OF LINEAR, TIME-INVARIANT CIRCUITS

We have developed the general procedures for handling step and impulse forcing functions. Quite often the forcing functions include derivatives of such functions. The procedures in such cases are still the same. The highest singular term of the forcing function is used to determine the nature of discontinuities in the derivatives of the unknown variable. These assumed discontinuities are substituted in the differential equation and the values of the discontinuities determined. The process is quite straightforward but somewhat laborious. A conceptually simple procedure is discussed in this section.

The step response is determined first in the manner discussed for constant excitations. The response is zero prior to the step excitation. After the step, the excitation is constant. The initial state is continuous and yields the necessary initial conditions at $t = 0+$. The step response is then determined in a simple way.

The properties of linearity and time-invariance for zero-state circuits are then utilized to derive the impulse response. Let $e_k(t)$ be the excitation and $r_k(t)$ the (zero-state) response of any circuit. Then linearity (Superposition II) implies that the response to excitation $e(t) = c_1e_1(t) + c_2e_2(t)$ is given by $r(t) = c_1r_1(t) + c_2r_2(t)$. Time-invariance implies that a time translation of the excitation, say $e_k(t-\Delta)$, results in an identical translation of the response to $r_k(t-\Delta)$. Table 4.2.1 shows these excitation-response pairs together with some consequences.

From properties 1–3 in Table 4.2.1 it follows that for an excitation $[e(t)-e(t-\Delta)]/\Delta$ the response is given by $[r(t)-r(t-\Delta)]/\Delta$. In the limit as $\Delta \to 0$, these quantities become derivatives and state that for an excitation $e'(t)$ the response is $r'(t)$. This is shown in properties 4 and 5 of the table.

Property 6 follows from similar arguments. The lower limit of the integral is $0-$ in order to include any singular functions at the origin.

The property 5 of Table 4.2.1 is very useful in determining the impulse response. Since the unit impulse is the derivative of the unit step, the impulse response is simply the derivative of the step response. Thus, the

Table 4.2.1

Excitation	(Zero-state) Response
1. $e_k(t)$	$r_k(t)$
2. $c_1e_1(t) + c_2e_2(t)$	$c_1r_1(t) + c_2r_2(t)$
3. $e_k(t-\Delta)$	$r_k(t-\Delta)$
4. $\lim\limits_{\Delta\to 0} \dfrac{e(t)-e(t-\Delta)}{\Delta}$	$\lim\limits_{\Delta\to 0} \dfrac{r(t)-r(t-\Delta)}{\Delta}$
5. $\dfrac{d}{dt}e(t)$	$\dfrac{d}{dt}r(t)$
6. $\int_{0-}^{t} e(\tau)d\tau$	$\int_{0-}^{t} r(\tau)d\tau$

impulse response can be determined very easily at least conceptually. Conversely, the step response is the integral of the impulse response from property 6. Similarly, the response to ramp functions is obtained by integration of the step response, and so on.

4.3 STEP AND IMPULSE RESPONSE

Let us consider the problem of obtaining the step and impulse responses for any voltage or current in the network. The problem of obtaining the step and impulse response is stated as follows: Given a network with *zero initial state*, we are required to solve for a specified response (current or voltage) due to a given excitation function $u(t)$ or $\delta(t)$, which either can be a current or a voltage source. If the excitation is a step of voltage, the physical analogy is that of a switch—closing at time $t = 0$—which connects a $1V$ battery to a circuit. The physical analogy of an impulse excitation is that of a very short (compared to the time constants of the circuit) pulse with large amplitude. Since our discussion now involves the possible presence of singular (impulse and its derivatives) functions at the origin, all lower limits are taken to be $0-$ for integrals arising in branch constraints and Kirchhoff's laws.

The problems involved are best illustrated by means of examples. Consider the series RC circuit in Fig. 4.1 with the initial state $v_C(0-) = 0$. The KVL equation for the circuit is

$$Ri(t) + \frac{1}{C}\int_{0-}^{t} i(\tau)d\tau = u(t) \tag{4.27}$$

This equation is solved as in Chapter 2. For $t < 0$, the response $i(t)$ is zero since the initial capacitor voltage is zero and the excitation is zero.

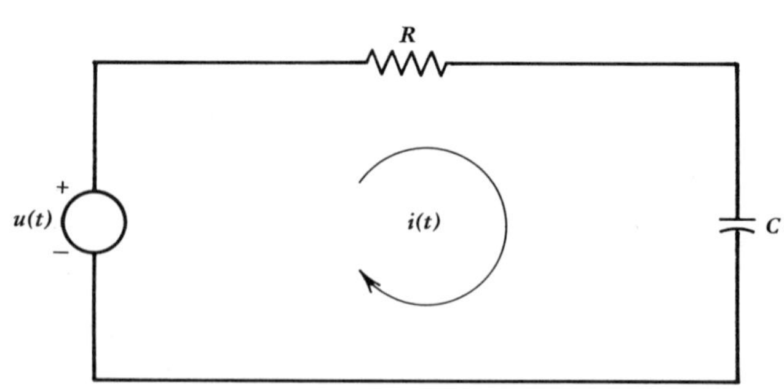

Fig. 4.1

For $t \geqslant 0$, Eq. 4.27 can be written as

$$Ri(t) + \frac{1}{C}\int_{0-}^{t} i(\tau)d\tau = 1 \tag{4.28}$$

which upon differentiation gives

$$Ri'(t) + \frac{1}{C}i(t) = 0 \tag{4.29}$$

The initial state $v_C(0-) = v_C(0+) = 0$ and so the initial current $(0+)$ from Eq. 4.28 is

$$i(0+) = \frac{1}{R} \tag{4.30}$$

The general solution of Eq. 4.29 is

$$i(t) = Ae^{-t/RC} \tag{4.31}$$

and with the initial condition of Eq. 4.30 gives

$$i(t) = \frac{1}{R}e^{-t/RC} \qquad t \geqslant 0 \tag{4.32}$$

Since the solution is zero for $t < 0$, we write the step response as

$$i_u(t) = \frac{1}{R}e^{-t/RC}u(t) \tag{4.33}$$

The impulse response of the circuit is obtained by differentiating Eq. 4.33:

$$i_\delta(t) = \frac{1}{R}\left[\delta(t)e^{-t/RC} - \frac{1}{RC}e^{-t/RC}u(t)\right] \tag{4.34}$$

since $\delta(t)$ is zero for every $t \neq 0$, its coefficients need only be evaluated for $t = 0$. The exponential at $t = 0$ is unity and the impulse response becomes

$$i_\delta(t) = \frac{1}{R}\left[\delta(t) - \frac{1}{RC}e^{-t/RC}u(t)\right] \tag{4.35}$$

Alternatively, we can use the procedures of Sec. 4.1 to determine the step response. We differentiate Eq. 4.27 to obtain

$$Ri'(t) + \frac{1}{C}i(t) = \delta(t) \tag{4.36}$$

Using the procedure for an impulse forcing function we integrate both sides between $0-$ and $0+$ to give

$$i(0+) = \frac{1}{R} \tag{4.37}$$

since $i(0-) = 0$. Equation 4.36 is then solved as a homogeneous equation to give

$$i(t) = Ae^{-t/RC}u(t) \tag{4.38}$$

from the initial condition in Eq. 4.37 we obtain the step response as

$$i(t) = \frac{1}{R}e^{-t/RC}u(t) \tag{4.39}$$

The impulse response is obtained in a similar manner or by differentiation of the step response. Either way we arrive at the same step and impulse responses. These responses are shown in Fig. 4.2.

In a second example consider the parallel *RC* circuit in Fig. 4.3 driven by a current source $i(t) = I_0u(t)$, where I_0 is a constant. The initial zero state implies $v_C(0-) = v_C(0+) = 0$. The circuit differential equation is

$$Cv'(t) + Gv = I_0u(t) \tag{4.40}$$

with initial condition $v(0+) = 0$. For $t < 0$, the solution is zero. For $t \geqslant 0$, the solution is obtained easily as

$$v(t) = I_0R(1 - e^{-t/RC}) \tag{4.41}$$

The response for all t may be expressed as

$$v(t) = I_0R(1 - e^{-t/RC})u(t) \tag{4.42}$$

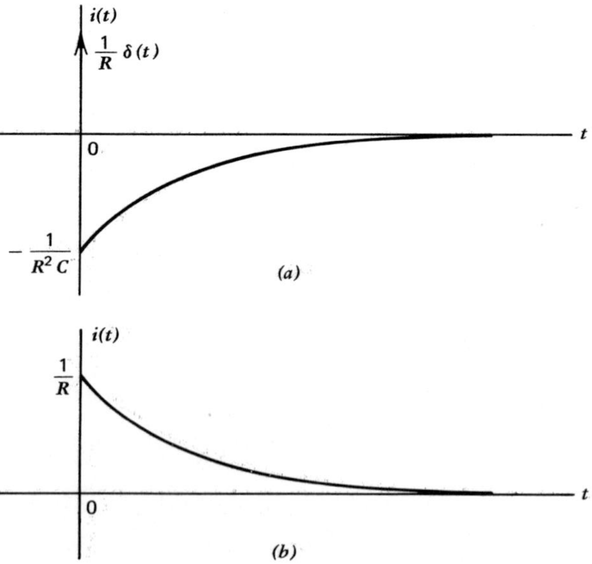

Fig. 4.2 (*a*) Impulse response of *RC* circuit. (*b*) Step response of *RC* circuit.

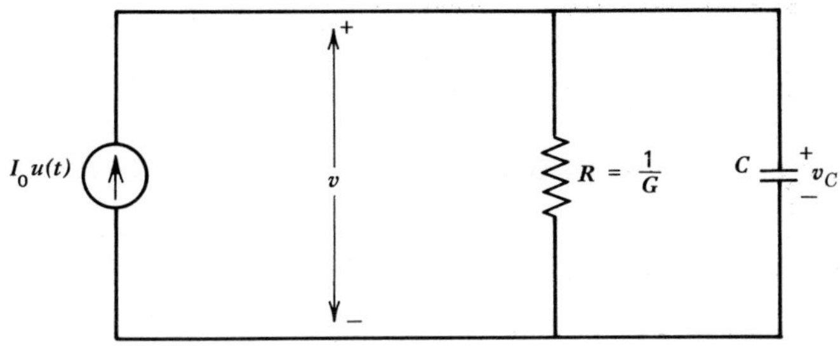

Fig. 4.3

The response to an impulse $I_0\delta(t)$ is obtained by differentiation:

$$v'(t) = \frac{I_0}{C} e^{-t/RC} u(t) \tag{4.43}$$

The responses to these nonunit step and impulse functions are plotted in Fig. 4.4.

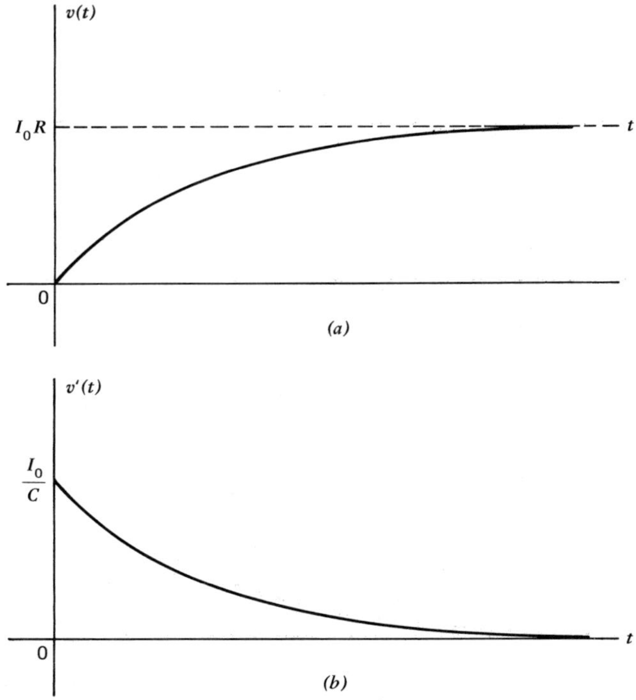

Fig. 4.4 (*a*) Response of parallel *RC* circuit to $I_0u(t)$. (*b*) Impulse response of parallel *RC* circuit to $I_0\delta(t)$.

Note that the terms step response and impulse response signify the responses to unit-step and unit-impulse excitations only.

4.4 CONVOLUTION: PIECEWISE-CONSTANT FUNCTIONS

In this section, we use the properties of linear, time-invariant systems described in Table 4.2.1 to develop their response to a special class of excitations in terms of the unit-step response. It must be understood at all times that we are referring to the zero-state response. The class of excitations in this section have finite numbers of discontinuities or jumps between which their values are constant. Such functions are called piecewise constant since they have constant values defined over certain intervals, and the values are different over different intervals. As we shall see, piecewise-constant functions may be expressed as linear combinations of unit-step functions; hence, the response to such excitations can be expressed as corresponding linear combinations of the step responses.

Let us consider a few simple examples of piecewise-constant functions and use the unit step as a building block to express these functions. Consider the pulse of unit height and width T shown in Fig. 4.5. This is called the unit pulse (*not* impulse) and defined by

$$\begin{aligned} w_1(t) &= 1 \qquad 0 \leqslant t < T \\ &= 0 \qquad \text{otherwise} \end{aligned} \tag{4.44}$$

A pulse of width T and height K_0 is given by $K_0 w_1(t)$. The unit pulse is represented as a linear combination of two unit-step functions as shown in Fig. 4.6. The function $u(t)$ provides the unit value for $t \geqslant 0$ and the

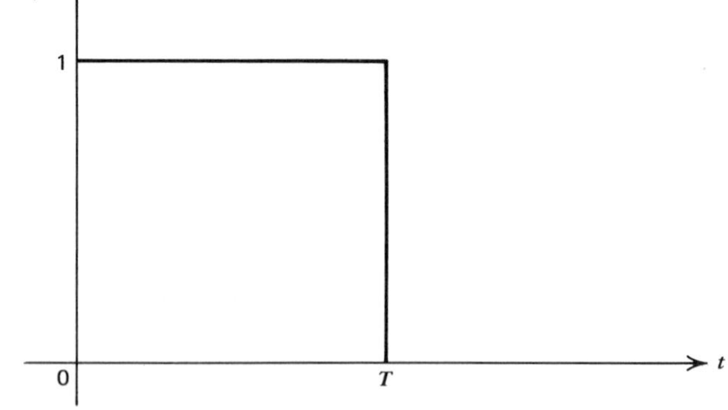

Fig. 4.5 A unit pulse.

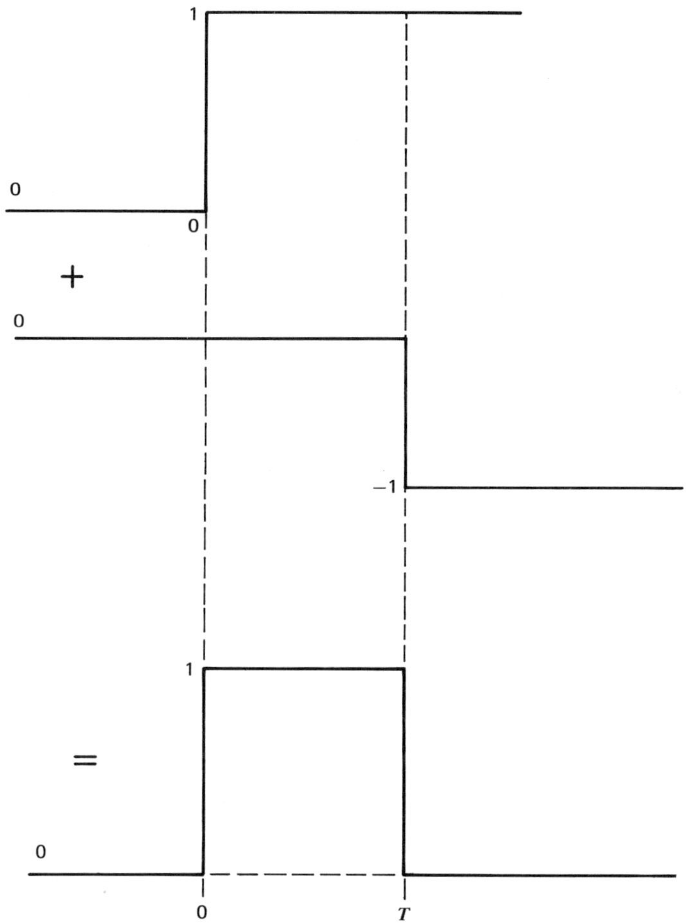

Fig. 4.6 A unit pulse as a combination of two step functions.

function $-u(t-T)$ adds to it the value -1 for all $t \geqslant T$. Thus

$$w_1(t) = u(t) - u(t-T) \tag{4.45}$$

Similarly a unit pulse with value unity in the interval from T_1 to T_2 and value zero elsewhere is given by

$$w_2(t) = u(t-T_1) - u(t-T_2) \tag{4.46}$$

where $0 \leqslant T_1 \leqslant T_2$. This pulse is shown in Fig. 4.7.

Now consider the response of a circuit to the pulse excitations $w_1(t)$ or $w_2(t)$. Let $r_0(t)$ be the step response of the circuit. Then by properties 2

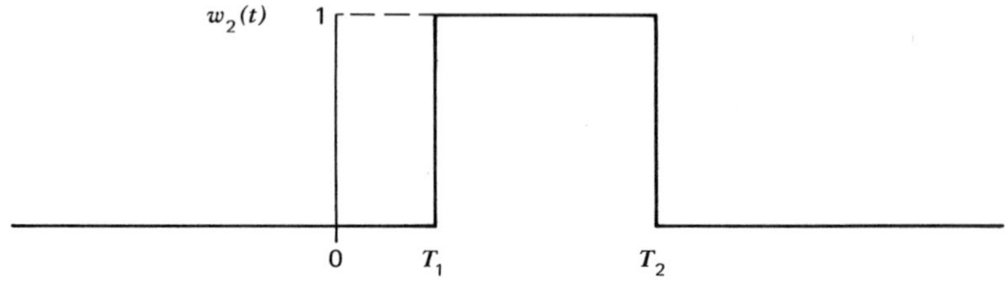

Fig. 4.7

and 3 of Table 4.1, the response of the circuit to $w_1(t)$ is given by

$$g_1(t) = r_0(t) - r_0(t-T) \tag{4.47}$$

Similarly, the response of the circuit to $w_2(t)$ is given by

$$g_2(t) = r_0(t-T_1) - r_0(t-T_2) \tag{4.48}$$

Also, the response to an excitation $c_1w_1(t)$ where c_1 is a constant is given by $c_1g_1(t)$, and so on.

For the next example, consider the piecewise-constant function with three discontinuities as shown in Fig. 4.8. The function is decomposed into a step at $t = T_0$ of height K_1 to which is added a step at $t = T_1$ of height $(K_2 - K_1)$, and finally a negative step of height K_2 is added on at $t = T_2$. That is,

$$w_3(t) = K_1u(t-T_0) + (K_2-K_1)u(t-T_1) - K_2u(t-T_2) \tag{4.49}$$

Alternatively, the function may be viewed as a linear combination of unit pulses rather than unit steps. For example, the function $w_3(t)$ in Fig. 4.8 is a pulse of height K_1 in the interval from T_0 to T_1 to which is added a pulse of height K_2 in the interval from T_1 to T_2. The terms in Eq. 4.49

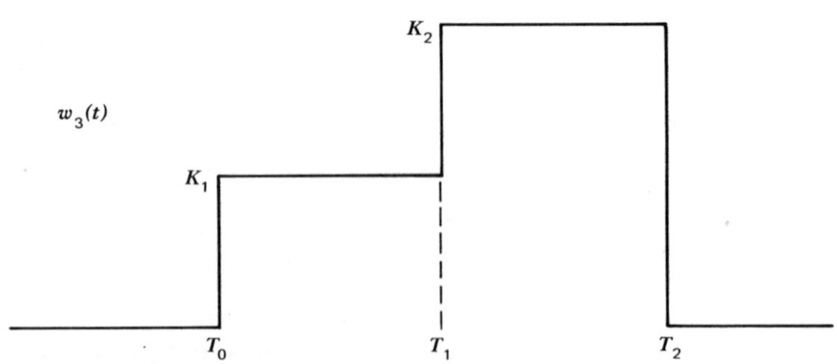

Fig. 4.8 Double step pulse.

can be rearranged to reflect this interpretation:

$$w_3(t) = K_1[u(t-T_0) - u(t-T_1)] + K_2[u(t-T_1) - u(t-T_2)] \tag{4.50}$$

For a circuit with step response $r_0(t)$, the response to $w_3(t)$ is given by

$$g_3(t) = K_1 r_0(t-T_0) + (K_2 - K_1) r_0(t-T_1) - K_2 r_0(t-T_2) \tag{4.51}$$

using Eq. 4.49. If we group terms as in Eq. 4.50, the response may also be written

$$g_3(t) = K_1[r_0(t-T_0) - r_0(t-T_1)] + K_2[r_0(t-T_1) - r_0(t-T_2)] \tag{4.52}$$

Figures 4.9 and 4.10 show the two interpretations of $w_3(t)$ in Eqs. 4.49 and 4.50, respectively.

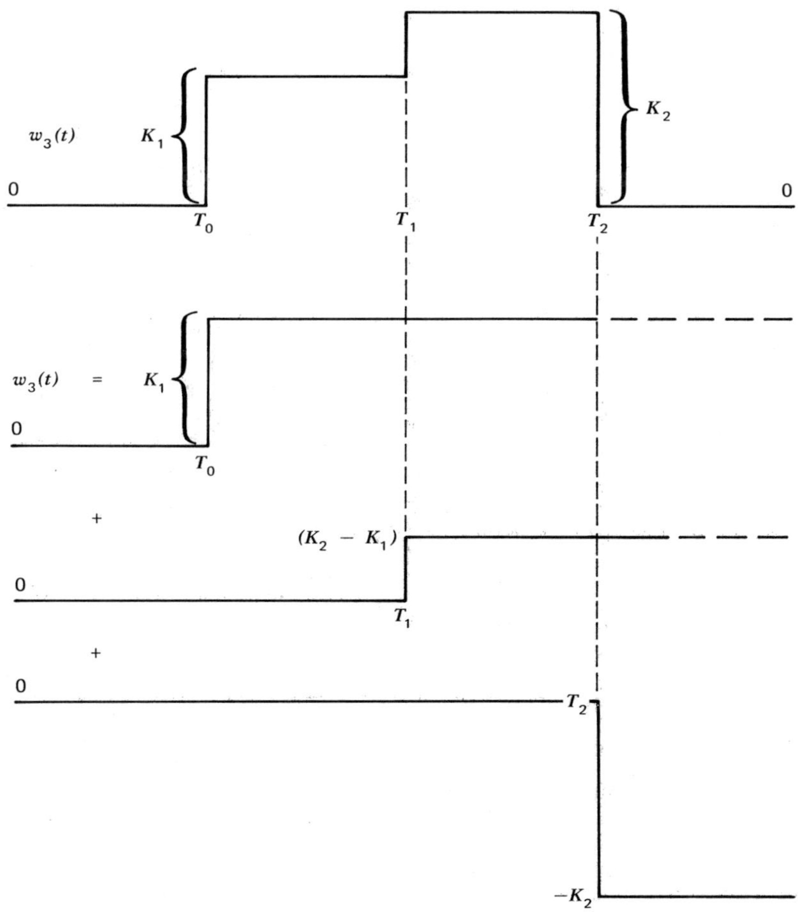

Fig. 4.9 $W_3(t)$ decomposed into step functions.

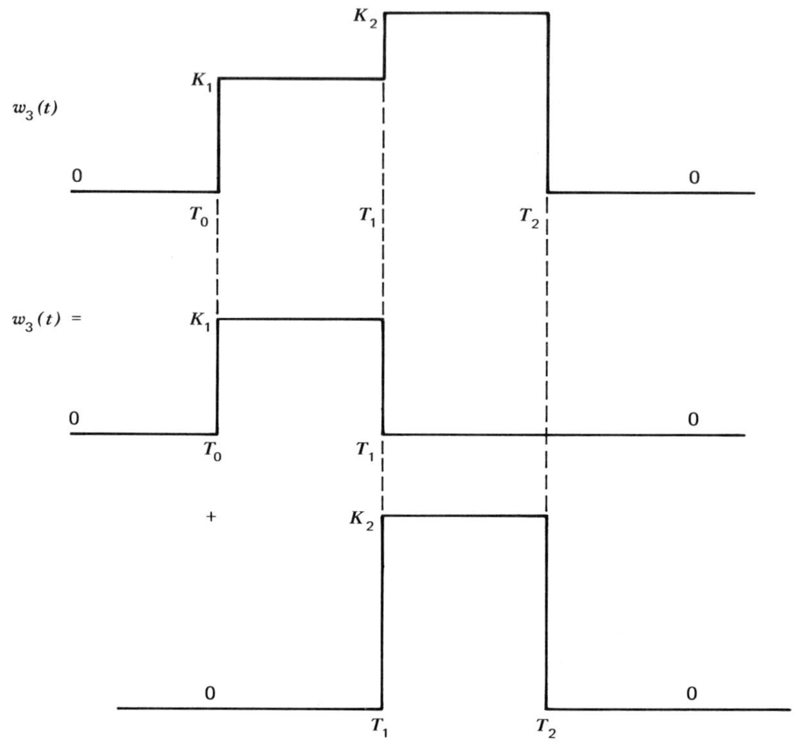

Fig. 4.10 $w_3(t)$ decomposed into pulses.

Example 4.4.1 Consider the *RC* circuit in Fig. 4.11 excited by a current source $i(t)$, where $i(t)$ takes on the values of the functions $w_1(t)$, $w_2(t)$, and $w_3(t)$ in Eqs. 4.45, 4.46, and 4.50. The response $v(t)$ is to be determined in each case. The step voltage response of the circuit has already been developed in Sec. 4.3 and is given by

$$v_u(t) = r_0(t) = (1 - e^{-t/RC})u(t) \tag{4.53}$$

where we have assumed $R = 1\Omega$. The response is plotted in Fig. 4.12. The

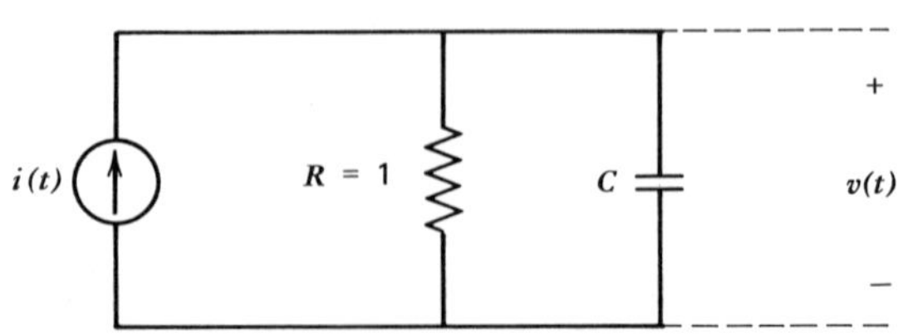

Fig. 4.11

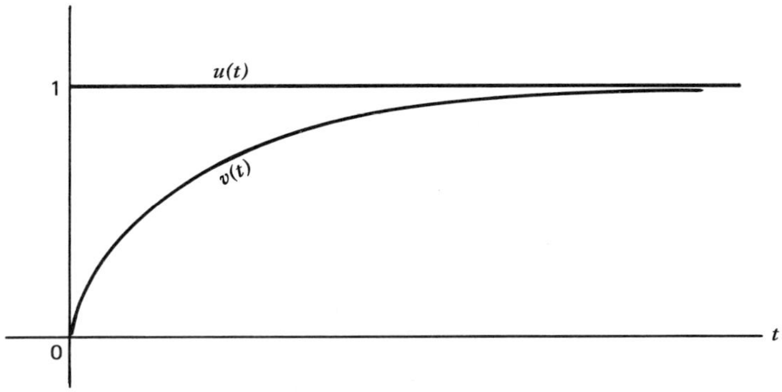

Fig. 4.12 Step voltage response.

response of the circuit to $w_1(t)$ is given by Eq. 4.47:

$$\begin{aligned} g_1(t) &= r_0(t) - r_0(t-T) \\ &= [1-e^{-t/RC}]u(t) - [1-e^{-(t-T)/RC}]u(t-T) \end{aligned} \tag{4.54}$$

Similarly, the response to $w_2(t)$ is

$$\begin{aligned} g_2(t) &= r_0(t-T_1) - r_0(t-T_2) \\ &= [1-e^{-(t-T_1)/RC}]u(t-T_1) - [1-e^{-(t-T_2)/RC}]u(t-T_2) \end{aligned} \tag{4.55}$$

and the response to $w_3(t)$ is by Eq. 4.52:

$$\begin{aligned} g_3(t) = {} & K_1[(1-e^{-(t-T_0)/RC})u(t-T_0) - (1-e^{-(t-T_1)/RC})u(t-T_1)] \\ & + K_2[(1-e^{-(t-T_1)/RC})u(t-T_1) - (1-e^{-(t-T_2)/RC})u(t-T_2)] \end{aligned} \tag{4.56}$$

The responses in Eqs. 4.54–4.56 are plotted in Figs. 4.13–4.15. The response $g_3(t)$ may, of course, also be expressed as in Eq. 4.51 by step

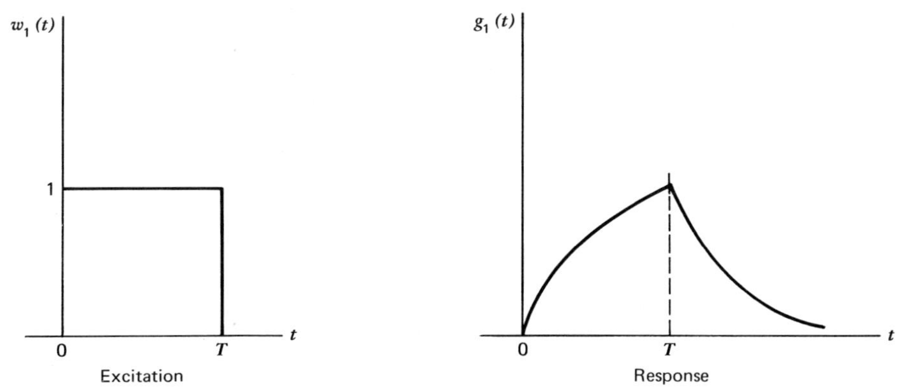

Fig. 4.13 Response to a unit pulse.

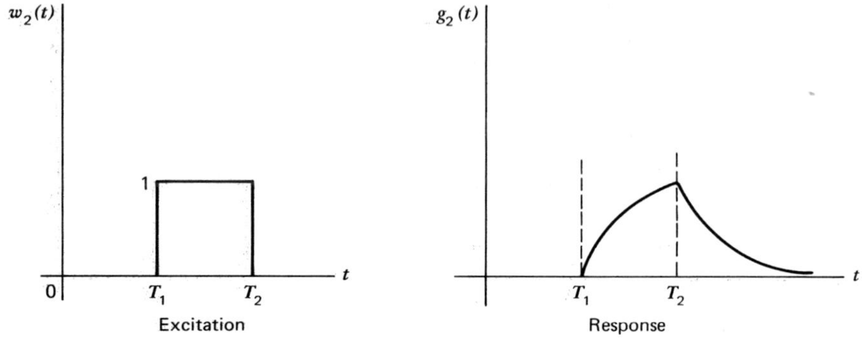

Fig. 4.14 Response to $w_2(t)$.

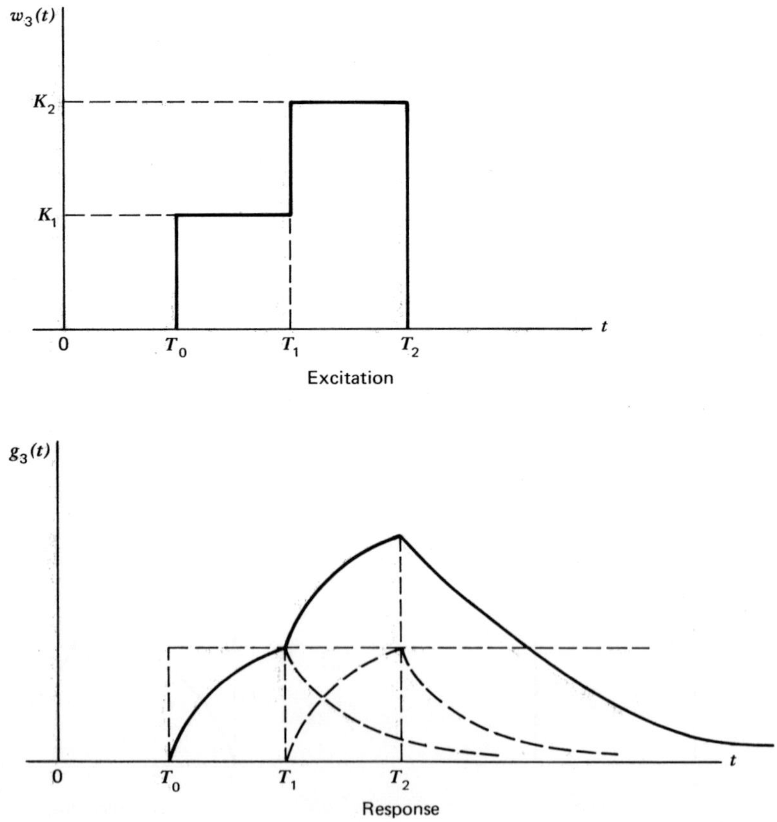

Fig. 4.15 Response to Excitation $w_3(t)$.

responses instead of grouping terms to indicate pulse responses. The pulse approach is more convenient since the excitation is expressed in terms of finite duration pulses. This concept will generalize our discussion of piecewise-constant excitation functions.

Let us reexamine the function $w_3(t)$ expressed in Eq. 4.50. From the definition of the unit-step function, it is clear that

$$w_3(T_0) = K_1$$

and

$$w_3(T_1) = K_2 \tag{4.57}$$

The same can be observed from Eq. 4.49. Then Eq. 4.50 may be written using Eq. 4.57 as

$$w_3(t) = w_3(T_0)[u(t-T_0)-u(t-T_1)]+w_3(T_1)[u(t-T_1)-u(t-T_2)]. \tag{4.58}$$

Similarly, the response $g_3(t)$ in Eq. 4.52 is written as

$$g_3(t) = w_3(T_0)[r_0(t-T_0)-r_0(t-T_1)]+w_3(T_1)[r_0(t-T_1)-r_0(t-T_2)]. \tag{4.59}$$

From Eq. 4.58, we observe that any *piecewise-constant* (p.c.) function of finite duration can be expressed as a finite linear combination of unit pulses with coefficients given by the values of the p.c. function at the start of each pulse. In general, then, a p.c. function of finite duration is expressed as

$$e(t) = e(T_0)[u(t-T_0)-u(t-T_1)]+e(T_1)[u(t-T_1)-u(t-T_2)]$$
$$+e(T_2)[u(t-T_2)-u(t-T_3)]+\cdots+e(T_N)[u(t-T_n)-u(t-T_{N+1})]$$

or

$$e(t) = \sum_{\nu=0}^{N} e(T_\nu)[u(t-T_\nu)-u(t-T_{\nu+1})] \tag{4.60}$$

The response of any circuit to $e(t)$ is then given by

$$r(t) = \sum_{\nu=0}^{N} e(T_\nu)[r_0(t-T_\nu)-r_0(t-T_{\nu+1})] \tag{4.61}$$

where $r_0(t)$ is the step response of the circuit.

If the p.c. function is of infinite duration, then it can be represented as a step function plus a finite combination of unit pulses. The step function provides the required infinite duration. Of course, a unit-step function may be interpreted as an infinite combination (sum) of unit pulses. Therefore, any p.c. excitation function is a linear combination of unit pulses as expressed in Eq. 4.60 with the response given by Eq. 4.61. The only difference is that N is no longer finite.

As we have thus seen, the response of any circuit to an arbitrary piecewise constant excitation is obtained as a linear combination of its response to the unit step or, equivalently, to the unit pulse. The coefficients in the linear combination are given by the values of the p.c. excitation at the starting instant of each pulse. The concept developed here is the piecewise-constant version of *convolution*. This concept forms the basis of our discussion of convolution in the next section.

4.5 CONVOLUTION

The idea of developing the response of a circuit (to a piecewise-constant excitation) in terms of the responses to unit steps or pulses appropriately translated in time may be generalized with little effort to arbitrary excitations. Any excitation, which may not be piecewise constant, may be *approximated* by a piecewise-constant function. An *approximate response* is then obtained for the *p.c. approximate excitation*. An arbitrary excitation with the approximating p.c. function is shown in Fig. 4.16. The steps or discontinuities occur at equally spaced intervals of width Δ, that is, $T_\nu = \nu\Delta$, where $\nu = 0, 1, 2, \ldots$. The smaller the value of Δ, the better the approximation. In the limit as $\Delta \to 0$, the p.c. function approaches the arbitrary excitation. Similarly, the approximate response approaches the exact response in the limit as Δ approaches zero. This is the concept of convolution developed below.

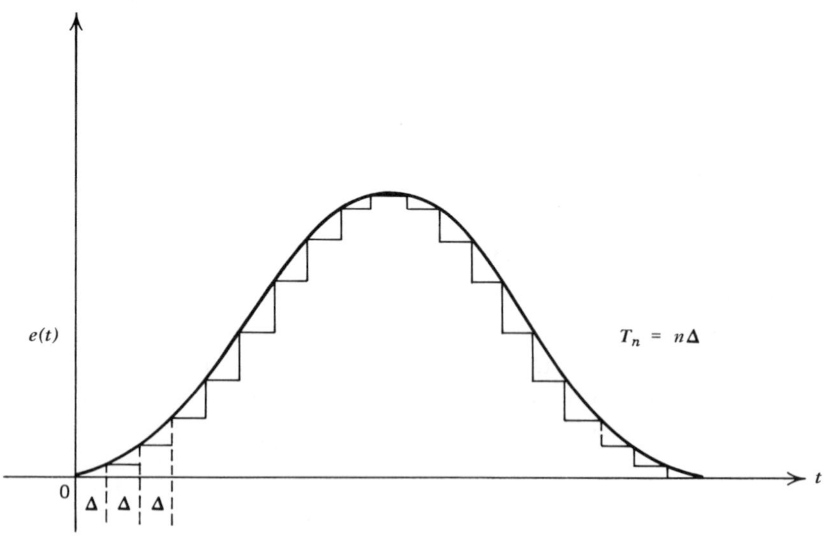

Fig. 4.16 An arbitrary function.

The excitation $e(t)$ in Fig. 4.16 may be expressed in an approximate manner by a p.c. function of the form in Eq. 4.60:

$$e(t) \doteq \sum_{\nu=0}^{[t/\Delta]} e(\nu\Delta)[u(t-\nu\Delta) - u(t-\nu\Delta-\Delta)] \tag{4.62}$$

where $T_\nu = \nu\Delta$. Since $u(t-\nu\Delta)$ is zero for $(t-\nu\Delta) < 0$, only the nonzero terms need be included in the summation. Of course, a good approximation is obtained even if we include a term such as $u(t-\nu\Delta)$ when $(t-\nu\Delta) < 0$, provided $t \doteq \nu\Delta$, that is, the integer nearest to t/Δ is ν. Therefore, the upper limit on the summation is specified by $[t/\Delta]$, the nearest integer to t/Δ. The expression in Eq. 4.62 approximates the value of $e(t)$ at any instant of time t.

The response $r(t)$ is approximated in a similar manner by the response to the approximate p.c. excitation of Eq. 4.62. From Eq. 4.61, the approximate response is

$$r(t) \doteq \sum_{\nu=0}^{[t/\Delta]} e(\nu\Delta)[r_0(t-\nu\Delta) - r_0(t-\nu\Delta-\Delta)] \tag{4.63}$$

where the upper limit of the summation is again given by $[t/\Delta]$ due to similar reasoning.

The approximate excitation and response of Eqs. 4.62 and 4.63 approach the actual excitation and response in the limit as Δ approaches zero. As $\Delta \to 0$, the variable $\nu\Delta$ becomes a continuous variable of integration, say $\nu\Delta = \tau$, with the integration from $\nu\Delta = 0$ to $\nu\Delta = t$. The infinitesimal Δ becomes the differential element, $\Delta = d\tau$. The quantity in the brackets in Eq. 4.62 is

$$[u(t-n\Delta) - u(t-n\Delta-\Delta)] = \left[\frac{u(t-n\Delta) - u(t-n\Delta-\Delta)}{\Delta}\right]\Delta \tag{4.64}$$

which in the limit as $\Delta \to 0$ becomes

$$\lim_{d\tau\to 0}\left[\frac{u(t-\tau) - u(t-\tau-d\tau)}{d\tau}\right]d\tau = \left[\frac{d}{d\tau}u(t-\tau)\right]d\tau \tag{4.65}$$

Similarly, the quantity in the brackets in Eq. 4.63 becomes in the limit as $\Delta \to 0$

$$[r_0(t-n\Delta) - r_0(t-n\Delta-\Delta)] = \left[\frac{d}{d\tau}r_0(t-\tau)\right]d\tau \tag{4.66}$$

Upon substitution of Eqs. 4.65 and 4.66 into Eqs. 4.62 and 4.63 in the limit as $\Delta \to 0$, and replacing $\nu\Delta$ by τ, Δ by $d\tau$, we have

$$e(t) = \int_0^t e(\tau)u'(t-\tau)d\tau \tag{4.67}$$

and,

$$r(t) = \int_0^t e(\tau) r_0'(t-\tau)\, d\tau \tag{4.68}$$

The above expression in Eq. 4.68 is the continuous version of the convolution that expresses the response to an arbitrary excitation $e(t)$ in terms of the derivative of the step response. The coefficient $e(\tau)$ is the value of the excitation $e(t)$ observed, approximated, or sampled by the infinitesimal pulse $u'(t-\tau)d\tau$ located at $t=\tau$ as in Eq. 4.67. The derivative of the step function $u(t)$ is the impulse function $\delta(t)$, and the derivative of the step response $r_0(t)$ is the impulse response $r_0'(t)$. The impulse response $r_0'(t)$ is commonly denoted by $h(t)$. The Eqs. 4.67 and 4.68 are thus written as

$$e(t) = \int_0^t \delta(t-\tau) e(\tau)\, d\tau \tag{4.69}$$

and

$$r(t) = \int_0^t h(t-\tau) e(\tau)\, d\tau \tag{4.70}$$

The first integral expression in Eq. 4.69 represents the sampling property of the impulse function. The impulse at $\tau = t$ yields the value of the function $e(\tau)$ at $\tau = t$. The second integral expression in Eq. 4.70 is the *convolution integral* that yields the response to $e(t)$ in terms of the impulse response of the system or the circuit.

The convolution integral process is also referred to as the convolution of two functions $e(t)$ and $h(t)$. This usage is a consequence of the fact that the convolution result is unchanged if the impulse response and the excitation were interchanged. To show this, let $\xi = t-\tau$. Then $d\xi = -d\tau$, and Eq. 4.70 becomes

$$\begin{aligned} r(t) &= \int_0^t h(t-\tau) e(\tau)\, d\tau \\ &= -\int_t^0 h(\xi) e(t-\xi)\, d\xi \\ &= \int_0^t e(t-\xi) h(\xi)\, d\xi \end{aligned} \tag{4.71}$$

Since ξ and τ are merely the variables of integration we may write Eq. 4.71 as

$$r(t) = \int_0^t h(t-\tau) e(\tau)\, d\tau = \int_0^t h(\tau) e(t-\tau)\, d\tau \tag{4.72}$$

that is,

$$r = h * e = e * h \tag{4.73}$$

where $*$ represents *convolved with.*

The limits in the convolution integral are a consequence of our particular situation. The excitation is assumed to be zero for negative time, and so the lower limit starts at $\tau = 0$. The impulse response for a zero-state circuit is also zero for negative time. Therefore, the upper limit must not exceed $\tau = t$. If these restrictions were not explicitly included in our derivation, the convolution integral would have taken its most general form

$$r(t) = \int_{-\infty}^{\infty} h(t-\tau)e(\tau)d\tau = \int_{-\infty}^{\infty} e(t-\tau)h(\tau)d\tau \tag{4.74}$$

which, of course, reduces to Eq. 4.72 for the case of $e(t) = h(t) = 0$ for $t < 0$.

The convolution integral is a powerful tool in analysis of linear, time-invariant systems since it provides the response to any excitation once the impulse response is known. Consequently, the impulse response of a system completely describes the system behavior through the convolution integral.

Example 4.5.1 Convolve $e^{-t}u(t)$ and 1.

Solution. In this example are two functions, one of which is zero for negative time. The other function is a constant for *all* time. We include this example to show the use of the general convolution integral of Eq. 4.74. First, let us try

$$\begin{aligned} r(t) &= \int_{-\infty}^{\infty} e^{-(t-\tau)}u(t-\tau)d\tau \\ &= \int_{-\infty}^{t} e^{-(t-\tau)}d\tau \\ &= e^{-t}[e^{\tau}]_{-\infty}^{t} \\ &= e^{-t}[e^{t}] = 1 \end{aligned}$$

Next, let us try

$$\begin{aligned} r(t) &= \int_{-\infty}^{\infty} e^{-\tau}u(\tau)d\tau \\ &= \int_{0}^{\infty} e^{-\tau}d\tau = 1 \end{aligned}$$

The above result is a purely mathematical one and no physical conclusions should be drawn from it. It is developed here to demonstrate the use of the general convolution integral. The general integral allows one to introduce appropriate limits without fixing them as in Eq. 4.72 where we specifically assume $h(t)$ and $e(t)$ zero for negative time.

Example 4.5.2 Convolve $e^{-t}u(t)$ and $u(t)$.

Solution. In this case, we have

$$r(t) = \int_{-\infty}^{\infty} e^{-\tau}u(\tau)u(t-\tau)d\tau$$

$$= \int_{0}^{t} e^{-\tau}d\tau \qquad t \geqslant 0$$

$$= (1-e^{-t})u(t)$$

Example 4.5.3 Convolve $e^{-t}u(t)$ and $tu(t)$.

Solution. Both functions are zero for negative time, and hence the convolution integral is

$$r(t) = u(t)\int_{0}^{t} e^{-(t-\tau)}\tau\, d\tau$$

We have not included the step function in the integrand since the limits imply the functions are zero for negative arguments. We have multiplied the integral by $u(t)$ to indicate that for negative time the $r(t)$ is zero. Then

$$\int_{0}^{t} e^{-(t-\tau)}\tau d\tau = e^{-t}[e^{\tau}\tau - e^{\tau}]_{0}^{t}$$

$$= e^{-t}[e^{t}t - e^{t} + 1]$$

$$= t - 1 + e^{-t}$$

Or,

$$r(t) = (t + e^{-t} - 1)u(t)$$

Example 4.5.4 Find the voltage response of the series RC circuit in Fig. 4.17 when excited by the unit pulse, $e(t) = [u(t) - u(t-T)]$.

The step voltage response of the circuit is easily derived as

$$r_0(t) = (1 - e^{-t/RC})u(t)$$

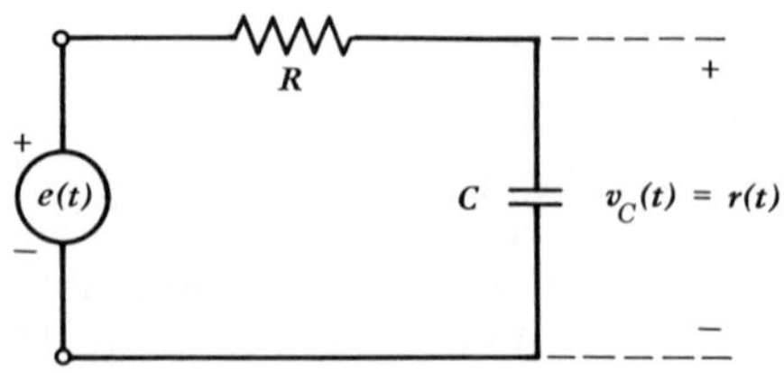

Fig. 4.17 Series RC circuit.

and the impulse voltage response is

$$h(t) = r_0'(t) = \frac{1}{RC} e^{-t/RC} u(t)$$

Then by convolution (Eq. 4.72)

$$\begin{aligned} r(t) &= \int_0^t \frac{1}{RC} e^{-(t-\tau)/RC} u(t-\tau)[u(\tau) - u(\tau - T)]d\tau \\ &= u(t) \int_0^t \frac{1}{RC} e^{-(t-\tau)/RC} d\tau - u(t-T) \int_T^t \frac{1}{RC} e^{-(t-\tau)/RC} d\tau \\ &= u(t)[e^{-(t-\tau)/RC}]_{\tau=0}^{t} - u(t-T)[e^{-(t-\tau)/RC}]_{\tau=T}^{t} \\ &= [1 - e^{-t/RC}]u(t) - [1 - e^{-(t-T)/RC}]u(t-T) \end{aligned}$$

In this case, of course, the response could have been obtained for the p.c. excitation by Eq. 4.54.

$$r(t) = [r_0(t) - r_0(t-T)]$$

which is the same as above.

An intuitive understanding of convolution interprets the excitation function as sampled by impulses at certain intervals, that is, impulse at

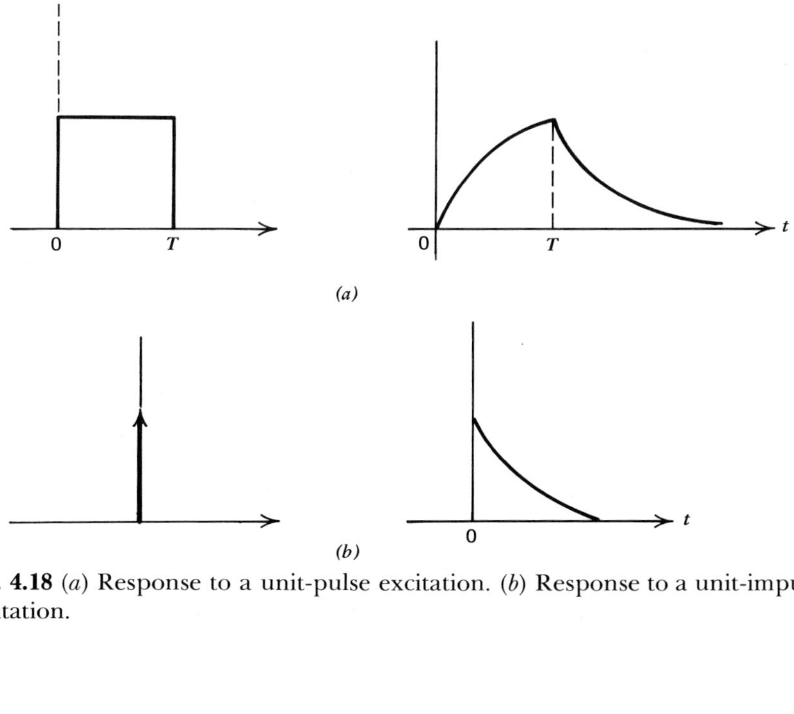

Fig. 4.18 (*a*) Response to a unit-pulse excitation. (*b*) Response to a unit-impulse excitation.

$t = \tau$ is weighted by the value $e(\tau)$ and the sampling interval $d\tau$. Then the excitation is an integration or a sum of the impulses $\delta(t-\tau)e(\tau)d\tau$ over the interval from $\tau = 0$ to t. The response to each weighted impulse is the impulse response times the weight, that is, $h(t-\tau)e(\tau)d\tau$. The response is then obtained by a summation or integration of these weighted responses. The finer the sampling (i.e., more frequent sampling) of the excitation, the more accurate is the response. In the limit as the sampling interval approaches zero, the actual response is obtained.

An example illustrates the intuitive concepts. Consider the circuit in Fig. 4.17 excited by the unit pulse. The excitation and the response are shown in Fig. 4.18*a* and the impulse response is shown in Fig. 4.18*b*. Fig. 4.19 shows the unit pulse sampled at different frequencies and the corresponding responses. When the unit pulse is sampled crudely at only three points (Fig. 4.19*a*), the response is correspondingly a rough approximation. When the unit pulse is sampled at frequent intervals (Fig. 4.19*c*), the response is a pretty good approximation and approaches the actual response in the limit.

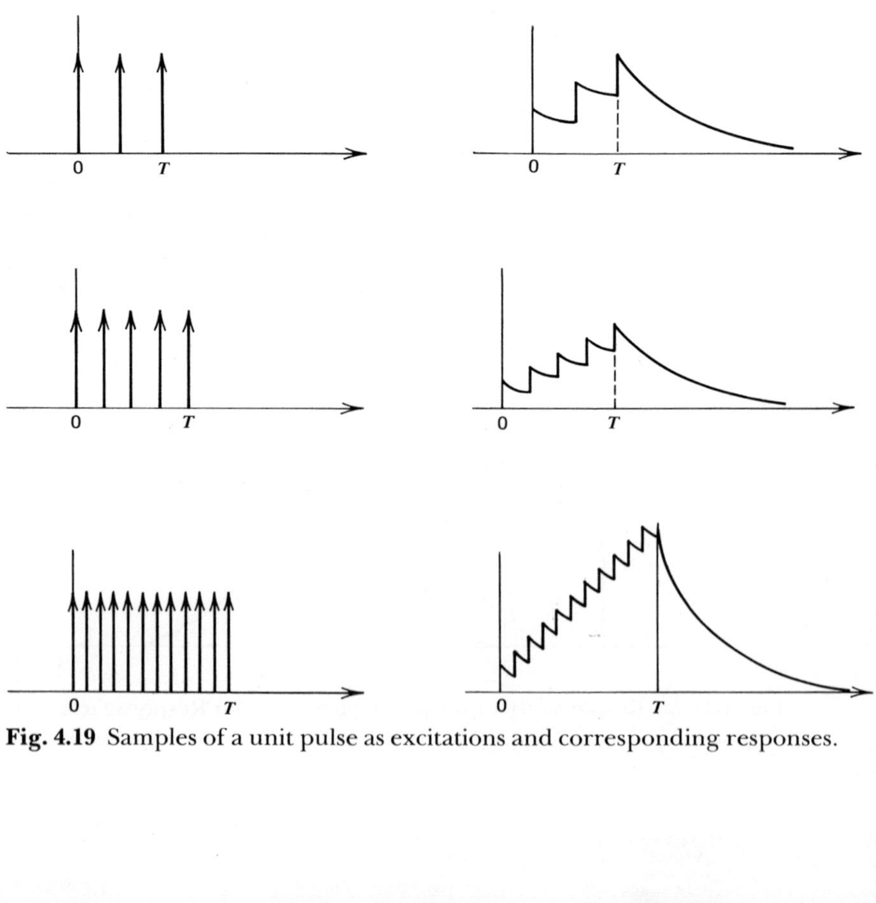

Fig. 4.19 Samples of a unit pulse as excitations and corresponding responses.

4.6 GRAPHICAL INTERPRETATION OF CONVOLUTION

The convolution integral gives the value of the function $r(t)$ at the desired instant of time t. An integral is an area under the curve represented by the integrand between certain limits. The convolution integral can also be interpreted as an area under a curve between the desired limits. Consider

$$r(t) = \int_0^t h(t-\tau)e(\tau)d\tau$$

where we are assuming $e(t)$ and $h(t)$ zero for $t < 0$. The value of the integral is evaluated at the desired instant of time to give $r(t)$ at that instant. The value of the integral at any t is the area under the curve $h(t-\tau)e(\tau)$ between the limits $\tau = 0$ and $\tau = t$. We now examine this curve and the area under it.

Let $\phi(\tau) = h(t-\tau)e(\tau)$ be the curve as a function of τ for any given t. Then we must develop $h(t-\tau)$, multiply it by $e(\tau)$, and find the area under the resulting product curve $\phi(\tau)$ between the limits *0* and t. The functions $e(t)$ and $h(t)$ are given, and the process is straightforward as described below for an illustrative example.

1. The functions $e(\cdot)$ and $h(\cdot)$ are given and shown in Figs. 4.20 and 4.21 as functions of τ.

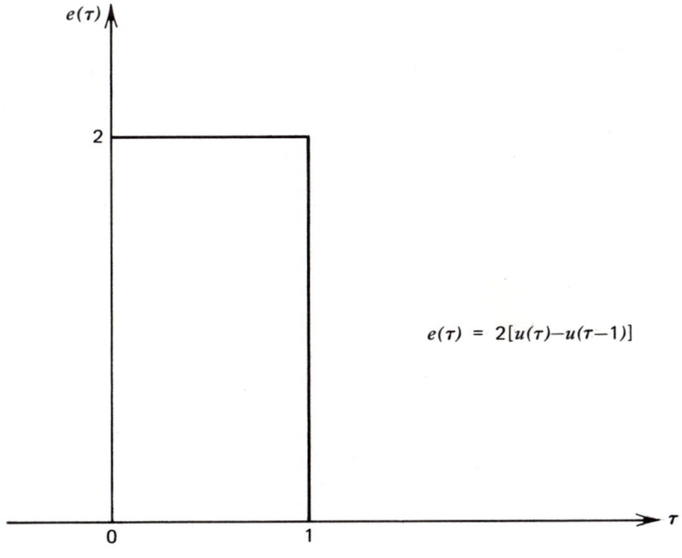

Fig. 4.20

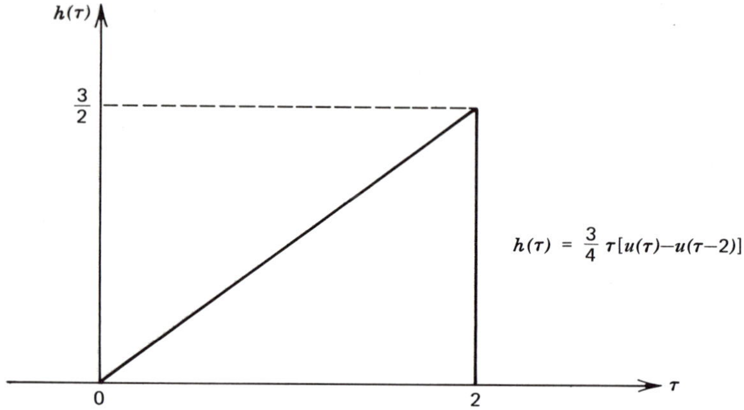

Fig. 4.21

2. Obtain $h(-\tau)$ from $h(\tau)$ by replacing τ by $-\tau$, that is, by *flipping* the curve $h(\tau)$ about the vertical axis. This is shown in Fig. 4.22.
3. Shift the curve $h(-\tau)$ to the right by an amount $\tau = t$ secs. A translation or delay of t seconds amounts to replacing τ by $(\tau - t)$. The resulting function is described by $h(-(\tau - t)) = h(t - \tau)$. The process is shown in Fig. 4.23.
4. Multiply $h(t-\tau)$ by $e(\tau)$ to obtain $\phi(\tau) = h(t-\tau)e(\tau)$. Fig. 4.24 shows this for $1 \leqslant t \leqslant 2$.
5. Determine the area under $\phi(\tau)$ between *0* and t. This value of the area is the value of $r(t)$ at t. Fig. 4.25 shows this for $1 \leqslant t \leqslant 2$. In this case $r(t) = (3t/2 - 3/4)$ for $1 \leqslant t \leqslant 2$.

The procedure described above must be followed for all values of t to obtain $r(t)$ as a function of time. The figures show the result only for

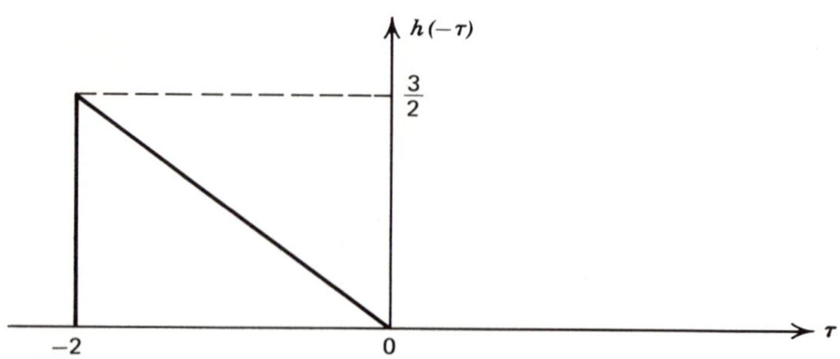

Fig. 4.22

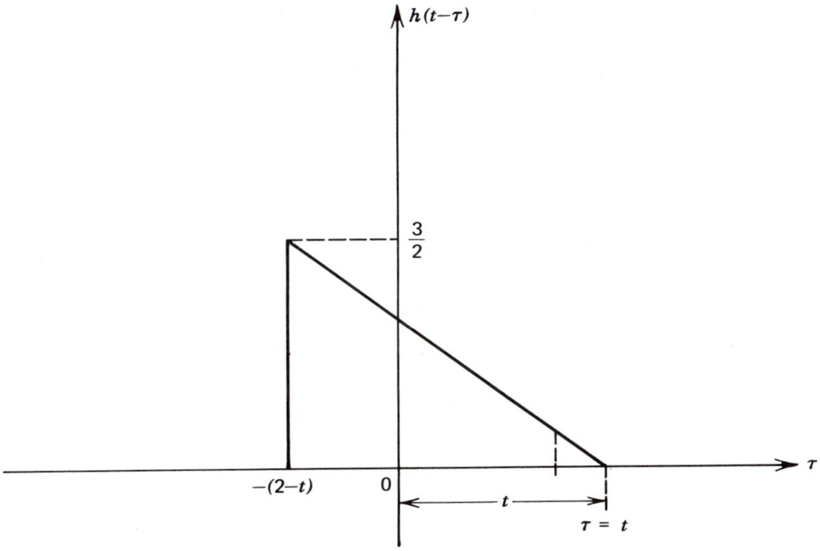

Fig. 4.23

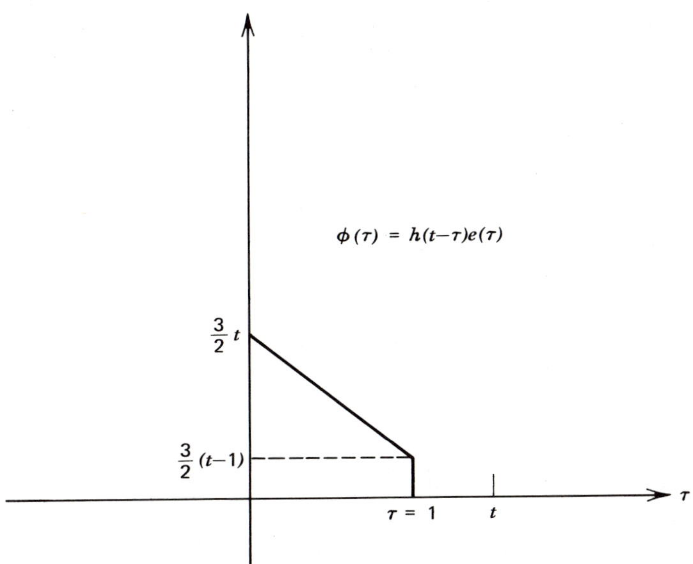

Fig. 4.24

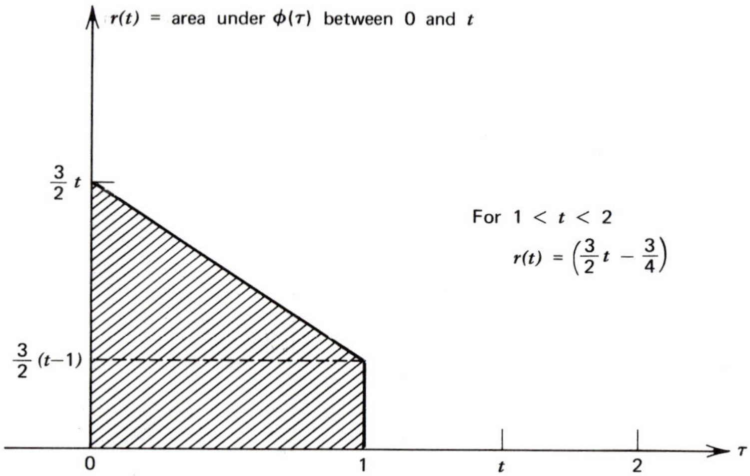

Fig. 4.25

$1 \leqslant t \leqslant 2$. From Figs. 4.20 and 4.22, it should be observed that $r(t) = 0$ for $t < 0$ and $t > 3$ since $\phi(\tau) = 0$ for these ranges of t. Thus, to complete our discussion of the example we need to determine $r(t)$ for $0 \leqslant t \leqslant 1$ and $2 \leqslant t \leqslant 3$.

For $0 \leqslant t \leqslant 1$, the curve $\phi(\tau)$ is shown in Fig. 4.26. The area under $\phi(\tau)$ between *0* and t is given by $r(t) = 3t^2/4$ for $0 \leqslant t \leqslant 1$. Finally, the curve $\phi(\tau)$ for $2 \leqslant t \leqslant 3$ is shown in Fig. 4.27, and $r(t) = 3-3(t-1)^2/4$ for this range.

The response function is now determined for all possible values of t

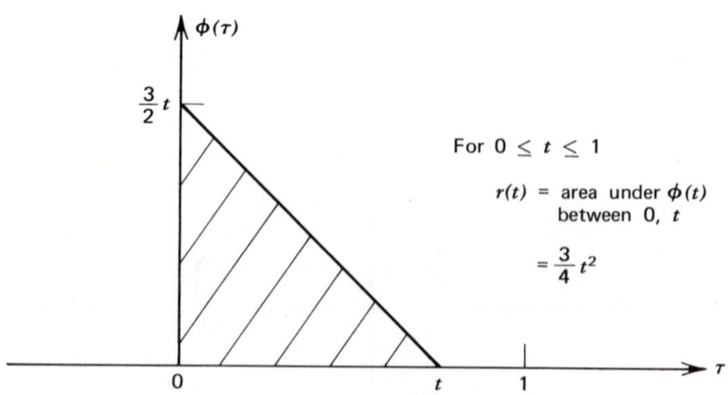

Fig. 4.26

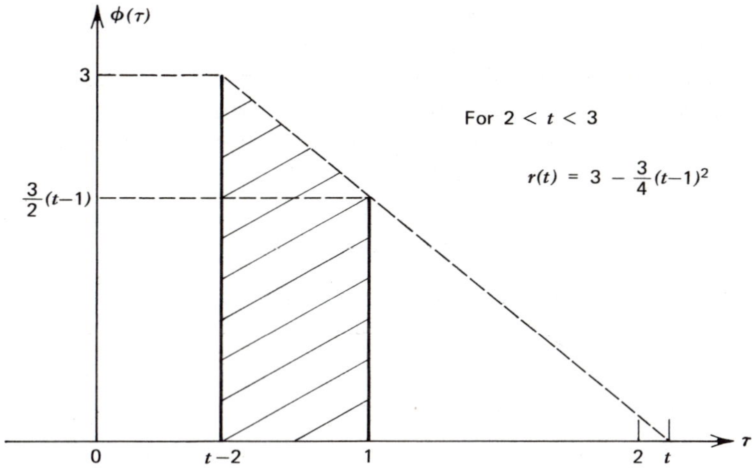

Fig. 4.27

and given by

$$
\begin{aligned}
r(t) &= 0 \qquad t < 0 \\
&= \tfrac{3}{4}t^2 \qquad 0 \leqslant t \leqslant 1 \\
&= \tfrac{3}{2}t - \tfrac{3}{4} \qquad 1 \leqslant t \leqslant 2 \\
&= 3 - \tfrac{3}{4}(t-1)^2 \qquad 2 \leqslant t \leqslant 3 \\
&= 0 \qquad t > 3
\end{aligned}
$$

The response function is plotted in Fig. 4.28.

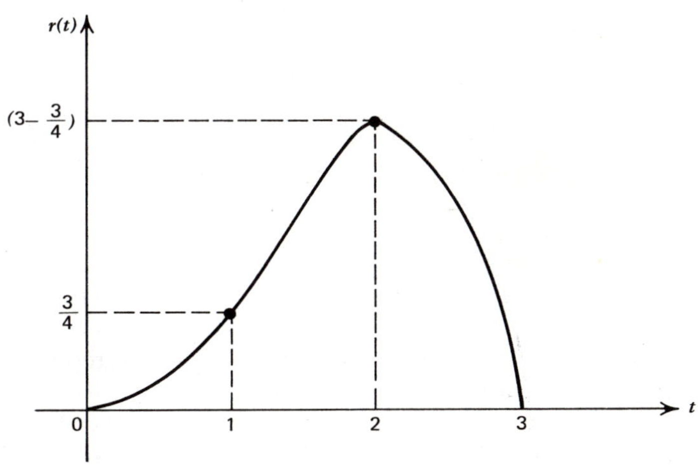

Fig. 4.28

The graphical procedure of this section is an aid in getting a rough picture of the response function or convolution integral. The results are simple and algebraic only for very simple functions, as in our example. Even the integration procedure (as distinct from the graphical procedure) for complex functions becomes intractable analytically. Numerical integration techniques must then be used for convolution, as discussed in the next section.

4.7 NUMERICAL INTEGRATION METHODS

We have seen that given the excitation $e(t)$ and the impulse response (derivative of the step response) $h(t)$ of a system, the response of the system is given by the convolution of $e(t)$ and $h(t)$:

$$r(t) = \int_0^t e(\tau)h(t-\tau)d\tau \tag{4.75}$$

The response at a certain time t or $r(t)$ is equal to the integration of $e(\tau)h(t-\tau)$ from 0 to t.

If e and h are given in very simple analytical forms, then the evaluation of Eq. 4.75 may not be very difficult. However, it is generally difficult to find an explicit analytical form for $r(t)$. Furthermore, if e and h are given as experimental or numerical data sets, then it is not possible at all to find an analytical form for $r(t)$. In these circumstances we can only satisfy ourselves by finding an approximate value for the integral $r(t)$ in Eq. 4.75.

This leads to the general integration problem of finding the numerical value of

$$I = \int_a^b k(x)dx \tag{4.76}$$

where a, b are some constants. The function $k(x)$ is assumed known or given at a set of $n+1$ *equally spaced* points in the interval $[a,b]$, or

$$x_i = a + ih_0 \qquad i = 0, 1, 2, \cdots, n \tag{4.77}$$

where $h_0 = (b-a)/n$ is the step size.

The approach we will take is essentially similar to the multistep method used in Chapter 3 to solve a first order differential equation: namely, an approximation function $\phi(x)$ that is easily integrated replaces the actual function $k(x)$ within the interval $[a,b]$ and then the integration in Eq. 4.76 is carried out to get an approximation for the desired I. However, since the function $k(x)$ is known for the region $[a,b]$ whereas the derivative function $f(t,r)$ in Sec. 3.6 was unknown for the new interval t_m to t_{m+1}, it is easier to solve the integration problem in Eq. 4.76 than to solve the differential equation problem.

Rectangular Rule

For simplicity, let us first restrict ourselves to the problem of finding the integral of $k(x)$ over the *single* interval $[x_0, x_1]$ of width h_0. Since $k(x_0)$ and $k(x_1)$ are known, a simple function may be used to approximate $k(x)$ over the interval $[x_0, x_1]$. Let us choose the simplest possible function $\phi = k(x_0) = k_0$ to approximate $k(x)$ in the interval $[x_0, x_1]$. Then we have,

$$\int_{x_0}^{x_1} k(x)dx \approx h_0k_0 \tag{4.78}$$

This is pictured in Fig. 4.29 as the area of the inscribed rectangle of height k_0 and width h_0.

The approximate integration process is called the *rectangular rule*. The error of this approximation is

$$\begin{aligned}\epsilon &= \int_{x_0}^{x_1} k(x)dx - \int_{x_0}^{x_1} k_0dx \\ &= \int_{x_0}^{x_1} [k(x) - k_0]dx \end{aligned} \tag{4.79}$$

Using Taylor's expansion for $k(x)$ with remainder, we have

$$k(x) = k_0 + \frac{x - x_0}{1} k'(\xi) \qquad x_0 \leqslant \xi \leqslant x$$

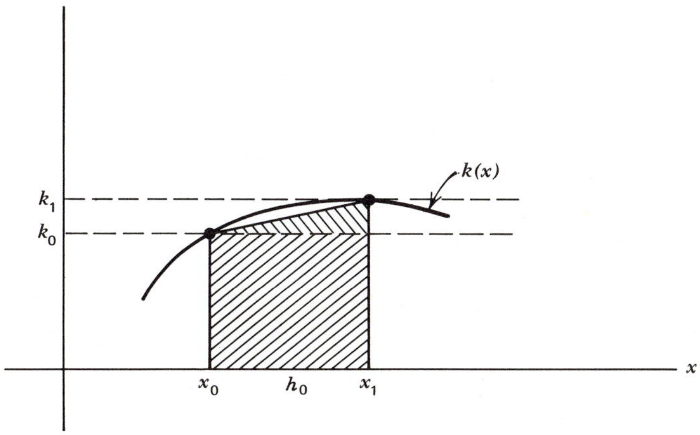

Fig. 4.29 Rectangular and trapezoidal rules.

The error becomes

$$\begin{aligned}\epsilon &= k'(\xi)\int_{x_0}^{x_1}(x-x_0)dx \\ &= k'(\xi)\frac{{h_0}^2}{2} \qquad x_0 \leqslant \xi \leqslant x_1 \qquad (4.80)\end{aligned}$$

From Eq. 4.80 the error of the rectangular rule is proportional to ${h_0}^2$.

A more accurate result for the integral could be obtained if we approximated the integrand by a straight line connecting (x_0, k_0) and (x_1, k_1). The integral is then given approximately by the area under the inscribed trapezoid as shown in Fig. 4.29. This procedure is called the *trapezoidal rule*.

Trapezoidal Rule

The approximating function $\phi(x)$ in this case is

$$\phi(x) = \frac{(k_1-k_0)}{h_0}(x-x_1)+k_1 \qquad (4.81)$$

Notice that $\phi(x_0) = k_0$ and $\phi(x_1) = k_1$. Using Eq. 4.81 as the approximation of $k(x)$ in the interval $[x_0, x_1]$, we have

$$\begin{aligned}\int_{x_0}^{x_1} k(x)dx &\approx \int_{x_0}^{x_1}\phi(x)dx \\ &= \frac{h_0}{2}(k_0+k_1) \qquad (4.82)\end{aligned}$$

Referring again to Fig. 4.29, we see that Eq. 4.82 is simply the area of the trapezoid inscribed under the curve $k(x)$. It can be shown that the error of this method is

$$\begin{aligned}\epsilon &= \int_{x_0}^{x_1} k(x)dx - \int_{x_0}^{x_1}\phi(x)dx \\ &= -\frac{{h_0}^3}{12}k''(\xi_0) \qquad x_0 \leqslant \xi_0 \leqslant x_1\end{aligned}$$

Hence this is a better approximation than the rectangular rule as the error is proportional to ${h_0}^3$. Using Eq. 4.82 for each of the n subdivisions $[x_i, x_{i+1}]$, $i = 0, 1, \cdots, n-1$, we have

$$I = \int_a^b k(x)dx \approx T_n = h_0\left(\frac{k_0}{2}+k_1+k_2+\cdots+k_{n-1}+\frac{k_n}{2}\right) \qquad (4.83)$$

The total error is therefore the sum of the individual errors

$$I - T_n = -\frac{h_0^3}{12}[k''(\xi_0) + k''(\xi_1) + k''(\xi_2) + \cdots + k''(\xi_{n-1})]$$

where $x_i \leqslant \xi_i \leqslant x_{i+1}$, $i = 0, 1, 2, \cdots, n-1$. If $k''(x)$ is continuous over the interval $[a, b]$, there exists a point ξ in $[a, b]$ for which

$$\sum_{i=0}^{n-1} k''(\xi_i) = nk''(\xi)$$

Hence we have the total error

$$I - T_n = -\frac{h_0^3}{12} nk''(\xi) = -\frac{(b-a)h_0^2}{12} k''(\xi),\ a \leqslant \xi \leqslant b \qquad (4.84)$$

Example 4.7.1 Let us consider a numerical example. For the *RC* series circuit in Fig. 4.30, the impulse voltage response, $v_C(t) = h(t)$ is

$$h(t) = \frac{1}{RC} e^{-t/RC} u(t) \qquad (4.85)$$

For $RC = 1$

$$h(t) = e^{-t} u(t) \qquad (4.86)$$

Suppose now we have an excitation voltage $v(t) = tu(t)$; then the response is given as the convolution of $v(t)$ and $h(t)$:

$$v_C(t) = \int_0^t v(\tau)h(t-\tau)d\tau$$

$$= \int_0^t \tau u(\tau) e^{-(t-\tau)} d\tau \qquad t \geqslant 0 \qquad (4.87)$$

The analytic solution for Eq. 4.87 is

$$v_C(t) = (t + e^{-t} - 1)u(t) \qquad (4.88)$$

Let us now apply the trapezoidal rule to Eq. 4.87 to find $v_C(t)$ at $t = 1$.

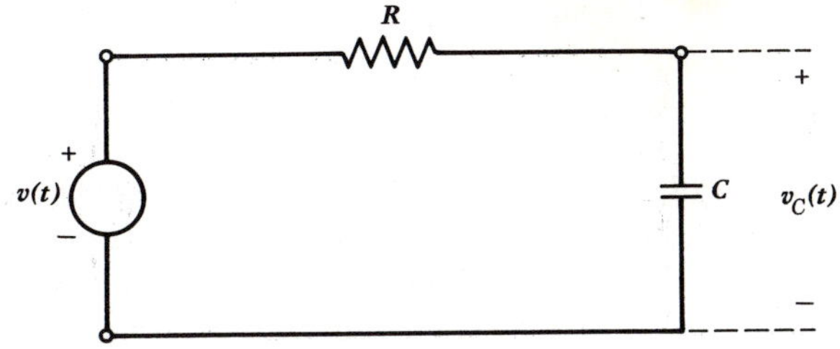

Fig. 4.30 *RC* series circuit.

We use $v(t-\tau)h(\tau)$ as our integrand to obtain:

$$v_C(t) = \int_0^1 (1-\tau)e^{-t}d\tau = \int_0^1 k(\tau)d\tau$$

where $k(\tau) \stackrel{\Delta}{=} (1-\tau)e^{-\tau}$. Using only four subdivisions, we have

$$\begin{aligned}
&k(0) = k_0 = e^{-0} = 1.00000 \\
&k(0.25) = k_1 = 0.75\, e^{-0.25} = 0.58410 \\
&k(0.5) = k_2 = 0.5\, e^{-0.5} = 0.30326 \\
&k(0.75) = k_3 = 0.25\, e^{-0.75} = 0.11809 \\
&k(1) = k_4 = 0.00000
\end{aligned}$$

These values are correct to five decimal places. Substituting these values into Eq. 4.83, we have

$$v_C(1) \approx T_4 = 0.25(\tfrac{1}{2} + 0.58410 + 0.30326 + 0.11809) = 0.37636$$

whereas the exact value of $v_C(1)$ from Eq. 4.88 is equal to 0.3678794. The error is about 1×10^{-2}.

Using eight subdivisions, we have the following results:

$$\begin{aligned}
&k(0) = k_0 = 1.00000 \\
&k(0.125) = k_1 = 0.875\, e^{-0.125} = 0.77218 \\
&k(0.25) = k_2 = 0.58410 \\
&k(0.375) = k_3 = 0.625\, e^{-0.375} = 0.42955 \\
&k(0.5) = k_4 = 0.30326 \\
&k(0.625) = k_5 = 0.375\, e^{-0.625} = 0.20072 \\
&k(0.75) = k_6 = 0.11809 \\
&k(0.875) = k_7 = 0.125\, e^{-0.875} = 0.05210 \\
&k(1) = k_8 = 0
\end{aligned}$$

$$v_C(1) \approx T_8 = 0.125\left[\tfrac{1}{2}k_0 + \sum_{i=1}^{7} k_1 + \tfrac{1}{2}k_8\right] = 0.37000$$

The error in this case is about 0.2×10^{-2}. From these results we see that we need even finer subdivisions in order to find a better approximation.

Simpson's Rule

The rectangular rule uses only one value of $k(x)$ each time to approximate the function $k(x)$ by a constant. The trapezoidal rule gives us a better approximation function $\phi(x)$ by connecting two known points (x_0, k_0) and (x_1, k_1) with a straight line. Now if we are given three points in the $(x, k(x))$ plane, then a parabola could be constructed to pass through, say, (x_0, k_0), (x_1, k_1), and (x_2, k_2). Then the integral $I = \int_{x_0}^{x_2} k(x)dx$ could be even better approximated by the area under the parabola from $x = x_0$ to $x = x_2$. This is shown in Fig. 4.31 and the approximation is called *Simpson's rule*.

Let us assume that the parabola $\phi(x)$ is of the following second-degree polynomial:

$$\phi(x) = l_0 + (x - x_0)l_1 + (x - x_0)(x - x_1)l_2 \tag{4.89}$$

The parabola $\phi(x)$ must pass through (x_0, k_0), (x_1, k_1), and (x_2, k_2); hence,

$$\begin{aligned} \phi(x_0) &= l_0 = k_0 \\ \phi(x_1) &= l_0 + h_0 l_1 = k_1 \\ \phi(x_2) &= l_0 + 2h_0 l_1 + 2h_0^2 l_2 = k_2 \end{aligned} \tag{4.90}$$

The solution to Eq. 4.90 is easily seen to be

$$\begin{aligned} l_0 &= k_0 \\ l_1 &= \frac{k_1 - k_0}{h_0} \\ l_2 &= \frac{k_2 - 2(k_1 - k_0) - k_0}{2h_0^2} = \frac{k_2 - 2k_1 + k_0}{2h_0^2} \end{aligned} \tag{4.91}$$

With the coefficients l_0, l_1, and l_2 known, the integration from x_0 to x_2 is approximated by

$$\int_{x_0}^{x_2} k(x)dx \approx \int_{x_0}^{x_2} \phi(x)dx = \int_{x_0}^{x_2} [l_0 + l_1(x - x_0) + l_2(x - x_0)(x - x_1)]dx \tag{4.92}$$

Simplifying Eq. 4.92, we have

$$\int_{x_0}^{x_2} k(x)dx \approx \frac{h_0}{3}[k_0 + 4k_1 + k_2] \tag{4.93}$$

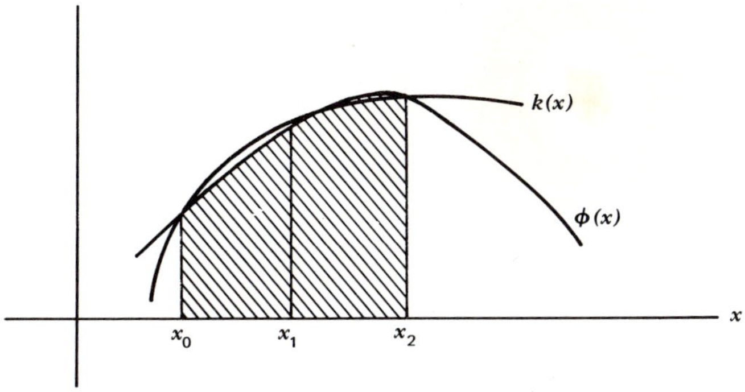

Fig. 4.31 Simpson's rule.

and the error in Eq. 4.93 can be shown to be

$$\epsilon = \int_{x_0}^{x_2} k(x)dx - \int_{x_0}^{x_2} \phi(x)dx = -\frac{h_0^5}{90} k^{(4)}(\eta) \qquad x_0 \leqslant \eta \leqslant x_2 \quad (4.94)$$

To extend Simpson's rule for integration over an interval $[a, b]$ we now divide $[a, b]$ into an even number, $2n$, of subdivisions of width h_0 and apply Simpson's rule over two subdivisions at a time. Let

$$x_i = a + ih_0 \qquad i = 1, 2, \cdots, 2n,$$

where

$$h_0 = \frac{(b-a)}{2n}$$

Using Simpson's rule over the interval $[x_i, x_{i+2}]$ we have

$$\int_{x_i}^{x_{i+2}} k(x)dx = \frac{h}{3}[k_i + 4k_{i+1} + k_{i+2}] - \frac{h_0^5}{90} k^{(4)}(\eta_i)$$

where $x_i \leqslant \eta_i \leqslant x_{i+2}$. If we now sum over the n subgroups of two intervals each,

$$\int_a^b k(x)dx = \frac{h_0}{3}(k_0 + 4k_1 + 2k_2 + \cdots + 4k_{2n-1} + k_{2n}) - \frac{n}{90} h_0^5 k^{(4)}(\eta)$$

$$a \leqslant \eta \leqslant b \qquad (4.95)$$

where we again used the continuity of $k^{(4)}(x)$ over $[a, b]$ to simplify the error term. Simpson's rule for integration over $[a, b]$ that has been subdivided into $2n$ subintervals of length h_0 then is

$$S_{2n} = \frac{h_0}{3}[k_0 + 4k_1 + 2k_2 + 4k_3 + \cdots + 4k_{2n-1} + k_{2n}] \qquad (4.96)$$

Example 4.7.2 Solve Example 4.7.1 using Simpson's rule.

Solution. The values of $k(x)$ for four subdivisions and eight subdivisions are given in Example 4.7.1. The results using Simpson's rule are as follows:

Four subdivisions

$$\begin{aligned} v_C(1) \approx S_4 &= \frac{0.25}{3}[k_0 + 4k_1 + 2k_2 + 4k_3 + k_4] \\ &= \frac{0.25}{3}[1 + 2.33640 + 0.60652 + 0.47236 + 0] \\ &= 0.36794 \end{aligned}$$

Eight subdivisions

$$v_C(1) \approx S_8 = \frac{0.125}{3}[k_0 + 4k_1 + 2k_2 + 4k_3 + 2k_4 + 4k_5 + 2k_6 + 4k_7 + k_8]$$

$$= 0.36788$$

Comparing these results with the exact solution $i(1) = 0.3678794$, we see that Simpson's rule gives us a result with an error of 1×10^{-4} or 1×10^{-6} by using four subdivisions or eight subdivisions respectively. These results are far more accurate than those obtained by using the trapezoidal rule. Naturally, if we further reduce the interval length h_0, we would have more accurate results. However, if we have two approximations such as S_{2n} and S_{4n} resulting from applying Simpson's rule with $2n$ and $4n$ subdivisions respectively, then there is an elegant refinement procedure called the *Richardson extrapolation* where two approximations are used to find a hopefully better third approximation.

Richardson Extrapolation

Simpson's rule in Eq. 4.96 can be rewritten in the following form:

$$I = \int_a^b k(x)dx = S_{2n} - \frac{n}{90} h_0{}^5 k^{(4)}(\eta) \qquad a \leqslant \eta \leqslant b \tag{4.97}$$

In terms of a, b, and n, since $h_0 = (b-a)/2n$, we have

$$I = S_{2n} - \frac{(b-a)^5}{2880n^4} k^{(4)}(\eta) \qquad a \leqslant \eta \leqslant b \tag{4.98}$$

Similarly, for $4n$ subdivisions, we have

$$I = S_{4n} - \frac{(b-a)^5}{2880(2n)^4} k^{(4)}(\xi) \qquad a \leqslant \xi \leqslant b \tag{4.99}$$

Multiply Eq. 4.99 by 2^4 and then subtract Eq. 4.98 from it to obtain

$$15\,I = 16\,S_{4n} - S_{2n} - \frac{(b-a)^5}{2880n^4}[k^{(4)}(\xi) - k^{(4)}(\eta)] \qquad a \leqslant \eta, \qquad \xi \leqslant b \tag{4.100}$$

In general we hope that $k^{(4)}(\xi)$ and $k^{(4)}(\eta)$ are very close together so that the difference in the error terms of Eq. 4.100 is negligible. Indeed Eq. 4.100 can be simplified to

$$I = \frac{16\,S_{4n} - S_{2n}}{15} - \frac{(b-a)^7 k^{(6)}(\sigma)}{(n^6)\cdot 1935360} \qquad a \leqslant \sigma \leqslant b \tag{4.101}$$

provided that $k(x)$ has a continuous and bounded sixth derivative on the interval $[a,b]$. We see that the error term is negligible if $(b-a)/n$ is less than 1. Thus, Eq. 4.101 with the error term neglected should give a very good approximation to I, and this is called Richardson extrapolation.

Let us use this method to refine the results of S_{2n} and S_{4n} in our numerical example. We have

$$S_{2n} = S_4 = 0.36794$$

$$S_{4n} = S_8 = 0.36788$$

Hence the exact integral $v_C(1)$ is estimated as

$$I \approx \frac{16\,S_8 - S_4}{15} = \frac{1}{15}(16 \times 0.36788 - 0.36794)$$

$$= \frac{1}{15}(5.51814) = 0.36788$$

This result is identical with the result of S_8, therefore we could conclude that 0.36788 is indeed a very good approximation to $v_C(1)$. While in this case Richardson extrapolation is no improvement on S_8, it gives much better estimates as a rule and is well worth the minor additional computation. For this reason, it is included in the computer program described next.

Computer Program

The implementation of Simpson's rule is quite simple. We need the interval parameters a, b, and the number of subdivisions $2n$ where n is any integer.

If the function $k(x)$ is given in analytical form, then a FORTRAN statement FUNCTION can be used to define $k(x)$. In this case Simpson's rule could be used twice and the Richardson extrapolation used for a final refinement. If the function $k(x)$ is given as a set of data points, then either Simpson's rule is employed once by using all points available, or it can be employed twice by utilizing first data points with $2h_0$ spacing between adjacent points and then using all points available for Simpson's rule. In the latter case the Richardson extrapolation could be used as a final refinement for the approximations.

In case the number of data points given is not an even number, then we could use the trapezoidal rule for the first subinterval integration and then employ Simpson's rule for the remaining even-numbered subintervals.

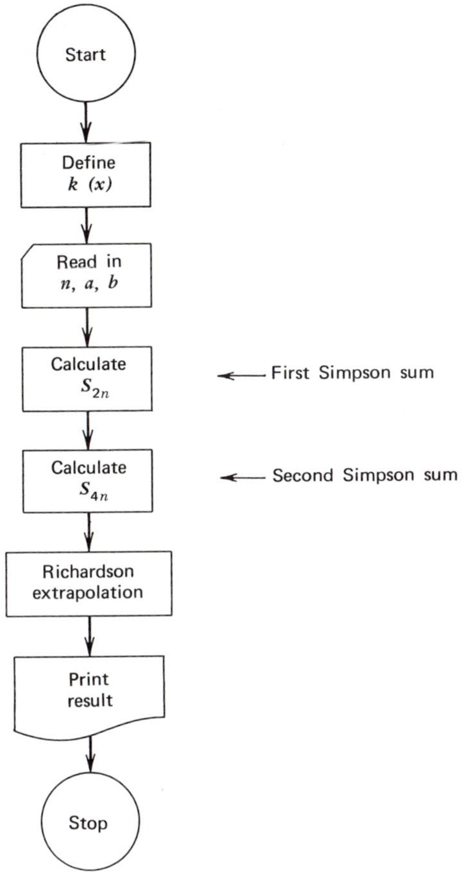

Fig. 4.32 Simpson's rule integration.

A flowchart using Simpson's rule for integration is given in Figure 4.32.

Example 4.7.3 Solve the problem in Example 4.7.1 using Simpson's rule.

Solution. The desired voltage response is

$$v_C(t) = \int_0^1 (1-\tau)e^{-\tau}d\tau$$

Dividing the interval $[0, 1]$ into $2n$ and $4n$ subintervals where $n = 6$, we can obtain S_{12} and S_{24} respectively. The Richardson extrapolation is then used for final refinement.

The FORTRAN program, and a sample computer output are shown in Figs. 4.33 and 4.34. The parameters are the following:

$FK(X)$	Integrand function
N, XN	Number of $2h_0$ subintervals
A	Integration lower limit
B	Integration upper limit
H	Subinterval step size h_0
$H2$	Twice of H
$Y(1)$	First Simpson sum, S_{2n}
$Y(2)$	Second Simpson sum, S_{4n}
R	Integration result after Richardson extrapolation

```
      DIMENSICN Y(2)
    1 FORMAT(I5,2F10.2)
    3 FORMAT(10X,4E14.5,I5)
      FK(X)=(1.-X)*EXP(-X)
      READ 1, N,A,B
      DO 10 J=1,2
         XN=N
         H=(B-A)/(XN*2.)
         H2=H*2.
         X=A
         Y(J)=0.
         DO 2 I=1,N
            Y(J)=Y(J)+FK(X)+4.*FK(X+H)+FK(X+H2)
            X=X+H2
    2    CONTINUE
         Y(J)=Y(J)*H/3.
         PRINT 3, Y(J),A,B,H,N
         N=N*2
   10 CONTINUE
      R=(16.*Y(2)-Y(1))/15.
      PRINT 3, R
      END
```

Fig. 4.33 FORTRAN Program for Example 4.7.3

S_{2N}	A	B	H	N
0.36788E 00	0.0	0.10000E 01	0.83333E-01	6
0.36788E 00	0.0	0.10000E 01	0.41667E-01	12

```
     R
0.36788E 00
```

Fig. 4.34 Sample computer output.

PROBLEMS

4-1. For the given initial conditions $x'(0-) = x''(0-) = 0$, find the solutions for the equations:

(i) $x''(t) + 2x'(t) + 2x(t) = 3\delta(t)$

(ii) $x''(t) + 7x'(t) + 12x(t) = 5\delta'(t)$

4-1a. Repeat 4-1 for the following equations:

(i) $x''(t) + 4x'(t) + 3x(t) = \delta(t) + u(t), x(0-) = x'(0-) = 0.$

(ii) $x''(t) + 6x'(t) + 25x(t) = \delta'(t), x(0-) = x'(0-) = 1.$

(iii) $x''(t) + 8x'(t) + 16x(t) = 2u(t) + 2\delta(t), x(0-) = -x'(0-) = 1.$

(iv) $x'(t) + 2x(t) + 2\int_0^t x(\tau)d\tau = (3/4)e^{-t} + \delta(t), x(0-) = -2.$

4-2. Given the set of equations

$$x'(t) + x(t) + y(t) = u(t)$$
$$x(t) + y'(t) + 2y(t) = \delta(t)$$

with $x(0-) = x'(0-) = y(0-) = y'(0-) = 0$

(i) Find $x(0+)$, $x'(0+)$, $y(0+)$, and $y'(0+)$.

(ii) Solve for $x(t)$ and $y(t)$ for $t \geqslant 0$.

4-2a. Repeat 4-2 for the following set of equations:

$$2x'(t) + 2x(t) + y'(t) - y(t) = 2u(t)$$
$$x'(t) + x(t) + y'(t) + 2y(t) = 3e^{-t} + \delta(t)$$

with $x(0-) = x'(0-) = y(0-) = y'(0-) = 0.$

4-3. An *RC* differentiator circuit is shown in Fig. P4-3.

(i) Find the response voltage $v_0(t)$ due to an exponential input voltage $v_i(t) = e^{jwt}$ where w is a constant.

(ii) Find the requirement for the time constant *RC* such that $v_0(t)$ is approximately proportional to the derivative of $v_i(t)$.

(iii) Can this circuit be used to generate a unit impulse function? If so, determine how large must the step input be for a unit impulse output.

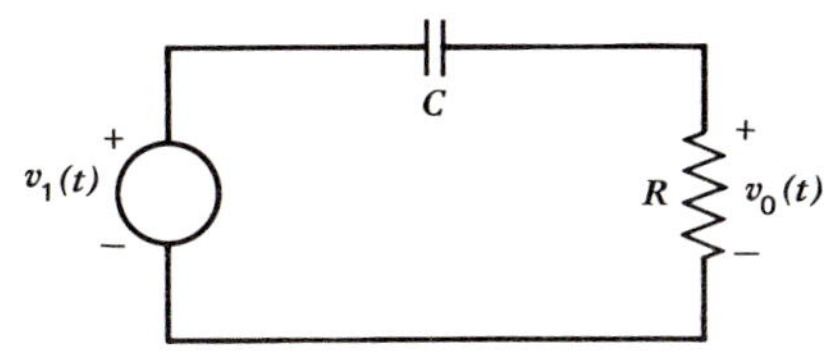

Fig. P4-3

4-4. An *RC* integrator circuit is shown in Fig. P4-4.

(i) For an exponential input voltage $v_i(t) = e^{j\omega t}$ find the requirement for the time constant *RC* such that $v_0(t)$ is proportional approximately to the integral of the input voltage.

(ii) Specify the magnitude of a step-input voltage such that the output voltage is approximately a unit ramp function.

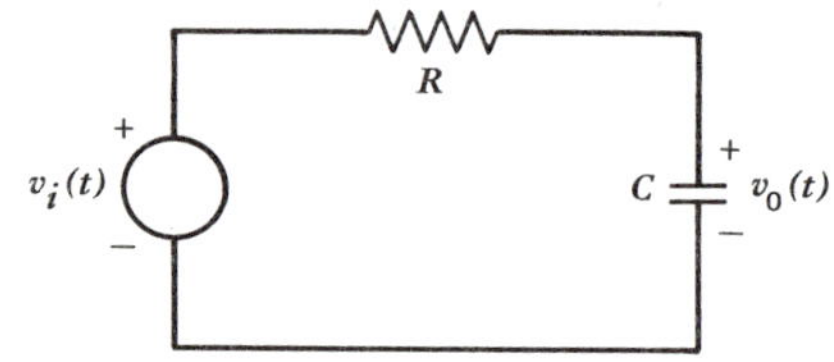

Fig. P4-4

4-5. The circuit shown in Fig. P4-5 is approximately analogous to a loudspeaker in which the cone velocity is proportional to the voltage $v_0(t)$ shown. Let $i(0-) = 0$. Find and sketch $v_0(t)$ if the input voltage $v_i(t)$ is given by

(i) $v_i(t) = u(t)V$

(ii) $v_i(t) = \delta(t)V$

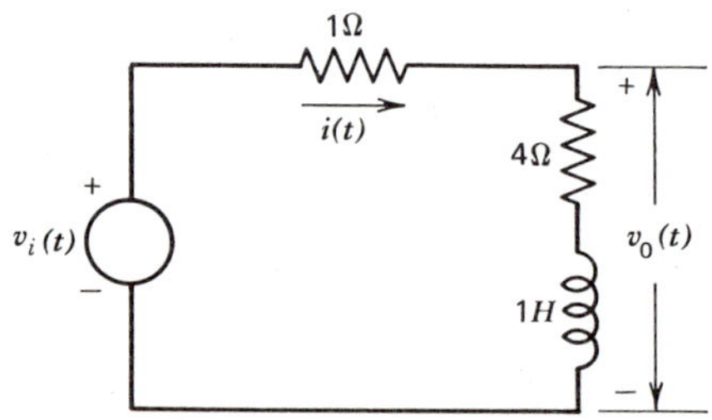

Fig. P4-5

4-6. From the results of 4-5, find $v_0(t)$ of the circuit of Fig. P4-5 by convolution if

(i) $v_i(t) = e^{-t}u(t)V$

(ii) $v_i(t) = e^{-10t}u(t)V$

(iii) $v_i(t) = (e^{-t}\cos 10t)u(t)V$

4-7. Using the linearity principle and the results of 4-5, find $v_0(t)$ of the circuit of Fig. P4-5 subject to the following excitations (you do not have to use convolution):

(i) $v_i(t) = e^{-(t-1)}u(t-1)V$

(ii) $v_i(t) = e^{-t}u(t-1)V$

4-7a. For the zero-state circuit shown in Fig. P4-7*a*, find the voltage

response $v(t)$ subject to each of the following excitations:
(i) $i(t) = 10u(t)$
(ii) $i(t) = 10\delta(t)$

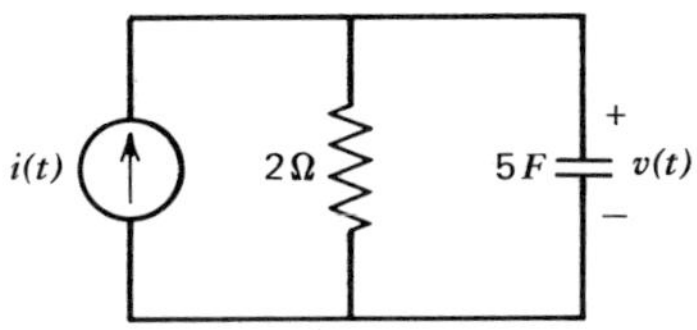

Fig. P4-7*a*

4-7b. For the zero-state circuit of Fig. P4-7*a*, find the response voltage $v(t)$ subject to each of the following excitations:
(i) $i(t) = 10u(t) - 5u(t-1)$
(ii) $i(t) = tu(t) - (t-1)u(t-1)$
(iii) $i(t) = e^{-0.1t}[u(t) - u(t-1)]$

4-7c. Repeat 4-7b for the semiperiodic excitations shown in Fig. P4-7*c*.

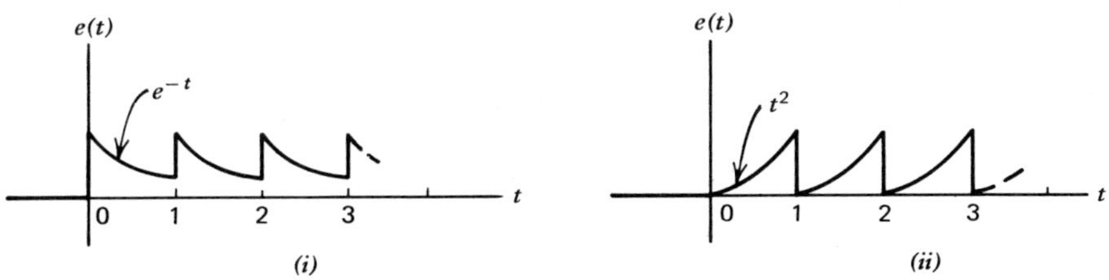

Fig. P4-7*c*

4-8. The unit step response of a zero-state linear system is

$$r(t) = (2e^{-2t} - 1)u(t)$$

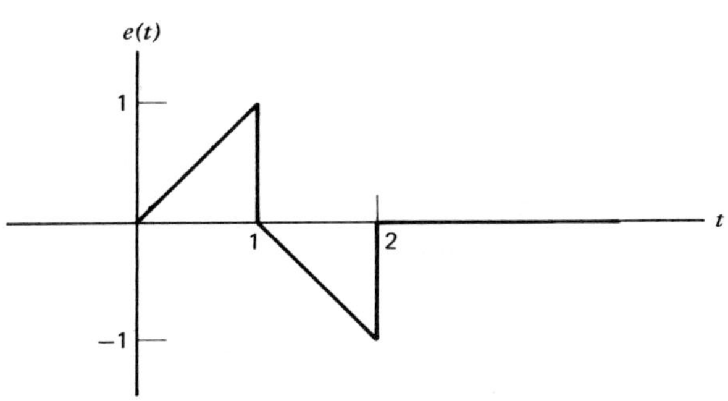

Fig. P4-8

(i) Find the response $r(t)$ due to a ramp excitation

$$e(t) = tu(t)$$

(ii) Find the response $r(t)$ due to the excitation shown in Fig. P4-8.

4-8a. Find the convolution $r(t) = h(t)*e(t)$ of the following set of functions:

(i) $h(t) = \epsilon^{-t}u(t)$, $e(t) = tu(t)$.
(ii) $h(t) = \sin 2\pi t[u(t) - u(t-1)]$, $e(t) = u(t)$.
(iii) $h(t) = t^2[u(t) - u(t-1)]$, $e(t) = e^{-t}u(t)$.

4-8b. Show that the convolution shares the following algebraic properties with ordinary multiplication:
If f_1, f_2, and f_3 are integrable functions (so that $f_1*f_2 = \int_0^t f_1(x) \times f_2(t-x)dx$ is defined and so is f_1*f_3), then we have

(i) $f_1*f_2 = f_2*f_1$ (commutative law)
(ii) $f_1*(f_2*f_3) = (f_1*f_2)*f_3$ (associative law)
(iii) $u(t)*f_3 = f_3*u(t) = f_3$ (identity)
(iv) $f_1*(f_2+f_3) = f_1*f_2 + f_1*f_3$ (distributive law)

4-8c. Demonstrate the validity of the identities of 4-8b using functions of 4-8a. Specifically, let $f_1 = h(t)$, $f_2 = e(t)$, $f_3 = \sin t$.

4-9. The excitation of a linear system is $e(t)$ and the system impulse response $h(t)$ are shown in Fig. P4-9. Sketch the system response to $e(t)$. (No equations need be written. A neat, carefully dimensioned sketch will suffice.)

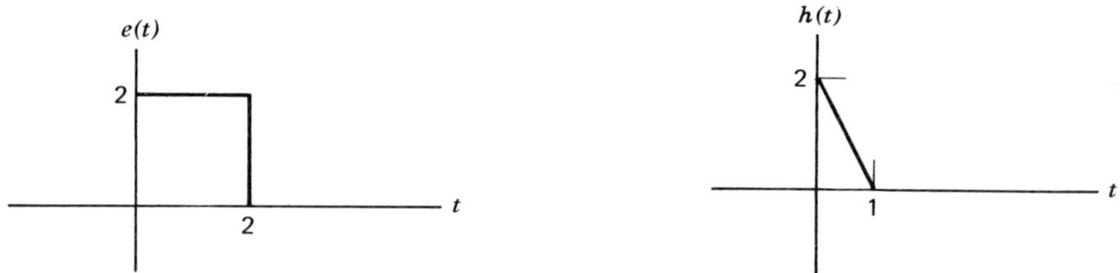

Fig. P4-9

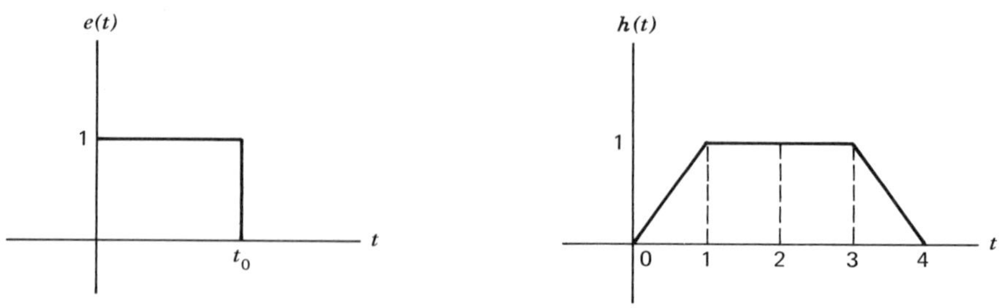

Fig. P4-9a

4-9a. Repeat 4-9 for the pair of $e(t)$ and $h(t)$ shown in Fig. P4-9*a* under the following conditions: (i) $t_0 = 3$, (ii) $t_0 = 4$, and (iii) $t_0 = 1$.

4-9b. Repeat 4-9a for the $h(t)$ shown in Fig. P4-9*b*.

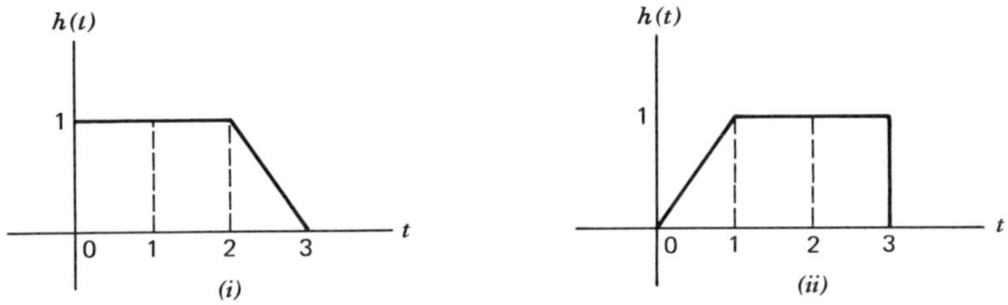

Fig. P4-9*b*

4-10. The impulse response $h(t)$ of a linear system is shown in Fig. P4-10. If the excitation is $e(t) = 3e^{-2t}u(t)$, determine the response values $r(1)$ and $r(4)$ using graphical convolution.

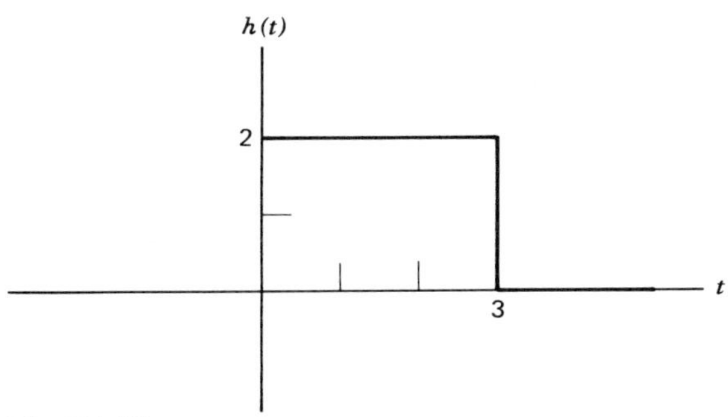

Fig. P4-10

4-11. For a given circuit, the total response $r(t)$ due to an arbitrary excitation $e(t)$ can be expressed as

$$r(t) = r_{ZI}(t) + r_{ZS}(t)$$

where $r_{ZI}(t)$ and $r_{ZS}(t)$ are zero-input and zero-state responses respectively. Furthermore from linearity principle the zero-state response of the circuit due to an excitation $e'(t)$, which is the derivative of $e(t)$, is $r'_{ZS}(t)$. Since $r_{ZI}(t)$ is independent of the external excitation, the total response due to $e'(t)$ for arbitrary initial states is given as

$$r_1(t) = r_{ZI}(t) + r'_{ZS}(t)$$

(i) Verify the above statement by finding the response due to $di(t)/dt$ for the circuit of Fig. P2-1.
(ii) Repeat (i) for problem 3-4 if the excitation $v_1(t)$ is changed to $10\delta(t)V$.

4-12. From 4-11 we see that if the response $r(t)$ due to an excitation $e(t)$ is known as

$$r(t) = r_{ZI}(t) + r_{ZS}(t)$$

then the response due to $e'(t)$ can be obtained easily by

$$r_1(t) = r_{ZI}(t) + r'_{ZS}(t)$$

Hence in order to obtain the response $r_1(t)$ of a circuit due to impulse excitation $\delta(t)$, we need the response $r(t)$ due to step excitation $u(t)$. However, since an arbitrary circuit under impulse excitation is also equivalent to the same circuit under zero-input excitation with its initial states modified at $t=0+$, all we need to know in order to find the impulse response $r_1(t)$ for $t \geqslant 0+$ is the zero-input response due to the modified initial state at $t=0+$.
(i) Verify the above statement for the circuit of Fig. P2-1.
(ii) What is the difference between the result of (i) and that of problem 4-11(i)?
(iii) Describe what happens at $t=0$. (Is the capacitor short-circuited at $t=0$? How much is the capacitor voltage build up from $t=0-$ to $t=0+$? And so on.)

4-13. The following approximation for convolution was suggested by E. A. Guillemin. To find the convolution of $e(t)$ and $h(t)$, approximate $e(t)$ by $e_a(t)$ that consists of straight-line segments jointed together (piecewise linear approximation). The derivative of $e_a(t)$ consists of a staircase curve and the second derivative of $e_a(t)$ is a train of impulses of various delays and various areas. The response to $e''_a(t)$ is obtained easily. By linearity principle, the response due to $e'_a(t)$ is obtained by integrating $e''_a(t)$; the response due to $e_a(t)$ is obtained by integrating $e'_a(t)$. Suppose we choose $n+1$ points on the curve of $e(t)$, $t_0, t_1, \cdots, t_n$ and form $e_a(t)$ by joining the points $e(t_0), e(t_1), \cdots, e(t_n)$ by straight lines. Find a formula for the response of a circuit due to an excitation $e(t)$ with impulse response given by $h(t)$.

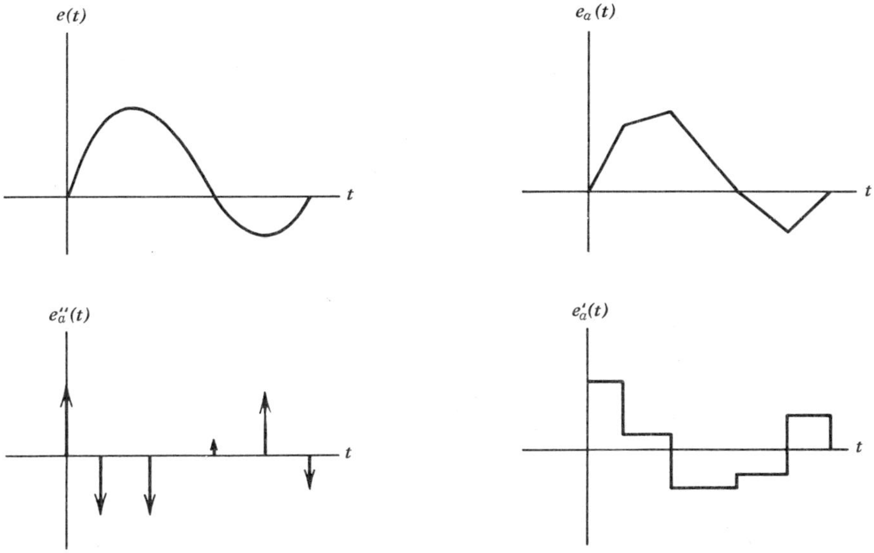

Fig. P4-13

4-14. Using the results of 4-13 to solve 4-9 and 4-10. Compare the results with the exact answers found in 4-9 and 4-10.

4-14a. Using the results of 4-13 to solve 4-9a and 4-9b. Compare with the exact answers.

4-14b. Apply Simpson's rule to find the approximation S_{2n} for the convolutional integral $r(t) = h^*e$ of the functions of 4-8a for $t = 0$, 0.2, 0.4, $\cdots$, 2.0. For each t, set $n = 16$. Compare with the exact results.

4-14c. Repeat 4-14b for $n = 4$ and 8 and then apply Richardson extrapolation. Are the results as good as the ones of 4-14b?

4-14d. Apply Simpson's rule with $n = 1$ and 2 to determine the approximations S_{2n} for each of the following definite integrals: (Use 3-digit chopped floating-point arithmetic.)

(i) $I = \int_0^1 t\,dt$; (ii) $I = \int_0^1 t^3 dt$.

(iii) $I = \int_1^2 \frac{1}{t}\,dt$; (iv) $I = \int_0^1 \frac{1}{1+t^2}\,dt$.

Answers: (i) 1/2, (ii) 1/4, (iii) $\log_e 2$, (iv) $\pi/4$.

4-14e. Apply Richardson extrapolation to the results of 4-14d. Are the new approximations better?

4-14f. Repeat 4-14d with $n = 4$. Compare with the results of 4-14e.

4-14g. Repeat 4-14d with $n = 16$.

4-15. To evaluate the integral

$$I = \int_a^b k(x)\, dx$$

approximately, we can use Simpson's rule. An even better approximation for I can also be obtained if a third or higher order polynomial $\phi(x)$ is used to approximate $k(x)$ by having $\phi(x)$ passing through four or more points in the $(x, k(x))$ plane. Let us assume that the approximation polynomial $\phi(x)$ is of the following third-degree polynomial:

$$\phi(x) = L_0 + (x - x_0)L_1 + (x - x_0)(x - x_1)L_2 + (x - x_0)(x - x_1)(x - x_2)L_3$$

(i) Find L_0, L_1, L_2, L_3 if $\phi(x)$ is required to pass through $(x_0, k(x_0))$, $(x_1, k(x_1))$, $(x_2, k(x_2))$, and $(x_3, k(x_3))$.

(ii) If the integration interval is divided into multiples of 3 or $3n$ subdivisions of width h_0, develop a formula using this $\phi(x)$ to approximate I.

(iii) Write a FORTRAN subroutine to implement this formula.

4-15a. Find approximations to the definite integrals of 4-14d using the subroutine of 4-15 with 33 subdivisions. Are the results better than that of 4-14g?

4-16. Let S_{2n} denote the approximate integral obtained by using the Simpson's rule with $2n$ subdivisions. For the problem of Example 4.7.1:

(i) Evaluate S_{2n} for $n = 2^0, 2^1, 2^2$ by chopped floating-point arithmetic with 5-digit mantissa.

(ii) With S_2 and S_4 known, apply Richardson extrapolation to get a new approximation. Compare this result with S_8. Can this new approximation be considered as S_8? If so, then S_8 need not be computed. It can be obtained directly from S_2 and S_4. This new "S_8" can then be used with S_4 to obtain "S_{16}" and so on. Write a subroutine to implement this recursive procedure. Limit the number of recursions to 20. This procedure belongs to the so-called Romberg integration methods.

4-16a. Write a computer program to find the definite integrals of 4-14d by the following steps:

(i) Generate S_2 and S_4 using Simpson's rule.

(ii) Generate new Simpson approximation S_{2n} by Romberg's method discussed in 4-16.

(iii) Stop computation when $S_{2n} = S_{2(n-1)}$ or when $n \geqq 20$.

4-17. Apply Simpson's rule with 32 subdivisions to solve 4-8(i) for $r(5)$.

4-18. Solve problem 4-8(i) for $r(5)$ using the subroutine of 4-15 with 33 subdivisions.

4-19. Solve 4-8(i) for $r(5)$ using the subroutine of 4-16 with 3 applications of Richardson extrapolation.

4-20. Compare the results of 4-17 and 4-18 with the exact one of 4-8(i). In terms of number of operations needed and the accuracy of the results, which algorithm is better? How about the result of problem 4-19?

CHAPTER 5
Resistive Networks: Linear Algebraic Systems

The convolution integral completes the general discussion of linear, time-invariant circuits. The time-domain (i.e., a function of time) behavior of zero-state circuits for arbitrary excitations is entirely determined by the step or impulse responses. The impulse response is an important concept in describing linear systems and is commonly used as the system specification. The impulse excitation, however, destroys the physically or intuitively satisfying notion that the state of a system does not change instantly. This continuity of the state variables is preserved under all bounded excitations. It is usually preferable to derive the step response of a system with the continuity of the state variables assured, and then differentiate the step response to obtain the system impulse response. Convolution then gives the zero-state response to any excitation, and the response of a circuit whose initial state is nonzero is obtained by adding the zero-input response to the zero-state response.

The step excitation for zero-state circuits is equivalent to constant excitation for the appropriate time interval ($t \geqslant 0$). Circuit response to constant and other elementary or exponential excitations is determined in a fairly simple manner from the circuit differential equations. The solution of the differential equation is governed by the initial conditions that must be satisfied by the solution variable and its derivatives. These initial conditions are derived from the specified initial state of the circuit. The state variables are continuous for bounded excitations at the instant of excitation. The postexcitation initial conditions are thus obtained from the known initial state values for bounded excitations. The impulse response may be obtained from the step response by differentiation. Or, the impulse response is obtained directly as discussed in Sec. 4.1 by first determining the resulting discontinuities in the initial conditions. The implied assertion is that one can determine any voltage or current in the circuit from the given value of the state and the circuit equilibrium equations.

The determination of the voltages and currents anywhere in the circuit from the known circuit state involves only routine manipulations. Once the capacitor voltages and the inductor currents are known at any instant, only the resistors and sources must be considered in determining the desired voltages or currents. For complex circuits, even this routine manipulation can be difficult, and systematic procedures are necessary for the analysis of purely resistive circuits or networks. The capacitor voltages and the inductor currents are represented at the given instant by voltage and current sources, respectively, of corresponding values at that instant. The resulting resistive circuit is then solved.

Alternatively, circuits in steady-state conditions (i.e., when all transients have vanished) under constant excitations may be represented as resistive networks. Steady-state conditions under constant excitations imply that all voltages and currents have reached constant values (cf. steady-state solution for constant forcing functions). Constant voltage across a capacitor implies zero current through it and constant current through an inductor implies zero voltage across it. In this situation, capacitors may be replaced by open circuits (i.e., zero currents) and inductors by short circuits (i.e., zero voltages). The resulting network of resistors and sources may then be solved to determine steady-state voltages and currents.

Resistive networks also arise in other situations and constitute frequent occurences. The equilibrium equations for such circuits contain neither derivatives or integrals since the branch constraints governing resistive networks are given by Ohm's law. The resulting circuit equations are therefore algebraic equations, *not* differential equations. A large network of resistors yields a large, simultaneous set of algebraic equations.

The solution of a system or set of algebraic equations thus constitutes the analysis of resistive networks. The solution of a large set of algebraic equations can be a difficult computational problem and efficient computational methods are necessary to handle such problems. Sets of algebraic equations occur frequently in other circuit analysis problems as well. In particular, we shall see that sinusoidal steady-state analysis problems also yield sets of algebraic equations. It is no exaggeration to say that the solution of a set of algebraic equations is one of the most important computational problems of circuit analysis.

In this chapter we discuss resistive networks and their analysis using node equations. A brief discussion of loop analysis is also included. State equations do not exist for resistive networks since there are no reactive elements present. Finally, we describe an efficient and widely used method to handle sets of algebraic equations.

5.1 NODE ANALYSIS: CURRENT SOURCES

The use of KCL node equations to analyze arbitrary circuits has been thoroughly discussed in Chapter 3. General formalism in terms of simultaneous sets of node equations was described. The same approach with simplifications arising from the purely resistive nature of our circuits is used here. Each separate part electrically connected together is treated as a separate entity for writing node equations. One reference node is needed for each separate part. The resulting equations are then coupled together by the appropriate coupling constraints. We shall concentrate, therefore, on a single electrically connected part of the circuit.

For a singly connected circuit with n nodes, one of the nodes is chosen as the reference node. A KCL node equation is written for each of the remaining $(n-1)$ nodes. The equations are then solved for the unknown node voltages with respect to the reference node. Whenever a node voltage is already known or constrained with respect to any other node, obviously no KCL node equation is necessary for that node. The presence of voltage sources constrains the node voltages across their terminals to specified source signal values. In such cases, the number of independent node equations is reduced by the number of voltage sources. Thus, if n_V is the number of independent voltage sources in a network with n nodes, the number of node equations needed to describe the circuit is $(n-n_V-1)$; one equation is at each of the independent (unconstrained) nodes. This is usually done by inspection for simple circuits as discussed in Chapters 2 and 3. For large networks, systematic procedures can be developed to account for voltage sources and are described in Sec. 5.2.

We describe the systematic procedure for writing the KCL node equations for resistive networks with current sources in this section. The case of mixed current and voltage sources is then discussed in the next section. The procedure is best developed through the use of several examples.

Consider the circuit with two current sources and resistors designated as R_1, R_2, R_3, and R_4 shown in Fig. 5.1. Let us choose node 4 as the reference node and apply KCL to nodes 1, 2, and 3. For consistency, let us assign positive signs to the unknown currents flowing out of a node.

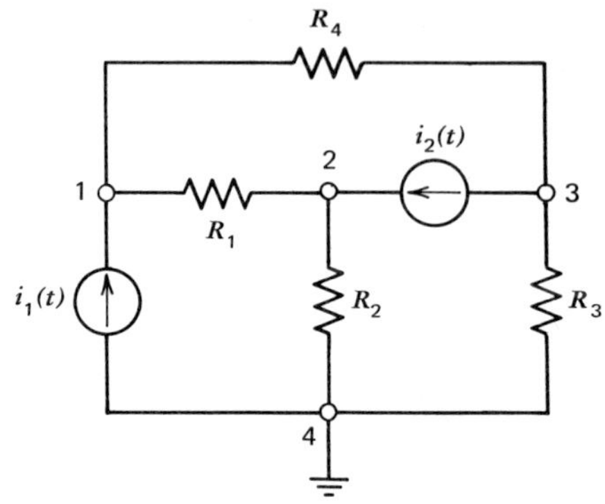

Fig. 5.1 A resistive network.

Then, we have the following equations for the three nodes:

$$\begin{aligned}
\left(\frac{1}{R_1}+\frac{1}{R_4}\right)v_1-\left(\frac{1}{R_1}\right)v_2-\left(\frac{1}{R_4}\right)v_3 &= i_1(t)\\
-\left(\frac{1}{R_1}\right)v_1+\left(\frac{1}{R_1}+\frac{1}{R_2}\right)v_2 \qquad\qquad &= i_2(t)\\
-\left(\frac{1}{R_4}\right)v_1 \qquad\qquad +\left(\frac{1}{R_3}+\frac{1}{R_4}\right)v_3 &= -i_2(t)
\end{aligned} \tag{5.1}$$

where v_1, v_2, and v_3 are the voltages of the nodes 1, 2, and 3, respectively with respect to the reference. From Eq. 5.1 we see that the coefficient associated with each unknown node voltage is the sum of reciprocal values of certain resistors in the circuit. Let us use the *conductances*, defined by

$$G_i=\frac{1}{R_i} \tag{5.2}$$

to write the node equations. Using this notation, Eq. 5.1 becomes

$$\begin{aligned}
(G_1+G_4)v_1-(G_1)v_2-(G_4)v_3 &= i_1(t)\\
-(G_1)v_1+(G_1+G_2)v_2 \qquad &= i_2(t)\\
-(G_4)v_1 \qquad +(G_3+G_4)v_3 &= -i_2(t)
\end{aligned} \tag{5.3}$$

The right-hand side quantity in each equation represents the *current source* signal entering the node. Thus in Fig. 5.1 the *current source* entering node 1 is $i_1(t)$. For node 2 it is $i_2(t)$; however, the *current source* entering node 3 is $-i_2(t)$. The minus sign indicates that $i_2(t)$ is leaving the node instead of entering the node.

Secondly, the sign for the ith node voltage coefficient is positive in the equation formed by applying KCL at that particular node. Otherwise, the signs are negative. Thus in the first equation written for node 1, the sign for v_1 is positive and those for v_2 and v_3 are negative. The same principle applies to the remaining equations in Eq. 5.3.

Finally, observe from the figure that the coefficient for the ith node voltage in the ith node equation is simply the sum of all the conductances connected to the ith node whereas the coefficient for the jth node voltage in the ith node equation is the conductance directly connecting nodes i and j. Thus in the first node equation above, the coefficient for v_1 is G_4+G_1, or the sum of the conductances connected to node 1. The coefficient for v_2 is G_1 since node 2 and node 1 are connected by G_1. Notice that the coefficient for v_3 in the second equation is zero because there is no direct connection between nodes 2 and 3.

In general for a circuit of $(N+1)$ nodes driven by *current sources only*, we form N node equations by fixing one node as the reference node. We have set the nodes $n = N+1$ for convenience in writing the subscripts in our equations. The general format is:

$$
\begin{array}{cccc}
y_{11}v_1 + y_{12}v_2 + \cdots + y_{1N}v_N = i_1 \\
y_{21}v_1 + y_{22}v_2 + \cdots + y_{2N}v_N = i_2 \\
\cdot \quad \cdot \quad \cdot \quad \cdot \\
\cdot \quad \cdot \quad \cdot \quad \cdot \\
\cdot \quad \cdot \quad \cdot \quad \cdot \\
y_{N1}v_1 + y_{N2}v_2 + \cdots + y_{NN}v_N = i_N
\end{array}
\tag{5.4}
$$

where i_k denotes the *current source entering* the kth node. The coefficient y_{ij} is determined using the following rules:

(a) If $i=j$, y_{ii} is obtained by adding all of the conductances incident to (touching) the node i.

(b) If $i \neq j$, y_{ij} is equal to the negative of the sum of the conductances between nodes i and j.

The conductance of a *current source* is considered zero in determining the value of y_{ii} or y_{ij}.

As an additional example, let us write a set of node equations for the circuit of Fig. 5.2. Let node 4 be the reference node. Using the format in Eq. 5.4, we immediately arrive at the following set of equations:

$$
\begin{array}{llllll}
\text{Node 1:} & (3+2+6)v_1 - & 2v_2 & - & 6v_3 & = 10 \\
\text{Node 2:} & -2v_1 & +(2+5+1)v_2 - & & 5v_3 & = 0 \\
\text{Node 3:} & -6v_1 & - & 5v_2 & +(6+5+4)v_3 & = -8
\end{array}
\tag{5.5}
$$

Observe that the current source *entering* node 3 is -8 and the current source entering node 2 is zero.

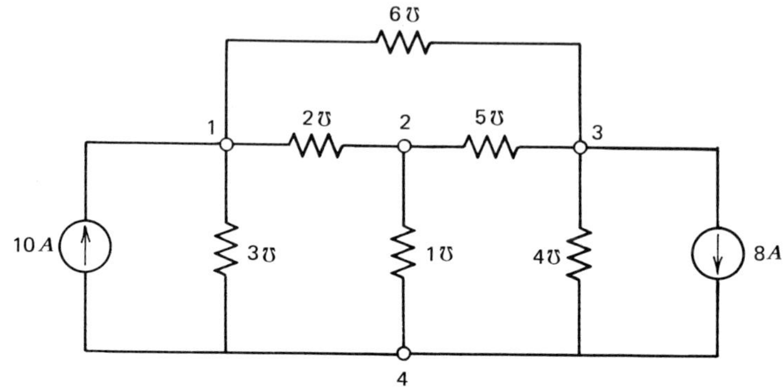

Fig. 5.2 A 4-node resistive network.

5.2 NODE ANALYSIS: MIXED SOURCES

When the voltage sources are present in a resistive network, the node equations are written with the same relative ease except there are modifications due to the constraints on some of the node voltages. If the presence of a voltage source at a node specifies that node voltage with respect to the reference, then obviously there is no need to determine the node voltage or to write the node equation for that node. If the voltage source at a node specifies the node voltage with respect to some other node, not the reference node, then the determination of the voltage of one of the two nodes also gives the voltage of the other node. A node equation is then required only at any one of the two nodes.

The procedure for writing the node equations as shown below, consists of simply writing the KCL equations at each node and assuming an unknown source current for each voltage source. This unknown source current will appear in each of the two constrained equations provided the reference node is not one of the voltage source terminals. The two equations are then combined into one and the unknown source current eliminated. If the reference node is a voltage source terminal, the other terminal of that source is a constrained node. The unknown source current appears only once in this case, and the equation is deleted since the constrained node voltage is already known.

The procedure is routine and straightforward for any number or locations of voltage sources. Some examples illustrate the procedure.

Consider the circuit of Fig. 5.3 with one current source and one voltage source. There are four nodes and we choose node 4 as the reference. Since the voltage source s_1 constrains node 1 with respect to the reference node, the node voltage $v_1 = s_1$ is known, and the node equation for node

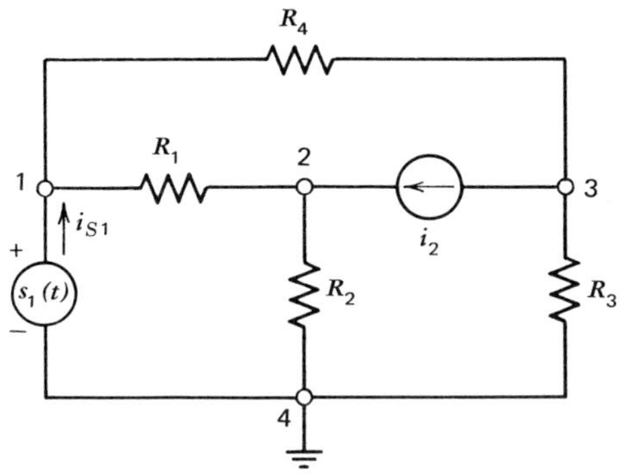

Fig. 5.3 A resistive network with mixed sources.

1 must be deleted. Let us proceed systematically and write the KCL node equations at each of the three nodes using the procedure described in Eq. 5.4. We assume the unknown source current i_{s1} through the voltage source. Then the equations are:

$$\begin{aligned} &\text{Node 1:} \quad (G_1+G_4)v_1 - G_1v_2 - G_4v_3 = i_{s1} \\ &\text{Node 2:} \quad -G_1v_1 + (G_1+G_2)v_2 = i_2 \\ &\text{Node 3:} \quad -G_4v_1 + (G_3+G_4)v_3 = -i_2 \end{aligned} \tag{5.6}$$

The unknown source current i_{si} appears in only one equation representing a constrained node whose voltage is known. This equation is deleted and the value of the constrained voltage $v_1 = s_1$ substituted to obtain the desired node equations:

$$\begin{aligned} (G_1+G_2)v_2 &= i_2 + G_1s_1(t) \\ (G_3+G_4)v_3 &= -i_2 + G_4s_1(t) \end{aligned} \tag{5.7}$$

Next, consider the same circuit driven by two voltage sources, as shown in Fig. 5.4. Again, we choose node 4 as the reference and write the KCL equations for the other nodes assuming the unknown source currents i_{s1} and i_{s2} as shown. The equations are

$$\begin{aligned} &\text{Node 1:} \quad (G_1+G_4)v_1 - G_1v_2 - G_4v_3 = i_{s1} \\ &\text{Node 2:} \quad -G_1v_1 + (G_1+G_2)v_2 = i_{s2} \\ &\text{Node 3:} \quad -G_4v_1 + (G_3+G_4)v_3 = -i_{s2} \end{aligned} \tag{5.8}$$

The unknown source current i_{S1} appears only once and the first equation is deleted and the constraint $v_1 = s_1$ introduced in the other equations. The unknown source current i_{s2} appears in the two remaining equations and is eliminated by adding the two equations to give

$$(G_1+G_2)v_2 + (G_3+G_4)v_3 = (G_1+G_4)s_1(t) \tag{5.9}$$

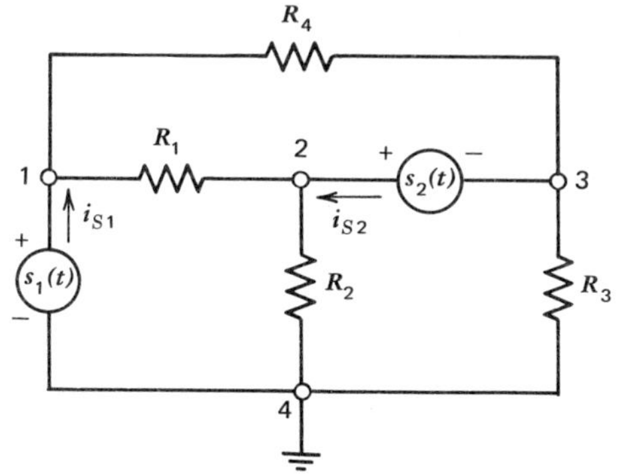

Fig. 5.4 A resistive network.

where we have put $v_1 = s_1$. The voltage variables v_2 and v_3 are constrained by the voltage source s_2 so that $v_2 - v_3 = s_2$. This constraint is now introduced in Eq. 5.9 by choosing the desired node 2 or 3 as the independent node. Let us choose node 3 as independent. Then we substitute $v_2 = v_3 + s_2$ into Eq. 5.9 to obtain

$$(G_1 + G_2 + G_3 + G_4)v_3 = (G_1 + G_4)s_1(t) - (G_3 + G_4)s_2(t) \tag{5.10}$$

as the only node equation required to describe the circuit. The equation is solved for v_3 and v_2 obtained from the given constraint. All the node voltages are then known and the circuit behavior is determined.

Sometimes a whole set of nodes are constrained by a succession of voltage sources, but the procedure remains unchanged. Elimination of an unknown source current eliminates one of the node equations and the corresponding voltage source constraint eliminates one of the voltage variables. Thus the process is applied for each voltage source to reduce the number of equations and the voltage variables by one.

Consider a fairly complicated circuit shown in Fig. 5.5 with four voltage sources. We choose node 8 as the reference and write the node equations for the remaining nodes using the unknown source currents as shown:

$$\begin{array}{ll}
\text{Node 1:} & (G_1+G_2)v_1 - G_2v_4 - G_1v_7 = -i_{s1} \\
\text{Node 2:} & (G_3+G_7)v_2 - G_3v_5 = i_{s1} - i_{s2} \\
\text{Node 3:} & G_4v_3 - G_4v_6 = i_{s2} - i_{s3} \\
\text{Node 4:} & -G_2v_1 + (G_2+G_8)v_4 - G_8v_5 = 0 \\
\text{Node 5:} & -G_3v_2 - G_8v_4 + (G_3+G_8)v_5 = i_{s4} \\
\text{Node 6:} & -G_4v_3 + G_4v_6 = -i_{s4} \\
\text{Node 7:} & -G_1v_1 + (G_1+G_6)v_7 = 0 \\
\text{Node 9:} & G_5v_9 = i_{s3}
\end{array} \tag{5.11}$$

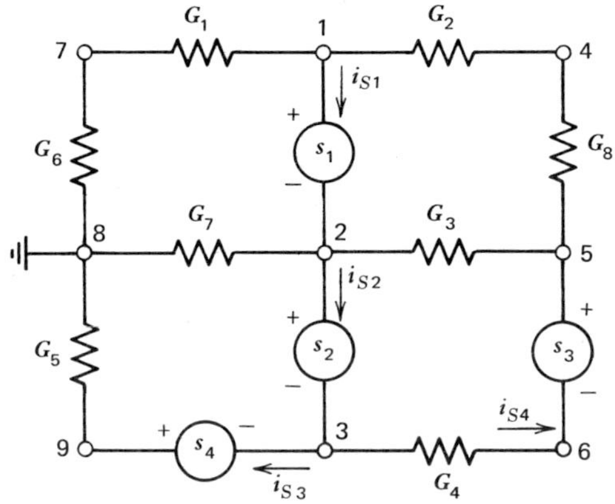

Fig. 5.5 A resistive network.

We must now eliminate the unknown source currents i_{s1}, i_{s2}, i_{s3}, and i_{s4} and thereby eliminate four equations and four constrained variables. Observe that i_{s1} can be eliminated by adding equations 1 and 2, i_{s2} by adding Eqs. 2 and 3, i_{s3} by adding Eqs. 3 and 9, and i_{s4} by adding Eqs. 5 and 6. This means that nodes 1, 2, 3 and 9 are constrained by the three sources and only one independent node variable exists in this set. Let us choose node 1 as the independent node and nodes 2, 3, and 9 as the constrained nodes. Similarly, we choose node 5 as the independent node in the set 5 and 6. The constrained voltages are given by

$$\begin{aligned} v_2 &= v_1 - s_1 \\ v_3 &= v_1 - s_1 - s_2 \\ v_9 &= v_1 - s_1 - s_2 + s_4 \\ v_6 &= v_5 - s_3 \end{aligned} \tag{5.12}$$

We substitute these constraints into Eq. 5.11, eliminate i_{s1}, i_{s2}, i_{s3} by combining Eqs. 1, 2, 3, 9, and eliminate i_{s4} by combining Eqs. 5 and 6. The final node equations are:

$$\begin{aligned} &\text{Node 1:} \quad (G_1+G_2+G_3+G_4+G_5+G_7)v_1 - G_2v_4 - (G_3+G_4)v_5 - G_1v_7 \\ &\qquad = (G_3+G_4+G_5+G_7)s_1 + (G_4+G_5)s_2 - G_4s_3 - G_5s_4 \\ &\text{Node 4:} \quad -G_2v_1 + (G_2+G_8)v_4 - G_8v_5 = 0 \\ &\text{Node 5:} \quad -(G_3+G_4)v_1 - G_8v_4 + (G_3+G_4+G_8)v_5 \\ &\qquad\qquad = -(G_3+G_4)s_1 - G_4s_2 + G_4s_3 \\ &\text{Node 7:} \quad -G_1v_1 + (G_1+G_6)v_7 = 0 \end{aligned} \tag{5.13}$$

Thus, we have four node equations in four unknown voltages v_1, v_4, v_5, and v_7. These voltages are determined from Eq. 5.13, and all other node voltages are obtained from the constraints in Eq. 5.12.

The general procedure for writing the node equations for resistive (as well as nonresistive) networks with n nodes may be summarized as follows.

1. Choose a reference node and write standard form node equations for all other nodes assuming unknown source currents through the voltage sources.
2. If an unknown source current appears in only one node equation, that node voltage is constrained by a voltage source. Delete the corresponding equation and substitute the constraint in the other equations.
3. An unknown current appearing in two node equations implies mutually constrained nodes. A set of mutually constrained nodes arises when any two of the node equations has a common unknown source current. Choose one node as independent for each constrained set and express the constrained node voltages in terms of the independent node voltage and the source voltages.
4. Combine the constrained set of node equations into one equation by eliminating the unknown source currents.
5. Substitute the constraints into the resulting equations to obtain $(n - n_V - 1)$ node equations.

5.3 LOOP ANALYSIS

The KVL loop equations are not unique, as we have seen in Chapter 1, and the choice of independent loops requires careful consideration. Once the independent loops are identified, however, the procedure is straightforward. For a class of circuits called *planar*, however, it is indeed easy to identify the independent loops. Planar circuits are defined as circuits that can be placed on a plane surface with no overlapping of any connecting wires.

This section describes the procedure for writing loop equations for resistive circuits containing either voltage or current sources. Loop identification, on the other hand, is stated only for planar circuits. The form of the equations is, of course, valid for arbitrary circuits.

Consider the planar circuit of Fig. 5.6. The circuit when placed on a plane surface (paper) has no overlapping wires and the interconnection creates *windows* as shown. The loops identified by these windows with the unknown loop currents $i_1, i_2, \ldots, i_6$ assigned give the independent loops for the circuit. This choice obviously satisfies the requirements of an independent and interrelated complete set of KVL loop equations. Thus, the window loops are the simplest and unique choice of indepen-

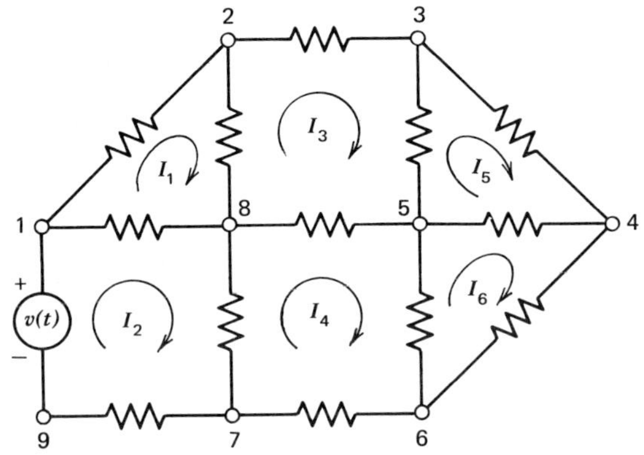

Fig. 5.6 A window-type circuit.

dent loops for all planar circuits. For *nonplanar* circuits, loop equations should be avoided as a rule. The example chosen is a singly connected circuit, but the window idea applies to any planar circuit.

Once the loops are identified, let us examine the writing of the loop equations. The procedure is similar to that described for node analysis. A few examples will illustrate the point.

Consider the circuit of Fig. 5.7 with two voltage sources; the window loops with unknown loop currents are shown. The KVL loop equations are:

$$\begin{aligned}
&\text{Loop 1:} \quad R_1(i_1 - i_3) + R_2(i_1 - i_2) = v_1 \\
&\text{Loop 2:} \quad R_3 i_2 + R_2(i_2 - i_1) = -v_2 \\
&\text{Loop 3:} \quad R_4 i_3 + R_1(i_3 - i_1) = v_2
\end{aligned} \tag{5.14}$$

Regrouping terms in Eq. 5.14, we have

$$\begin{aligned}
(R_1 + R_2)i_1 - R_2 i_2 - R_1 i_3 &= v_1 \\
-R_2 i_1 + (R_2 + R_3)i_2 &= -v_2 \\
-R_1 i_1 + (R_1 + R_4)i_3 &= v_2
\end{aligned} \tag{5.15}$$

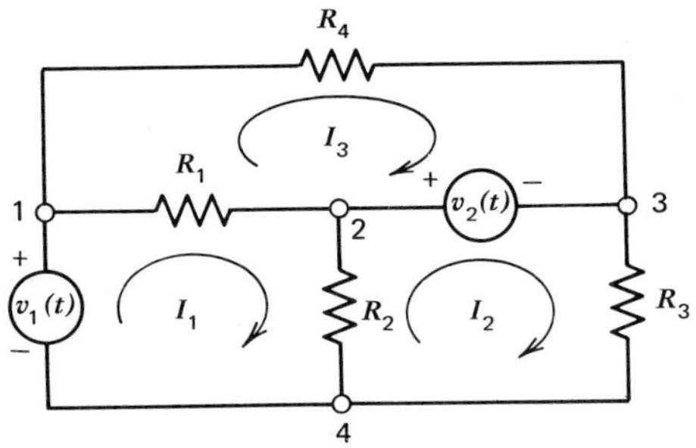

Fig. 5.7 A simple resistive network.

Observe in Eq. 5.15 that the consistent choice of clockwise direction for the flow of loop currents in the planar circuits gives us a symmetry in the signs of coefficients. The KVL loop equation for loop 1 gives a positive sign for the coefficient of i_1 and negative signs for all other coefficients—for i_2 and i_3. Similarly, the equation for loop 2 gives a positive coefficient for i_2 and negative coefficients for all other currents, and so on. In loop 1 equation, the coefficient of loop current i_1 is the sum of all resistors in loop 1, the coefficient of i_2 is the negative of the resistor common to loop 1 and loop 2, and the coefficient of i_3 is the negative of the resistor common to loop 1 and loop 3. The other loop equations have similar coefficients.

Generally, a single electrical part of a circuit with n nodes and b branches containing voltage sources only must give $l = (b - n + 1)$ *KVL* loop equations.

$$
\begin{array}{l}
a_{11}i_1 + a_{12}i_2 + \cdots + a_{1l}i_l = v_1 \\
a_{21}i_1 + a_{22}i_2 + \cdots + a_{2l}i \;\; = v_2 \\
\quad \cdot \qquad\quad \cdot \qquad\qquad\quad \cdot \quad\;\; \cdot \\
\quad \cdot \qquad\quad \cdot \qquad\qquad\quad \cdot \quad\;\; \cdot \\
\quad \cdot \qquad\quad \cdot \qquad\qquad\quad \cdot \quad\;\; \cdot \\
a_{l1}i_1 + a_{l2}i_2 + \cdots + a_{ll}i_l = v_l
\end{array}
\tag{5.16}
$$

The coefficients a_{ij} for planar circuits with consistent loop current directions are given by

$$
\begin{aligned}
a_{ii} &= \text{sum of resistances in loop } i \\
a_{ij} &= \text{negative sum of resistances common to loops } i \text{ and } j
\end{aligned}
\tag{5.17}
$$

where $i, j = 1, 2, \ldots, l$.

When current sources are present in loops, some of the loop currents are constrained. If the current source is present in a single loop, that loop current is specified by the source current, and no loop equation is necessary for that loop. If a current source has two loops in common, these loops have mutually constrained currents. One of the loop currents is selected as independent and the other as constrained. Loop equations are then written as usual for all loops with unknown voltage drops assumed across current sources. These unknown source voltage drops are then eliminated by combining the corresponding equations as in the case of node equations. The constrained loop current is also eliminated by introducing the appropriate constraint. If there are n_i current sources, the number of independent loops is $l = (b - n - n_i + 1)$. The procedure is identical to the case of voltage sources in node equations. A simple example illustrates the procedure.

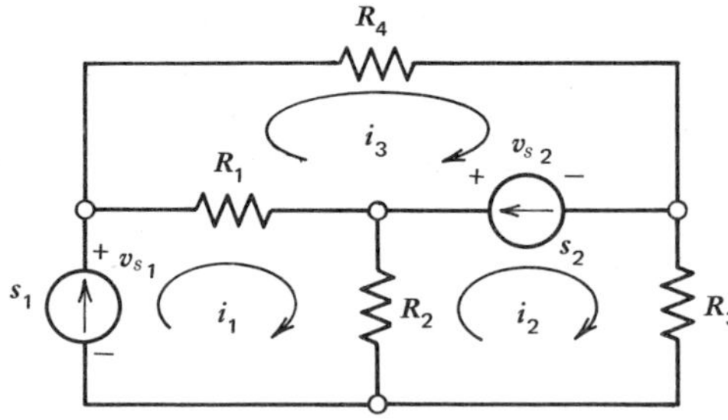

Fig. 5.8

Consider the circuit of Fig. 5.8 with two current sources. The loop equations are:

$$\begin{array}{llllll} \text{Loop 1:} & (R_1+R_2)i_1 - & R_2 i_2 & - & R_1 i_3 & = v_{s1} \\ \text{Loop 2:} & -R_2 i_1 & +(R_2+R_3)i_2 & & & = -v_{s2} \\ \text{Loop 3:} & -R_1 i_1 & & + & (R_1+R_4)i_3 & = v_{s2} \end{array} \tag{5.18}$$

The unknown source voltage v_{s1} occurs in one equation only and the equation is therefore deleted and $i_1 = s_1$ substituted. The other two equations are combined to eliminate v_{s2}. We choose loop 3 as the independent loop and substitute the constraint $i_2 = i_3 - s_2$ to give

$$-(R_1+R_2)s_1 + (R_1+R_2+R_3+R_4)i_3 - (R_2+R_3)s_2 = 0$$

or

$$(R_1+R_2+R_3+R_4)i_3 = (R_1+R_2)s_1 + (R_2+R_3)s_2 \tag{5.19}$$

This is the only loop equation necessary to analyze the circuit. From i_3, we obtain $i_2 = i_3 - s_2$ and i_1 is known already. All the loop currents are determined.

The procedure for writing loop equations for planar circuits is thus quite similar to that for writing node equations. In both cases, the resulting equations are simultaneous algebraic equations. The solution of these equations is discussed in the next section.

5.4 NUMERICAL SOLUTION OF LINEAR ALGEBRAIC EQUATIONS

Resistive networks have been described by systems (sets) of linear algebraic equations. The general solution of such equations can be obtained using determinants by Cramer's rule if the system is nonsingular. Equivalently, the solution can be obtained by inversion of the matrix of coefficients. Determinants and matrices are discussed in the appendix. They are fine for formal expressions of the solutions, but are quite impossible to use for practical numerical solutions of large sets of equations. When the number of equations becomes large, the amount of computation and the computational errors are prohibitive for direct applications of determinants or matrix inversions. In this section, we describe an efficient and widely used numerical method or algorithm to solve a set of algebraic equations.

In the previous sections resistive networks are described, in general, by a set of n linear algebraic equations in n unknowns:

$$\begin{aligned} a_{11}x_1 + a_{12}x_2 + \cdots + a_{1n}x_n &= b_1 \\ a_{21}x_1 + a_{22}x_2 + \cdots + a_{2n}x_n &= b_2 \\ \cdot \qquad \cdot \qquad \quad \cdot \quad &\quad \cdot \\ \cdot \qquad \cdot \qquad \quad \cdot \quad &\quad \cdot \\ \cdot \qquad \cdot \qquad \quad \cdot \quad &\quad \cdot \\ a_{n1}x_1 + a_{n2}x_2 + \cdots + a_{nn}x_n &= b_n \end{aligned} \tag{5.20}$$

where the coefficients a_{ij} and the constants b_i $(i, j = 1, 2, \ldots, n)$ are given. The problem is to find the values of $x_i (i = 1, 2, \ldots, n)$, if they exist, which satisfy Eq. 5.20. In matrix notation Eq. 5.20 takes the form

$$A\mathbf{x} = \mathbf{b} \tag{5.21}$$

where A is the matrix of coefficients, and $\mathbf{x}$, $\mathbf{b}$ are column vectors whose components are respectively x_i, b_i $(i = 1, 2, \ldots, n)$; that is,

$$A = \begin{bmatrix} a_{11} & a_{12} & \cdots & a_{1n} \\ a_{21} & a_{22} & \cdots & a_{2n} \\ \cdots & \cdots & \cdots & \cdots \\ a_{n1} & a_{n2} & \cdots & a_{nn} \end{bmatrix}, \quad \mathbf{x} = \begin{bmatrix} x_1 \\ x_2 \\ \cdots \\ x_n \end{bmatrix}, \quad \mathbf{b} = \begin{bmatrix} b_1 \\ b_2 \\ \cdots \\ b_n \end{bmatrix} \tag{5.22}$$

If the constants b_i in Eq. 5.20 are all zero, the system of equations is said to be homogeneous. On the other hand, if not all of the b_i are zero, then it is called an inhomogeneous system. It can be shown that Eq. 5.20 has a unique solution if and only if the coefficient matrix A is nonsingular or, equivalently, that the determinant of A is not equal to zero. In such circumstance, the solution is given by

$$\mathbf{x} = A^{-1}\mathbf{b} \tag{5.23}$$

where A^{-1} is the inverse of A. In the appendix it is shown that the inverse matrix A^{-1} is given by

$$A^{-1} = \frac{1}{\Delta} \begin{bmatrix} A_{11} & A_{21} & \cdots & A_{n1} \\ A_{12} & A_{22} & \cdots & A_{n2} \\ \cdots & \cdots & \cdots & \cdots \\ A_{1n} & A_{2n} & \cdots & A_{nn} \end{bmatrix} \tag{5.24}$$

where Δ is the determinant of matrix A and A_{ij} are the cofactors of the matrix A defined by $A_{ij} = (-1)^{i+j} M_{ij}$. The minors M_{ij} are determinants of the matrix A with ith row and jth column removed, that is,

$$M_{ij} = \begin{vmatrix} a_{11} & a_{12} & \cdots & a_{1j} & \cdots & a_{1n} \\ \cdots & \cdots & \cdots & \cdots & \cdots & \cdots \\ \not{a}_{i1} & \not{a}_{i2} & \cdots & \not{a}_{ij} & \cdots & \not{a}_{in} \\ \cdots & \cdots & \cdots & \cdots & \cdots & \cdots \\ a_{n1} & a_{n2} & \cdots & a_{nj} & \cdots & a_{nn} \end{vmatrix} \leftarrow i\text{th row} \tag{5.25}$$

$\uparrow$
jth column

and

$$A_{ij} = (-1)^{i+j} M_{ij} \tag{5.26}$$

for $i, j = 1, 2, \ldots, n$. Cramer's rule for solving simultaneous algebraic equations is equivalent to matrix inversion. By Cramer's rule given in the appendix,

$$\begin{aligned} x_i &= \frac{1}{\Delta}(A_{1i}b_1 + A_{2i}b_2 + \cdots + A_{ni}b_n) \\ &= \frac{1}{\Delta}\sum_{j=1}^{n} A_{ji}b_j \end{aligned} \tag{5.27}$$

which is equivalent to Eqs. 5.23 and 5.24.

The solution of a set of n algebraic equations thus requires the evaluation of several determinants; in general, we require the determinant of A and for each unknown we must evaluate n cofactors in addition. When the value of n is large, the computational problem is formidable and the errors can be quite large. An idea of the computational problem involved can be illustrated by an example for $n = 3$. Consider the set

$$\begin{aligned} 11x_1 - 5x_2 \phantom{{}-3x_3} &= 5 \\ -10x_1 + 13x_2 - 3x_3 &= 1 \\ -x_2 + 6x_3 &= 1 \end{aligned} \tag{5.28}$$

or

$$\begin{bmatrix} 11 & -5 & 0 \\ -10 & 13 & -3 \\ 0 & -1 & 6 \end{bmatrix} \begin{bmatrix} x_1 \\ x_2 \\ x_3 \end{bmatrix} = \begin{bmatrix} 5 \\ 1 \\ 1 \end{bmatrix} \tag{5.29}$$

The solution is given by

$$\begin{bmatrix} x_1 \\ x_2 \\ x_3 \end{bmatrix} = \begin{bmatrix} 11 & -5 & 0 \\ -10 & 13 & -3 \\ 0 & -1 & 6 \end{bmatrix}^{-1} \begin{bmatrix} 5 \\ 1 \\ 1 \end{bmatrix} \tag{5.30}$$

$$= \frac{1}{525}\begin{bmatrix} 75 & 30 & 15 \\ 60 & 66 & 33 \\ 10 & 11 & 93 \end{bmatrix}\begin{bmatrix} 5 \\ 1 \\ 1 \end{bmatrix} \tag{5.31}$$

$$= \frac{1}{525}\begin{bmatrix} 420 \\ 399 \\ 154 \end{bmatrix} \tag{5.32}$$

To obtain the inverse matrix in Eq. 5.31 we have to evaluate nine (2×2) determinants for the cofactors and one (3×3) determinant.

By Cramer's rule the solutions are obtained as

$$x_1 = \frac{1}{\Delta}\begin{vmatrix} 5 & -5 & 0 \\ 1 & 13 & -3 \\ 1 & -1 & 6 \end{vmatrix} = \frac{420}{525} \tag{5.33}$$

$$x_2 = \frac{1}{\Delta}\begin{vmatrix} 11 & 5 & 0 \\ -10 & 1 & -3 \\ 0 & 1 & 6 \end{vmatrix} = \frac{399}{525} \tag{5.34}$$

$$x_3 = \frac{1}{\Delta}\begin{vmatrix} 11 & -5 & 5 \\ -10 & 13 & 1 \\ 0 & -1 & 1 \end{vmatrix} = \frac{154}{525} \tag{5.35}$$

where

$$\Delta = \begin{vmatrix} 11 & 5 & 0 \\ -10 & 13 & -3 \\ 0 & -1 & 6 \end{vmatrix} = 525 \tag{5.36}$$

In this case, we have evaluated four (3×3) determinants or twelve (2×2) determinants, the same as before.

In general, the solution of n equations requires the evaluation of one $(n \times n)$ determinant and n^2 $(n-1) \times (n-1)$ determinants. When n is of the order of a hundred, it is easy to see that the number of determinants becomes huge and the evaluation of each determinant of order 100 is formidable in itself. Efficient numerical methods thus become essential for the solution of such equations.

We now discuss an elementary and yet very efficient numerical method known as Gaussian elimination method to solve the nonhomogeneous system of Equations, that is, Eq. 5.20.

Gaussian Elimination with Pivoting

The objective of the Gaussian elimination is to transform the given system by sequential elimination of the unknown variables in Eq. 5.20 into an equivalent triangular system whose solution is easily obtained.

For illustration, let us consider a general system of three linear equations

$$\begin{aligned} a_{11}x_1 + a_{12}x_2 + a_{13}x_3 &= b_1 \\ a_{21}x_1 + a_{22}x_2 + a_{23}x_3 &= b_2 \\ a_{31}x_1 + a_{32}x_2 + a_{33}x_3 &= b_3 \end{aligned} \tag{5.37}$$

Assume that $a_{11} \neq 0$. (We call a_{11} the first pivot.) To eliminate the x_1 terms from the second and third equations, we multiply the first equation by a_{21}/a_{11} and a_{31}/a_{11} and subtract it respectively from the second and third equations (leaving the first equation unchanged). We then have an equivalent system of equations given by

$$\begin{aligned} a_{11}x_1 + a_{12}x_2 + a_{13}x_3 &= b_1 \\ a_{22}^{(1)}x_2 + a_{23}^{(1)}x_3 &= b_2^{(1)} \\ a_{32}^{(1)}x_2 + a_{33}^{(1)}x_3 &= b_3^{(1)} \end{aligned} \tag{5.38}$$

where

$$a_{22}^{(1)} = a_{22} - \frac{a_{21}}{a_{11}} a_{12} \qquad a_{23}^{(1)} = a_{23} - \frac{a_{21}}{a_{11}} a_{13}$$

$$a_{32}^{(1)} = a_{32} - \frac{a_{31}}{a_{11}} a_{12} \qquad a_{33}^{(1)} = a_{33} - \frac{a_{31}}{a_{11}} a_{13}$$

$$b_2^{(1)} = b_2 - \frac{a_{21}}{a_{11}} b_1 \qquad b_3^{(1)} = b_3 - \frac{a_{31}}{a_{11}} b_1$$

If $a_{22}^{(1)} \neq 0$, we call $a_{22}^{(1)}$ the second pivot. The x_2 term in the third equation is eliminated by first multiplying the second equation by $a_{32}^{(1)}/a_{22}^{(1)}$ and then subtracting it from the third equation in Eq. 5.38. We thus have the equivalent triangular system of equations

$$\begin{aligned} a_{11}x_1 + a_{12}x_2 + a_{13}x_3 &= b_1 \\ a_{22}^{(1)}x_2 + a_{23}^{(1)}x_3 &= b_2^{(1)} \\ a_{33}^{(2)}x_3 &= b_3^{(2)} \end{aligned} \tag{5.39}$$

where

$$a_{33}^{(2)} = a_{33}^{(1)} - \frac{a_{32}^{(1)}}{a_{22}^{(1)}} a_{23}^{(1)}$$

$$b_3^{(2)} = b_3^{(1)} - \frac{a_{32}^{(1)}}{a_{22}^{(1)}} b_2^{(1)}$$

The solution of the triangular system of Eq. 5.39 is the required solution of the original system of Eq. 5.37. If none of the pivots a_{11}, $a_{22}^{(1)}$, $a_{33}^{(2)}$ vanish, the solution of the triangular system is obtained in reverse order by back-substitution as follows:

$$
\begin{aligned}
x_3 &= \frac{b_3^{(2)}}{a_{33}^{(2)}} \\
x_2 &= \frac{1}{a_{22}^{(1)}}\left(b_2^{(1)} - a_{23}^{(1)}x_3\right) \\
x_1 &= \frac{1}{a_{11}}\left(b_1 - a_{12}x_2 - a_{13}x_3\right)
\end{aligned}
\tag{5.40}
$$

The process of elimination can be carried out more conveniently in matrix form. Thus let the augmented matrix of Eq. 5.20 be defined as the matrix A with the column vector $\mathbf{b}$ adjoined. Therefore, for the system in Eq. 5.37 the augmented matrix is

$$
\begin{matrix} m \\ \\ \dfrac{a_{21}}{a_{11}} \\ \dfrac{a_{31}}{a_{11}} \end{matrix}
\quad
\begin{bmatrix}
a_{11} & a_{12} & a_{13} & b_1 \\
a_{21} & a_{22} & a_{23} & b_2 \\
a_{31} & a_{32} & a_{33} & b_3
\end{bmatrix}
\tag{5.41}
$$

where the column m indicates the multipliers needed during the first stage of elimination. Performing the elimination, we obtain for the second stage

$$
\begin{matrix} m \\ \\ \\ \dfrac{a_{32}^{(1)}}{a_{22}^{(1)}} \end{matrix}
\quad
\begin{bmatrix}
a_{11} & a_{12} & a_{13} & b_1 \\
0 & a_{22}^{(1)} & a_{23}^{(1)} & b_2^{(1)} \\
0 & a_{32}^{(1)} & a_{33}^{(1)} & b_3^{(1)}
\end{bmatrix}
\tag{5.42}
$$

Eliminating x_2 in the second column using the multiplier $a_{32}^{(1)}/a_{22}^{(1)}$, we arrive at the triangular system

$$
\begin{bmatrix}
a_{11} & a_{12} & a_{13} & b_1 \\
 & a_{22}^{(1)} & a_{23}^{(1)} & b_2^{(1)} \\
 & & a_{33}^{(2)} & b_3^{(2)}
\end{bmatrix}
\tag{5.43}
$$

which corresponds to Eq. 5.39. This is now solved using Eq. 5.40.

Example 5.4.1 Let us now solve Eq. 5.28 using Gaussian elimination. In matrix notation we have the augmented matrix of Eq. 5.28 as

$$\begin{matrix} m \\ \\ -\frac{10}{11} \\ 0 \end{matrix} \begin{bmatrix} 11 & -5 & 0 & 5 \\ -10 & 13 & -3 & 1 \\ 0 & -1 & 6 & 1 \end{bmatrix} \tag{5.44}$$

Multiplying the first row by $-10/11$ and subtracting from the second row, we eliminate the x_1 term from the second equation. The third equation has no x_1 term as indicated by the zero multiplier and the third row is thus unchanged. Performing the elimination, we have

$$\begin{matrix} m \\ \\ \\ -\frac{11}{93} \end{matrix} \begin{bmatrix} 11 & -5 & 0 & 5 \\ 0 & \frac{93}{11} & -3 & \frac{61}{11} \\ 0 & -1 & 6 & 1 \end{bmatrix}$$

The x_2 term is now eliminated from the third row by multiplying the second row by $-11/93$ and subtracting it from the third row. We thus have the triangular system.

$$\begin{bmatrix} 11 & -5 & 0 & 5 \\ 0 & \frac{93}{11} & -3 & \frac{61}{11} \\ 0 & 0 & \frac{525}{93} & \frac{154}{93} \end{bmatrix}$$

The solution is obtained by back-substitution as in Eq. 5.40:

$$\begin{aligned} x_3 &= \frac{154}{525} \\ x_2 &= \frac{11}{93}\left[\frac{61}{11} + 3\left(\frac{154}{525}\right)\right] = \frac{399}{525} \\ x_1 &= \frac{1}{11}\left[5 + 5\left(\frac{399}{525}\right)\right] = \frac{420}{525} \end{aligned} \tag{5.45}$$

The solution is thus obtained fairly easily with minimal computation.

This procedure of elimination and back-substitution can be applied to a system of equations of any order. However, the elimination process would fail if, for our 3×3 system, either a_{11} or $a_{22}^{(1)}$ vanishes since the multipliers necessary for elimination would be indeterminate in this case. Hence, we should rearrange the rows (i.e., the ordering of the equations) so as to obtain a nonvanishing pivot. If this is impossible, then the matrix A is singular, and the system has no solution. The same conclusion holds if $a_{33}^{(2)}$ vanishes.

Furthermore, to ensure minimum error during the process of elimination or back-substitution, small multipliers are required as is indicated in Chapter 1. Therefore, we arrange to have, at any stage of elimination, a pivot with magnitude larger than or equal to any remaining element in the column below the pivot element. This would then give us all multipliers with magnitudes less than one. This elimination process with pivots chosen as the largest elements in the lower column is called *Gaussian elimination with pivoting*.

The above example of Eq. 5.44 had the pivot elements that were the largest elements for the remainder of the columns and the pivoting was satisfied. Let us now consider an example for which pivoting needs rearranging.

Example 5.4.2 Solve the following system by elimination without pivoting:

$$\begin{aligned} 2x_1 + 3x_2 - x_3 &= 5 \\ 4x_1 + 4x_2 - 3x_3 &= 3 \\ 2x_1 - 3x_2 + x_3 &= -1 \end{aligned} \tag{5.46}$$

Solution. In matrix notation the augmented matrix of Eq. 5.46 is

$$\begin{matrix} m \\ \\ 2 \\ 1 \end{matrix} \quad \begin{matrix} \\ \begin{bmatrix} 2 & 3 & -1 & 5 \\ 4 & 4 & -3 & 3 \\ 2 & -3 & 1 & -1 \end{bmatrix} \end{matrix}$$

Multiplying the first row by 2 and 1 and subtracting it from the second row and third row, respectively, we have eliminated the x_1 terms from the second and third equations. The results are

$$\begin{matrix} m \\ \\ \\ 3 \end{matrix} \quad \begin{matrix} \\ \begin{bmatrix} 2 & 3 & -1 & 5 \\ & -2 & -1 & -7 \\ & -6 & 2 & -6 \end{bmatrix} \end{matrix}$$

The x_2 term is then eliminated from the third equation by multiplying the second row by 3 and subtracting it from the third row. Hence we have the triangular system

$$\begin{bmatrix} 2 & 3 & -1 & 5 \\ & -2 & -1 & -7 \\ & & 5 & 15 \end{bmatrix}$$

And the solution of Eq. 5.46 is obtained as follows:

$$x_3 = \frac{15}{5} = 3$$

$$x_2 = -\tfrac{1}{2}[-7-(-1)(3)] = 2 \tag{5.47}$$

and

$$x_1 = \tfrac{1}{2}[5-(3)(2)-(-1)(3)] = 1$$

Example 5.4.3 Solve the example of Eq. 5.46 using Gaussian elimination with pivoting.

Solution. The augmented matrix for the system is

$$\begin{bmatrix} 2 & 3 & -1 & 5 \\ 4 & 4 & -3 & 3 \\ 2 & -3 & 1 & -1 \end{bmatrix}$$

Step 1. Interchange rows 1 and 2 so that the pivot is the largest number in column 1. We have

$$\begin{matrix} m \\ \\ \tfrac{1}{2} \\ \tfrac{1}{2} \end{matrix} \begin{bmatrix} 4 & 4 & -3 & 3 \\ 2 & 3 & -1 & 5 \\ 2 & -3 & 1 & -1 \end{bmatrix}$$

Step 2. Eliminating x_2 terms from the second and third equations, we have

$$\begin{bmatrix} 4 & 4 & -3 & 3 \\ & 1 & \tfrac{1}{2} & \tfrac{7}{2} \\ & -5 & \tfrac{5}{2} & -\tfrac{5}{2} \end{bmatrix}$$

Step 3. Interchange rows 2 and 3. We have

$$\begin{matrix} m \\ \\ \\ -\tfrac{1}{5} \end{matrix} \begin{bmatrix} 4 & 4 & -3 & 3 \\ & -5 & \tfrac{5}{2} & -\tfrac{5}{2} \\ & 1 & \tfrac{1}{2} & \tfrac{7}{2} \end{bmatrix}$$

Step 4. Eliminating x_2 term from the third equation, we arrive at

$$\begin{bmatrix} 4 & 4 & -3 & 3 \\ & -5 & \frac{5}{2} & -\frac{5}{2} \\ & & 1 & 3 \end{bmatrix}$$

Step 5. Solve by back-substitution,

$$\begin{aligned} x_3 &= 3 \\ x_2 &= -\tfrac{1}{5}[-\tfrac{5}{2}-\tfrac{5}{2}(3)] \qquad = 2 \\ x_1 &= \tfrac{1}{4}[3-4(2)-(-3)(3)] = 1 \end{aligned}$$

The solution for the above simple example is performed with exact numbers. In general, of course, the computational results are given approximately. Let us consider another example to illustrate the actual computation. We choose an example for which the exact answers are easily obtained.

Example 5.4.4 Use Gaussian elimination with pivoting to solve the equations in Eq. 5.48.

Solution. The original equations are:

$$\begin{aligned} 11v_1 - 5v_2 \qquad &= 5 \\ -10v_1 + 13v_2 - 3v_3 &= 0 \\ -v_2 + 6v_3 &= 0 \end{aligned} \tag{5.48}$$

The exact solution is given by

$$\begin{bmatrix} v_1 \\ v_2 \\ v_3 \end{bmatrix} = \begin{bmatrix} 11 & -5 & 0 \\ -10 & 13 & -3 \\ 0 & -1 & 6 \end{bmatrix}^{-1} \begin{bmatrix} 5 \\ 0 \\ 0 \end{bmatrix} \tag{5.49}$$

The inverse matrix is given in Eq. 5.31 and we have

$$\begin{aligned} \begin{bmatrix} v_1 \\ v_2 \\ v_3 \end{bmatrix} &= \frac{1}{525} \begin{bmatrix} 75 & 30 & 15 \\ 60 & 66 & 33 \\ 10 & 11 & 93 \end{bmatrix} \begin{bmatrix} 5 \\ 0 \\ 0 \end{bmatrix} \\ &= \frac{1}{105} \begin{bmatrix} 75 \\ 60 \\ 10 \end{bmatrix} \end{aligned} \tag{5.50}$$

The exact answers are

$$\begin{aligned} v_1 &= 0.71428 \\ v_2 &= 0.57142 \\ v_3 &= 0.09523 \end{aligned} \tag{5.51}$$

correct to five decimal places.

Let us now use Gaussian elimination with pivoting. For this system the augmented matrix is

$$\begin{bmatrix} 11 & -5 & 0 & 5 \\ -10 & 13 & 3 & 0 \\ 0 & -1 & 6 & 0 \end{bmatrix}$$

Since 11 is already the largest element in the first column, no row interchange is necessary. If we perform the computation retaining five decimal places, we have the following results:

Step 1. No row interchange is necessary; we have

$$\begin{matrix} m \\ \\ -0.90909 \\ 0.00000 \end{matrix} \quad \begin{matrix} \\ \begin{bmatrix} 11 & -5 & 0 & 5 \\ -10 & 13 & -3 & 0 \\ 0 & -1 & 6 & 0 \end{bmatrix} \end{matrix}$$

Step 2. Eliminating v_2 from the second and third equation, we have

$$\begin{bmatrix} 11 & -5 & 0 & 5 \\ 0 & 8.45455 & -3 & 4.54545 \\ 0 & -1 & 6 & 0 \end{bmatrix}$$

Step 3. No row interchange is necessary; we have

$$\begin{matrix} m \\ \\ \\ -0.11828 \end{matrix} \quad \begin{matrix} \\ \begin{bmatrix} 11 & -5 & 0 & 5 \\ 0 & 8.45455 & -3 & 4.54545 \\ 0 & -1 & 6 & 0 \end{bmatrix} \end{matrix}$$

Step 4. Eliminating v_3 from the third equation, we have

$$\begin{bmatrix} 11 & -5 & 0 & 5 \\ 0 & 8.45455 & -3 & 4.54545 \\ 0 & 0 & 5.64516 & 0.53764 \end{bmatrix}$$

Step 5. Solve the resulting triangular system by back-substitution. We have

$$v_3 = \frac{0.53764}{5.64516} = 0.09523$$

$$v_2 = \frac{1}{8.45455}(4.54545 - (-3)v_3) = 0.57142$$

$$v_1 = \frac{1}{11}(5 - (-5)v_2 - 0v_3) = 0.71428$$

A final note on the elimination method. Sometimes we wish to find the inverse of a matrix, and the elimination method can be used for that also. It turns out to be computationally more advantageous to use the elimination method rather than the matrix inversion approach. The inverse matrix is defined by

$$AA^{-1} = \mathrm{I}$$

where I is an nth order identity matrix. By the rule of matrix operation, the multiplication of A and the first column of A^{-1} is equal to the first column of I; the multiplication of A and the second column of A^{-1} is equal to the second column of I; and so forth. Hence, if we assign the **b** vector in the system $A\mathbf{x} = \mathbf{b}$ as any column of the identity matrix, then we would obtain a corresponding column of the A^{-1} as the solution vector **x**.

Example 5.4.5 Find the first column of the inverse of the matrix

$$A = \begin{bmatrix} 2 & 3 & -1 \\ 4 & 4 & -3 \\ 2 & -3 & 1 \end{bmatrix}$$

Solution. By definition the required column is the solution to the system $A\mathbf{x} = \mathbf{b}$ with $\mathbf{b} = [1 \;\; 0 \;\; 0]^T$. Applying Gaussian elimination without pivoting, we have the following result:

$$\begin{matrix} \\ 2 \\ 1 \end{matrix} \begin{bmatrix} 2 & 3 & -1 & 1 \\ 4 & 4 & -3 & 0 \\ 2 & -3 & 1 & 0 \end{bmatrix}$$

$$\begin{matrix} \\ \\ 3 \end{matrix} \begin{bmatrix} 2 & 3 & -1 & 1 \\ & -2 & -1 & -2 \\ & -6 & 2 & -1 \end{bmatrix}$$

$$\begin{bmatrix} 2 & 3 & -1 & 1 \\ & -2 & -1 & -2 \\ & & 5 & 5 \end{bmatrix}$$

Hence we have

$$x_3 = \tfrac{5}{5} = 1$$

$$x_2 = \tfrac{1}{-2}(-2 + 1 \times 1) = \tfrac{1}{2}$$

$$x_1 = \tfrac{1}{2}[1 - 3(\tfrac{1}{2}) - 1\,(1)] = \tfrac{1}{4}$$

As a check, we carry out the multiplication of $A\mathbf{x}$ and obtain

$$A\mathbf{x} = \begin{bmatrix} 2 & 3 & -1 \\ 4 & 4 & -3 \\ 2 & -3 & 1 \end{bmatrix} \begin{bmatrix} \frac{1}{4} \\ \frac{1}{2} \\ 1 \end{bmatrix} = \begin{bmatrix} 1 \\ 0 \\ 0 \end{bmatrix}$$

Computer Program

To implement the Gaussian elimination procedure, we employ a two dimensional array $A(I, J)$, where I and J ranges from 1 to n, to store the respective element a_{ij} of the system matrix A. For the **b** vector, a vector $B(k)$ is also used where $k = 1, 2, \cdots, n$. A subroutine GE(N, A, B) is written to solve up to a system of 20 linear equations based on the flow-chart in Fig. 5.9. In order to use GE, the system order N, the matrix A,

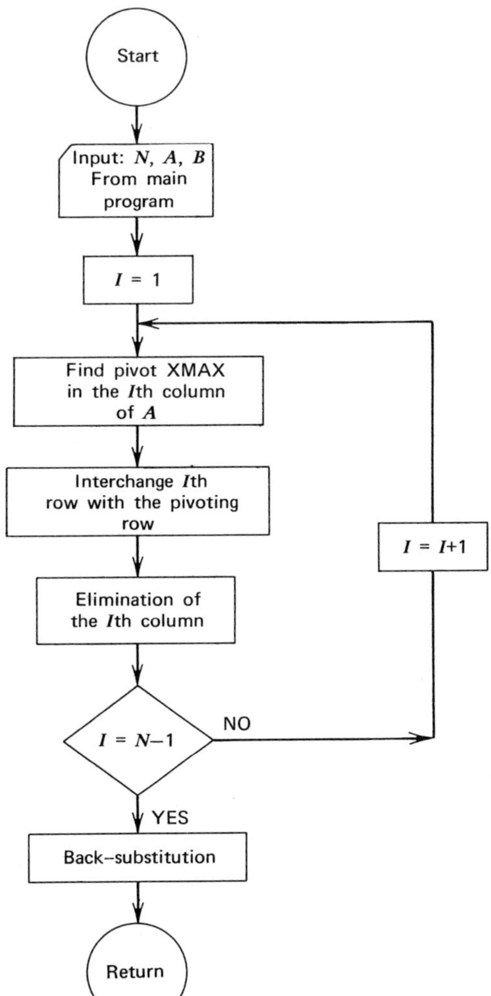

Fig. 5.9 Gaussian elimination subroutine GE.

and the vector B have to be defined. Then a calling to GE(N, A, B) will give us the result $\mathbf{x} = A^{-1}\mathbf{b}$ stored in B. In case we have more than 20 linear equations in 20 variables, then the dimension of each vector or array has to be changed. Thus the dimension statement in GE should be

$$\text{DIMENSION } A(50{,}50), B(50)$$

if we have a system of 50 equations in 50 unknowns.

Example 5.4.6 Solve Example 5.4.4 using the subroutine GE. The exact solution is given as

$$v_1 = 0.71428$$
$$v_2 = 0.57142$$
$$v_3 = 0.09523$$

The matrix A and vector $\mathbf{b}$ are

$$A = \begin{bmatrix} 11 & -5 & 0 \\ 10 & -13 & 3 \\ 0 & 1 & -6 \end{bmatrix} \qquad \mathbf{b} = \begin{bmatrix} 5 \\ 0 \\ 0 \end{bmatrix}$$

Solution. The FORTRAN program, the subroutine GE, input data listing and the computer output are illustrated in Figs. 5.10 through 5.12. The parameters are the following:

N	Number of equations
A	N by N coefficient matrix
B	N vector, right-hand side of $A\mathbf{x} = \mathbf{b}$, also the solution $\mathbf{x} = A^{-1}\mathbf{b}$
$N1$	$N-1$
$XMAX$	Maximum element in a column
L	Location indicator of $XMAX$
$SAVE$	Temporary storage used when interchanging rows

```
      DIMENSICN A(10,10),B(10)
    1 FCRMAT(I5)
    2 FCRMAT(F10.5)
    3 FORMAT(10X,E12.5)
    4 READ 1,N
      READ 2, ((A(I,J), I=1,N), J=1,N)
      PRINT 3, ((A(I,J), I=1,N), J=1,N)
      READ 2, (B(I), I=1,N)
      PRINT 3, (B(I), I=1,N)
      CALL GE(N,A,B)
      PRINT 3, (B(I), I=1,N)
      GC TO 4
      END
```

Fig. 5.10 Solving a system of N linear equations.

```
      SUBROUTINE GE(N,A,B)
      DIMENSION A(10,10),B(10)
      N1=N-1
      DO 300 I=1,N1
        XMAX=0.
        DO 40 K=I,N
          IF(XMAX-ABS(A(K,I))) 50,40,40
 50       XMAX=A(K,I)
          L=K
 40     CONTINUE
        IF(ABS(XMAX)-0.0) 60,60,70
 60     PRINT 80
 80     FORMAT(10X,19H SYSTEM IS SINGULAR)
        RETURN
 70     DO 90 J=I,N
          SAVE=A(I,J)
          A(I,J)=A(L,J)
          A(L,J)=SAVE
 90     CONTINUE
        SAVE=B(I)
        B(I)=B(L)
        B(L)=SAVE
        KK=I+1
        DO 100 K=KK,N
          A(K,I)=A(K,I)/XMAX
100     CONTINUE
        DO 120 K=KK,N
          DO 110 JK=KK,N
            A(K,JK)=A(K,JK)-A(K,I)*A(I,JK)
110       CONTINUE
          B(K)=B(K)-A(K,I)*B(I)
120     CONTINUE
300   CONTINUE
      B(N)=B(N)/A(N,N)
      DO 510 I=1,N1
        IB=N-I
        IC=N
        DO 500 K=1,I
          B(IB)=B(IB)-A(IB,IC)*B(IC)
          IC=IC-1
500     CONTINUE
        B(IB)=B(IB)/A(IB,IB)
510   CONTINUE
      RETURN
      END
```

Fig. 5.11 Subroutine GE (N,A,B).

```
First column of A:        0.11000E 02
                          0.10000E 02
                          0.0

Second column of A:      -0.50000E 01
                         -0.13000E 02
                          0.10000E 01

Third column of A:        0.0
                          0.30000E 01
                         -0.60000E 01

Vector b:                 0.50000E 01
                          0.0
                          0.0

Solution x = A^-1 b:      0.71429E 00
                          0.57143E 00
                          0.95238E-01
```

Fig. 5.12 Input data and computer output.

PROBLEMS

5-1. Using chopped floating-point arithmetic operations with 5-digit mantissa, solve the following system of linear equations by Gaussian elimination with pivoting.

(i) $A\mathbf{x} = \mathbf{b}$ where

$$A = \begin{bmatrix} 9 & -36 & 30 \\ -36 & 192 & -180 \\ 30 & -180 & 180 \end{bmatrix} \qquad \mathbf{b} = \begin{bmatrix} 1 \\ 0 \\ 0 \end{bmatrix}$$

(ii) $A\mathbf{x} = \mathbf{b}$ where

$$A = \begin{bmatrix} 33 & 16 & 72 \\ -24 & -10 & -57 \\ -8 & -4 & -17 \end{bmatrix} \qquad \mathbf{b} = \begin{bmatrix} 359 \\ -281 \\ -85 \end{bmatrix}$$

5-1a. Repeat 5-1 for the following systems.

(i) $A\mathbf{x} = \mathbf{b}$ where

$$A = \begin{bmatrix} 3 & 5 & 1 \\ 2 & 4 & 5 \\ 1 & 2 & 2 \end{bmatrix} \qquad \mathbf{b} = \begin{bmatrix} -4 \\ 9 \\ 3 \end{bmatrix}$$

(ii) $A\mathbf{x} = \mathbf{b}$ where

$$A = \begin{bmatrix} 1 & 2 & 3 \\ 2 & 3 & 4 \\ 5 & 6 & 6 \end{bmatrix} \qquad \mathbf{b} = \begin{bmatrix} 2 \\ 3 \\ 5 \end{bmatrix}$$

5-1b. Repeat 5-1 using 3-digit mantissa. Is the answer less accurate?

5-1c. Repeat 5-1b for the systems of 5-1a.

5-2. For the circuit of Fig. P5-2, find the steady-state capacitor voltages v_{C1}, v_{C2} if $v_1(t) = 2V$ and $v_2(t) = 4V$.

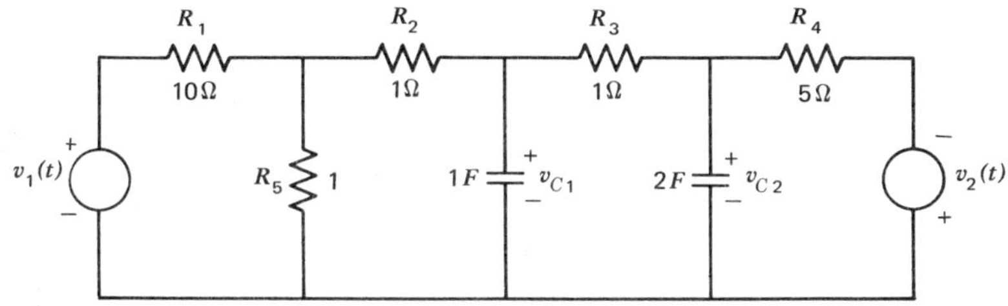

Fig. P5-2

5-3. For the circuit of Fig. P5-2, if initially $v_{C1}(0-) = 5V$ and $v_{C2}(0-) = 10V$ while $v_1(t)$ and $v_2(t)$ are given as in problem 5-2, determine the currents through the resistors at $t = 0$. Will v_{C1} and v_{C2} settle to the steady-state values found in 5-2? Why?

5-4. For the circuit of Fig. P5-4, find the final states of the reactive elements.

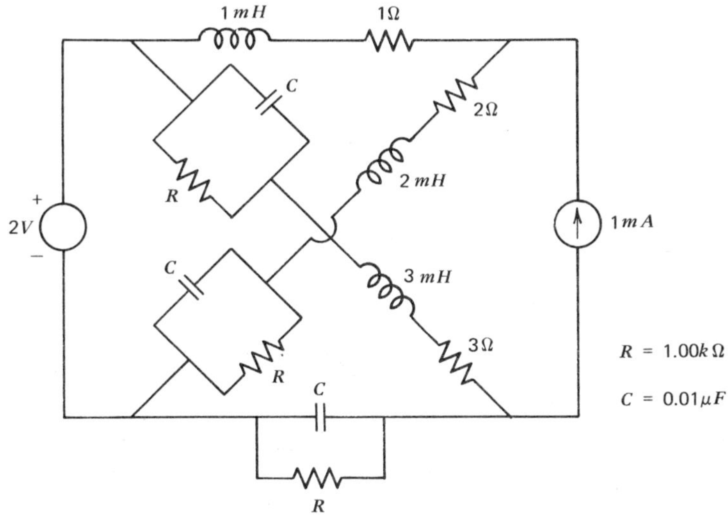

Fig. P5-4

5-4a. Formulate a set of independent node equations for the zero-state circuit of Fig. P5-4.

5-5. Find the currents through the resistors of the circuit shown in Fig. P5-5.

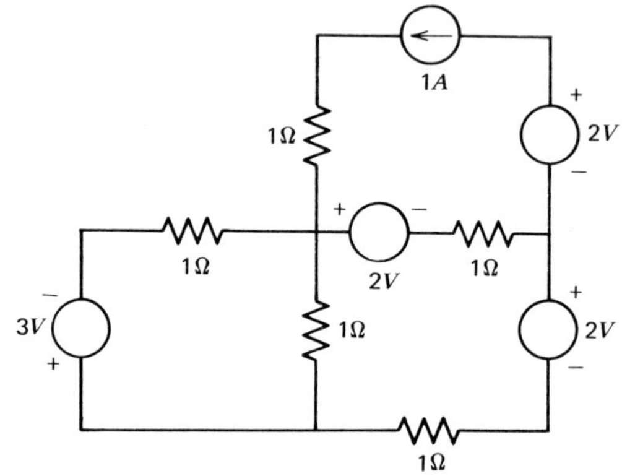

Fig. P5-5

5-6. Repeat 5-5 for the circuit of Fig. P5-6, with $i_1(t) = 10A$ and $i_2(t) = 5A$.

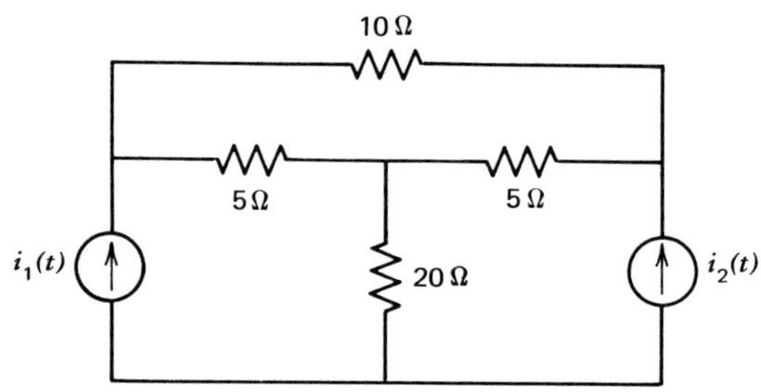

Fig. P5-6

5-7 Repeat problem 5-6 for $i_1(t) = i_2(t) = 10A$.

5-8. For the circuit of Fig. P5-8, formula a set of independent loop equations.

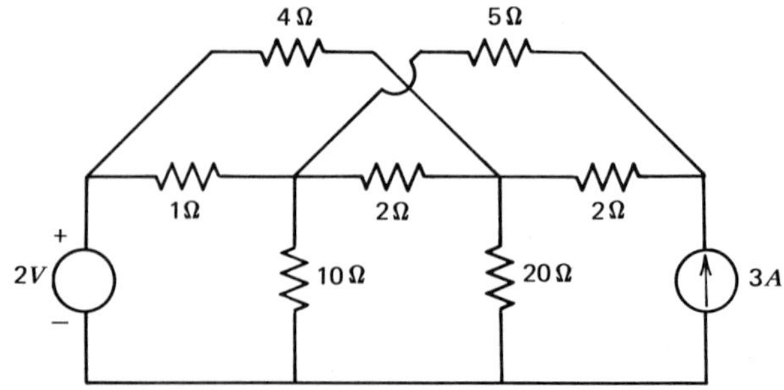

Fig. P5-8

5-9. Form a set of independent node equations for the circuit of Fig. P5-8.

5-10. Determine the final states of the reactive elements in the circuit of Fig. P5-10. State your reasonings.

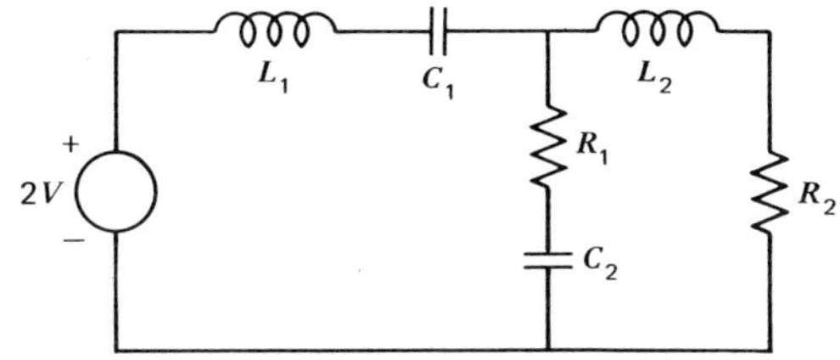

Fig. P5-10

5-11. Find the node voltages of the circuit of Fig. P5-11 by using the subroutine GE(N, A, B). The reference node is set at node 7.

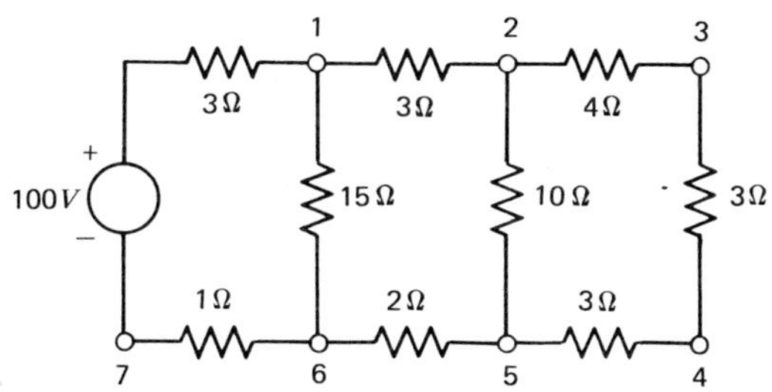

Fig. P5-11

5-12. Find the currents through the resistors of the circuit of Fig. P5-12 by using the subroutine GE(N, A, B).

5-12a. Solve 5-1 using the subroutine GE(N, A, B).

5-12b. Solve 5-2 using the subroutine GE(N, A, B).

5-12c. Solve the following systems using the subroutine GE(N, A, B).

(i)
$$\begin{bmatrix} 5 & 7 & 6 & 5 \\ 7 & 10 & 8 & 7 \\ 6 & 8 & 10 & 9 \\ 5 & 7 & 9 & 10 \end{bmatrix} \begin{bmatrix} x_1 \\ x_2 \\ x_3 \\ x_4 \end{bmatrix} = \begin{bmatrix} 23 \\ 32 \\ 33 \\ 31 \end{bmatrix} \qquad \text{Ans: } x_1 = x_2 = x_3 = x_4 = 1.$$

(ii)
$$\begin{bmatrix} -120 & 16 & -140 & 240 \\ 1200 & -120 & 1680 & -2700 \\ -2700 & 240 & -4200 & 6480 \\ 1680 & -140 & 2800 & -4200 \end{bmatrix} \begin{bmatrix} x_1 \\ x_2 \\ x_3 \\ x_4 \end{bmatrix} = \begin{bmatrix} 1 \\ 0 \\ 0 \\ 0 \end{bmatrix} \qquad \text{Ans: } \begin{aligned} x_1 &= 1/2 \\ x_2 &= 1 \\ x_3 &= 1/4 \\ x_4 &= 1/3. \end{aligned}$$

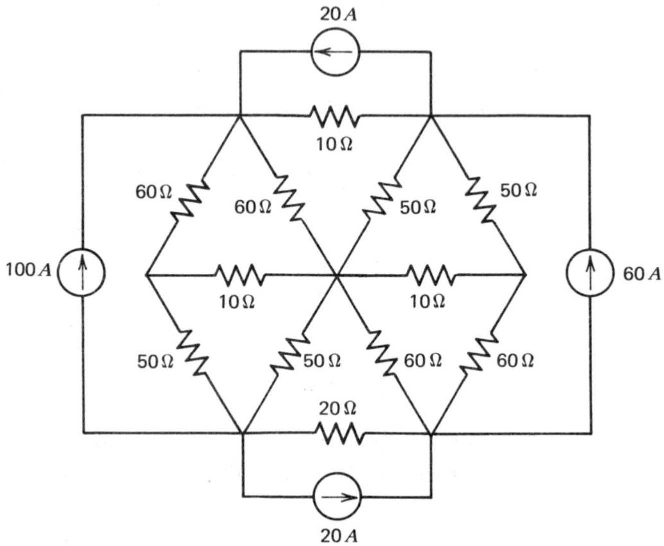

Fig. P5-12

5-12d. Solve 5-3 using the subroutine GE(N, A, B).
5-12e. Solve 5-4 using the subroutine GE(N, A, B).
5-12f. Solve 5-4a using the subroutine GE(N, A, B).
5-12g. Solve 5-8 using the subroutine GE(N, A, B).
5-12h. Solve 5-9 using the subroutine GE(N, A, B).

CHAPTER 6
Sinusoidal Steady-State Analysis

The transient time-domain analysis of linear, time-invariant circuits is now complete both in theory and in computation. The analysis has been developed to give the circuit behavior in the time domain (i.e., as a function of time) from the moment of excitation to the steady-state conditions. The discrepancy between the initial state at the instant of excitation and the steady-state gives rise to transients. The transient (response), governed by the circuit-element values and the initial-state values, eventually vanishes leaving only the driven response under steady-state conditions. The transient part of the response may also not arise if the initial-state values matched the steady-state values at the instant of excitation. The steady-state response is what remains as $t \to \infty$, that is, after the transient has vanished. It is governed by the driving force or excitation and therefore is called the driven response as well as the steady-state response. The steady-state or the driven response is present as long as the excitation is present and is consequently of ultimate importance in most situations. In this chapter, we discuss the steady-state response (i.e., response under steady-state conditions) in depth.

The excitations that we shall consider for the steady-state analysis are the sinusoidal signals. As we have seen in previous chapters, while the computational techniques are applicable to arbitrary excitations, the theory can be developed explicitly for elementary or exponential signals only. The convolution integral then provided the zero-state response to arbitrary excitations in terms of the response to the elementary signal (step response or impulse response). The response for nonzero initial state is then obtained by adding the zero-input response to the zero-state response. The formal theory of circuits is thus developed in terms of the response to elementary excitations only (a step signal is a constant signal initiated at a given instant). The formal theory, however, could provide explicit steady-state response to exponential excitations with arbitrary complex exponents. The only practical signals of interest representable in terms of *simple* exponentials are the sinusoidal and the constant excitations. The constant excitation may also be viewed as a sinusoid of zero frequency, that is, $A \sin (\omega t + \theta)|_{\omega=0} = A \sin \theta =$ constant. The sinusoidal steady-state analysis thus includes sinusoidal and constant excitation.

The sinusoidal excitations are also the most widely encountered signals in practice. For example, the power supply in our homes, offices, and factories is sinusoidal. Many of the oscillators (or, sources) in electronic circuit applications also generate sinusoidal signals. Radio and television broadcasts transmit sinusoidal (carrier) signals that are appropriately modified (modulated) to carry the desired information such as picture, music, or speech. As we shall also see, the concepts developed for sinusoidal signals are easily generalized to other signals. The sinusoidal steady-state analysis thus represents a major milestone in the study of circuits both from a practical and a theoretical viewpoint.

The approach we follow is what is called the *frequency-domain analysis*. It is based on the fact that the steady-state response is completely determined for any given *frequency* of sinusoidal excitation by the amplitude and phase of the response. These quantities depend on the frequency of excitation—hence the name frequency-domain analysis. As we shall see, the sinusoids are represented as real or imaginary parts of complex exponentials. The analysis is then performed for the simpler functions, the exponentials. Finally, the response to sinusoids is obtained by taking the appropriate real or imaginary part of the response to the exponentials. This approach not only simplifies the analysis, but it introduces many important concepts such as *impedance*, *network functions*, etc. to describe the frequency dependent properties of circuits.

6.1 FREQUENCY-DOMAIN ANALYSIS

The steady-state response of a linear, time-invariant circuit excited by a sinusoid of a given frequency is also a sinusoid of the same frequency as seen in Chapter 2. The amplitude and phase of the response sinusoid relative to the excitation are determined by the circuit through its equilibrium or differential equations. The steady-state analysis thus constitutes the determination of the phase and amplitude of the response when the phase and amplitude of the excitation are given for the sinusoid of a *specified frequency*. The specified frequency and the circuit parameters determine the response relative to the excitation. Quite often the response of the circuit is determined for all possible frequencies of interest and the circuit behavior is completely specified for steady-state sinusoidal excitation by the response specified for each frequency of interest. The analysis is thus performed with frequency as the main parameter rather than time, and we call this the frequency-domain analysis. Therefore, steady-state sinusoidal analysis initiates the frequency-domain analysis of circuits.

Frequency-domain analysis of circuits is based on the fact that the steady-state response is a sinusoid of the same frequency as the excitation. Let us clarify this assertion. We have already seen that each node voltage and each loop current is a sum of sinusoids of the same frequency as the excitation since each node voltage or loop current is determined by the circuit differential equation with the sinusoidal excitation (plus its derivatives) as the forcing functions. When the forcing function is the excitation plus its derivatives, the conclusion follows by Superposition II applied to steady-state solutions. The derivatives of a sinusoid are also sinusoids of the same frequency. Thus, the steady-state solution of such a differential equation is a sum of sinusoids of the same frequency. The response, a voltage or a current somewhere in the circuit, is then a linear combination of the appropriate node voltages or loop currents. We thus find that the steady-state response is a sum of sinusoids of the same frequency as the excitation. We show below that this sum is also a sinusoid of the same frequency as the excitation.

Let us first show that the sum of a sine wave and a cosine wave of the same frequency ω can be expressed as a sine wave of the same frequency. Let

$$\begin{aligned} v_0(t) &= A_1 \sin(\omega t + \phi_1) + A_2 \cos(\omega t + \phi_2) \\ &= A_1 \cos\phi_1 \sin\omega t + A_1 \sin\phi_1 \cos\omega t + A_2 \cos\phi_2 \cos\omega t - A_2 \sin\phi_2 \sin\omega t \\ &= V_1 \sin\omega t + V_2 \cos\omega t \end{aligned} \tag{6.1}$$

We now wish to show that $v(t)$ is a sine wave of the same frequency ω:

$$v_0(t) = V \sin(\omega t + \theta) \tag{6.2}$$

We would like to determine the values of V and θ such that $v_0(t)$ in Eq. 6.2 equals the function in Eq. 6.1. Expand the function in Eq. 6.2 and obtain:

$$v_0(t) = V\cos\theta \sin\omega t + V \sin\theta \cos\omega t$$

Equation 6.1 is the same as Eq. 6.2 if

$$\begin{aligned} V_1 &= V\cos\theta \\ V_2 &= V\sin\theta \end{aligned} \tag{6.3}$$

Now square V_1 and V_2 and add them together to obtain:

$$V_1^2 + V_2^2 = V^2 \tag{6.4}$$

and if we take the ratio of V_2 and V_1, we get

$$\frac{V_2}{V_1} = \tan\theta \tag{6.5}$$

This then gives us the rules for finding the values of V and θ.

$$V = \sqrt{V_1^2 + V_2^2}$$

and

$$\theta = \tan^{-1}\frac{V_2}{V_1} \tag{6.6}$$

We have combined a sine and a cosine function of a given frequency with different amplitudes and phases into a single sine function of the same frequency. We now determine the sum of n such sine functions of the same frequency but with different amplitudes and phase angles. Consider

$$v(t) = \sum_{n=1}^{N} V_n \sin(\omega t + \theta_n) \tag{6.7}$$

To combine this function into a single sine function, we proceed by expanding the right-hand side of Eq. 6.7 as:

$$v(t) = \sum_{n=1}^{N} (V_n \cos\theta_n \sin\omega t + V_n \sin\theta_n \cos\omega t) \tag{6.8}$$

Rearranging the sum, we have

$$v(t) = \left(\sum_{n=1}^{N} V_n \cos\theta_n\right) \sin\omega t + \left(\sum_{n=1}^{N} V_n \sin\theta_n\right) \cos\omega t$$

We can see that this is simply

$$v(t) = B_1 \sin\omega t + B_2 \cos\omega t \tag{6.9}$$

From Eq. 6.6, this reduces to

$$v(t) = V\sin(\omega t + \theta) \tag{6.10}$$

Equations 6.1–6.6 show that the sum of a sine function and a cosine function of the same frequency is a sine function of that frequency. Equations 6.7–6.10 show that the sum of any number of sine functions of a given frequency is a sine function of the same frequency. Thus, the sum of any number of sinusoids of frequency ω is also a sinusoid of the same frequency ω. The steady-state response and the excitation are thus sinusoids of the same frequency.

6.2 COMPLEX NUMBERS

The steady-state solution for a sinusoidal excitation was particularly simplified in Chapters 2 and 3 by the use of the complex exponentials. An extensive discussion of steady-state analysis is best carried out in terms of these complex exponentials. Recall that

$$
\begin{aligned}
e^{j\omega t} &= \cos \omega t + j \sin \omega t \\
e^{-j\omega t} &= \cos \omega t - j \sin \omega t
\end{aligned}
\tag{6.11}
$$

where $j = \sqrt{-1}$. The expression for sine and cosine functions in terms of complex exponentials are obtained by the difference and the sum of these equations respectively:

$$
\begin{aligned}
\sin \omega t &= \frac{1}{2j}\left(e^{j\omega t} - e^{-j\omega t}\right) \\
\cos \omega t &= \frac{1}{2}\left(e^{j\omega t} - e^{-j\omega t}\right)
\end{aligned}
\tag{6.12}
$$

Derivatives and integrals of complex exponentials give rise to complex coefficients of these exponentials. In general, the use of complex exponentials for sinusoidal analysis requires considerable manipulation of complex numbers. A thorough familiarity with complex numbers is therefore essential to our purposes. We proceed with a discussion of the complex number arithmetic.

A complex number, represented as the sum of a real number and an imaginary number, is given by

$$
c = a + jb \tag{6.13}
$$

where a and b are real.

We call the real number a, the real part of the complex number c, and we represent it as $a = \text{Re}(c)$. The imaginary number is jb and b is called the imaginary part of the complex number c, and it is represented as $b = \text{Im}(c)$.

To represent a complex number uniquely, we need to specify the real part a and the imaginary part b of c. What we have then are pairs of numbers, a and b, which we can show graphically by letting one of the axes on a plane be the real axis and the other the imaginary axis. Then we can represent a number c on the plane by the real and imaginary parts of the complex number c. This may be represented in rectangular or Cartesian form as shown in Fig. 6.1.

We can also represent the point c uniquely on the complex plane by $|c|$, the (radial) distance of c from the origin, and θ, the counterclock-

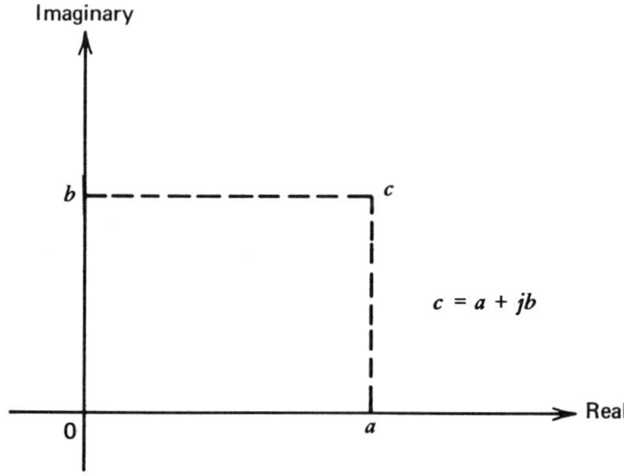

Fig. 6.1 Rectangular form of a complex number.

wise angle of the ray (from the origin to the point c) with respect to the positive horizontal axis (Fig. 6.2*a*). The complex number is usually shown by a vector from the origin to the point c. Remember that the angle is measured *counter-clockwise* with respect to the *positive real or horizontal* axis.

This form is called the *polar form* representation of a complex number and is denoted by

$$c = |c| \angle \theta. \tag{6.14}$$

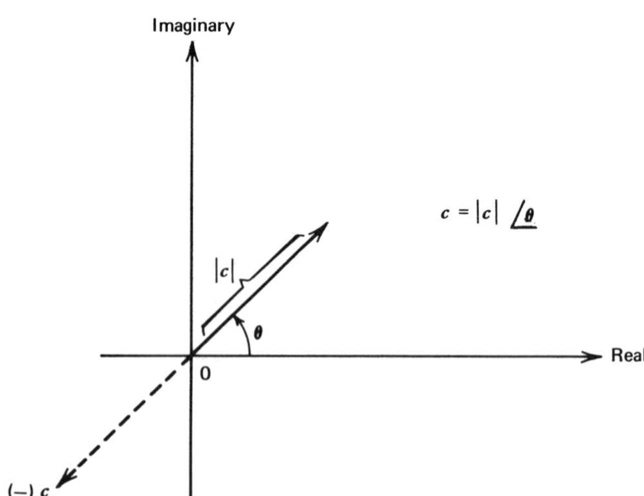

Fig. 6.2*a* Polar form of a complex number.

Let us examine the relationship between the Cartesian form and the polar form since they both represent the same point or number c. Observe that for a right angle triangle with sides a and b, and hypotenuse $|c|$,

$$a = |c| \cos \theta$$

and

$$b = |c| \sin \theta \tag{6.15}$$

where a is the projection of $|c|$ on the horizontal axis and b is the projection of $|c|$ on the vertical axis.

Squaring a and b and adding, we get

$$a^2 + b^2 = |c|^2$$

or

$$|c| = \sqrt{a^2 + b^2} \tag{6.16}$$

Dividing b by a we get:

$$\frac{b}{a} = \tan \theta$$

or

$$\theta = \tan^{-1} \frac{b}{a} \tag{6.17}$$

Thus, Eqs. 6.16 and 6.17 give the relationship between the rectangular and the polar forms of representations.

Complex numbers may also be represented by complex exponentials. We develop this as follows. In the complex number

$$c = a + jb$$

substitute the values of a and b from Eq. 6.15 with $|c|$ and θ defined by Eqs. 6.16 and 6.17. Then

$$\begin{aligned} c &= |c| \cos \theta + j\,|c| \sin \theta \\ &= |c|\,(\cos \theta + j \sin \theta) \\ &= |c|\, e^{j\theta} \end{aligned} \tag{6.18}$$

since by Eq. 6.11 $e^{j\theta} = \cos \theta + j \sin \theta$. This representation is called the *exponential form* and is directly related to the polar form. Observe that the rotation of the vector by 180° (i.e., π radians) gives the negative of the original number since $|c|e^{j(\theta+\pi)} = -|c|e^{j\theta}$. This is shown in Fig. 6.2*a*. Finally, the counterclockwise angle θ is equivalent to the negative clockwise angle ϕ with respect to the positive real axis. This is shown in Fig. 6.2*b* where $(\theta + \phi) = 2\pi$ radians. Then

$$\begin{aligned} c = |c|e^{j\theta} &= |c|e^{j(2\pi - \phi)} \\ &= |c|e^{j2\pi}e^{-j\phi} \\ &= |c|e^{-j\phi} \end{aligned}$$

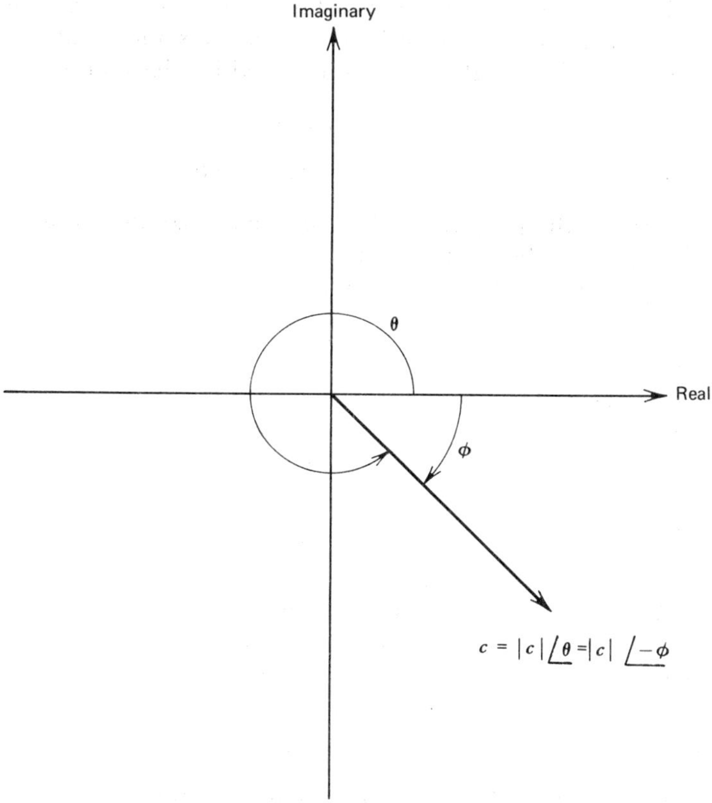

Fig. 6.2b

These different representations of complex numbers are now used to add, subtract, multiply, and divide these complex numbers.

Addition and Subtraction

First, let us examine addition and subtraction.

$$\begin{aligned} c &= c_1 \pm c_2 \\ &= (a_1 \pm a_2) + j\,(b_1 \pm b_2) \end{aligned} \tag{6.19}$$

We see that in adding and subtracting the complex numbers we merely add or subtract the corresponding real and imaginary parts of the complex numbers.

Graphically, addition may be represented as shown in Fig. 6.3. The

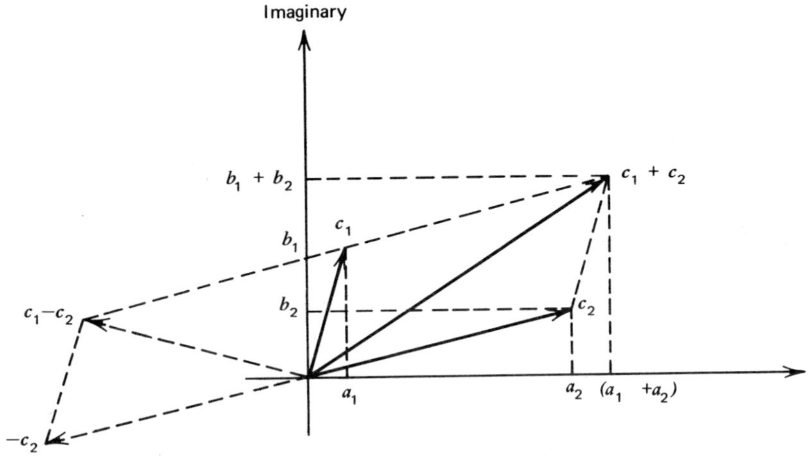

Fig. 6.3 Addition and subtraction of two complex numbers.

figure shows that the addition of two complex numbers is the same as the addition of two vectors by the parallelogram law, where the sum vector is from the origin to the opposite corner of the parallelogram. Subtraction is merely the addition of the negative of a complex number, and the results are obtained as shown.

Multiplication

Multiplication of two complex numbers

$$c_1 = a_1 + jb_1$$

and

$$c_2 = a_2 + jb_2 \tag{6.20}$$

is performed by going through the multiplication in the usual manner and replacing j^2 by -1. Thus

$$\begin{aligned} c_1c_2 &= (a_1 + jb_1)(a_2 + jb_2) \\ &= a_1a_2 + j(b_1a_2 + a_1b_2) + j^2b_1b_2 \end{aligned}$$

Recalling that $j^2 = -1$, we have

$$c_1c_2 = \underbrace{(a_1a_2 - b_1b_2)}_{\text{real}} + j\underbrace{(a_1b_2 + a_2b_1)}_{\text{imaginary}} \tag{6.21}$$

So now we have a new complex number with its real and imaginary parts as shown. We can also represent the product in the exponential form:

$$c_1c_2 = (|c_1|e^{j\theta_1})(|c_2|e^{j\theta_2}) \tag{6.22}$$

Simplifying Eq. 6.22, we have

$$c_1 c_2 = |c_1||c_2| e^{j(\theta_1 + \theta_2)} \tag{6.23}$$

We see that it is very easy to multiply two complex numbers if they are in an exponential form. The same is true of division as shown below.

Division

In the exponential form, the quotient of two complex numbers becomes simply:

$$\frac{c_1}{c_2} = \frac{|c_1| e^{j\theta_1}}{|c_2| e^{j\theta_2}} = \frac{|c_1|}{|c_2|} e^{j(\theta_1 - \theta_2)} \tag{6.24}$$

Examining division in the rectangular form, we have

$$\frac{c_1}{c_2} = \frac{(a_1 + jb_1)}{(a_2 + jb_2)} \tag{6.25}$$

To divide, we first multiply the numerator and denominator by a complex number c_2^*, where $c_2^* = a_2 - jb_2$.

$$\frac{c_1}{c_2} = \frac{c_2^* c_1}{c_2^* c_2} = \frac{(a_2 - jb_2)(a_1 + jb_1)}{(a_2 - jb_2)(a_2 + jb_2)} \tag{6.26}$$

Multiplying we have

$$\frac{c_1}{c_2} = \frac{(a_1 a_2 + b_1 b_2) + j(b_1 a_2 - b_2 a_1)}{a_2{}^2 + b_2{}^2} \tag{6.27}$$

Note that the denominator now has a real number so we can represent our quotient as the sum of a real part and an imaginary part:

$$\frac{c_1}{c_2} = \underbrace{\frac{a_1 a_2 + b_1 b_2}{a_2{}^2 + b_2{}^2}}_{\text{real}} + j\underbrace{\frac{(b_1 a_2 - b_2 a_1)}{a_2{}^2 + b_2{}^2}}_{\text{imaginary}}. \tag{6.28}$$

The key to division is to make the denominator real by using c_2^*. This type of operation is used extensively where the number c^* is given the name *complex conjugate* of c. A complex conjugate is obtained by changing the sign of the imaginary part of c. Also,

$$\begin{aligned} cc^* &= (a + jb)(a - jb) \\ &= a^2 + b^2 = |c|^2 \end{aligned} \tag{6.29}$$

.3 PHASORS AND ROTATING PHASORS

The representations of complex numbers and rules for their arithmetic operations lead us naturally to the discussion of complex functions. The complex functions that concern us are the complex exponentials. In this section, we describe the representation of complex exponentials and some of the operations on these functions. We start by reexamining some of the operations on complex numbers.

As shown in the previous section, the sum of two complex numbers is obtained by drawing a parallelogram formed by the vectors representing c_1 and c_2 and performing the vector addition of c_1 and c_2. The sum vector is given by the diagonal from the origin. The addition may also be performed as shown in Fig. 6.4 by appropriately moving one of the vectors to adjoin the other vector. The sum vector is then drawn from the origin to the end point. That is, one of the vectors is moved and placed such that its tail is at the arrowhead of the other vector, and the angle of the displaced vector with the horizontal is unchanged. The sum is the vector from the origin to the arrowhead of the displaced vector. Such displaced

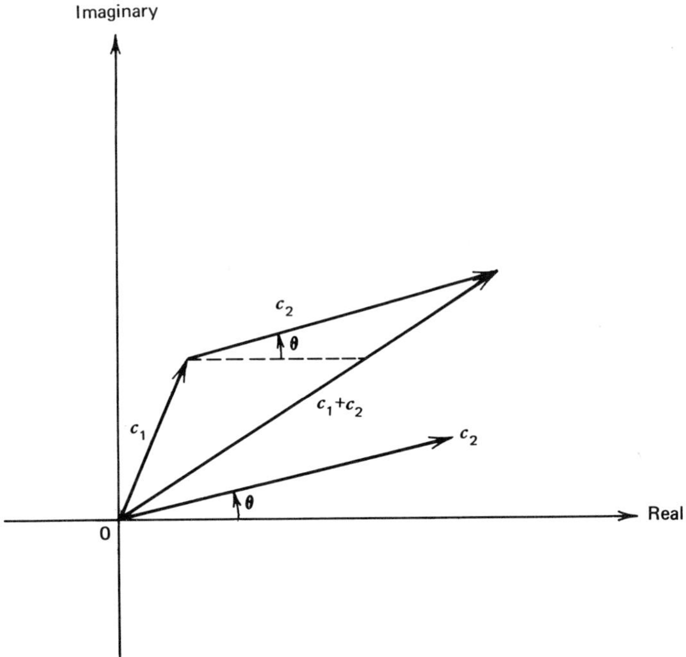

Fig. 6.4 Phasors.

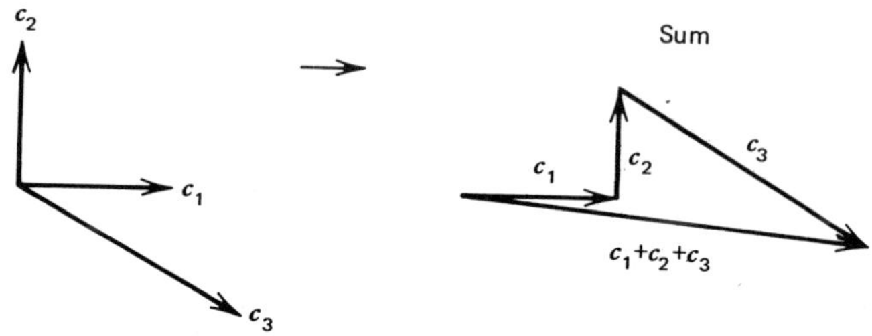

Fig. 6.5 Sum of three phasors.

vectors can thus facilitate the addition of complex numbers. The displaced vectors are called *phasors*. Phasors have lengths equal to the magnitudes of the complex numbers they represent, and their orientation is determined by the angles of the complex numbers. If a horizontal line in the positive real direction is drawn at the tail of a phasor, the *counterclockwise* angle measured between the positive horizontal line and the phasor represents the angle of the complex number.

Phasor addition is stated as follows: in any order, arrange all phasors, except the end ones, so that the tail of one sits on the arrowhead of another. Then the sum is the phasor obtained by joining the tail of the first phasor and the arrowhead of the last phasor with the arrow of the sum phasor appearing also at the arrowhead of the last phasor. Figure 6.5 shows the addition of three phasors. The complex number represented by the sum phasor has the magnitude given by the length of the phasor and the angle given by the counterclockwise angle the phasor makes with the positive horizontal. The complex number is also determined by a vector from the *origin* whose length is the same as that of the sum phasor and which is parallel to the sum phasor.

Note that the counterclockwise angle between the positive horizontal and the phasor determines the angle of the complex number. This follows from the polar representation of the complex number discussed in Sec. 6.2 where the angle was measured with respect to the positive horizontal axis in the counterclockwise direction. It is sometimes also convenient to use the equivalent negative clockwise angle.

Rotating Phasors

Let us now examine the complex exponential $|V|e^{j\omega t}$. At any given value of time t, this is a complex number with the polar representation or phasor magnitude $|V|$ and angle $\psi = \omega t$ (Fig. 6.6). When $t = 0$, the angle $\psi = \omega t = 0$, and the phasor is along the positive real axis. As the value of t

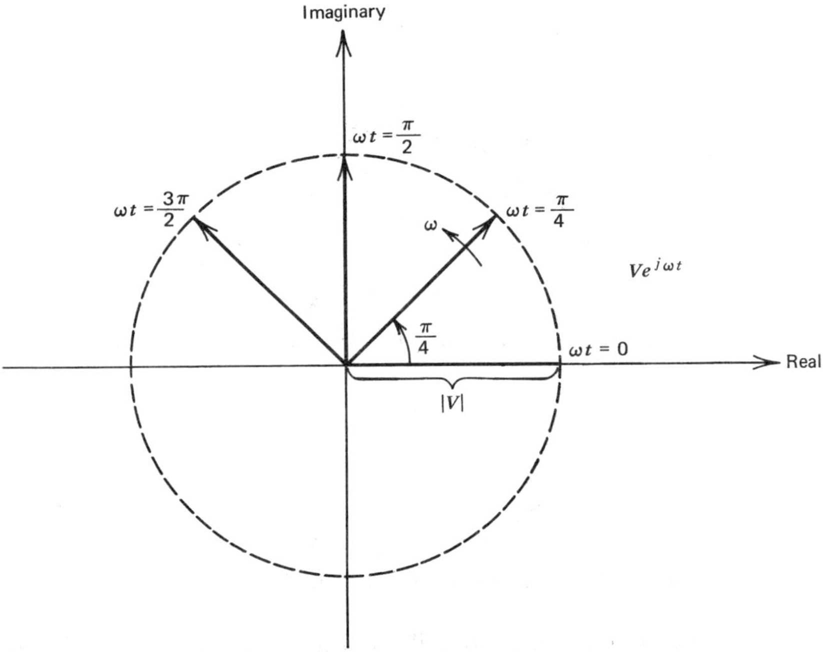

Fig. 6.6 Rotating phasors.

increases, angle $\psi = \omega t$ also increases, and the phasor rotates in a *counterclockwise* direction with its magnitude unchanged. Such a phasor with constant magnitude and rotating at the constant speed of ω radians per second is called a *rotating phasor* to distinguish it from the *stationary phasors* of the last section. If the value of t is fixed, the rotating phasor becomes a stationary phasor at that instant of time.

Sometimes, the exponential function may have an initial phase angle as in

$$v(t) = |V|e^{j(\omega t + \theta)}. \tag{6.30}$$

This function can be written as:

$$\begin{aligned} |V|e^{j(\omega t+\theta)} &= |V|e^{j\theta}e^{j\omega t} \\ &= Ve^{j\omega t} \end{aligned} \tag{6.31}$$

where V is a complex constant. The complex constant $V = |V|e^{j\theta}$ is a phasor also, but it is a stationary phasor since it does not rotate with time. On the other hand, $Ve^{j\omega t}$ is a rotating phasor. The representation for such a rotating phasor is shown in Fig. 6.7. At $\omega t = 0$ the angle is θ, at $\omega t = \pi/2$ the angle is $(\theta + \pi/2)$, and so on. Thus $|V|e^{j(\omega t+\theta)}$ is a rotating

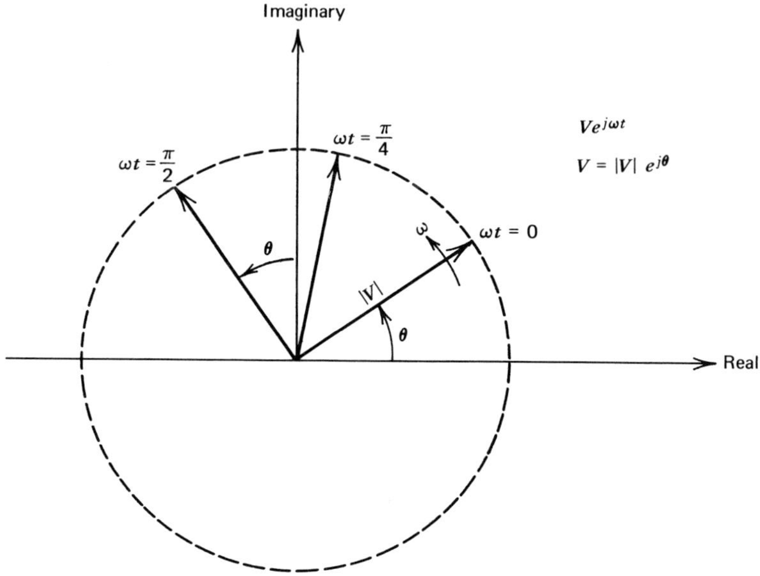

Fig. 6.7 Rotating phasor with initial phase angle.

phasor, but it starts initially at the phase angle θ and rotates in the counterclockwise direction at the speed of ω radians per second.

One of the problems frequently encountered in steady-state circuit analysis is the addition of different voltages or currents that are rotating phasors. We know that the sum of several sine waves of the same frequency adds up to a single sine wave of the same frequency. We should expect, therefore, that rotating phasors, rotating at the same speed, should add up to a single phasor rotating also at the same speed. Examining this algebraically, we have

$$v_1(t) = V_1 e^{j\omega t} \qquad v_2(t) = V_2 e^{j\omega t} \tag{6.32}$$

where V_1 and V_2 are complex constants. Adding

$$V_1 e^{j\omega t} + V_2 e^{j\omega t} = (V_1 + V_2) e^{j\omega t} = V e^{j\omega t} \tag{6.33}$$

where V is a new complex constant. So we see that the addition of two rotating phasors is merely the addition of two complex constants or, stationary phasors, multiplied by the rotating phasor $e^{j\omega t}$. Graphically, as shown in Fig. 6.8, we add together the stationary phasors at any instant of time, say at $\omega t = 0$, and the resulting phasor when multiplied by $e^{j\omega t}$ is the required rotating phasor. We see that $Ve^{j\omega t}$ rotates in the counterclockwise direction as do $V_1 e^{j\omega t}$ and $V_2 e^{j\omega t}$ with the same angular speed of ω radians per second.

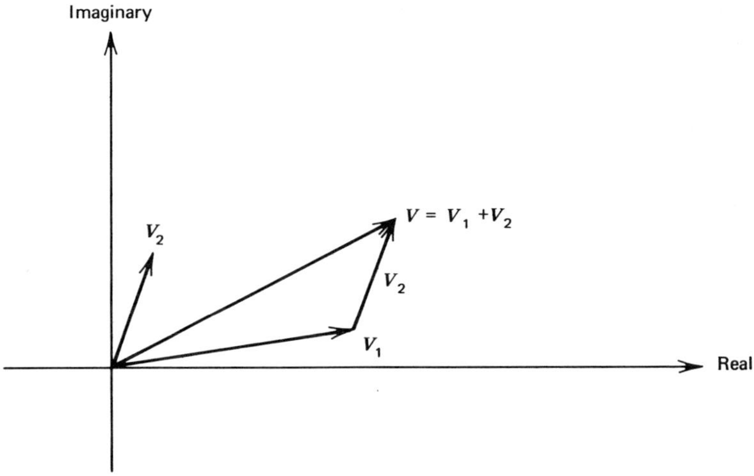

Fig. 6.8 Sum of rotating phasors at $\omega t = 0$.

Rotating Phasors and Sinusoids

Recall that sinusoidal excitations may be expressed as sums of two complex exponentials. Let us now examine the phasor representation of sinusoids. Consider a cosine function

$$v(t) = |V| \cos \omega t$$
$$= \frac{|V|}{2} e^{j\omega t} + \frac{|V|}{2} e^{-j\omega t} \tag{6.34}$$

The cosine function is thus represented by two phasors both of magnitude $|V|/2$ but rotating in opposite directions, counterclockwise and clockwise as shown in Fig. 6.9. The sum of these two rotating phasors gives the desired cosine function. This is shown in Fig. 6.10 for $\omega t = \pi/4$.

Observe in Fig. 6.10 that the vertical projections of the two phasors cancel out; the horizontal projections are identical and of like sign. The sum is, therefore, twice the horizontal projection of one of the phasors. Thus

$$|V| \cos \omega t = 2 \times \text{horizontal projection of either } \frac{|V|}{2} e^{j\omega t} \text{ or } \frac{|V|}{2} e^{-j\omega t}$$
$$= \text{horizontal projection of } |V| e^{j\omega t} \tag{6.35}$$

In a similar fashion the sine function is given by

$$|V| \sin \omega t = \text{vertical projection of } |V| e^{j\omega t} \tag{6.36}$$

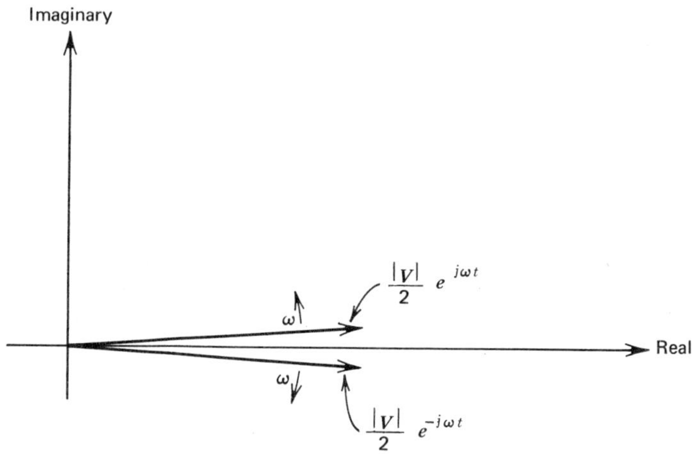

Fig. 6.9 Cosine function.

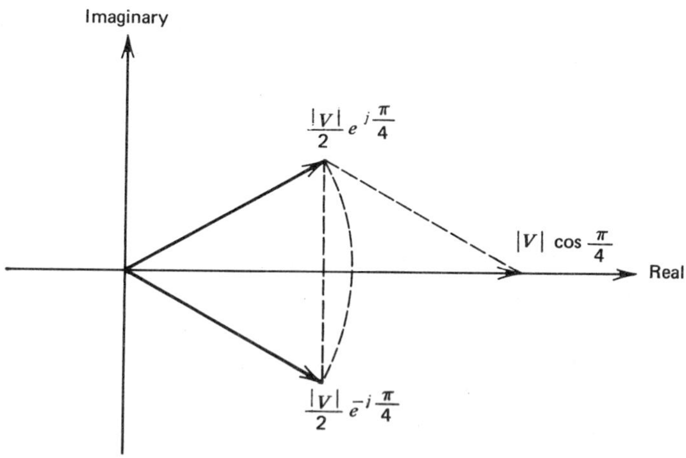

Fig. 6.10 Cosine as sum of horizontal projections.

The sine wave as a vertical projection of a rotating phasor is shown in Fig. 6.11. The horizontal projection of the rotating phasor can be plotted similarly to give the cosine wave.

The sine or cosine excitations can be represented thus by the horizontal or vertical projections, respectively, of rotating phasors. Algebraically, they can be represented by the real or the imaginary parts, respectively,

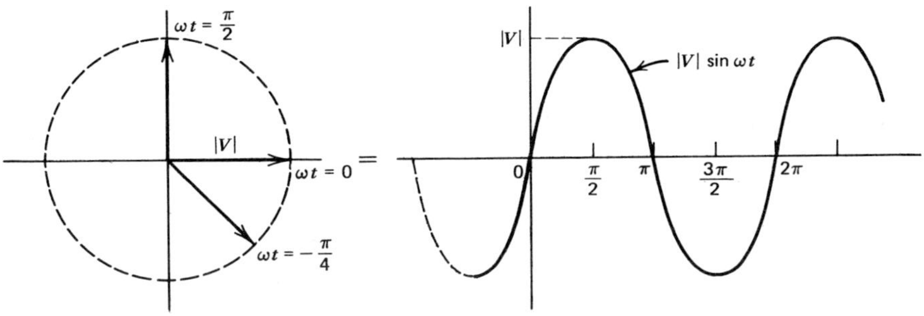

Fig. 6.11 Vertical projection of a rotating phasor.

of complex exponentials. Thus,

$$|V|\, e^{j\omega t} = |V|\,(\cos \omega t + j \sin \omega t) \tag{6.37}$$

and

$$\begin{aligned} |V| \cos \omega t &= \text{Re}\, |V| e^{j\omega t} \\ |V| \sin \omega t &= \text{Im}\, |V| e^{j\omega t} \end{aligned} \tag{6.38}$$

Similarly, it is easy to see that sinusoids with phase angles can also be represented by rotating phasors as follows:

$$\begin{aligned} |V| \cos (\omega t + \theta) &= \text{Re}\, V e^{j\omega t} \\ &= \text{horizontal projection of } V e^{j\omega t} \end{aligned} \tag{6.39}$$

$$\begin{aligned} |V| \sin (\omega t + \theta) &= \text{Im}\, V e^{j\omega t} \\ &= \text{vertical projection of } V e^{j\omega t} \end{aligned} \tag{6.40}$$

where

$$V = |V| e^{j\theta} \tag{6.41}$$

Finally, observe that the real part of a complex exponential is represented as the imaginary part of j times the same exponential. That is

$$\text{Re}\, V e^{j\omega t} = \text{Im}\, (jV)\, e^{j\omega t}$$

and

$$\text{Im}\, V e^{j\omega t} = \text{Re}\, (-jV)\, e^{j\omega t} \tag{6.42}$$

Thus all real parts may be converted to imaginary parts and vice versa. This is merely a restatement of the fact that sine functions may be represented by cosine functions and vice versa by introducing appropriate phase shifts of $\pi/2$ radians or 90°.

The solution of equilibrium equations for steady-state conditions and sinusoidal excitations involves sums of many voltages or currents that are themselves sinusoids of the same frequency. A single sinusoid can be represented as a real part or an imaginary part of a complex exponential. Sums of sinusoids, therefore, involve sums of real parts or sums of imaginary parts of complex exponentials. Since sine functions may be converted to cosine functions and vice versa, let us consider the sum of only cosine functions (i.e., the sum of real parts of several exponentials) with same frequency ω but different amplitudes and phases.

We use the following obvious real-part property of a complex function:

$$\text{Re}\,(V_1 + V_2) = \text{Re}\,V_1 + \text{Re}\,V_2$$

or, for the addition of N such terms,

$$\text{Re}\left[\sum_{n=1}^{N} V_n\right] = \sum_{n=1}^{N} (\text{Re}\,V_n)$$

With this property then, the sum of several cosine functions is obtained by adding the real parts of the corresponding complex exponentials or rotating phasors. For example,

$$\begin{aligned}\sum_{n=1}^{N} |V_n| \cos(\omega t + \theta_n) &= \sum_{n=1}^{N} [\text{Re}\,|V_n| e^{j(\omega t + \theta_n)}] \\ &= \text{Re}\left[\sum_{n=1}^{N} |V_n| e^{j(\omega t + \theta_n)}\right] \\ &= \text{Re}\left[\left(\sum_{n=1}^{N} |V_n| e^{j\theta_n}\right) e^{j\omega t}\right]\end{aligned} \tag{6.43}$$

Thus, the sum of real parts of rotating phasors is given by the real part of the sum of these rotating phasors. The concept is very important and forms the basis of steady-state sinusoidal analysis. The sinusoids are expressed as real (or imaginary) parts of rotating phasors, the rotating phasors are summed together into a new rotating phasor, and the real (or imaginary) part of the sum rotating phasor gives the sum of the sinusoids.

Phasor Diagrams

Sinusoidal signals at different points in the circuit can be represented by the real (or imaginary) parts of rotating phasors with different amplitudes and phases. The interrelationship between such signals or rotating phasors is shown by the magnitudes of the phasors and the mutual angular displacements or relative phase angles between them. For example, if there are two rotating phasors with angle θ between them at any instant of time, the same relative phase angle θ is between them at all times since they are both rotating with the same speed ω in the same counterclockwise direction. It is therefore necessary to consider only their corresponding stationary phasors to define the interrelationships between them. The stationary phasors are obtained at any instant of time by freezing the rotating phasors at that time. Since only the phasor magnitudes and relative phase angles are important, the specific instant of time chosen to fix the rotation is not relevant.

Thus, the mutual relationship between rotating phasors is described by a *phasor diagram* showing the corresponding stationary phasors. The choice of instant for freezing the rotation is irrelevant and is selected for convenience. The positive horizontal axis is the reference with respect to which phase angles are measured in the counterclockwise direction. The phase angle of a phasor along the reference line is therefore zero. Usually, the stationary phasors are shown so that one selected phasor called the *reference* is along the positive horizontal axis. All other relative phase angles are then measured with respect to this reference. It should be noted that any one of the phasors may be selected as the reference and relative phase angles measured with respect to it.

Consider two signals as shown in Fig. 6.12. We could choose either one to serve as the reference. The signals and the corresponding rotating phasors are

$$\begin{aligned} &① \quad |V_1| \sin(\omega t + \theta_1) \leftrightarrow |V_1| e^{j\theta_1} \cdot e^{j\omega t} \\ &② \quad |V_2| \sin(\omega t + \theta_2) \leftrightarrow |V_2| e^{j\theta_2} \cdot e^{j\omega t} \end{aligned} \qquad (6.44)$$

In Eq. 6.44, the signals are the imaginary parts of the respective rotating phasors. We may say that the sinusoids are *converted* or *transformed* into the corresponding rotating phasors. Conversely, the rotating phasors are *converted* or *transformed back* into the sinusoids. Let us consider the phasor diagram for these rotating phasors. Let $\theta_1 = \pi/3$ and $\theta_2 = \pi/6$. Then the mutual phase angle between the two rotating phasors is $\pi/6$ with the phasor V_1 ahead of or *leading* V_2 in the counterclockwise rota-

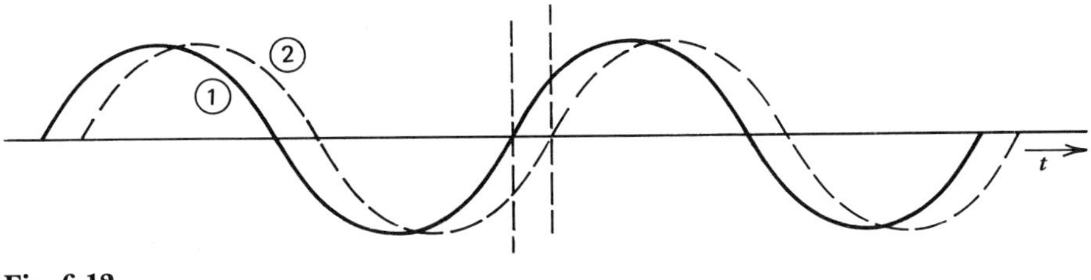

Fig. 6.12

tion. The phasors V_1 and V_2 are shown at some arbitrary instant in Fig 6.13. The arrow with ω indicates their rotation direction and the angular speed ω.

Now, the phasor diagram could just as validly be shown with one or the other phasors along the horizontal reference. In the steady-state conditions, the mutual relationship is all that counts. The sinusoids repeat indefinitely in time and any absolute clock starting at zero time is meaningless. The reference time and, therefore, the reference phasor is chosen arbitrarily without destroying the mutual relationships between the sinusoids.

In Fig. 6.12, for example, the signals remain in the same mutual positions regardless of where the reference time is chosen. Similarly in Eq. 6.44, a different choice of time origin changes only the values of θ_1 and θ_2 but not the value of $(\theta_1 - \theta_2)$. If the time origin is displaced by t_0 seconds, the new rotating phasors are given by

$$
\begin{aligned}
|V_1| e^{j\theta_1} e^{j\omega(t - t_0)} &= |V_1| e^{j(\theta_1 - \omega t_0)} e^{j\omega t} \\
V_2 e^{j\theta_2} e^{j\omega(t - t_0)} &= |V_2| e^{j(\theta_2 - \omega t_0)} e^{j\omega t}
\end{aligned}
\tag{6.45}
$$

The absolute phase angles of the two rotating phasors at $\omega t = 0$ are now given by $(\theta_1 - \omega t_0)$ and $(\theta_2 - \omega t_0)$, but the relative phase angle is still given by $(\theta_1 - \theta_2)$. Thus, the reference point in time or in the phasor diagram can be selected at will without altering the steady-state mutual relationships of the sinusoidal signals or rotating phasors.

For our example in Fig. 6.13, we may choose V_1 as the reference phasor, Then

$$
V_1 = |V_1| \angle 0
$$

and

$$
V_2 = |V_2| \angle -\pi/6 \tag{6.46}
$$

as shown in Fig. 6.14*a*. Alternatively, we may choose V_2 as the reference in which case

$$
V_2 = |V_2| \angle 0
$$

and

$$
V_1 = |V_1| \angle \pi/6 \tag{6.47}
$$

This is shown in Fig. 6.14*b*. Observe that with either phasor as reference V_1 *leads* V_2 by $\pi/6$ radians, or V_2 *lags* (behind) V_1 in the counterclockwise direction of rotation. Finally, a change in the choice of the reference merely rotates all phasors by the same amount so that the new phasor becomes the reference. In the figure all the phasors are rotated clockwise by $\pi/6$ radians to change the reference from V_2 to V_1. This is also equivalent to subtracting $\pi/6$ radians from all angles of the several phasors.

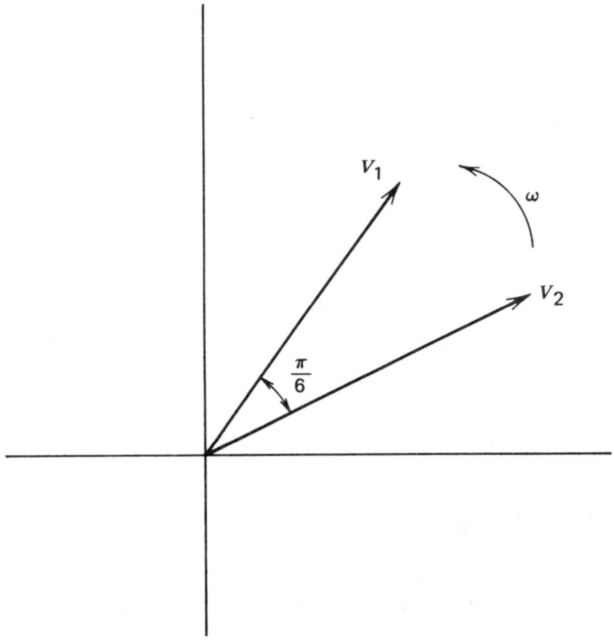

Fig. 6.13

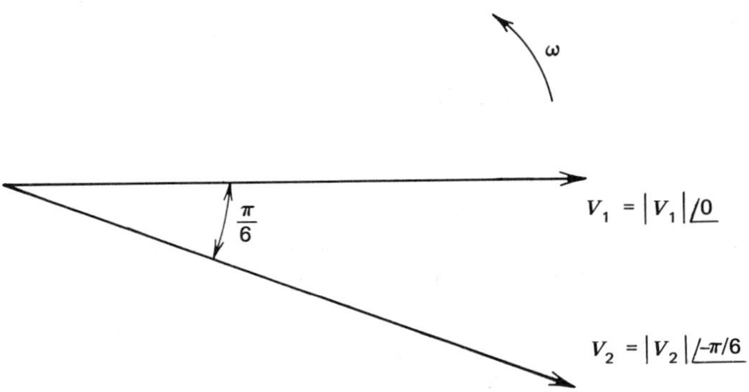

(a)

Fig. 6.14*a*

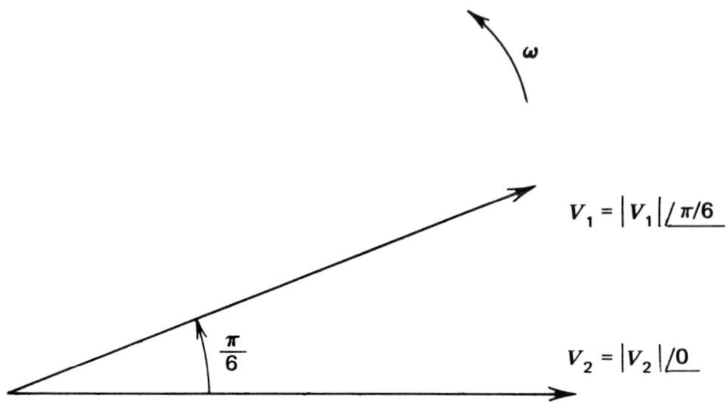

(b)

Fig. 6.14b

6.4 STEADY-STATE RESPONSE

Let us now develop the steady-state response under sinusoidal excitation in terms of steady-state response to the corresponding rotating phasor or complex exponential excitation. We have seen in Chapter 3 that the most general circuit response is a solution of the circuit differential equation

$$(a_nD^n + a_{n-1}D^{n-1} + \cdots + a_1D + a_0)x = (b_mD^m + b_{m-1}D^{m-1} + \cdots + b_1D + b_0)y \tag{6.48}$$

where x is the response and y is the excitation. The response when several excitations are present is obtained in terms of the response to a single excitation by superposition. We will consider, therefore, Eq. 6.48 to be the most general situation and develop its steady-state solution when y is a sinusoid.

First, let us assume that y is a cosine function

$$y = y_1 = |Y| \cos(\omega t + \theta) \tag{6.49}$$

Then, the steady-state solution of Eq. 6.48 is also a sinusoid of the same frequency that may also be expressed as a cosine function but with different amplitude and phase, say,

$$x = x_1 = |X| \cos(\omega t + \phi) \tag{6.50}$$

Here, $|X|$ and ϕ are determined by substitution Eqs. 6.49 and 6.50 into Eq. 6.48 and comparing coefficients of sine and cosine functions. Next, we know that a time translation of excitation merely gives the same amount of time translation in the response for all time-invariant circuits.

Therefore, if the excitation in Eq. 6.49 is shifted by $\omega t = \pi/2$, that is,

$$\begin{aligned} y = y_2 &= |Y| \cos\left(\omega t + \theta - \frac{\pi}{2}\right) \\ &= |Y| \sin(\omega t + \theta) \end{aligned} \tag{6.51}$$

then the response in Eq. 6.50 is also shifted by the same amount:

$$\begin{aligned} x = x_2 &= |X| \cos\left(\omega t + \phi - \frac{\pi}{2}\right) \\ &= |X| \sin(\omega t + \phi) \end{aligned} \tag{6.52}$$

Finally, if the excitation is multiplied by a constant, the response is also scaled by the same amount by linearity. Thus, if the excitation is jy_2, the response is jx_2. This can also be verified by substitution in Eq. 6.48. Superposition applied to steady-state response then says that if the excitation is

$$y_e = y_1 + jy_2 \tag{6.53}$$

the response is

$$x_r = x_1 + jx_2 \tag{6.54}$$

Equations 6.49–6.54 give the excitation-response pair as

$$\begin{aligned} y_e &= Ye^{j\omega t} \\ x_r &= Xe^{j\omega t} \end{aligned} \tag{6.55}$$

where

$$\begin{aligned} Y &= |Y| e^{j\theta} \\ X &= |X| e^{j\phi} \end{aligned} \tag{6.56}$$

Equation 6.55 merely states that the response to a complex exponential excitation is also a complex exponential with the same exponent. In phasor terminology, the response is a rotating phasor with the same speed of rotation ω as the excitation. There is nothing new here as already discussed in Chapter 3. What is significant in this development is that the steady-state response to a sinusoidal excitation is obtained very simply from the (steady-state) response to a rotating phasor. Let us clarify this statement!

Assume that the response x_r is determined for the excitation $y_e = Ye^{j\omega t}$. This is obtained by substituting $x_r = Xe^{j\omega t}$, with X unknown, into Eq. 6.48. The value of X is given by

$$X = \frac{b_m(j\omega)^m + b_{m-1}(j\omega)^{m-1} + \cdots + b_1(j\omega) + b_0}{a_n(j\omega)^n + a_{n-1}(j\omega)^{n-1} + \cdots + a_1(j\omega) + a_0} Y \tag{6.57}$$

Note that both Y and X are complex in general. We thus know th excitation and response pair in Eq. 6.55 for any given Y and ω. Next observe that the real part of y_e is y_1 and the real part of x_r is x_1. Thus, th excitation-response pair y_1, x_1 is obtained by taking the real parts of y_e x_r. Similarly, the excitation-response pair y_2, x_2 is obtained by taking th imaginary parts of y_e, x_r.

The steady-state response to the cosine excitation of Eq. 6.49 is thu obtained by taking the real part of x_r given by Eqs. 6.55 and 6.57 as

$$x_1 = \text{Re}\left[\frac{b_m(j\omega)^m + \cdots + b_1(j\omega) + b_0}{a_n(j\omega)^n + \cdots + a_1(j\omega) + a_0} Ye^{j\omega t}\right]. \qquad (6.58$$

Similarly, the response to the sine excitation of Eq. 6.51 is given by takin the imaginary part of the quantity in the brackets above.

The procedure for the steady-state sinusoidal response may be sum marized as follows:

1. Transform the sinusoidal excitation into the rotating-phasor excita tion. Remember whether the sinusoid is a real part (Re) or an imagi nary part (Im) of the rotating phasor.
2. Obtain the steady-state response to the rotating-phasor excitation The response is also a rotating phasor with the same ω.
3. Transform back the rotating-phasor response into the sinusoida steady-state response by using the (Re) or (Im) operation as specifie in 1 above. This is the steady-state response to the original sinusoida excitation. Figure 6.15 shows the steps in the procedure.

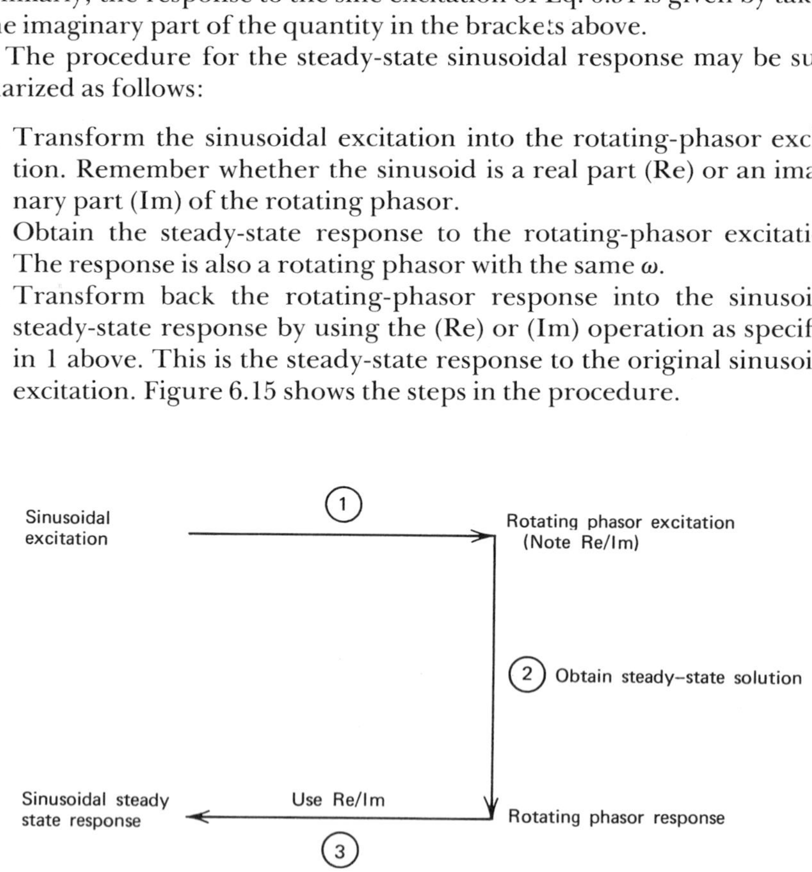

Fig. 6.15

Example 6.4.1 Find the steady-state solution to

$$\frac{dx}{dt}+x=\cos t$$
$$=\text{Re}\,[e^{jt}]$$

Solution. Rotating phasor e^{jt}, operation Re.
Consider

$$\frac{dx_r}{dt}+x_r=e^{jt}$$

Substituting $x_r=Xe^{jt}$

$$X(j+1)\,e^{jt}=e^{jt}$$

The rotating-phasor solution is

$$x_r=\left(\frac{1}{1+j}\right)e^{jt}$$

Now transform back using the Re operation to obtain the steady-state solution:

$$x=\text{Re}\;x_r=\text{Re}\left[\frac{1}{1+j}\,e^{jt}\right]$$
$$=\text{Re}\left[\frac{1}{\sqrt{2}\,e^{j\pi/4}}\,e^{jt}\right]$$
$$=\frac{1}{\sqrt{2}}\cos\left(t-\frac{\pi}{4}\right)$$

Example 6.4.2 Find the steady-state solution to

$$\left[\left(\frac{d}{dt}\right)^2+\frac{d}{dt}+110\right]x_r=100\sqrt{2}\,e^{j(\pi/4)}\,e^{j\omega t}$$
$$=\text{Im}\;100\sqrt{2}\;e^{j(10t+\pi/4)}$$

Solution. Operation Im.
Consider

$$\left(\frac{d^2}{dt^2}+\frac{d}{dt}+110\right)x_r=100\sqrt{2}\,e^{j(\pi/4)}e^{j\omega t}$$

Substitute $x_r=Xe^{j10t}$ into the equation

$$[(j\,10)^2+j\,10+110]Xe^{j10t}=100\sqrt{2}\;e^{j(\pi/4)}\,.\,e^{j10t}$$
$$X=\frac{100\sqrt{2}\;e^{j(\pi/4)}}{10+j\,10}$$
$$=\frac{100\sqrt{2}\;e^{j(\pi/4)}}{10\sqrt{2}\;e^{j(\pi/4)}}=10\angle 0$$

Then $$x_r = 10\, e^{j10t}$$

and $$x = \text{Im}\, x_r = \text{Im}\, 10\, e^{j10t} = 10 \sin 10t$$

Example 6.4.3 Find the steady-state solution to

$$\left(\frac{d^3}{dt^3} + 3\frac{d^2}{dt^2} + 5\frac{d}{dt} + 7\right) x = \left(2\frac{d}{dt} + 1\right) \sin(\omega t + \theta)$$

$$= \left(2\frac{d}{dt} + 1\right) \text{Im}\, e^{j(\omega t + \theta)}$$

Solution. Operation is Im. The rotating-phasor differential equation is

$$\left(\frac{d^3}{dt^3} + 3\frac{d^2}{dt^2} + 5\frac{d}{dt} + 7\right) x_r = \left(2\frac{d}{dt} + 1\right) e^{j\theta} \cdot e^{j\omega t}$$

Let $x_r = Xe^{j\omega t}$. Then by substitution,

$$X = \frac{(2j\omega + 1)e^{j\theta}}{(j\omega)^3 + 3(j\omega)^2 + 5(j\omega) + 7}$$

For a given value of ω and θ, we can calculate

$$X = |X| e^{j\phi}$$

Then $$x_r = Xe^{j\omega t} = |X|\, e^{j\phi} \cdot e^{j\omega t}$$

and $$x = \text{Im}\, x_r = |X| \sin(\omega t + \phi)$$

6.5 IMPEDANCE AND ADMITTANCE

In the last section, we showed that the steady-state sinusoidal analysis i considerably simplified through the use of rotating phasors and th appropriate *Re*/*Im* (real or imaginary part) operation. The analysis i really performed entirely in terms of the rotating phasors (or exponen tials), and the steady-state response is determined also as a rotatin phasor with the same frequency ω but with different magnitude an phase. The *Re*/*Im* operation is used only when the actual sinusoidal func tion is needed. Otherwise, the entire analysis is performed as if all sig nals, excitation or response anywhere in the circuit, were rotating phasor with the same angular frequency ω.

The rotating-phasor approach gives the frequency-domain respons merely as a complex constant times the excitation exponential or rotatin phasor. The *Re*/*Im* operations transforms back the frequency domai response into the time-domain sinusoidal response. One difficulty sti remains, however. The circuit differential equation is needed before one can use Eq. 6.57 to obtain the rotating-phasor response. The derivation of the differential equation is no mean task for most circuits and, as we shall see, not necessary at all for steady-state analysis.

Let us review the principles of circuit analysis. The Kirchhoff Laws (KVL and KCL) govern the equilibrium relationships between certain voltage drops or certain branch currents in the circuit. They are valid at all times. The branch constraints for the circuit elements introduce the derivatives and the integrals, and the equilibrium equations thus become cast in the differential-equation form.

In the steady-state rotating phasor analysis, all signals (excitations and responses anywhere) are rotating phasors as shown in the last section. Thus, all node voltages and loop currents in a circuit are rotating phasors rotating with the same speed ω. The individual branch currents and voltages that are combinations (or sums) of node voltages or loop currents are also rotating phasors. All the signals in a given circuit are thus rotating phasors with the same ω and are consequently related by complex constants. The excitation and response are related by a constant as in Eq. 6.57 and so are the branch currents and voltages. The equilibrium equations for rotating-phasor analysis thus should involve only complex constants apart from the exponential common to all signals. The rotating-phasor analysis, therefore, yields algebraic equations (with complex constants) and not differential equations.

In this section we define these constant relationships between the steady-state rotating-phasor voltage and current at the same pair of terminals for any circuit as well as for the individual circuit elements. These constant relationships or branch constraints are then used in the next section to develop the algebraic equilibrium equations.

Henceforth, we assume all circuits to be in steady-state conditions and excited by sinusoidal signals unless otherwise stated. The sinusoidal signals are indicated by the time functions $v(t)$ or $i(t)$, and the corresponding rotating phasors are denoted by $v_p(t)$ and $i_p(t)$. The *Re*/*Im* operation then converts v_p into v and i_p into i. The stationary phasors are indicated by capital letters V and I.

Consider the circuit in Fig. 6.16*a* with no internal sources excited at the terminals where the voltage and the current signals are shown. The corresponding rotating-phasor signals shown in Fig. 6.16*b* are given by

$$\begin{aligned} v_p(t) &= Ve^{j\omega t} \\ i_p(t) &= Ie^{j\omega t} \end{aligned} \tag{6.59}$$

where V and I are complex constants in general. The ratio of the rotating

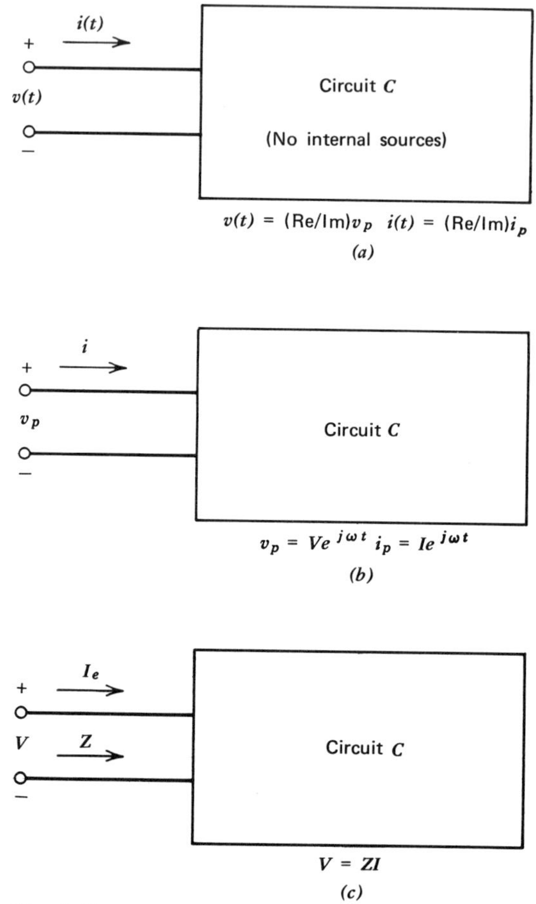

Fig. 6.16

phasor voltage and the rotating-phasor current is a constant, Z, called the *impedance*:

$$Z = \frac{v_p}{i_p} = \frac{Ve^{j\omega t}}{Ie^{j\omega t}} = \frac{V}{I} \tag{6.60}$$

The impedance Z is thus a ratio of the voltage and current rotating phasors or, more properly, the ratio of the associated stationary phasors. The impedance is a very useful concept and is used universally in describing circuits as we shall see. It is commonly used to relate the stationary voltage and current phasors; and $e^{j\omega t}$ is introduced whenever the rotating phasor is desired. For example for the circuit of Fig. 6.16*c*, with the polarities as shown, if the current (stationary) phasor I is given, then the voltage phasor $V = ZI$. The rotating voltage phasor may then be obtained by $v_p = Ve^{j\omega t}$, and the sinusoidal signal obtained by the *Re*/*Im* operation.

Note that Z is a constant and not dependent on time. Since V and I are usually complex constants, Z is also complex. Finally, *note that Z is a ratio of phasors either stationary or rotating, but not a ratio of sinusoids.*

The impedance as defined in Eq. 6.60 is a ratio of the voltage and current phasor signals at *the same terminal pair* of the circuit with polarities as shown in Fig. 6.16. One of the signals is considered the driving source or excitation and the other the response at the same driving terminals or points. For this reason, when the phasor signals are observed at the same terminals, the impedance is also called the *driving-point impedance.* This distinguishes it from other ratios defined later between voltage and current phasors at different terminals. We will mostly use the simpler term impedance when there is no chance of confusion.

Notice that by Eq. 6.60 $V = ZI$, a sort of a complex Ohm's law. Intuitively, one may view the impedance as impeding or resisting the flow of current. For a given phasor voltage V, the larger the magnitude of the impedance the smaller is the current (phasor) I. The phase angle θ of the impedance is the phase angle by which the current lags the voltage phasor. These properties will become clearer as we examine the various impedances of circuit elements.

The reciprocal of the impedance is also used frequently, and it is called the *admittance.* The admittance Y is given by

$$Y = \frac{1}{Z} = \frac{I}{V} \tag{6.61}$$

Finally, the complex impedance may be expressed as $Z = R + jX$, where the real part R is called the *resistance* of the circuit and the imaginary part X is called the *reactance.* Similarly, $Y = G + jB$, and the real part G is called the *conductance*; the imaginary part B is called the *susceptance.* These terms will become clear as we develop the impedances of circuits.

Example 6.5.1 Let the circuit impedance be given by

$$Z = 1 + j1$$

and the current by $i(t) = 10 \sin(30t + \pi/4)$. Find the voltage $v(t)$.

Solution. Convert $i(t)$ to $i_p(t)$, where

$$i_p(t) = 10\, e^{j(\pi/4)} e^{j\omega t}$$

and the operation Im $i_p = i(t)$. Then the current phasor $I = 10\, e^{j(\pi/4)}$. Th
voltage phasor is

$$
\begin{aligned}
V &= ZI \\
&= (\sqrt{2}\, e^{j(\pi/4)})(10\, e^{j(\pi/4)}) \\
&= 10\sqrt{2}\, e^{j(\pi/2)}
\end{aligned}
$$

The rotating phasor is given by

$$v_p = 10\sqrt{2}\, e^{j(\pi/2)} e^{j30t}$$

and

$$v(t) = \text{Im}\ v_p = 10\sqrt{2} \sin (30t + \pi/2)$$

The impedance or admittance of a circuit is thus of great value i determining the (steady-state) response of a circuit under (sinusoida excitation. The impedance is a property of the circuit parameters for given frequency and can be expressed simply. We determine below th impedances of the individual circuit elements. In later sections, thes impedances will be shown to serve as building blocks for impedances o arbitrary circuits.

The impedances of one-port circuit elements are determined simpl by applying the branch constraints to the rotating phasor signals an taking the ratios of the respective voltage and current phasors. Thus, th impedance of a one-port element serves to represent the (branch phasor constraint. For the two-port elements we give the phasor con straints and define impedances where appropriate.

Resistor

The voltage-current relationship for a resistor shown in Fig. 6.17 is

$$v_p = Ri_p \tag{6.62}$$

where

$$
\begin{aligned}
v_p &= Ve^{j\omega t} \\
i_p &= Ie^{j\omega t}
\end{aligned}
\tag{6.63}
$$

The impedance of a resistor is given by Eq. 6.62 to be

$$Z_R = \frac{v_p}{i_p} = \frac{V}{I} = R \tag{6.64}$$

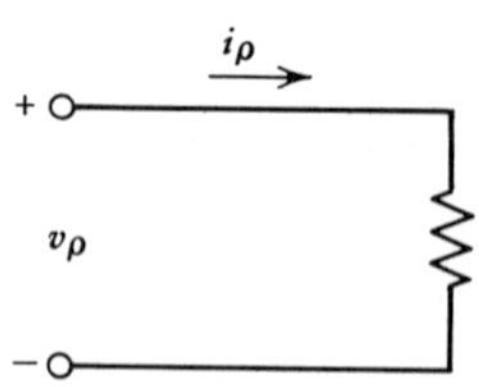

Fig. 6.17

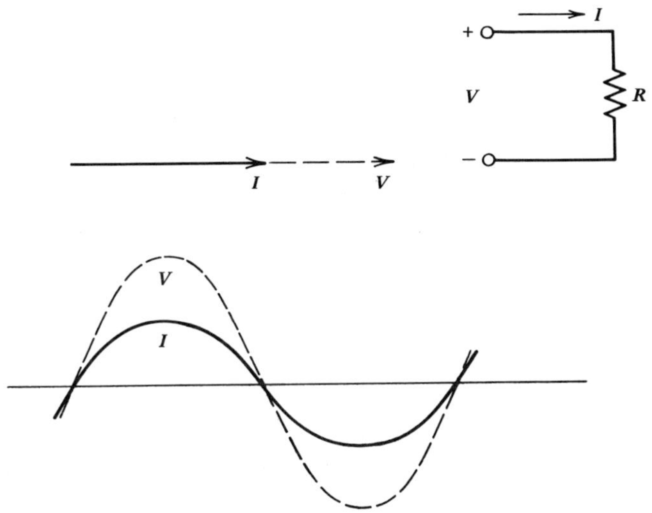

Fig. 6.18

The impedance is a real constant R and therefore the phasor V $(= RI)$ and the phasor I are in phase. If the phasor I is a reference, that is, $I = |I|\angle 0$, then $V = R|I|\angle 0$. The phasor constraint, the phasor diagram and the corresponding sinusoids are shown in Fig. 6.18. Note that the impedance relates the phasor-voltage polarities and current direction as shown. *Observe that for a resistor, the voltage and current are in phase.*

Inductor

The voltage-current relationship for the inductor in Fig. 6.19 is given by

$$v_\rho = L\frac{di_\rho}{dt} \tag{6.65}$$

Since $i_\rho = Ie^{j\omega t}$ and $v_p = Ve^{j\omega t}$, from Eq. 6.65

$$\begin{aligned} Ve^{j\omega t} = v_\rho &= L\frac{d}{dt}(Ie^{j\omega t}) \\ &= j\omega LIe^{j\omega t} \end{aligned}$$

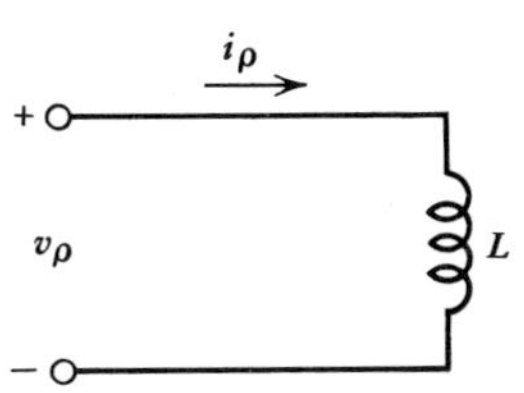

Fig. 6.19

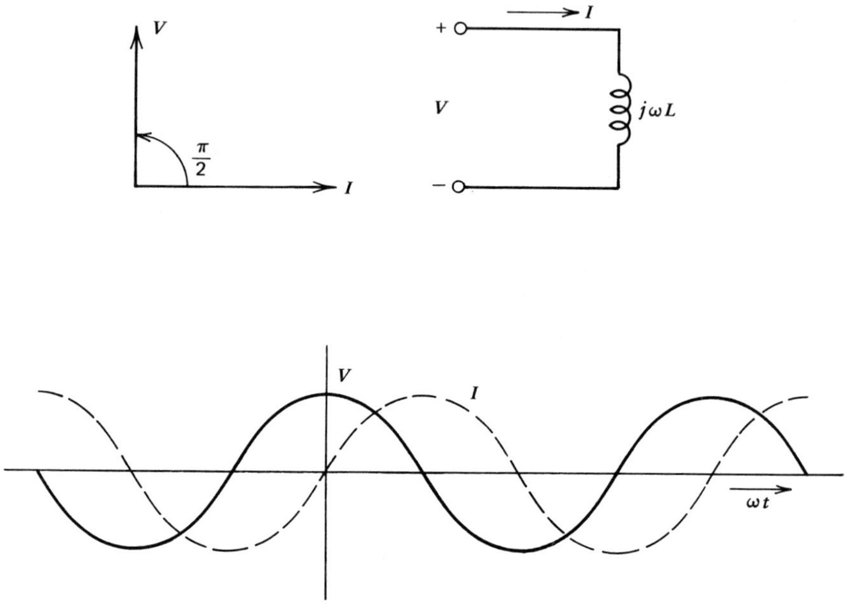

Fig. 6.20

The impedance of an inductor is thus

$$Z_L = \frac{v_p}{i_p} = \frac{V}{I} = j\omega L \tag{6.66}$$

The impedance of an inductor has magnitude (ωL) and phase angle $\pi/2$ radians. The phasor diagram with $I = |I|\angle 0$ as reference is shown in Fig. 6.20 with the corresponding sinusoids. *Observe that for an inductor, the current lags behind the voltage by $\pi/2$ radians or 90°.*

Capacitor

For the capacitor in Fig. 6.21,

$$i_p = C\frac{dv_p}{dt} \tag{6.67}$$

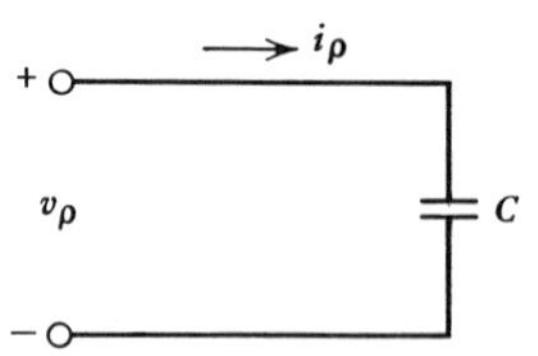

Fig. 6.21

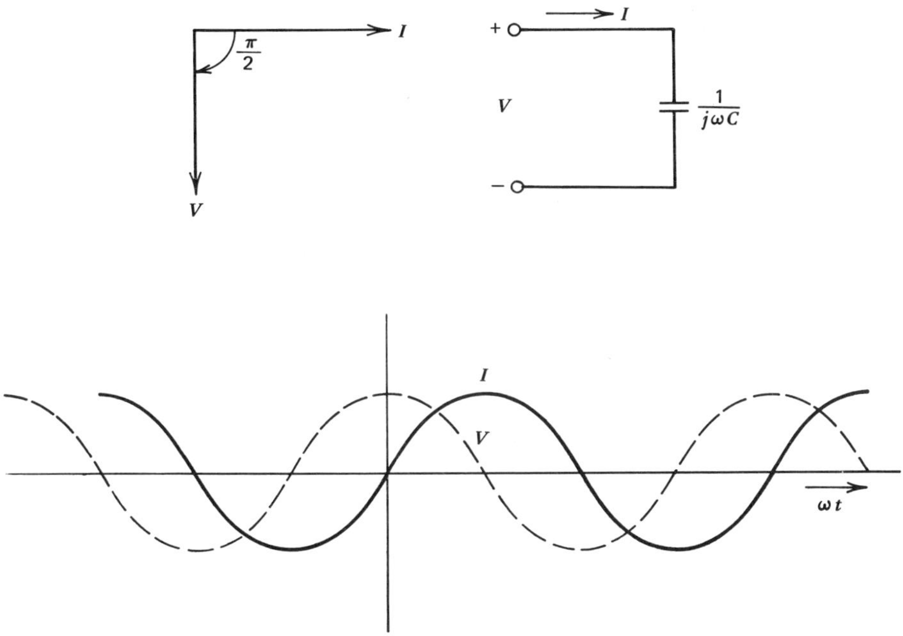

Fig. 6.22

where $v_\rho = Ve^{j\omega t}$ and $i_\rho = Ie^{j\omega t}$. Thus

$$\begin{aligned} i_\rho &= C\frac{d}{dt}(Ve^{j\omega t}) \\ &= j\omega CVe^{j\omega t} \\ &= Ie^{j\omega t} \end{aligned}$$

Therefore, the impedance of a capacitor is given by

$$Z_C = \frac{v_\rho}{i_\rho} = \frac{V}{I} = \frac{1}{j\omega C} \tag{6.68}$$

The impedance of a capacitor has magnitude $(1/\omega C)$ and phase angle $(-\pi/2)$ radians. The phasor diagram is shown in Fig. 6.22 with the current phasor $I = |I|\angle 0$ as the reference. *Observe that for a capacitor, the current leads the voltage by* $\pi/2$ *radians or* 90°.

Coupled Coils

The voltage-current relationships for the coupled coils in Fig. 6.23 are

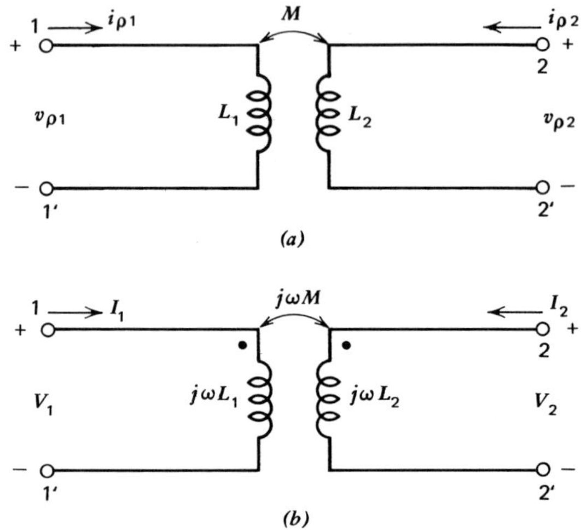

Fig. 6.23

given by

$$L_1 \frac{di_{\rho 1}}{dt} \pm M \frac{di_{\rho 2}}{dt} = v_{\rho 1}$$
$$\pm M \frac{di_{\rho 1}}{dt} + L_2 \frac{di_{\rho 2}}{dt} = v_{\rho 2} \tag{6.69}$$

where the rotating phasors are

$$v_{\rho 1} = V_1 e^{j\omega t}$$
$$v_{\rho 2} = V_2 e^{j\omega t}$$
$$i_{\rho 1} = I_1 e^{j\omega t} \tag{6.70}$$
$$i_{\rho 2} = I_2 e^{j\omega t}$$

The relationships involve voltages and currents at different terminal pairs and the constraints are two-part constraints. The signals at either terminal pair or port depends on the signals at the other port, and no impedance can be defined in the usual manner as before. Generally, we may substitute Eq. 6.70 into Eq. 6.69 to obtain

$$j\omega L_1 I_1 \pm j\omega\, M I_2 = V_1$$
$$\pm j\omega M I_1 + j\omega L_2 I_2 = V_2 \tag{6.71}$$

for the two-port phasor relations. Let us now consider the coupled coils driven by a source $i_{\rho 1} = I_1 e^{j\omega t}$ with the other end open-circuited, that is, $i_{\rho 2} = 0$ or $I_2 = 0$. Then, we can define the impedance at the terminal pairs 1–1′ or port 1:

$$Z_{11} = \left.\frac{v_{\rho 1}}{i_{\rho 1}}\right|_{i_{\rho 2}=0} = \frac{V_1}{I_1} = j\omega L_1 \tag{6.72}$$

The subscript of Z_{11} indicates voltage at port 1 and current at port 1. But the voltage at port 2 is present also in this case, and we can define a *transfer impedance* as the ratio of voltage phasor at port 2 to the current phasor at port 1, that is,

$$Z_{21} = \left.\frac{v_{\rho 2}}{i_{\rho 1}}\right|_{i_{\rho 2}=0} = \frac{V_2}{I_1} = \pm\, j\omega M \tag{6.73}$$

Similarly, it is possible to define impedances

$$Z_{22} = \left.\frac{v_{\rho 2}}{i_{\rho 2}}\right|_{i_{\rho 2}=0} = \frac{V_2}{I_2} = j\omega L_2 \tag{6.74}$$

$$Z_{12} = \left.\frac{v_{\rho 1}}{i_{\rho 2}}\right|_{i_{\rho 1}=0} = \frac{V_1}{I_2} = \pm\, j\omega M \tag{6.75}$$

The coupled coils phasor constraints and the relevant impedances are shown in Fig. 6.23*b*. The dot conventions are introduced to make our discussion more specific. Let us interpret the Eqs. 6.71–6.75. The impedance of the self-inductance is $j\omega L_1$, and so the self-induced phasor voltage in coil 1 due to phasor current I_1 is $j\omega L_1 I_1$. This is a voltage drop in the direction of the current flow I_1. The impedance of the mutual inductance is $\pm j\omega M$, and the voltage induced in coil 1 due to phasor current I_2 is $\pm j\omega M I_2$. The polarity of the induced voltage is determined by the dots. Since I_2 is flowing into the coil at the dot end, the mutually induced voltage is positive at the dot end of coil 1. Thus, in the direction of I_1 the mutually induced voltage is a voltage drop $+j\omega M I_2$ and the positive sign must be used in Eq. 6.71. Similar reasoning is used for the second coil phasor voltages.

Other two-port elements involve only constants as phasor constraints. No change is involved for phasor relationships except that the phasors are substituted for time functions. The controlled source and the ideal transformer-phasor constraints are shown in Figs. 6.24 and 6.25.

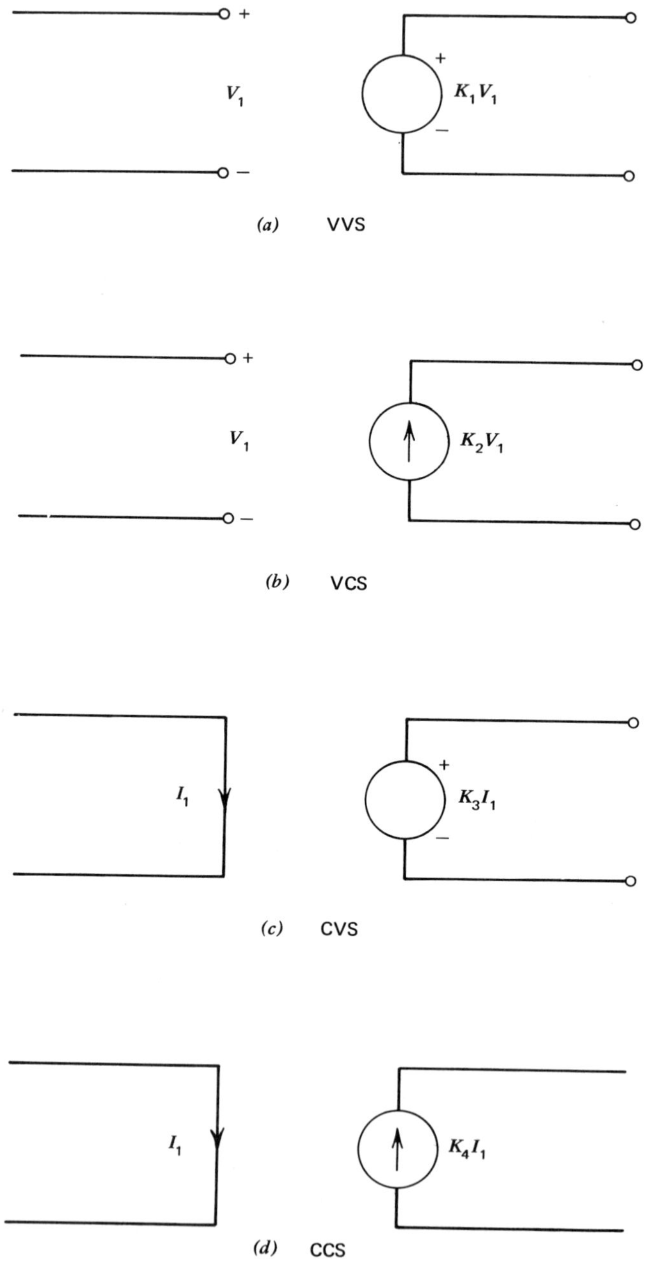

$v_{p1} = V_1 e^{j\omega t},\ i_{p1} = I_1 e^{j\omega t}$

Fig. 6.24 Phasor constraints for controlled sources.

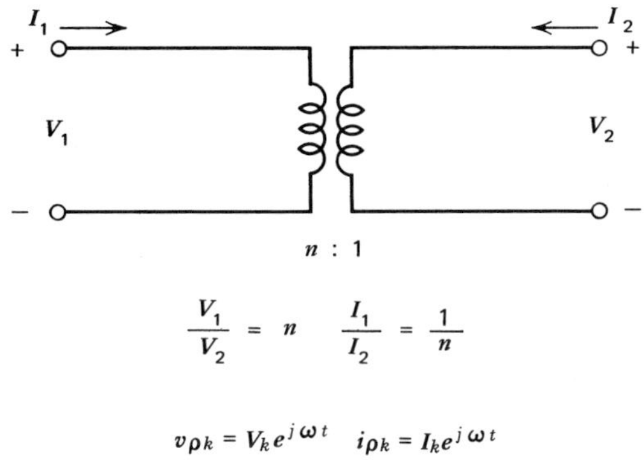

Fig. 6.25 Ideal transformers.

6.6 KIRCHHOFF'S *(PHASOR)* LAWS

The phasor branch constraints represented by the impedances ($R, j\omega L, 1/j\omega C$) for one-port elements with the corresponding two-port constraints further simplify the steady-state analysis. The branch constraints need be expressed only for the stationary phasors since $e^{j\omega t}$, common to all rotating phasors, cancels out. The rotating-phasor equilibrium equations also simplify to stationary phasor equations since again the exponential $e^{j\omega t}$, common to all rotating phasor terms, cancels out. Let us examine the equilibrium laws (KVL and KCL) for rotating phasors.

Kirchhoff's Voltage Law

Consider a closed loop of a circuit as shown in Fig. 6.26 and let us apply the KVL to the rotating-phasor voltages across the elements. Then

$$v_{\rho 1} + v_{\rho 2} + v_{\rho 3} = v_{\rho S} \tag{6.76}$$

where the rotating phasors $v_{\rho k} = V_k e^{j\omega t}$, $k = 1, 2, 3, S$. Equation 6.76,

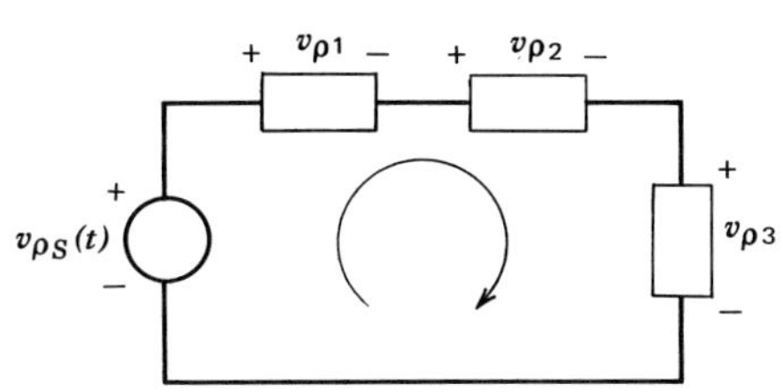

Fig. 6.26

upon substituting $v_{\rho k} = V_k e^{j\omega t}$, becomes

$$V_1 e^{j\omega t} + V_2 e^{j\omega t} + V_3 e^{j\omega t} = V_S e^{j\omega t}$$

or

$$V_1 + V_2 + V_3 = V_S \tag{6.77}$$

Thus, the KVL is simply an equilibrium equation for stationary-phasor voltages. These phasors may be expressed in terms of the stationary phasor loop currents by the use of impedances or phasor branch constraints. The resulting equations are thus simultaneous algebraic equations in phasor loop currents. The solution of these algebraic equations gives the phasor currents, and when they are multiplied by $e^{j\omega t}$, we have the rotating-phasor currents. The operation Re/Im may then be used to obtain the sinusoidal currents if needed.

Kirchhoff's Current Law

Consider the KCL applied to the rotating phasor currents in Fig. 6.27:

$$i_{\rho 1} + i_{\rho 2} + i_{\rho 3} = i_{\rho S} \tag{6.78}$$

where $i_{\rho k} = I_k e^{j\omega t}$. Substituting, we have

$$I_1 e^{j\omega t} + I_2 e^{j\omega t} + I_3 e^{j\omega t} = I_S e^{j\omega t}$$

or

$$I_1 + I_2 + I_3 = I_S \tag{6.79}$$

Once again the equilibrium equation is in terms of stationary phasors. The phasor branch constraints are used to express such equations in terms of phasor node voltages yielding simultaneous algebraic equations as before.

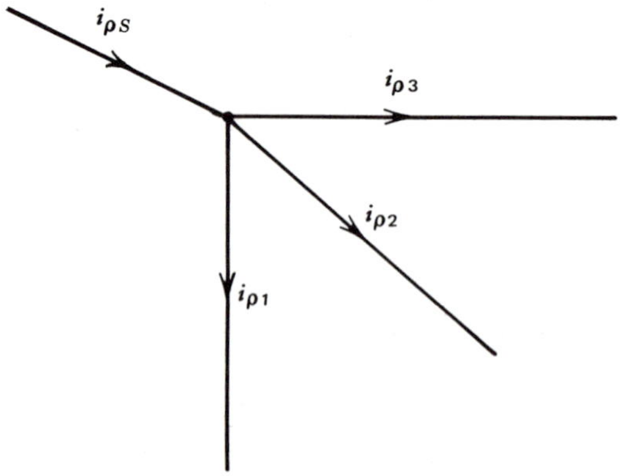

Fig. 6.27

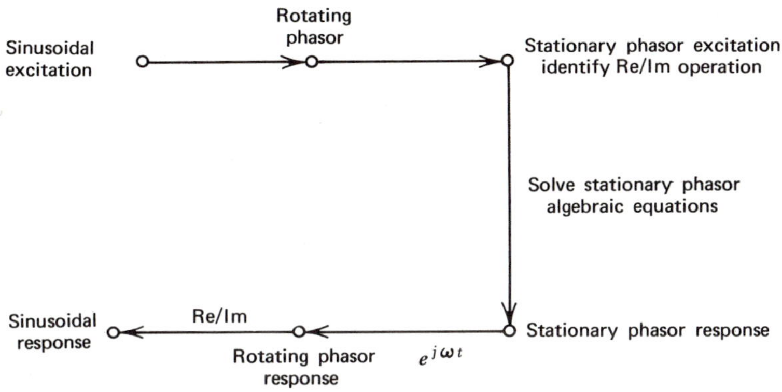

Fig. 6.28

A few examples illustrate the phasor analysis. The steps followed in the analysis are:

1. The source sinusoidal signal is replaced by the corresponding rotating phasor. The *Re*/*Im* operation to convert the rotating phasor into the sinusoid is identified.
2. The stationary source phasor is now used to write the phasor equilibrium equations in terms of stationary-phasor signals.
3. The algebraic phasor equations are solved for the unknown phasors.
4. The stationary response phasors are converted into rotating phasors.
5. The *Re*/*Im* operation is used to obtain the sinusoidal response.

Figure 6.28 shows the steps followed.

Example 6.6.1 Find the (steady-state) node voltage $v(t)$ for the circuit in Fig. 6.29 if $i_S(t) = |I_S| \sin \omega t$.

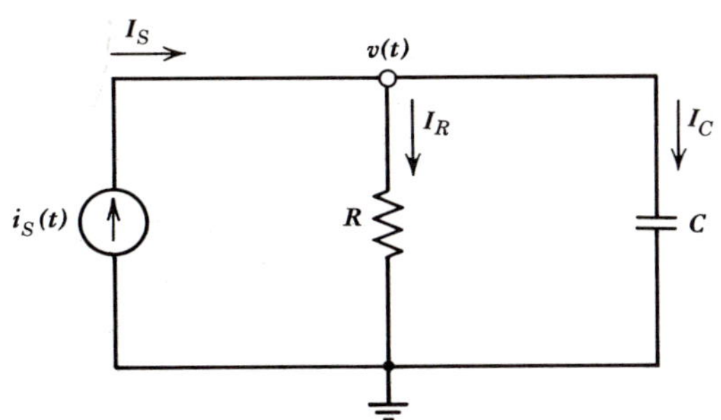

Fig. 6.29

Solution.

$$i_S(t) = |I_S| \sin \omega t = \text{Im}[\,|I_S|\,e^{j\omega t}]$$

or

$$I_S = |I_S|\angle 0$$

KCL: $I_R + I_C = I_S$

or

$$\frac{V}{R} + j\omega C\,V = I_S = |I_S| \tag{6.80}$$

$$V = \frac{|I_S|}{G + j\omega C} \qquad \left(G = \frac{1}{R}\right) \tag{6.81}$$

Then

$$v_\rho(t) = Ve^{j\omega t} = \frac{|I_S|}{G + j\omega C}\,e^{j\omega t}$$

and

$$v(t) = \text{Im}\;v_\rho(t) = \frac{|I_S|}{\sqrt{G^2 + \omega^2 C^2}} \sin\left(\omega t - \tan^{-1}\frac{\omega C}{G}\right) \tag{6.82}$$

Example 6.6.2 Find $i(t)$ for the circuit in Fig. 6.30 for

$$\begin{aligned} v_S(t) &= |V_S| \sin(\omega t + \theta) \\ &= \text{Im}\;V_S e^{j\omega t} \end{aligned}$$

where

$$V_S = |V_S|\,e^{j\theta}$$

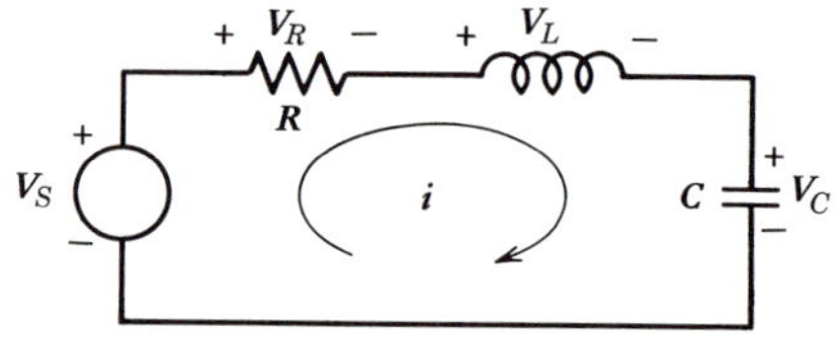

Fig. 6.30

Solution. Applying KVL to stationary-phasor voltages

$$\text{KVL:} \qquad V_R + V_L + V_C = V_S$$

or

$$RI + j\omega LI + \frac{1}{j\omega C}I = |V_S|\,e^{j\theta} \tag{6.83}$$

$$I = \frac{|V_S|\,e^{j\theta}}{R + j\left(\omega L - \dfrac{1}{\omega C}\right)} \tag{6.84}$$

Then $\quad i_p(t) = Ie^{j\omega t}$

and $\quad i(t) = \text{Im}\, i_p$

$$= \text{Im}\left[\frac{|V_S|e^{j\theta}}{R + j(\omega L - 1/\omega C)}e^{j\omega t}\right]$$

$$= \frac{|V_S|}{\sqrt{R^2 + (\omega L - 1/\omega C)^2}} \sin\left(\omega t + \theta - \tan^{-1}\frac{\omega L - 1/\omega C}{R}\right) \tag{6.85}$$

The phasor equilibrium equations for multinode and multiloop circuits are written in the same manner and lead to sets of simultaneous algebraic equations in phasor quantities. The equations are similar to those obtained for resistive networks except that now the coefficients are complex in general. The formal solution of these equations is straightforward and the numerical method of solution is only slightly modified as described in the computation section at the end of the chapter. We illustrate the procedure by examples before stating the general form of the equations. Only the phasor-signal quantities are shown in the circuit diagrams since the equilibrium equations are written for these signals.

Example 6.6.3 Consider the circuit of Fig. 6.31 with the reference node as shown. The phasor KCL's at the remaining nodes give:

$$\begin{aligned} \frac{V_1}{R_1} + j\omega C(V_1 - V_2) &= I \\ j\omega C(V_2 - V_1) + \frac{V_2}{j\omega L} + \frac{V_2}{R_2} &= 0 \end{aligned} \tag{6.86}$$

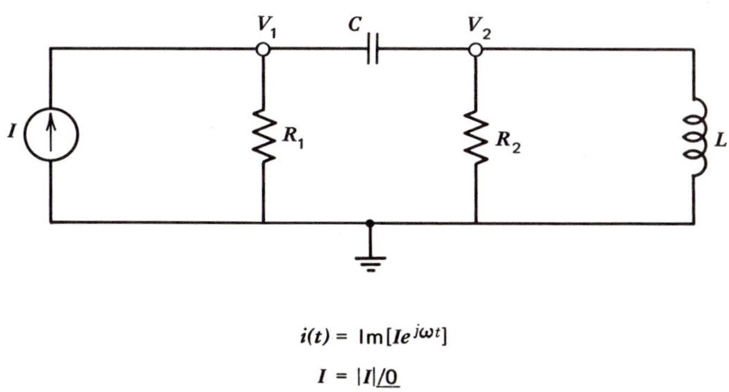

Fig. 6.31

Rearranging terms, we have

$$\begin{aligned} (G_1+j\omega C)V_1 - j\omega C\, V_2 &= I \\ -j\omega C\, V_1 + \left(G_2 + j\omega C + \frac{1}{j\omega L}\right)V_2 &= 0 \end{aligned} \tag{6.87}$$

Observe that these node equations have the same general form as the equations in Chapters 3 and 5. The coefficient of V_1 is positive in the first equation for node 1, and its value is given by the sum of the admittances of the elements incident (connected) at node 1. In the same equation the coefficient of V_2 is the negative of the admittance of the element connected between nodes 1 and 2. Similarly, in the equation for node 2, the coefficient of V_2 is the positive sum of all admittances incident at node 2; whereas, the coefficient of V_1 is the negative admittance of the element connected between nodes 2 and 1.

For resistive networks, the coefficients were conductances whereas now they are admittances. For transient analysis, the coefficients were $(G_{ij}+C_{ij}D+(1/L_{ij})D^{-1})$ whereas now the operation D is replaced by $(j\omega)$ and D^{-1} by $(j\omega)^{-1}$. The time variation for rotating-phasor analysis is $e^{j\omega t}$ which is common to all terms and $D(e^{j\omega t}) = j\omega(e^{j\omega t})$ and $D^{-1}(e^{j\omega t}) = (1/j\omega)(e^{j\omega t})$. The common exponentials are then cancelled out leaving D replaced by $j\omega$ and D^{-1} by $1/j\omega$. Otherwise, the procedure for writing the equations is the same. Remember, however, that the signals are stationary phasors now and not time functions.

Let us obtain the solution V_1 for Eq. 6.87 for $G_1 = G_2 = 2$ mhos, $C = 1/4\, f$, $L = 4/3\, h$, $\omega = 3$ rads/sec and $I = 12\ \angle 0$. The equations become

$$\begin{aligned} (2+j0.75)V_1 - j0.75\, V_2 &= 12 \\ -j0.75\, V_1 + (2+j0.50)V_2 &= 0 \end{aligned}$$

Then

$$V_1 = \frac{\begin{vmatrix} 12 & -j\ 0.75 \\ 0 & 2+j\ 0.50 \end{vmatrix}}{\begin{vmatrix} 2+0.75 & -j\ 0.75 \\ -j\ 0.75 & 2+j\ 0.50 \end{vmatrix}}$$

$$= \frac{24+j6}{4.1875+j2.5}$$

The sinusoidal time function is then obtained by evaluating Im $[V_1 e^{j3t}]$.

Example 6.6.4 Consider next the three loop circuit in Fig. 6.32 and write the phasor-loop equations. Observe that the mutual inductance is common between loops 1 and 2, loops 1 and 3, and loops 2 and 3. An appro-

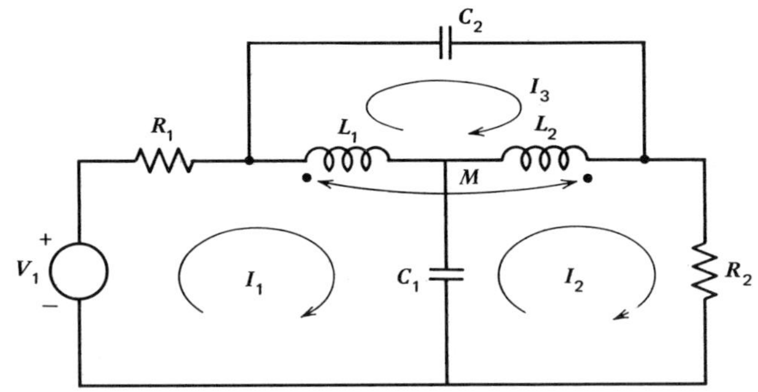

Fig. 6.32

priate impedance $\pm j\omega M$ must be introduced to represent the corresponding voltage drop due to the mutual impedance.

The loop equations are:

$$\left(R_1+j\omega L_1+\frac{1}{j\omega C_1}\right)I_1-\left(j\omega M+\frac{1}{j\omega C_1}\right)I_2-(j\omega L_1-j\omega M)I_3=V_1$$

$$-\left(j\omega M+\frac{1}{j\omega C_1}\right)I_1+\left(R_2+j\omega L_2+\frac{1}{j\omega C_1}\right)I_2-(j\omega L_2-j\omega M)I_3=0$$

$$-(j\omega L_1-j\omega M)I_1-(j\omega L_2-j\omega M)I_2+\left(\frac{1}{j\omega C_2}+j\omega L_1+j\omega L_2-2j\omega M\right)I_3=0 \qquad (6.88)$$

Observe that the coefficients in the phasor loop equations are impedances R_{ij}, $j\omega L_{ij}$, $1/j\omega C_{ij}$ and, where appropriate, $\pm j\omega M$. The conventions are as for general loop equations for planar networks with D replaced by $(j\omega)$ and D^{-1} by $1/j\omega$.

The general phasor equations are thus algebraic simultaneous equations, and we now describe the form of the equations for circuits with only one-port elements. The presence of two-port elements modifies the coefficients appropriately. For the node equations, the phasor equations are generally of the form

$$Y\mathbf{V}=\mathbf{I} \qquad (6.89)$$

where $\mathbf{V}$ is the vector of phasors $V_1, V_2, \ldots, V_N$ for an $(N+1)$ node circuit. The vector $\mathbf{I}$ is the source phasor vector and the matrix Y of coefficients y_{ij} has elements given by

$$y_{ij}=\pm\left(G_{ij}+j\omega C_{ij}+\frac{1}{j\omega L_{ij}}\right) \qquad (6.90)$$

where the plus sign is for y_{ii} and the negative sign for $i \neq j$. The elements y_{ii} represent the sum of admittances incident at node i and the y_{ij} represent the negative sum of the admittances connected between nodes i and j. The elements y_{ij} are modified when the voltage sources are present or when two-port elements are present. The voltage sources are handled in the same way as for resistive networks, and the two-port elements are handled as for general transient analysis.

Similarly, for phasor loop equations the general form is

$$Z\mathbf{I} = \mathbf{V} \tag{6.91}$$

where $\mathbf{V}$ is the phasor source vector, $\mathbf{I}$ is the phasor loop current vector, and Z for planar networks with no current sources or two-port elements has elements

$$z_{ij} = \pm\left(R_{ij} + j\omega L_{ij} + \frac{1}{j\omega C_{ij}}\right) \tag{6.92}$$

The consistent signs are obtained as before by assigning all loop currents in a consistent direction.

Both the Eqs. 6.89 and 6.91 are simultaneous algebraic equations, and their solutions are obtained in the same manner as for resistive networks. The fact that the matrices Z and Y are complex makes no difference in the general solution:

$$\begin{aligned} &\text{Node Equations:} \quad \mathbf{V} = Y^{-1}\mathbf{I} \\ &\text{Loop Equations:} \quad \mathbf{I} = Z^{-1}\mathbf{V} \end{aligned} \tag{6.93}$$

Equivalently, Cramer's rule can also be applied to obtain the solutions. Numerical algorithm modifications are discussed in the last section of this chapter.

The next section describes further simplifications that arise due to series and parallel connections of circuit elements. Impedances of such combinations are easily expressed and the phasor signals anywhere in the circuit are easily obtained.

6.7 SERIES AND PARALLEL CONNECTIONS

The phasor equilibrium equations and the concept of impedance considerably simplifies the sinusoidal steady-state analysis as we saw in the last section. Even the solution of phasor algebraic equations is unnecessary if the impedance of the entire circuit at the source terminal is known. The impedance gives the phasor excitation-response relationship at the source terminal, and consequently the response phasor is immediately determined. Thus, the knowledge of the circuit impedance can be very convenient in analysis, particularly if the impedance is readily determined.

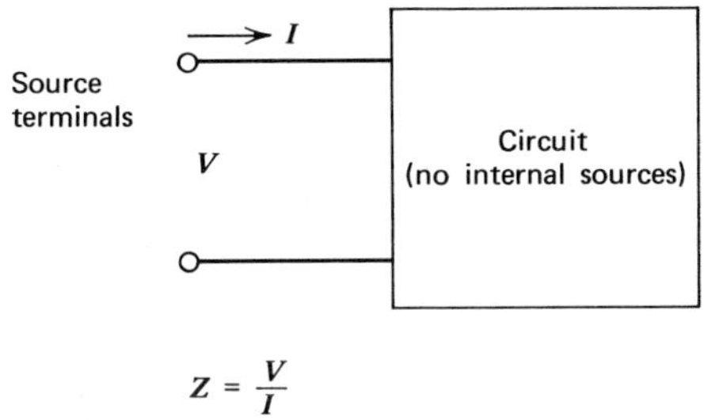

Fig. 6.33

Generally, the circuit impedance must be determined by the solution of the phasor equilibrium equations. Then, the voltage phasor at the source terminals divided by the current phasor at the same terminals gives the circuit impedance as seen by the source as shown in Fig. 6.33. The impedance can be derived assuming any arbitrary current phasor as excitation and determining the voltage phasor at the same terminals. The ratio of V to I is then the impedance as seen at these terminals, that is,

$$Z = \frac{V}{I} \tag{6.94}$$

Fortunately, the determination of the circuit impedance for many circuits can be considerably simplified and performed by inspection. Specifically, series and parallel connections of circuits or circuit elements have impedances obtained by simple rules. We now derive these rules.

Series Connection

Consider the series connection of two circuit impedances as shown in Fig. 6.34 with the phasor voltages and currents as shown. The series connection assures that the same current phasor I is passing through both

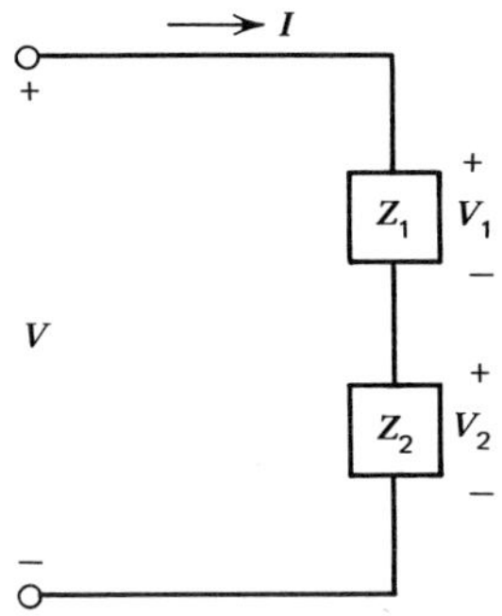

Fig. 6.34

the circuit impedances. The voltages across them are thus given by Z_1I and Z_2I, that is,

$$Z_1 = \frac{V_1}{I} \tag{6.95}$$

and

$$Z_2 = \frac{V_2}{I} \tag{6.96}$$

Now the total voltage phasor from KVL is

$$V = V_1 + V_2 \tag{6.97}$$

We find the total impedance Z to be

$$Z = \frac{V}{I} \tag{6.98}$$

$$= \frac{V_1 + V_2}{I} = \frac{V_1}{I} + \frac{V_2}{I}$$

$$Z = Z_1 + Z_2 \tag{6.99}$$

That is, if we put together two impedances, Z_1 and Z_2 in series, the total impedance is the sum of these two individual impedances. The idea can be generalized to n impedance branches in series. The total impedance is simply the sum of the individual impedances in series, that is,

$$Z = Z_1 + Z_2 + \cdots + Z_n \tag{6.100}$$

for n impedances connected in series.

Parallel Circuit

For impedances branches in parallel, consider only two impedances in parallel at first (Fig. 6.35). We will then generalize the idea to **n** imped-

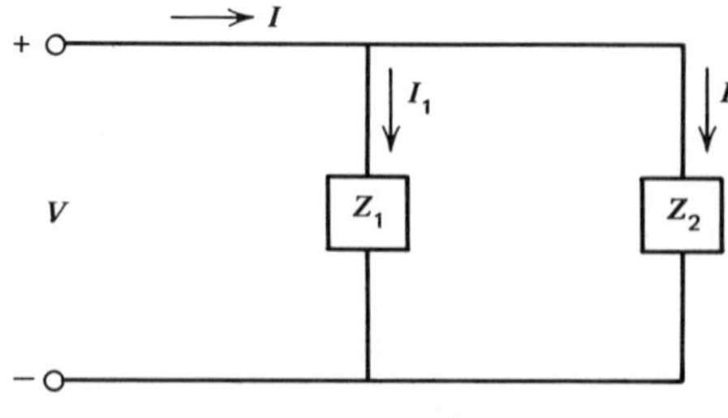

Fig. 6.35

ances in parallel. In this case the impedances are

$$Z_1 = \frac{V}{I_1} \qquad Z_2 = \frac{V}{I_2} \tag{6.101}$$

The total current from KCL is

$$I = I_1 + I_2$$

Dividing this equation by V, we get

$$\frac{I}{V} = \frac{I_1}{V} + \frac{I_2}{V} \tag{6.102}$$

which gives

$$\frac{1}{Z} = \frac{1}{Z_1} + \frac{1}{Z_2} \tag{6.103}$$

Thus, for a parallel combination of two impedances, the reciprocal of the total impedance is obtained by adding together the reciprocals of the individual impedances. We have already defined the reciprocal of impedance as admittance, that is,

$$\text{Admittance} = Y = \frac{1}{Z} = \frac{I}{V} \tag{6.104}$$

We now restate the impedance of parallel branches in terms of admittances. The total admittance of a parallel combination equals the sum of the admittances of the parallel branches. Generalizing to n branches, we have

$$Y = Y_1 + Y_2 + \cdots + Y_n \tag{6.105}$$

for parallel combination.

To summarize, we sum the branch impedances to determine the impedance of series circuits. To determine the admittance of parallel circuits, we sum the branch admittances. Let us consider some examples and also consider the phasor diagrams for these simple connections. Phasor diagrams give us an insight into the behavior of these simple circuits even when no specific excitation or reference phasor is given.

Example 6.7.1 Consider the series RL circuit of Fig. 6.36

The total impedance for this circuit is the sum of the individual

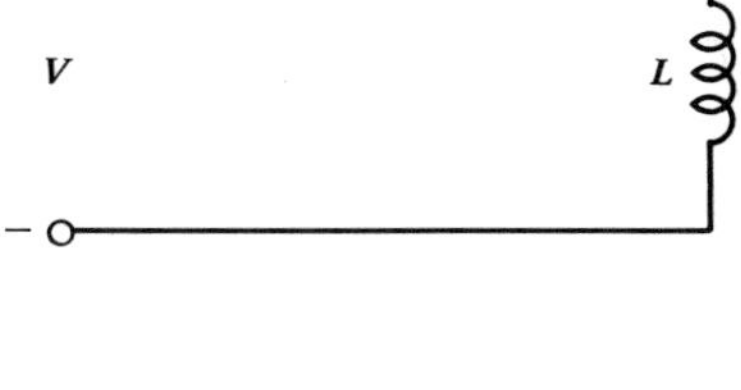

Fig. 6.36

impedances, that is,

$$Z = R + j\omega L \tag{6.106}$$

We see that the impedance is a complex number

$$Z = |Z|\angle\theta \tag{6.107}$$

where

$$|Z| = \sqrt{R^2 + \omega^2 L^2}$$

and

$$\theta = \tan^{-1}\frac{\omega L}{R} \tag{6.108}$$

The phasor voltage V across this circuit, in terms of the phasor I and the impedance Z, is

$$\begin{aligned} V &= ZI \\ &= (R + j\omega L)I \end{aligned} \tag{6.109}$$

If I is chosen as the reference vector, that is, $\mathbf{I} = |I|\ \angle 0$, we have the phasor diagram given in Fig. 6.37.

We see that the voltage phasor V is the vector sum of the voltages RI and $j\omega LI$, where j represents a 90° lead over the reference vector. Remember that when we have a pure inductance, the current in the inductance lags behind the voltage by 90° and the current through a resistance is in phase with the voltage across it. We see that in an inductive circuit, the current is still lagging behind the voltage but by an angle less than 90° or less than $\pi/2$ radians.

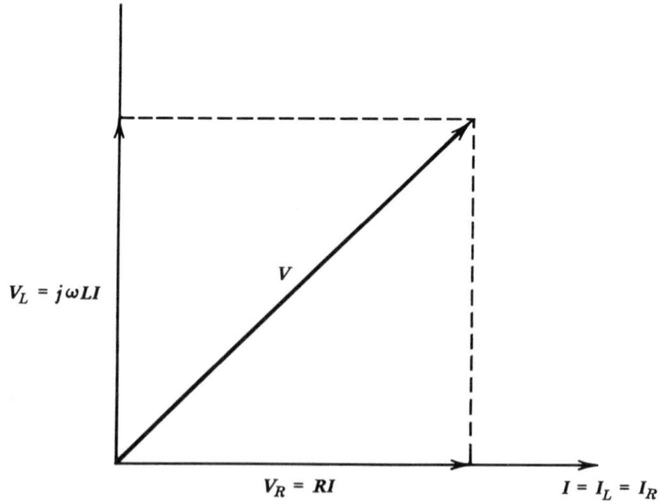

Fig. 6.37

The impedance $Z = R + j\omega L$ of the series circuit has a real part that corresponds to the resistive part and an imaginary part that corresponds to the reactive part. Therefore the real part of Z is termed the resistance and the imaginary part the reactance of the circuit. The reactance is the magnitude of the impedance of the reactive part of the circuit, that is, $X = \omega L$.

Example 6.7.2 Let us now take another series circuit with a resistor and a capacitor in series (Fig. 6.38).

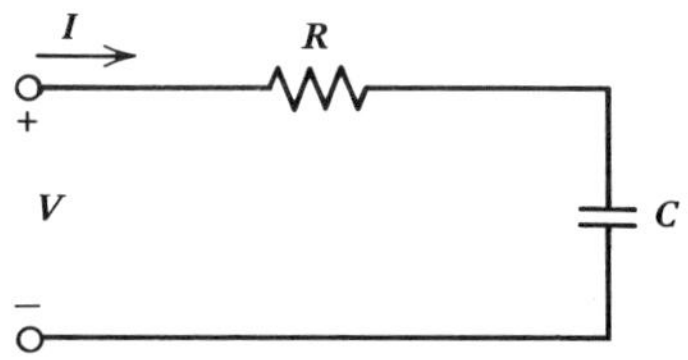

Fig. 6.38

In this case, the total impedance is

$$Z = R + \frac{1}{j\omega C} = R - \frac{j}{\omega C} \tag{6.110}$$

$$= \sqrt{R^2 + X^2} \angle \tan^{-1} - \frac{1}{\omega CR} \tag{6.111}$$

The reactance $X = (-1/\omega C)$ in this case. With current as our reference again, the phasor diagram is shown in Fig. 6.39. Recall that in a pure

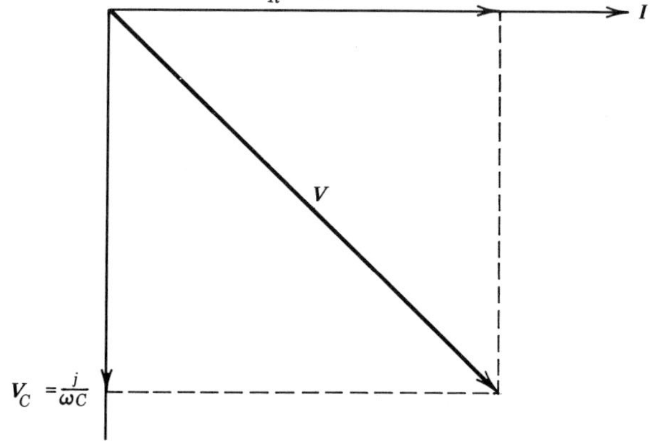

Fig. 6.39

capacitance, the current leads the voltage by 90°. In the present capacitive circuit, the current still leads the voltage; however, the phase angle depends on the amount of resistance in the circuit.

Example 6.7.3 Let us take the case where we have a resistor and a capacitor in parallel (Fig. 6.40).

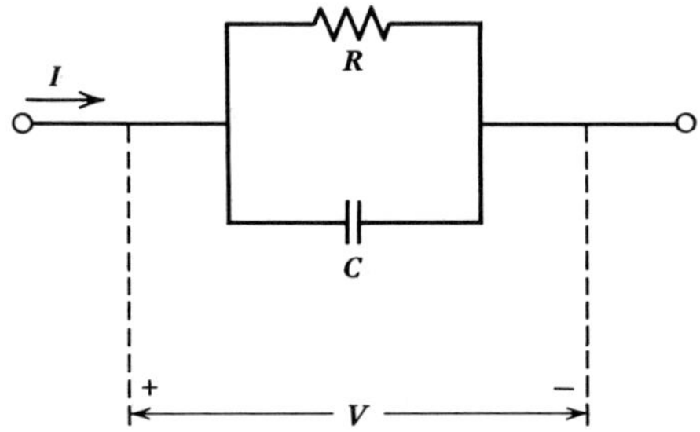

Fig. 6.40

In this case the admittance is obtained in terms of branch admittances

$$Y = \frac{1}{Z} = \frac{1}{R} + j\omega C \tag{6.112}$$

or

$$Y = G + j\omega c \tag{6.113}$$

where $G = 1/R$. This explains the term conductance for the real part of

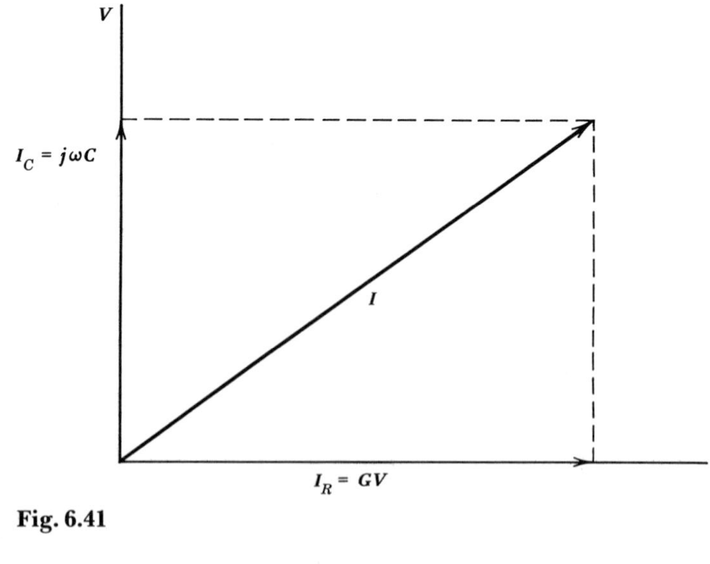

Fig. 6.41

the admittance. The susceptance is the magnitude of the reactive admittance, $B = \omega C$.

Now if we chose the voltage to be the reference phasor, the phasor diagram is as shown in Fig. 6.41.

Again we observe that the current in a capacitive circuit (whether series or parallel) leads the voltage, and the angle depends on the resistance involved.

Similarly, for a resistor and an inductor in parallel as in Fig. 6.42, we have the phasor diagram of Fig. 6.43.

We see thus that for inductive circuits the current is always lagging

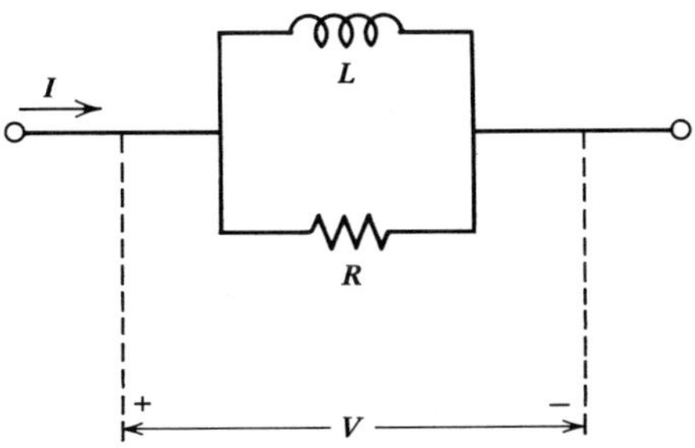

Fig. 6.42

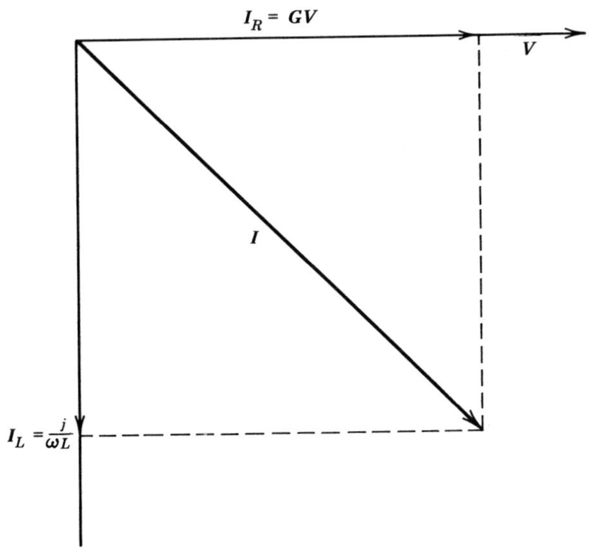

Fig. 6.43

behind the voltage phasor, and in capacitive circuits, the current is always leading the voltage phasor.

The above simple examples illustrate the power of the impedance concept and the phasor diagrams. Without any specified excitation we described the relationships between all the phasor signals in the circuit. We are also able to determine the total impedance of the circuit. The analysis for a given excitation is very simple once we know the impedance. Let us consider the series *RLC* circuit shown in Fig. 6.44. First, we examine the circuit generally and then obtain a specific response for a given excitation.

We will refer to the phasor voltages and currents for the most part as voltages and currents where there is no chance of confusion. Other signal descriptions will be specified as sinusoidal, rotating phasors, etc.

For the circuit of Fig. 6.44, the phasor quantities involved are the voltage drop across the resistor, V_R, the voltage drop across the inductor, V_L, and the voltage drop across the capacitor, V_C, in the direction of the current flow. For this circuit by KVL

$$V_R + V_L + V_C = V \tag{6.114}$$

The impedance of the series circuit is

$$\begin{aligned} Z &= R + j\omega L - \frac{j}{\omega C} \\ &= R + j\left(\omega L - \frac{1}{\omega C}\right) \end{aligned} \tag{6.115}$$

The real part of the impedance of a series *RLC* circuit is equal to the value of the resistance. The imaginary part of the total impedance is equal to the impedance of the reactive part of the circuit, that is, the difference of the magnitudes of the impedances of the inductance and the capacitance. If ωL is greater than $1/\omega C$, then the total impedance would be inductive; that is, the impedance would be R plus a positive reactance, and the circuit will be called *inductive.* If $1/\omega C$ is larger than ωL, the reactance would be negative, and the impedance would be *capacitive.* The relative magnitudes of ωL and $1/\omega C$ determine whether the current is leading or lagging behind the voltage.

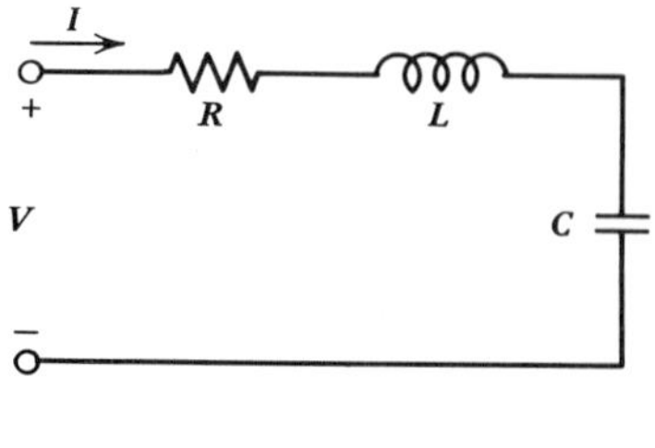

Fig. 6.44

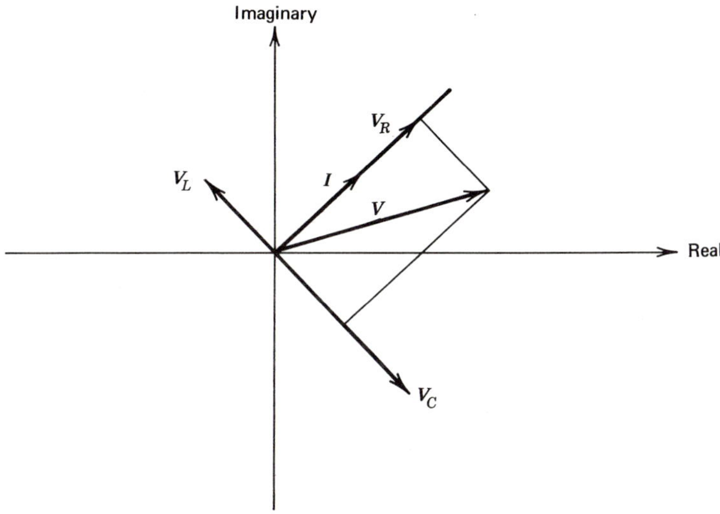

Fig. 6.45

A phasor diagram for a series *RLC* network is given in Fig. 6.45. Any one of the phasors may be chosen as a reference but none is chosen in the figure. Observe that V_R, the phasor for the voltage across the resistor, is in phase with I, the current phasor, and V_L, the voltage phasor across the inductor, leads the current phasor by 90° and V_C, the voltage phasor across the capacitor, is 90° behind the current phasor.

The magnitude of V_C is assumed to be larger than the magnitude of V_L; therefore, the circuit is capacitive. And the net voltage drop across the reactance, that is, the reactive part, of the circuit is a capacitive voltage drop. The total voltage across the circuit is the sum of V_R and the net voltage drop across the reactive part of the circuit. Since the circuit is capacitive, the current leads the total voltage across the circuit.

Example 6.7.4 Determine the sinusoidal current $i(t)$ in the series *RLC* circuit if $v(t) = 10 \cos 20t$, $R = 30\Omega$, $L = 1H$, $C = 1 \times 10^{-3}F$. The reference phasor is $V = 10 \angle 0$ and $v(t) = Re\,(10e^{j20t})$. The impedance is

$$\begin{aligned} Z &= R + j\left(\omega L - \frac{1}{\omega C}\right) \\ &= 30 + j\left(20 - \frac{1000}{20}\right) \\ &= 30 - j30 \\ &= 30\sqrt{2}\,e^{-j(\pi/4)} \end{aligned}$$

The current is

$$I=\frac{V}{Z}=\frac{10}{30\sqrt{2}\,e^{-j(\pi/4)}}$$
$$=\frac{1}{3\sqrt{2}}e^{j(\pi/4)}$$

and

$$i(t)=\frac{1}{3\sqrt{2}}\cos\left(20t+\frac{\pi}{4}\right)$$

Observe that the current is leading the voltage because the circuit is capacitive.

Similarly, for a parallel *RLC* circuit shown in Fig. 6.46*a* the *KCL* gives

$$I_R+I_L+I_C=I \tag{6.116}$$

The total admittance Y of the circuit is as follows:

$$\begin{aligned} Y&=\frac{1}{R}+j\omega C-j\frac{1}{\omega L}\\ &=\frac{1}{R}+j\left(\omega C-\frac{1}{\omega L}\right)\end{aligned} \tag{6.117}$$

If the quantity $(\omega C-1/\omega L)$ is positive, the admittance is capacitive; and if $(\omega C-1/\omega L)$ is negative, the admittance will be inductive in nature. The phasor diagram of a parallel *RLC* circuit is shown in Fig. 6.46*b*. The circuit is assumed to be inductive. The phasor diagram illustrates that the sum of the currents through each element in the circuit is equal to the current I. V is the common voltage for the parallel circuit. The current through the resistor, I_R, is equal to V divided by R, the value of the resistance, and is, therefore, in phase with the voltage phasor. The current in the capacitor, I_C, is V times the admittance of the capacitor and must lead by 90°. The current through the inductor, I_L, lags behind the voltage by 90°, And in our example, the values of ω, L, and C are such that the total current through the inductor is larger than the current in the capacitor. Hence, the net current through the reactive part looks inductive.

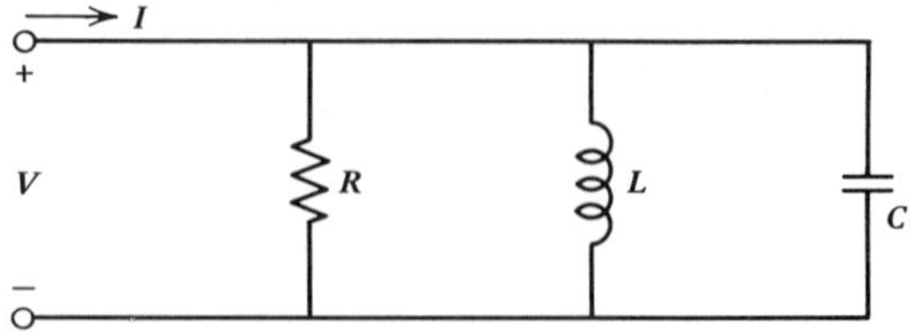

Fig. 6.46*a*

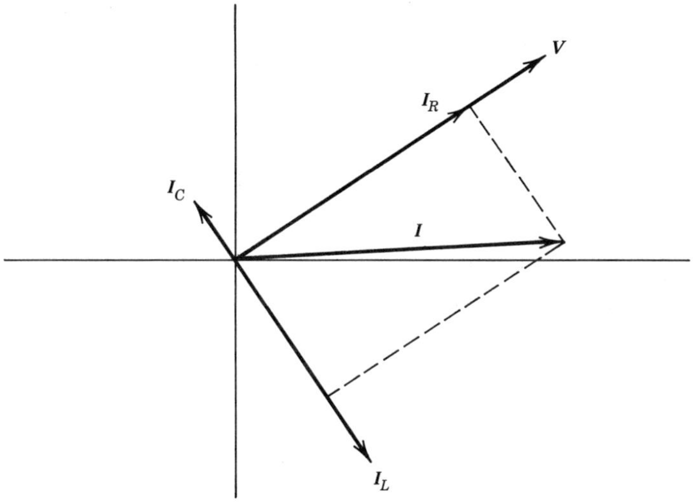

Fig. 6.46*b*

The sum of that current and the resistive current is the total current I. Since the reactive part of the circuit is inductive in nature, the current lags behind the voltage.

These examples of series and parallel circuits illustrate the ease with which we are able to analyze the circuits by using the impedance or admittance of the circuits. More complex circuits involving repeated series and parallel connections can also be handled in a similar manner. The impedance of such series-parallel circuits is also easily determined as discussed in the next chapter. The rule for the impedance of series circuits and the admittance of parallel circuits is repeatedly used to determine the circuit impedance. We will discuss the procedures in the next chapter not only for the impedance but for the other network or circuit parameters.

The phasor diagrams are used for a qualitative understanding of the simpler circuits, but not to obtain precise answers. They are sketched easily for arbitrary, unspecified excitations to examine the relationships between the different signals in a simple circuit.

Circuits that are so complex that they cannot be expressed as simple or repeated series and parallel connections can always be analyzed using the general techniques for writing the equilibrium equations described in Sec. 6.6. Impedances of such circuits must be obtained by the solution of the equilibrium equations. Once the impedance is obtained for a circuit, it describes the voltage-current relationship at its terminals for any excitation whatever at the specified frequency.

Before discussing the numerical solutions of equilibrium equations, we include in the next section an important property of circuits under sinusoidal steady-state conditions. We show that the power and energy delivered to a circuit under sinusoidal steady-state conditions can be described in a simple and meaningful way.

6.8 POWER AND ENERGY

The electrical power delivered to a load (i.e., a circuit) plays an important role in circuit analysis and design. The electrical signals must have sufficient energy to perform their functions effectively, whereas the circuit elements do not function properly (i.e., they burn out) if certain allowable power and energy levels are exceeded. Circuit design depends on a clear understanding of the concepts discussed here. As we shall see, we are able to give conceptually simple and concrete measures of power and energy for sinusoidal steady-state conditions. Other situations can be converted to the steady-state case through techniques to be discussed in later chapters. We concentrate here on sinusoidal steady-state analysis.

We indicated in the first chapter that the power delivered to a circuit is given by the product of the voltage across the circuit times the current flowing into the circuit with polarities as shown in Fig. 6.47,

$$p(t) = v(t)i(t) \tag{6.118}$$

The voltage and the current are functions of time and so the power is also time-dependent. The power as shown in Eq. 6.118 is therefore called *instantaneous power*.

Let us now examine the situation under steady-state conditions. First, let us examine the constant excitation steady-state or *direct current*, that is, *d.c.* case. The direct current term stems from the fact that the excitation-response do not alternate in sign with time. The sinusoidal steady-state case is called the *alternating current* or *a.c.* case. Of course, our discussion for the sinusoidal case covers both d.c. and a.c. The d.c. case is obtained by setting $\omega = 0$.

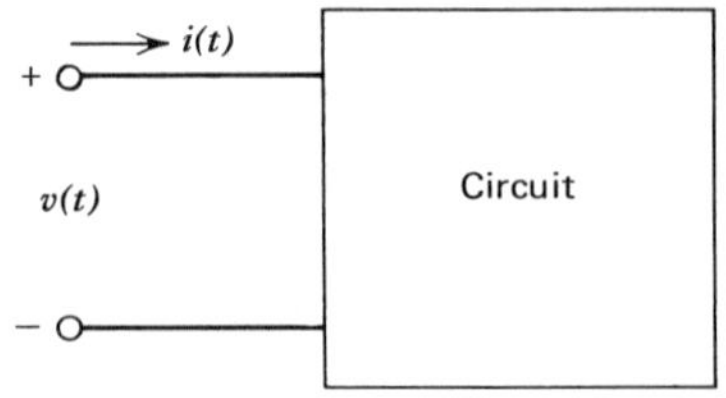

Fig. 6.47

Then for the steady-state analysis with $\omega = 0$, the impedances are all resistive. The impedances of the reactive elements are either zero ($j\omega L$, $j\omega M$) or infinite ($1/j\omega C$). Zero impedance represents short-circuit and infinite impedance represents open-circuit. The overall impedance results only from the resistors present in the circuit. The constant voltages and currents are thus real constants expressed by V and I at the circuit terminals. The power in Eq. 6.118 becomes

$$p(t) = P = VI \tag{6.119}$$

which is constant with time. Since the impedance of the circuit is purely resistive, say $Z = R$, the power becomes

$$P = \frac{V^2}{R} = I^2 R \tag{6.120}$$

For $\omega \neq 0$, the steady-state voltage and current are sinusoids of the same frequency and, therefore, the same period. Let us assume that the period of the sinusoids is $T = 2\pi/\omega$. If both $v(t)$ and $i(t)$ in Eq. 6.118 are periodic with period T, then so is $p(t)$, that is,

$$p(t) = p(t+T) \tag{6.121}$$

Thus, while the power is still a function of time for the a.c. case ($\omega \neq 0$), it is a periodic function that must be defined for any one period. For all other periods it merely repeats itself. The energy delivered to the circuit in any one period is

$$W_T = \int_{t_0}^{t_0+T} p(\tau) d\tau \tag{6.122}$$

where t_0 is any instant of time.

Since $p(t)$ is periodic with period T, the integral is the same over any period. Thus, the energy delivered to the circuit has the same value for each period. Moreover, this is a specific value for a given function $p(t)$ since it is integrated between definite limits.

A possible measure of power for sinusoidal a.c. case is the *average power*, P_{av}, which when multiplied by the period T gives the actual value of energy W_T. The rationale for the concept is, of course, the simplicity similar to the d.c. case. For the d.c. case, the energy over T seconds is P times T (i.e., PT) since P is constant:

$$\int_{t_0}^{t_0+T} P d\tau = PT \tag{6.123}$$

In terms of the average power for the a.c. case we have from Eq. 6.122

$$W_T = P_{av}T = \int_{t_0}^{t_0+T} v(\tau)i(\tau)d\tau$$

so that

$$P_{av} = \frac{1}{T}\int_{t_0}^{t_0+T} v(\tau)i(\tau)d\tau = \frac{1}{T}\int_{t_0}^{t_0+T} p(\tau)d\tau \tag{6.124}$$

P_{av} now behaves very like the d.c. power term to yield the energy in one period.

The expression in Eq. 6.124 for the average power is used for averaging functions in many contexts. The value of t_0 is arbitrary, but for simplicity we choose it to be zero.

Let us now determine P_{av} for some circuits assuming the voltage as the reference, that is,

$$\begin{aligned} v(t) &= V_m \cos \omega t = Re[Ve^{j\omega t}] \\ i(t) &= I_m \cos(\omega t + \theta) = Re[Ie^{j\omega t}] \end{aligned} \tag{6.125}$$

where the phasors

$$\begin{aligned} V &= |V|\angle 0 = V_m \\ I &= |I|\angle\theta = I_m\angle\theta \end{aligned} \tag{6.126}$$

Observe that V_m and I_m are the maximum values the sinusoids assume.

Power in a Resistor

Here V and I are in phase, θ is zero, and $V = RI$. So

$$\begin{aligned} p(t) &= v(t)i(t) = \frac{v^2(t)}{R} = i^2(t)R \\ &= \frac{V_m{}^2}{R}\cos^2\omega t = RI_m{}^2\cos^2\omega t. \end{aligned}$$

We now use Eq. 6.124 with $t_0 = 0$ to determine P_{av} for a resistor:

$$\begin{aligned} P_{av} &= \frac{1}{T}\int_0^T p(\tau)\,d\tau \\ &= \frac{1}{T}\int_0^T \frac{V_m{}^2}{R}\cos^2\omega t \\ &= \frac{V_m{}^2}{TR}\int_0^T [\tfrac{1}{2} + \tfrac{1}{2}\cos 2\omega\tau]\,d\tau \\ &= \frac{V_m{}^2}{2R} \end{aligned} \tag{6.127}$$

Observe that the integral of a cosine or sine function over an integral number of periods is zero. Since the period of $\cos 2\omega\tau$ is π/ω, and $T = 2\pi/\omega$, the cosine term integrates to zero. Thus the average power delivered to a resistor R is

$$P_{av} = \frac{V_m^2}{2R} = \frac{RI_m^2}{2} \tag{6.128}$$

Comparing Eqs. 6.120 and 6.128, we see that the expressions are similar except for a factor of two. Let us define an effective value of the a.c. phasor such that the expressions are identical for both d.c. and a.c. values. Thus, let

$$P_{av} = \frac{V_{eff}^2}{R} = I_{eff}^2 \tag{6.129}$$

where $V_{eff} = (1/\sqrt{2})V_m = 0.707\,V_m$ and $I_{eff} = 0.707\,I_m$. Alternatively, from Eqs. 6.127 and 6.129

$$P_{av} = \frac{V_{eff}^2}{R} = \frac{1}{T}\int_0^T \frac{V_m^2}{R}\cos^2\omega\tau\,d\tau$$

or

$$V_{eff} = \sqrt{\frac{1}{T}\int_0^T V_m^2\cos^2\omega\tau\,d\tau}$$

$$= \sqrt{\frac{1}{T}\int_0^T v^2(\tau)\,d\tau} \tag{6.130}$$

Similarly

$$I_{eff} = \sqrt{\frac{1}{T}\int_0^T i^2(\tau)\,d\tau} \tag{6.131}$$

From the above expressions, we see that the effective value is the square root of the average of the squared value of the signal. The effective value, therefore, is called the *root mean square* (RMS) value of the signal. So

$$\begin{aligned} V_{eff} &= V_{RMS} = \frac{V_m}{\sqrt{2}} \\ I_{eff} &= I_{RMS} = \frac{I_m}{\sqrt{2}} \end{aligned} \tag{6.132}$$

Power in an Impedance

We now examine the average power delivered to an arbitrary impedance. Here θ is not zero and

$$Z = |Z|\angle\theta \tag{6.133}$$

with

$$v(t) = V_m \cos \omega t \tag{6.134}$$

$$i(t) = \frac{V_m}{|Z|} \cos (\omega t - \theta)$$

The average power delivered to the impedance is

$$\begin{aligned} P_{\text{av}} &= \frac{1}{T} \int_0^T V_m \cos \omega t \frac{V_m}{|Z|} \cos (\omega t - \theta)\, dt \\ &= \frac{1}{T} \int_0^T \frac{V_m^2}{|Z|} \cos \omega t [\cos \omega t \cos \theta + \sin \omega t \sin \theta]\, dt \\ &= \frac{V_m^2}{|Z|T} \left[\cos \theta \int_0^T \cos^2 \omega t\, dt + \sin \theta \int_0^T \tfrac{1}{2} \sin 2\omega t\, dt \right] \\ &= \frac{V_m^2}{|Z|T} [\cos \theta (\tfrac{1}{2}T)] \\ &= \frac{V_m^2 \cos \theta}{2|Z|} \end{aligned} \tag{6.135}$$

since the second integral equals zero.

Since $\dfrac{V_m}{|Z|} = I_m$, an alternative expression is

$$P_{\text{av}} = \frac{V_m I_m}{2} \cos \theta \tag{6.136}$$

and using RMS values

$$P_{\text{av}} = V_{\text{RMS}} I_{\text{RMS}} \cos \theta \tag{6.137}$$

The expressions for average power are quite simple and depend on the phasor magnitudes and the phase angle between them. This makes it very convenient to use the concept of average power in circuit analysis. Moreover, in most problems our interest is to find the energy in a given integral number of periods, and we are able to find this from the average power. For periodic signals that are not sinusoidal, the same concept of average power can still be used. It is also possible to represent such periodic signals as sums of sinusoids and the power calculated in terms of the average power in each sinusoidal signal.

Let us close this section with a few examples illustrating the calculation of P_{av} in different circuits.

Example 6.8.1 Determine P_{av} in a resistor. $R = 100\,\Omega$ if $v(t) = 100 \times \cos(300t - \pi/6)$ volts.

Solution. Since $V_m = 100$,

$$P_{av} = \frac{100^2}{2 \times 100} = \frac{100}{2} = 50 \text{ watts}$$

Example 6.8.2 With the same voltage consider a pure reactance whose impedance

$$Z = j\,100 = 100\,\angle\pi/2$$

Solution

$$P_{av} = \frac{V_m{}^2 \cos\theta}{2|Z|}$$

but

$$\cos\theta = \cos\frac{\pi}{2} = 0$$

and

$$P_{av} = 0$$

So in a pure reactive circuit made up of inductors and capacitors, $P_{av} = 0$. A plot of $p(t)$ for one period shows that for half the period it is positive and for the other half negative. The total is zero. This means that for half the period, energy is stored in the circuit and it is returned in the next half period.

Example 6.8.3 With the voltage as before, let

$$Z = 100\,\angle\pi/4$$

So

$$P_{av} = \frac{V_m{}^2 \cos\pi/4}{2 \times 100}$$

$$= \frac{100^2 \times 1/\sqrt{2}}{2 \times 100}$$

$$= \frac{100}{2\sqrt{2}} \text{ watts}$$

We close this section by deriving the conditions under which a practical source delivers the maximum power to a load. A practical source differs from an ideal source in that it has an internal impedance. A practical voltage source is represented by an ideal voltage source with a series impedance as shown in Fig. 6.48*a*. Similarly, a practical current source is represented in Fig. 6.48*b* as an ideal current source with an impedance in parallel. The sources are not ideal any more since the circuit connected influences the source terminal signals. For example, the

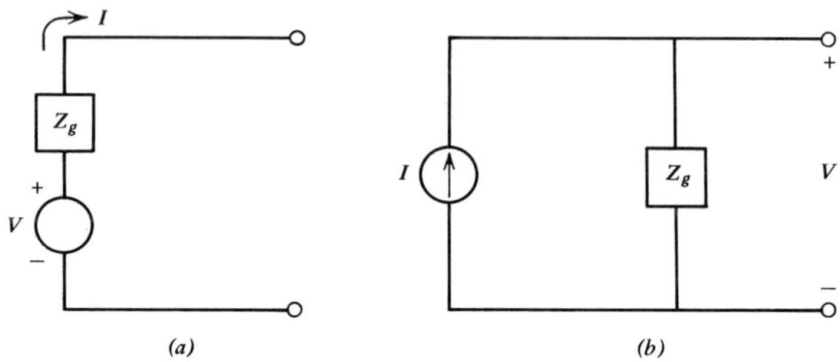

Fig. 6.48 Practical sources.

practical voltage source exhibits reduced voltage $(V-Z_gI)$ as the current drawn by the circuit increases.

Let us consider an impedance Z connected to a practical voltage source as in Fig. 6.49. Let $Z_g = R_g + jX_g$ and $Z = R + jX$. Then the average power delivered to the load Z is given by the power delivered to R since the average power in the reactive part X is zero. The average power in R is $\frac{1}{2}|I|^2R$, or

$$P_{\text{av}} = \frac{VR}{2|Z|^2}$$

$$= \frac{VR}{2[(R_g+R)^2+(X_g+X)^2]}$$

The average power attains its maximum value for those values of R and X for which

$$\frac{R}{(R_g+R)^2+(X_g+X)^2}$$

is a maximum. It is clear first that $(X+X_g)$ must equal zero, or $X=-X_g$. Next, we differentiate the expression with respect to R and set it to zero. Then

$$\frac{1}{(R_g+R)^2}-\frac{2R}{(R_g+R)^3}=0$$

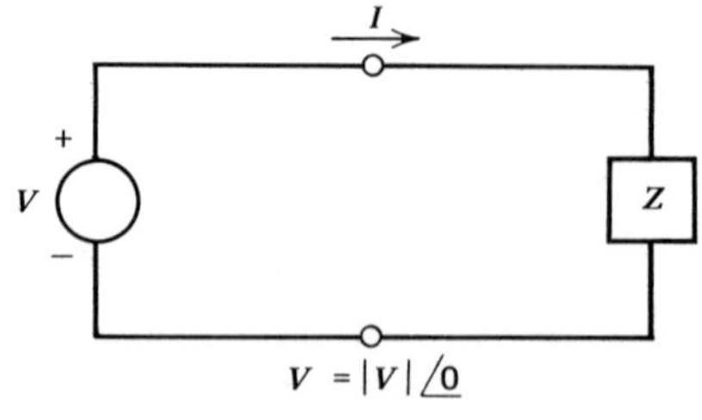

Fig. 6.49

or

$$R = R_g$$

The maximum value of P_{av}, that is, P_{max}, thus occurs when $R = R_g$ and $X = -X_g$. This may be also expressed as the condition that $Z = Z_g^* = (R_g + jX_g)^* = R_g - jX_g$. When the load Z is the complex conjugate of the generator impedance Z_g, the source delivers the maximum power. This condition is also called *conjugate matching*. Similar results are obtained for the current source.

6.9 NUMERICAL SOLUTION OF LINEAR ALGEBRAIC EQUATIONS: II COMPLEX COEFFICIENTS

In Sec. 6.6, we have shown that the phasor equilibrium equations for sinusoidal steady-state analysis give rise to a system of linear algebraic equations with complex coefficients. The equations are similar to those we discussed in Chapter 5. The only difference is that we now have complex coefficients rather than real coefficients. As we shall see, the numerical procedures discussed in Chapter 5 can be applied with minor modifications to the case of complex coefficients.

Consider the general system of n equations in n unknowns in the matrix form:

$$A\mathbf{x} = \mathbf{b} \tag{6.138}$$

where A is now a matrix of complex coefficients. A unique solution $\mathbf{x} = A^{-1}\mathbf{b}$ exists if and only if the determinant (magnitude) of the coefficient matrix A is nonzero. The Gaussian elimination procedure is once again the simplest and probably the most widely used method for numerical solution. However, we must use *complex* additions, multiplications, subtractions, and divisions for the elimination process. Pivoting is then interpreted at each elimination stage as the selection of a pivot whose *magnitude* (of a complex number) is larger than or equal to the remaining elements in the (lower) column that are to be eliminated. Let us consider a simple numerical example.

Example 6.9.1 For the circuit of Fig. 6.50 where phasor signals are shown, let $I_1 = |I_1| \angle 0$.

The node equations are

$$\left(G_1 + j\omega C + \frac{1}{j\omega L}\right)V_1 - \frac{1}{j\omega L}V_2 = |I_1|$$

$$-\frac{1}{j\omega L}V_1 + \left(G_2 + \frac{1}{j\omega L}\right)V_2 = 0$$

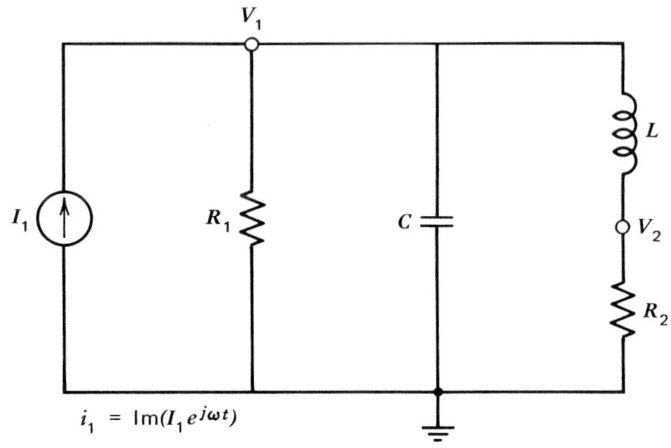

Fig. 6.50

Let $\omega = 10$ rads/sec., $R_1 = R_2 = 1\ \Omega$

$$C = 1F, L = 1H, \text{and } |I_1| = 1$$

Then the node equations become

$$\begin{aligned}(1+j9.9)V_1 + j0.1V_2 &= 1\\ +j0.1V_1 + (1-j0.1)V_2 &= 0\end{aligned}$$

The augmented matrix for Gaussian elimination process is obtained as

$$\begin{matrix} m & \\ \dfrac{j0.1}{1+j9.9} & \begin{bmatrix} 1+j9.9 & j0.1 & 1 \\ j0.1 & 1-j0.1 & 0 \end{bmatrix}\end{matrix}$$

Now the first row is multiplied by $j0.1/(1+j9.9)$ and subtracted from the second row; hence, we have

$$\begin{bmatrix} 1+j9.9 & j0.1 & 1 \\ & \dfrac{2+j9.8}{1+j9.9} & \dfrac{-j0.1}{1+j9.9} \end{bmatrix}$$

This triangular system could then be solved by back-substitution and we have

$$V_2 = -\frac{j0.1}{2+j9.8} = -0.009796 - j0.001999$$

$$V_1 = \frac{1}{1+j9.9}(1 - j0.1V_2) = 0.010196 - j0.099960$$

The results indicate that V_2 leads I by an angle of $\pi + \tan^{-1}$ (0.001999/0.009796) radians and V_1 lags I by $\tan^{-1}$ (0.099960/0.010196) radians.

Computer Program

To implement the Gaussian elimination method for systems with complex constant coefficients, all we need is an explicit COMPLEX declaration for the related variables. Thus, if the variable A is designated as the coefficient matrix and B as the right-hand side of the equation $A\mathbf{x} = \mathbf{b}$, then in the main program we need the following statement:

COMPLEX A,B

Of course, we should also use the related format for read-in and printout of the complex coefficients.

If we desire frequency response over a set of discrete frequencies, then the coefficients can best be simulated using a FORTRAN statement function in the main program. The response to each frequency can then be calculated by varying the frequency parameters in the coefficient matrix A and the B vector.

A subroutine CGE(N,A,B) for handling complex simultaneous equations is given in Fig. 6.53. Notice that the only difference between CGE(N,A,B) and GE(N,A,B) is an additional statement in CGE(N,A,B) to declare certain of its variables as complex variables as follows:

COMPLEX A,B,XMAX,SAVE

Also changed is the related function ABS(X) to find the absolute value of a real number X: we use the function CABS(Y) to find the magnitude of a complex variable Y.

The maximum number of linear equations to be solved is still set at 20. For more than a system of 20 equations, the dimension statement for the variables A and B has to be modified to the desired number of equations.

Example 6.9.2 Find the node voltages v_1 and v_2 of the circuit in Fig. 6.48 for $\omega = 1, 2, 3, \cdots, 10$.

Solution. The FORTRAN program, the sample computer output and the subroutine CGE, are shown in Figs. 6.51 to 6.53. The parameters are the following:

- N Number of equations
- S Complex frequency parameter
- XR Real part of S, equal to zero
- XI Imaginary part of S
- A Complex N by N coefficient matrix
- B Complex N vector, right-hand side of $A\mathbf{x} = \mathbf{b}$; also the solution $\mathbf{x} = A^{-1}\mathbf{b}$

*N*1	$N-1$
XMAX	Maximum element in a column
L	Location indicator of *XMAX*
SAVE	Temporary storage used when interchanging rows

```
      DIMENSION A(20,20),B(20)
      COMPLEX A,B,S
    2 FORMAT(F10.5)
    3 FORMAT(10X,2E14.5)
      N=2
      XR=0.0
    1 READ 2, XI
      S=CMPLX(XR,XI)
      A(1,1)=1.0+S+1.0/S
      A(1,2)=-1.0/S
      A(2,1)=-1.0/S
      A(2,2)=1.0+1.0/S
      PRINT 3, ((A(I,J), I=1,N), J=1,N)
      B(1)=(1.0,0.0)
      B(2)=(0.0,0.0)
      PRINT 3, (B(I), I=1,N)
      CALL CGE(N,A,B)
      PRINT 3, (B(I), I=1,N)
      GO TO 1
      END
```

Fig. 6.51 Steady-state response of a circuit.

	Real	Imaginary	
ω = 5	0.10000E 01	0.48000E 01	A matrix (columnwise)
	0.0	0.20000E 00	
	0.0	0.20000E 00	
	0.10000E 01	-0.20000E 00	
	0.10000E 01	0.0	B vector
	0.0	0.0	
	0.42925E-01	-0.19873E 00	$x = A^{-1}B$
	-0.36566E-01	-0.15898E-01	
ω = 6	0.10000E 01	0.58333E 01	
	0.0	0.16667E 00	
	0.0	0.16667E 00	
	0.10000E 01	-0.16667E 00	
	0.10000E 01	0.0	
	0.0	0.0	
	0.29231E-01	-0.16615E 00	
	-0.26154E-01	-0.92308E-02	
ω = 7	0.10000E 01	0.68571E 01	
	0.0	0.14286E 00	
	0.0	0.14286E 00	
	0.10000E 01	-0.14286E 00	
	0.10000E 01	0.0	
	0.0	0.0	
	0.21206E-01	-0.14262E 00	
	-0.19543E-01	-0.58212E-02	

Fig. 6.52 Sample computer output.

```
      SUBROUTINE CGE(N,A,B)
      DIMENSION A(20,20),B(20)
      COMPLEX A,B,XMAX,SAVE
      N1=N-1
      DO 300 I=1,N1
        XMAX=0.
        DO 40 K=I,N
          IF(CABS(XMAX)-CABS(A(K,I))) 50,40,40
50        XMAX=A(K,I)
          L=K
40      CONTINUE
        IF(CABS(XMAX)-0.0) 60,60,70
60      PRINT 80
80      FORMAT(10X,19H SYSTEM IS SINGULAR)
        RETURN
70      DO 90 J=I,N
          SAVE=A(I,J)
          A(I,J)=A(L,J)
          A(L,J)=SAVE
90      CONTINUE
        SAVE=B(I)
        B(I)=B(L)
        B(L)=SAVE
        KK=I+1
        DO 100 K=KK,N
          A(K,I)=A(K,I)/XMAX
100     CONTINUE
        DO 120 K=KK,N
          DO 110 JK=KK,N
            A(K,JK)=A(K,JK)-A(K,I)*A(I,JK)
110       CONTINUE
          B(K)=B(K)-A(K,I)*B(I)
120     CONTINUE
300   CONTINUE
      B(N)=B(N)/A(N,N)
      DO 510 I=1,N1
        IB=N-I
        IC=N
        DO 500 K=1,I
          B(IB)=B(IB)-A(IB,IC)*B(IC)
          IC=IC-1
500     CONTINUE
        B(IB)=B(IB)/A(IB,IB)
510   CONTINUE
      RETURN
      END
```

Fig. 6.53 Subroutine CGE for complex Gaussian elimination.

PROBLEMS

6-1. Express the following summations of sinusoids in the general form $K \sin(wt+\theta)$:

(i) $i(t) = 3 \sin t + 4 \cos t$.

(ii) $i(t) = 3 \cos t + 4 \sin t$.

(iii) $v(t) = 15 \sin(3t+\pi/4) + 7 \sin(3t+\pi/8)$.

(iv) $v(t) = 15 \cos(3t+\pi/4) + 7 \cos(3t+\pi/8)$.

6-2. Repeat 6-1 but expressing the results in the general form $K_1 \cos(wt+\phi)$. Is K_1 different from K? How about θ and ϕ?

6-3. We have seen that if we have

$$v(t) = \sum_{n=1}^{N} V_n \sin(wt+\theta_n)$$

then it can be simplified to

$$v(t) = B_1 \sin wt + B_2 \cos wt$$

or

$$v(t) = V \sin(wt+\theta)$$

(i) Find explicitly B_1, B_2, V, θ in terms of the known quantities V_n and θ_n.

(ii) Repeat (i) for

$$v(t) = \sum_{n=1}^{N} V_n \cos(wt+\theta_n)$$

What is the difference in phase angle between the results of (i) and (ii)? Is it independent of N, the number of terms?

6-4. Given $C_1 = 6{\cdot}5+j2$, $C_2 = 8+j13$, $C_3 = 12\angle 105°$, $C_4 = 6\angle 30°$, evaluate the following expressions:

(i) $C_1(C_3+C_4)$
(ii) $(C_2+C_3)/C_4$
(iii) $C_1C_2(C_1+C_3)$.

6-4a. Using the complex numbers specified in 6-4, evaluate the following expressions. Show each result in the complex plane.

(i) $(C_1+C_3)(C_2+C_4)$
(ii) $(C_1+C_2)/(C_3+C_4)$
(iii) $C_1/C_2(C_3+C_4)$

6-5. The conjugate of a complex number C is defined as one whose real part is equal to that of C and whose imaginary part is equal to the negative of that of C. It is denoted by C^*. Thus if $C = a+jb$, then we have $C^* = a-jb$. Verify the following:

(i) If $C = |C|e^{j\theta}$, then $C^* = |C|e^{-j\theta}$.
(ii) $C+C^* = 2a$.
(iii) $C = \frac{1}{2}(C+C^*) + \frac{1}{2}(C-C^*)$.
(iv) $C^*C = CC^* = a^2+b^2$.
(v) $(C_1C_2)^* = C_1^*C_2^*$ where C_1 and C_2 are complex numbers.

6-5a. Repeat 6-4a for the following expressions:

(i) $C_1C_2^*$
(ii) $(C_1C_2)^*$
(iii) $C_1C_2^*/C_3C_4^*$
(iv) $(C_1-C_1^*)/C_1C_2^*$.

6-5b. Evaluate the following expressions in the form of a complex number $a+jb$.

(i) $ln(j1)$
(ii) $ln(1+jw)$
(iii) $ln(-1-j1)$
(iv) $ln(1-w^2+j2\zeta w)$

6-5c. Evaluate the following expressions:

(i) $(-1)^{1/4}$
(ii) $(j32)^{1/5}$
(iii) $(3+j4)^{j6}$
(iv) $(2-j2)^{1+j1}$.

6-6. With the aid of a protractor and scale, graphically determine the following:

(i) $A = 5\angle 20° + 3\angle 60° + 8\angle 180° + 2\angle -240°$.
(ii) jA
(iii) $-2jA$
(iv) $(1-2j)A$
(v) A^*.

6-6a. Repeat 6-1 using phasors.

6-6b. Repeat 6-3 using phasors.

6-7. For the voltage sources of Fig. P6-7, $v_a(t) = 2 \sin w_0 t$. It is observed that $v_a(t) + v_b(t) = 10 \sin (w_0 t + \theta)$.

(i) Find $v_b(t)$ if $\theta = 0°$; if $\theta = 90°$; if $\theta = 180°$; if $\theta = -90°$.
(ii) Draw the appropriate phasors for $v_a(t), v_b(t), v_a(t) + v_b(t)$ for the values of θ in (i).

Fig. 6.7

6-8. For the linear time-invariant network of Fig. P6-8, the steady state voltage response is

$$v(t) = 45 \sin (377t + 20°) + 10 \cos (377t - 10°)$$

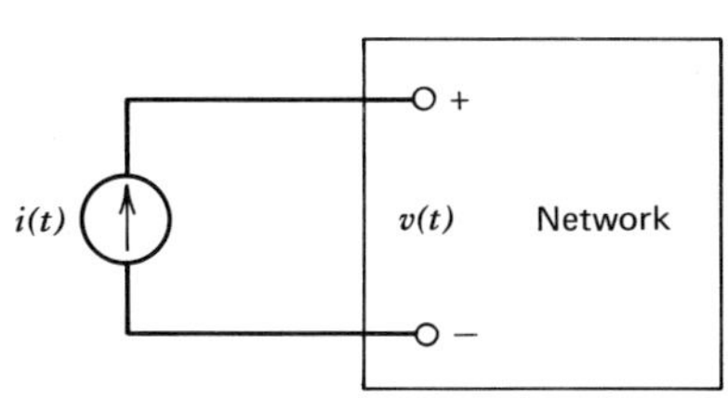

Fig. P6-8

due to a source $i(t) = \cos 377t$. Find $v(t)$ if $i(t)$ is changed to $3 \sin (377t - 30°)$.

6-9a. Find the steady state solution for each of the following equations:
(i) $3x'(t) + 2x(t) = \sin t$.
(ii) $5x'(t) + 6x(t) = 2 \cos 3t$.
(iii) $4x'(t) + x(t) = \sin (2t + \pi/6)$.

6-9b. Repeat 6-9a for each of the following equations:
(i) $x''(t) + 6x'(t) + 25x(t) = 2 \cos t$.
(ii) $x''(t) + 8x'(t) + 7x(t) = 2 \cos (7t + 1/4\pi)$.
(iii) $x''(t) + 5x'(t) = 3 \sin 5t$.

6-9 c. Repeat 6-9a for each of the following equations:

(i) $x'(t) + 5x(t) + \int_0^t x(\tau)d\tau = 2 \sin t$.

(ii) $x'(t) + 2x(t) + 2\int_0^t x(\tau)d\tau = 5 \sin (3t + \pi)$.

6-10a. Determine the driving-point impedances for the circuits shown in Fig. P6-10*a*.

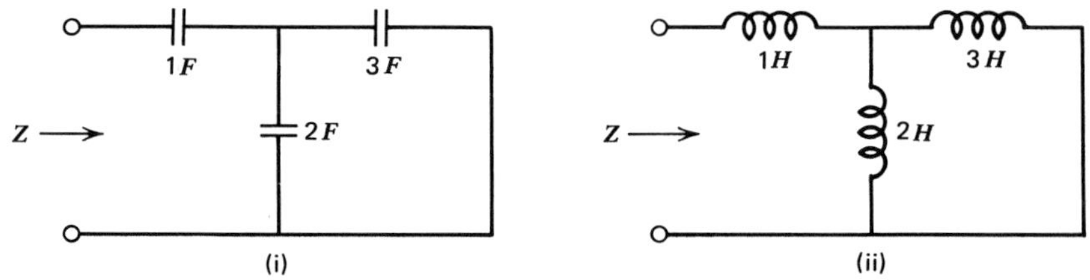

Fig. P6-10*a*

6-10b. Repeat 6-10a for the circuits of Fig. P6-10*b*. Are they equal?

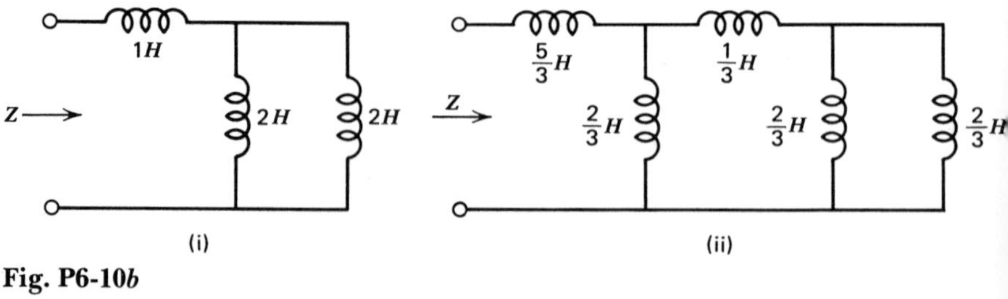

Fig. P6-10*b*

6-10c. Repeat 6-10a for the following circuits:

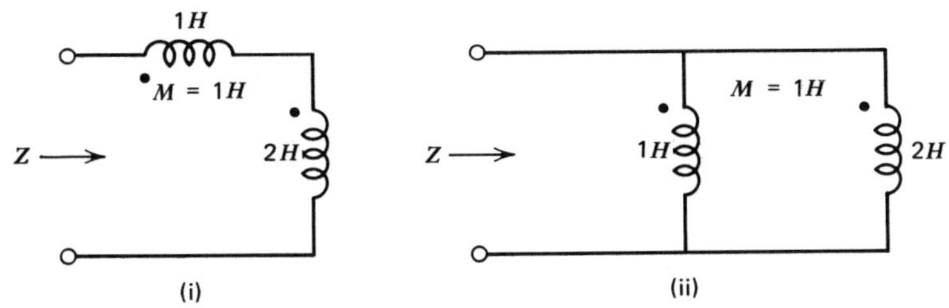

Fig. P6-10*c*

6-10d. Repeat 6-10a for the circuits shown in Fig. P6-10*d*.

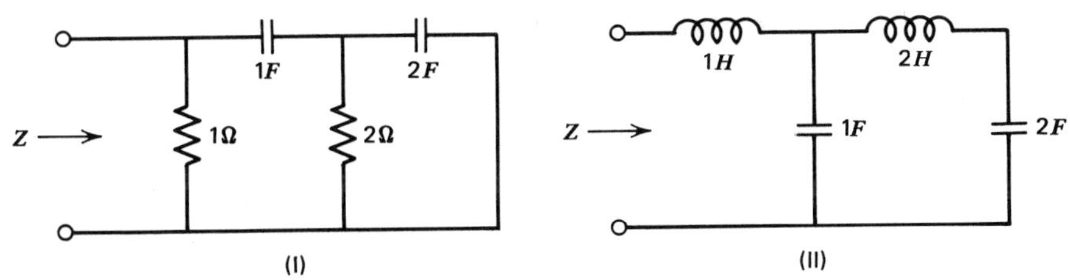

Fig. P6-10*d*

6-10e. Repeat 6-10a for the circuits shown in Fig. P6-10*e*.

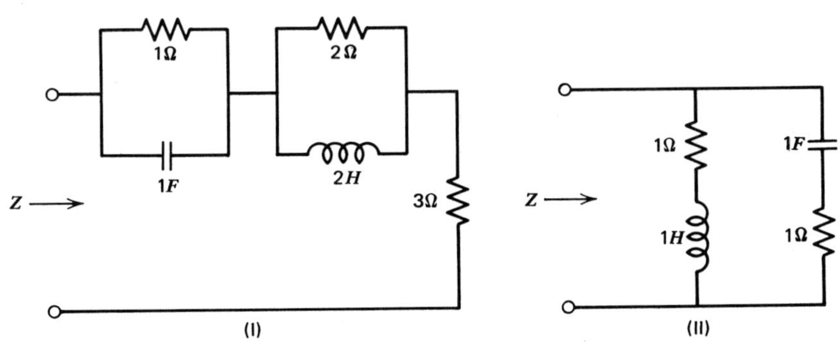

Fig. P6-10*e*

6-11. For the circuit of Fig. P6-11, find the steady state solutions of $i_2(t)$ and $i_L(t)$ for $v(t) = 2 \sin \pi t$.

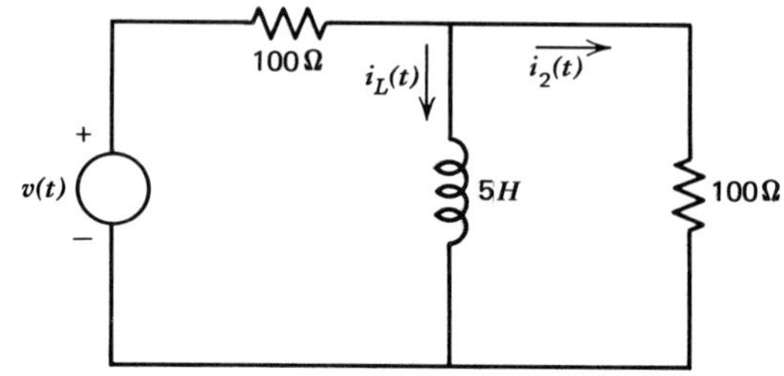

Fig. P6-11

6-12. For the circuit of Fig. P6-12, find the steady state solutions of $i_L(t)$ and $v_C(t)$ for $v(t) = 10 \cos wt$. How would you obtain the steady-state solutions for constant excitations? (Check problem 3-1.)

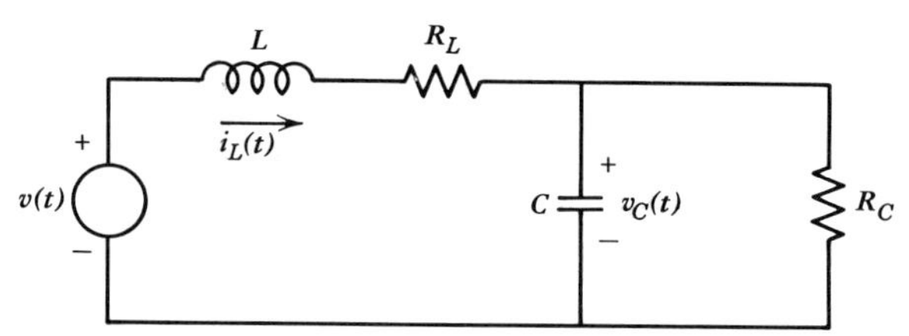

Fig. P6-12

6-13. Find the steady-state currents through the circuit elements of Fig. P6-10*a* for the following excitations:
(i) $v(t) = \sin t$
(ii) $i(t) = \sin 3t$.

6-14a. Repeat 6-13 for the circuits of Fig. P6-10*b*.

6-14b. Repeat 6-13 for the circuits of Fig. P6-10*c*.

6-14*c*. Repeat 6-13 for the circuits of Fig. P6-10*d*.

6-14d. Repeat 6-13 for the circuits of Fig. P6-10*e*.

6-15. For a two-port zero state linear network with two excitations $e_1(t)$ and $e_2(t)$ at the terminal pairs 1-1′ and 2-2′, the response $r(t)$ of a portion of the network can be obtained by applying the superposition principle, namely, the response $r(t)$ is obtained by adding $r_1(t)$ and $r_2(t)$ together where $r_1(t)$ is the response due to $e_1(t)$ with $e_2(t)$ removed (2-2′ is short-circuited if $e_2(t)$ is a voltage source and is open-circuited if $e_2(t)$ is a current source) and $r_2(t)$ is the response due to $e_2(t)$ with $e_1(t)$ removed. For the steady state response due to sinusoidal excitations, the phasor R of the response $r(t)$ can then be formulated in the following way:

$$R = k_1E_1 + k_2E_2$$

where

$$k_1 = \left.\frac{R_1}{E_1}\right|_{e_2(t)\text{ removed}}$$

$$k_2 = \left.\frac{R_2}{E_2}\right|_{e_1(t)\text{ removed.}}$$

And R_1, R_2, E_1, E_2 are the stationary phasors of $r_1(t)$, $r_2(t)$, $e_1(t)$, $e_2(t)$ respectively. In case that the responses and excitations are voltages, then k_1 and k_2 are dimensionless constants.
For the circuit of Fig. P6-15, find k_1 and k_2 for
(i) $e_1(t) = v_1(t), e_2(t) = v_2(t), r(t) = v_{ac}(t)$.
(ii) $e_1(t) = v_1(t), e_2(t) = v_2(t), r(t) = v_{bc}(t)$.
Check your answer by finding v_{ac} and v_{bc} with $v_1(t) = 2V$ and $v_2(t) = 4V$ and compare your results with the capacitor voltages found in problem 5-2.

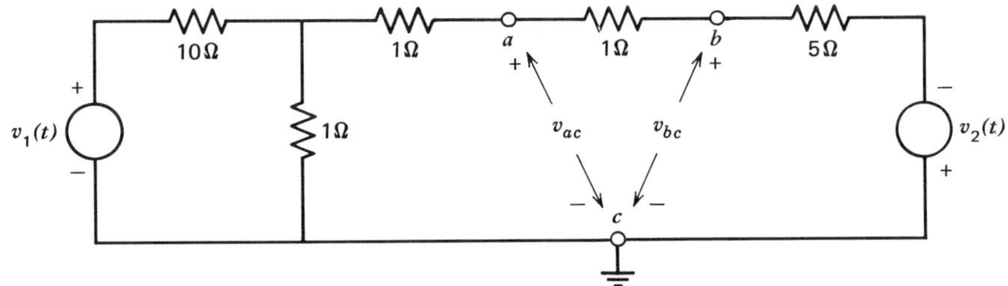

Fig. P6-15

6-16. Repeat 6-15 for the circuit of Fig. P6-16 where $e_1(t) = v_1(t)$, $e_2(t) = v_2(t)$ and $r(t) = v_C(t)$.

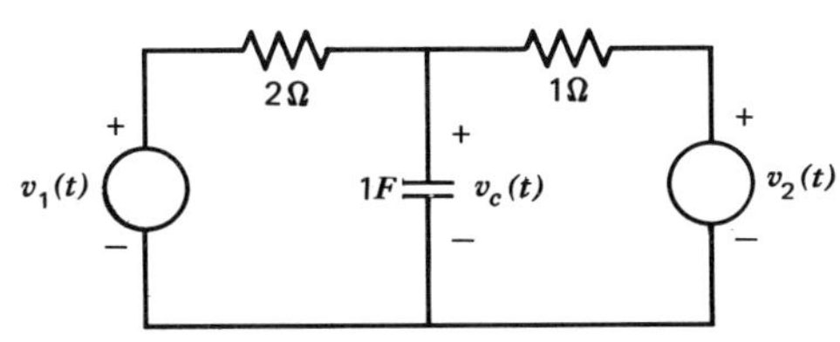

Fig. P6-16

6-17. For the circuit of Fig. P6-17, if the stationary phasors of $v_1(t)$ and $v_3(t)$ are V_1 and V_3 respectively, find the transfer ratio $H = V_3/V_1$.

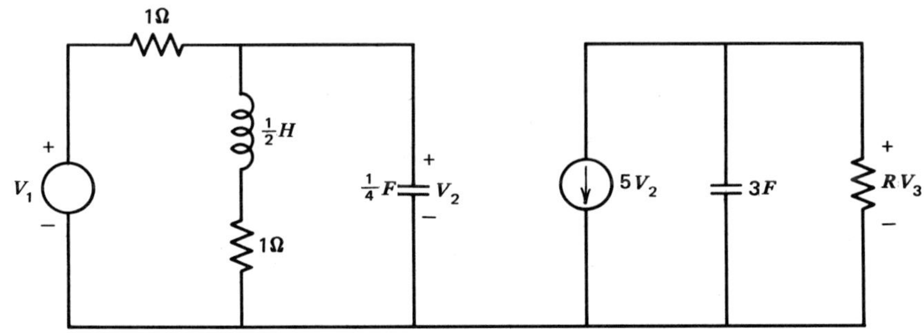

Fig. P6-17

6-18. With H found in 6-17, find V_3 if
(i) $v_1(t) = 2\cos w_0 t$
(ii) $v_1(t) = 2\cos 0t = 2$.
What happens to H as w approaches ∞?

6-19. For the circuit of Fig. P6-19, find the impedance of the network at terminals 1-1′ when $w = 1$ rad/sec.

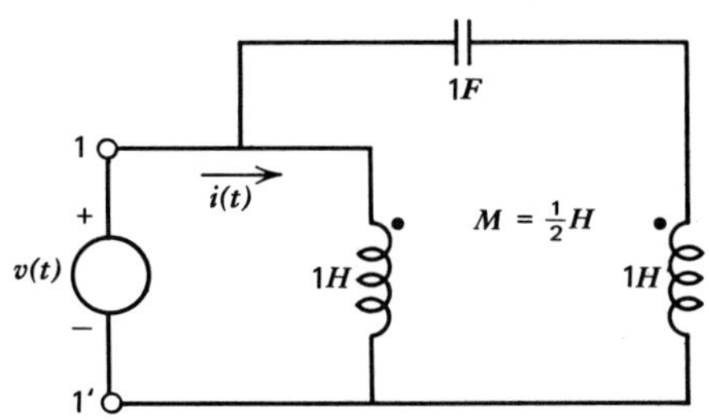

Fig. P6-19

6-20. For a certain network the voltage excitation is $v(t) = \cos t$. The steady state current response is found as

$$i(t) = \tfrac{1}{2}\sqrt{2}\cos(t - 45°)$$

(i) Find the admittance of this network.
(ii) Find a circuit to realize this network. Is the solution unique?

6-21. For the network shown in Fig. P6-21, the voltage sources are given phasors. Show that the output voltage phasor is given by

$$V = \frac{V_1Y_1 + V_2Y_2 + \cdots + V_nY_n}{Y_1 + Y_2 + \cdots + Y_n}$$

This is referred to as Millman's theorem.

Fig. P6-21

6-22. For the circuit of Fig. P6-22, $v_2(t) = 2 \sin 2t$. Determine $v_1(t)$ in the form $K \sin(2t + \theta)$.

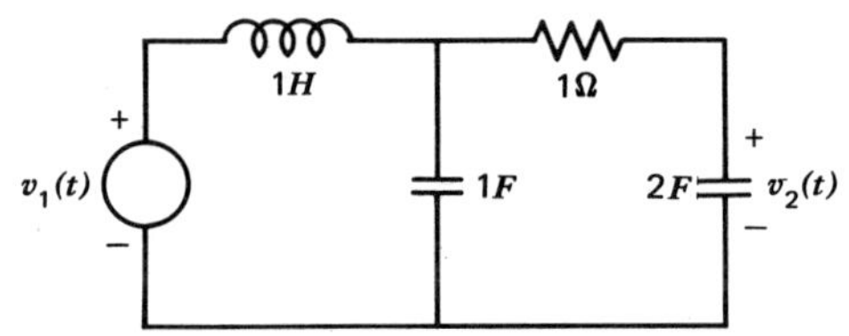

Fig. P6-22

6-23. For the circuit shown under steady state sinusoidal excitation, determine
(i) The phasor ratio $H = I_L/I_g$.
(ii) The frequency $w_{\max}$ such that the magnitude of H is a maximum.

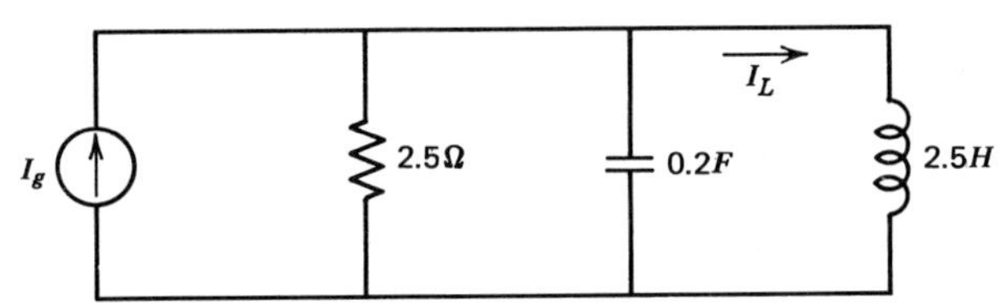

Fig. P6-23

6-24. Let $V = |V|\angle\alpha$ and $I = |I|\angle\beta$ be phasors representing the sinu-

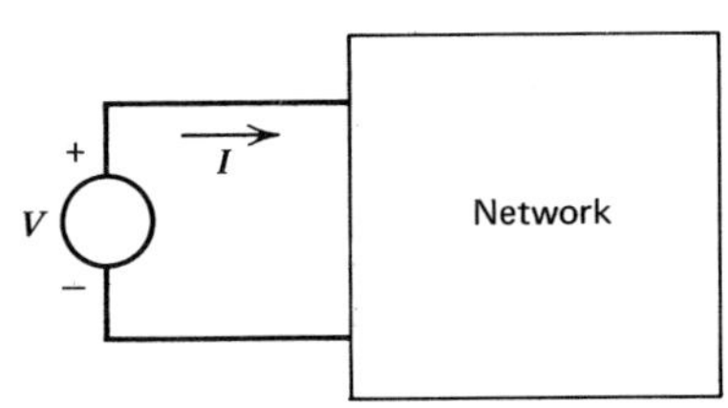

Fig. P6-24

soidal voltage and current as shown in Fig. P6-24. Let P_{av} be the average power entering the network. Show that $P_{av} = \frac{1}{2}Re(V^*I)$.

6-24a. For the circuits of Fig. P6-10*a*.

(i) Find the power delivered to each of the circuit elements under each of the excitations given in 6-13.

(ii) Find the power delivered by the excitation source. Is the sum of the results of (i) equal to the result of (ii)?

6-24b. Repeat 6-24a for the circuits of Fig. P6-10*b*.

6-24c. Repeat 6-24a for the circuits of Fig. P6-10*c*.

6-25. For the circuit of Fig. P6-24, it is given that $v(t) = 100 \sin(100t + 25°)$ and $i(t) = 2 \sin(100t + 65°)$.

(i) Determine the equivalent conductance of the network. Is the network capacitive or inductive?

(ii) Compute P_{av} by taking the average of the instantaneous power over a period.

(iii) Compute P_{av} by taking the average of $v^2 G_{eq}$ and compare with the answer in (ii). Is the average of $v^2 G_{eq}$ equal to $V_{rms}^2 G_{eq}$, where V_{rms} is the root-mean-square of $v(t)$?

6-25a. Solve each of the following systems of linear equations by complex arithmetic:

(i) $A\mathbf{x} = \mathbf{b}$ where

$$A = \begin{bmatrix} 1+j1 & 1-j1 \\ 2+j3 & 3+j4 \end{bmatrix} \qquad \mathbf{b} = \begin{bmatrix} 1+j1 \\ 0 \end{bmatrix}$$

(ii) $A\mathbf{x} = \mathbf{b}$ where

$$A = \begin{bmatrix} 3+j4 & 5-j6 \\ 7+j5 & 1-j2 \end{bmatrix} \qquad \mathbf{b} = \begin{bmatrix} 1 \\ 1 \end{bmatrix}$$

6-25b. Repeat 6-25b for the following systems:

(i) $A\mathbf{x} = \mathbf{b}$ where

$$A = \begin{bmatrix} 1 & 1+j1 \\ 2-j5 & 2 \end{bmatrix} \qquad \mathbf{b} = \begin{bmatrix} 2+j1 \\ 4-j5 \end{bmatrix}$$

(ii) $A\mathbf{x} = \mathbf{b}$ where

$$A = \begin{bmatrix} 10\angle 45° & 21\angle 30° \\ 1-j7 & 3+j5 \end{bmatrix} \qquad \mathbf{b} = \begin{bmatrix} 2\angle 10° \\ 3\angle 5° \end{bmatrix}$$

6-26. The analysis of steady-state response of a complex network at a given frequency w_0 often results in a system of linear algebraic equations. In matrix notation we have

$$A\mathbf{x} = \mathbf{b}$$

where A, $\mathbf{x}$, and $\mathbf{b}$ are all complex matrices.

Now if the real part and imaginary part of A, $\mathbf{x}$ and $\mathbf{b}$ are separated, then we have

$$(A_1+jA_2)(\mathbf{x}_1+j\mathbf{x}_2)=\mathbf{b}_1+j\mathbf{b}_2$$

or equivalently

$$A_1\mathbf{x}_1-A_2\mathbf{x}_2=\mathbf{b}_1$$
$$A_2\mathbf{x}_1+A_1\mathbf{x}_2=\mathbf{b}_2$$

If we denote the matrices G, $\mathbf{y}$ and $\mathbf{c}$ as

$$G=\begin{bmatrix}A_1 & -A_2\\ A_2 & A_1\end{bmatrix},\ \mathbf{y}=\begin{bmatrix}\mathbf{x}_1\\ \mathbf{x}_2\end{bmatrix},\ \text{and } \mathbf{c}=\begin{bmatrix}\mathbf{b}_1\\ \mathbf{b}_2\end{bmatrix}$$

then the equivalent system is a system of real algebraic equations

$$G\mathbf{y}=\mathbf{c}$$

Note that the dimension of $\mathbf{y}$ and $\mathbf{c}$ are twice of $\mathbf{x}$ or $\mathbf{b}$ and G is a $2n$ by $2n$ matrix if A is an n by n matrix.

(i) Formulate a system of linear equations for $v_C(t)$ of the circuit of Fig. P6-16. Find A, $\mathbf{b}$, G, and $\mathbf{c}$ with the phasors of $v_1(t)$ and $v_2(t)$ given as $V_1=2\angle 0°$ and $V_2=3\angle 90°$.

(ii) Solve the system $A\mathbf{x}=\mathbf{b}$ by complex arithmetic.

(iii) Solve the system $G\mathbf{y}=\mathbf{c}$ by real Gaussian elimination with pivoting. Compare with the answer in (ii).

6-26a. Solve the systems of 6-25a by the method of 6-26 using the subroutine GE(N, A, B). Compare with the exact results.

6-26b. Repeat 6-26a for the systems of 6-25b.

6-27. For Example 6.9.1, formulate a new set of real linear algebraic equations $G\mathbf{y}=\mathbf{c}$ for $w=5$ rad/sec. and solve the resulting 4 by 4 system by using the subroutine GE(N,A,B). Compare with the numerical results given in Fig. 6.52.

6-28. Solve the state equations, using the subroutine CGE(N,A,B), for the steady-state responses of v_{C1}, v_{C2}, v_{C3} of the circuit of Fig. P3-18 with $v_1(t)=2\cos(wt)$ and $v_2(t)=3\sin(wt+30°)$ and $w=0$, 10, 100, 1000 rad/sec.

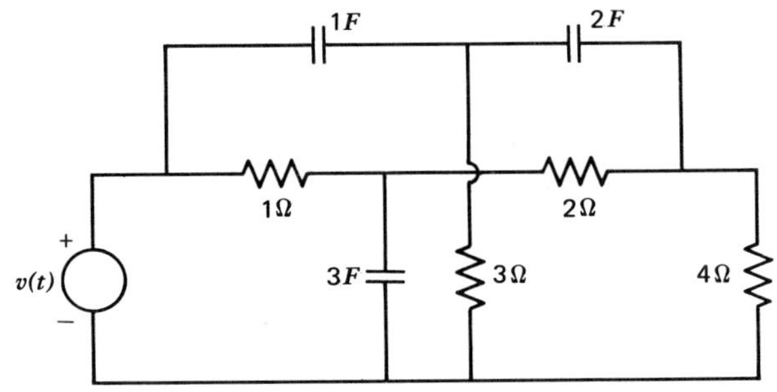

Fig. P6-28*a*

6-28a. For the circuit shown in Fig. P6-28a

(i) Write a set of node equations.

(ii) Using the subroutine CGE(N, A, B), find the steady-state node voltages of the equations of (i) for $v(t) = 10 \cos wt$ with $w = 0, 10, 100, 1000$ rad/sec.

6-28b. For the circuit shown in Fig. P6-28b

(i) Write a set of node equations.

(ii) Using the subroutine CGE(N, A, B), find the steady-state currents through each of the circuit elements for $i(t) = 2$ amp and $v_1 = v_2 = 3 \sin t$.

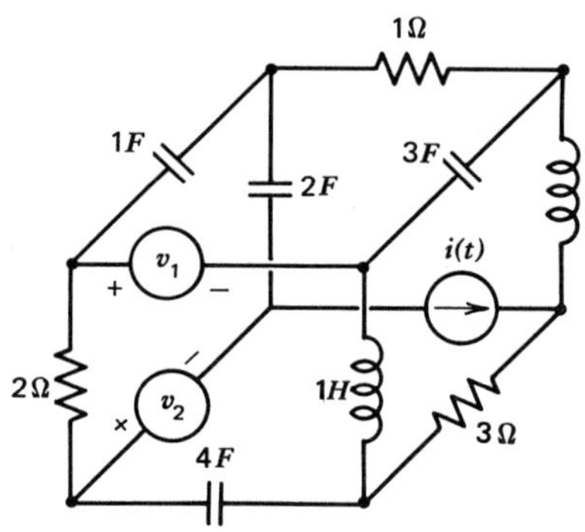

Fig. P6-28b

6-28c. Solve the equations of 6-28b using the method of 6-26.

6-29. Repeat problem 6-28 using the method of problem 6-26. Compare the results of 6-28 and 6-29 in terms of computing time and accuracy. (You can at least obtain a set of exact answers for v_{C1}, v_{C2}, v_{C3} at $w = 0$.) Which method do you prefer? Why?

CHAPTER 7
Network Functions

The steady-state sinusoidal analysis is considerably simplified by the use of the phasor concept as seen in Chapter 6. In fact, the entire analysis is performed for stationary phasors only. The sinusoids are transformed into the phasor representations and (stationary) phasor equilibrium equations for the circuits are written using Kirchhoff's Laws. The branch impedance constraints are then introduced to obtain the algebraic equations. These equations are easily solved for the desired unknowns by matrix inversion or numerical methods and the phasor quantities transformed back into sinusoids by introducing $e^{j\omega t}$ and using the (Re/Im) operation.

Alternatively, the circuit impedance may be defined at the circuit terminals. The response of the circuit at these terminals to a steady-state sinusoidal) excitation is related in terms of the corresponding phasors by the circuit impedance at the prescribed frequency.

The phasor analysis involves stationary phasors only with no time functions present. The impedances of the circuits (or circuit elements) depend on the particular frequency ω of interest. In our analysis, the numerical values may be obtained once the value of ω is specified. However, it is often convenient to express the impedance generally as a function of the frequency variable; that is, ω is left unspecified. The phasor equilibrium equations for any circuit are expressed in terms of the unknown frequency variable ω and the circuit impedance obtained as a function of ω. Whenever the impedance for a particular ω is needed, that value of ω is substituted in the impedance function. The response phasor may then be obtained for the desired value of ω. This approach has the merit that the impedance is known generally as a function of ω and does not have to be evaluated by a repetitive solution of the phasor equations for each value of ω.

Circuit description and behavior for linear, time-invariant circuits may thus be expressed by functions of frequency for steady-state sinusoidal analysis. No time description is involved, and the analysis is performed in the *frequency domain*. The *time-domain* analysis of Chapters 2 to 5, on the other hand, involves circuit description and behavior as a function of time. The frequency-domain analysis is the subject of this chapter and involves the determination and interpretation of the functions of frequency that characterize circuits under sinusoidal steady-state conditions. Later chapters will show how the frequency-domain analysis can be extended to nonsinusoidal excitations and transient analysis.

7.1 NETWORK OR SYSTEM FUNCTIONS

The *impedance function* expressed as a function of the frequency variable ω is a *network or system function*. A network or system function is the ratio of the response (phasor) to the excitation (phasor) expressed as a function of the frequency ω. The response and excitation may be at the same terminal pair or at different terminal pairs. When the response is a voltage and the excitation a current at the same driving terminal-pair, the network function is a driving-point impedance. Otherwise, the ratio is a transfer function if both the excitation and the response are voltages or currents. Transfer impedance is a ratio of response voltage and excitation current at different terminal pairs.

The frequency-domain analysis involves the determination of the pertinent network function as a function of ω. The function then defines the circuit behavior—that is, (phasor) excitation-response pair—for every ω. The response for any ω is then obtained by the value of the network function at ω times the excitation. The circuit behavior is thus completely determined for all ω by the network function.

The network function including the impedance is a property of the particular circuit. It depends solely on the circuit interconnections and the circuit parameters (i.e., element values). The network function is thus uniquely defined for a given circuit. Let us examine an arbitrary circuit (Fig. 7.1) for which we have the general phasor equations given by

$$A\mathbf{x} = \mathbf{b} \tag{7.1}$$

$\mathbf{x}$ is the vector of unknown node voltages or loop current (phasors) and $\mathbf{b}$ is the source (phasor) vector with only one nonzero source element, say b_1. The matrix A of the coefficients is made up of elementary branch impedances (R, $j\omega L$, $j\omega M$, $1/j\omega C$) expressed as functions of ω. The matrix A reflects the specific circuit parameters and the interconnections.

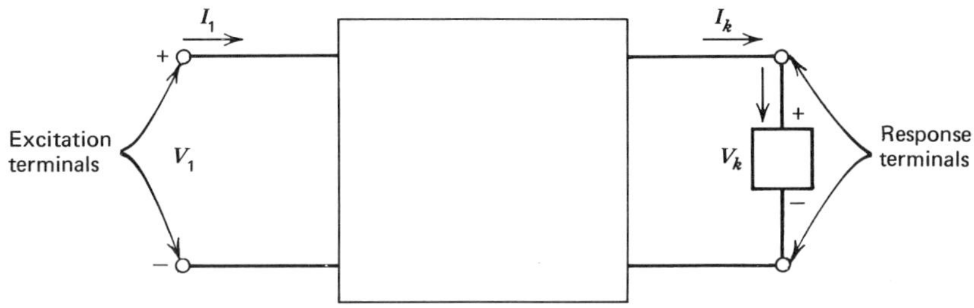

Fig. 7.1

The solution of the phasor Eq. 7.1 is formally given by

$$\mathbf{x} = A^{-1}\mathbf{b}. \tag{7.2}$$

Since only $b_1 \neq 0$, Eq. 7.2 becomes

$$\begin{bmatrix} x_1 \\ x_2 \\ \cdot \\ \cdot \\ \cdot \\ \cdot \\ x_n \end{bmatrix} = \frac{1}{|A|} \begin{bmatrix} A_{11} \\ A_{21} \\ \cdot \\ \cdot \\ \cdot \\ \cdot \\ A_{n1} \end{bmatrix} b_1 \tag{7.3}$$

where $|A|$ = determinant of A, and A_{ij} is the cofactor of A with the ith row and the jth column removed. The response phasor of interest may be some element x_k. Then the network function is

$$H(\omega) = \frac{x_k}{b_1} = \frac{A_{k1}}{|A|} \tag{7.4}$$

a function of ω which is completely determined by the specific circuit matrix A. The response phasor for any ω_0 and any excitation phasor b_1 is given by

$$x_k = H(\omega_0) b_1. \tag{7.5}$$

Let V_1 or I_1 be the possible excitation or input signal in Fig. 7.1, and let $k = 2$. Then, some examples of network functions are:

$$Z_{in} = \frac{V_1}{I_1} = \text{Input (or, driving-point) impedance}$$

$$Z_{21} = \frac{V_2}{I_1} = \text{Transfer impedance}$$

$$T_V = \frac{V_2}{V_1} = \text{Voltage transfer function}$$

$$T_I = \frac{I_2}{I_1} = \text{Current transfer function}$$

While generally the network functions must be derived by solving the phasor equilibrium equations, it is difficult at times to obtain explicit expressions as functions of ω. Equations 7.3 and 7.4 show that the network functions are determined by the determinant and cofactor of the circuit matrix. Though formally possible and straightforward, evaluation of determinants and cofactors as explicit functions of ω can be quite difficult when n is large as for moderately complex circuits. Fortunately, for a large and important class of circuits known as *series-parallel circuits*, it is quite easy to write down the network functions. We discuss, in the next section, these relatively simple procedures for developing network functions for series-parallel circuits. For all other cases, the equilibrium equations must be solved either explicitly as functions of ω or numerically for a whole range of values of ω as in the last chapter.

In this chapter, we also examine circuit behavior in the frequency domain and discuss some important signal processing circuits in terms of the network function properties.

7.2 SERIES-PARALLEL CIRCUITS

Successive series and parallel connections give rise to *series-parallel circuits*. The successive series and parallel connections start with the basic circuit element branches and continue with connections of more complex branches, each of which may be a series-parallel circuit. Let us examine a series-parallel circuit with arbitrary branch impedances as shown in Fig. 7.2. Each of the branch impedances may itself be the impedance of a complex circuit. In any case, the circuit is a combination of series and parallel connections of these branch impedances. The term branch here should not be confused with a *single* element branch of a circuit. The branch impedance represents rather a two-terminal circuit.

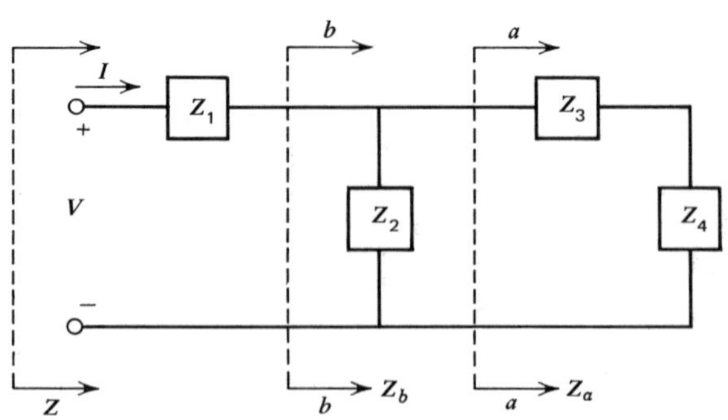

Fig. 7.2

Driving-Point or Input Impedance

The driving-point impedance, $Z = V/I$, at the circuit terminals (at the left end) is obtained by using the rules for combining impedances in series or in parallel at each series or parallel connection. The process is most conveniently developed by starting at the end farthest from the driving-point or input circuit terminals, that is, where the circuit may be excited and where the impedance is desired.

At the farthest end from the input terminals, Z_4 is in series with Z_3. Therefore, the impedance of this series combination is the impedance of the circuit looking to the right of the broken line *aa*. Let us denote this impedance by Z_a that is the sum of the series impedances Z_3 and Z_4:

$$Z_a = Z_3 + Z_4 \tag{7.6}$$

Next, observe that the series combination of Z_3 and Z_4 is connected in parallel with Z_2. The impedance of the circuit to the right of the line *bb*, Z_b, is thus given by the parallel combination of the Z_2 branch and the $Z_a = (Z_2 + Z_3)$ branch. That is,

$$\begin{aligned} Y_b = \frac{1}{Z_b} &= \frac{1}{Z_2} + \frac{1}{Z_a} \\ &= \frac{1}{Z_2} + \frac{1}{Z_3 + Z_4} \end{aligned} \tag{7.7}$$

Finally, the impedance Z_1 is in series with the impedance Z_b. Therefore, the total input impedance $Z = Z_{in}$ is given by the sum of Z_1 and Z_b:

$$\begin{aligned} Z = Z_{in} &= Z_1 + Z_b \\ &= Z_1 + \frac{1}{\dfrac{1}{Z_2} + \dfrac{1}{Z_3 + Z_4}} \end{aligned} \tag{7.8}$$

Thus, the total impedance for a series parallel circuit is simply determined by adding impedances for series connections and adding admittances for parallel connections. An example illustrates the steps involved.

Example 7.2.1 Consider the circuit in Fig. 7.3. The impedance, Z_3, of the series combination of R_3 and C_3, labeled "branch 3", is

$$Z_3 = R_3 + \frac{1}{j\omega C_3} \tag{7.9}$$

Similarly, the impedance, Z_2, of branch 2 is

$$Z_2 = R_2 + j\omega L_2 \tag{7.10}$$

The total admittance, Y_{23}, of the parallel combination of branches 2 and

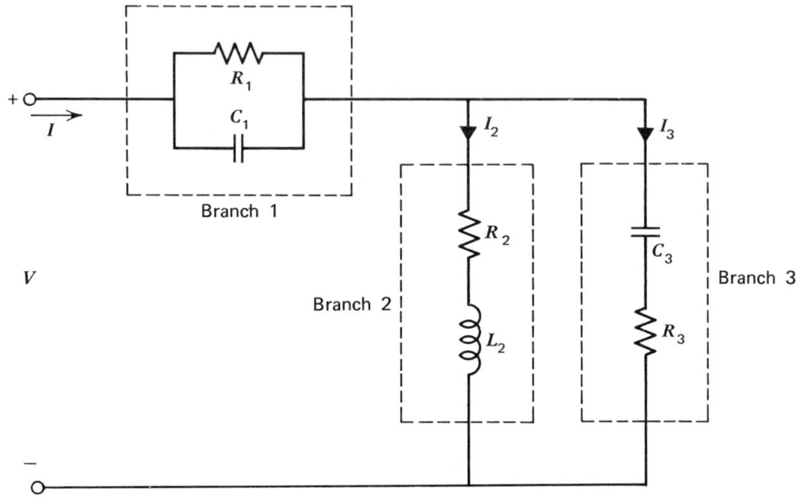

Fig. 7.3

3 is equal to the sum of the admittances of branches 2 and 3.

$$Y_{23} = Y_2 + Y_3 \tag{7.11}$$
$$= \frac{1}{Z_2} + \frac{1}{Z_3}$$

$$Z_{23} = \frac{1}{\dfrac{1}{Z_2} + \dfrac{1}{Z_3}} = \frac{Z_2 Z_3}{Z_2 + Z_3} \tag{7.12}$$

The admittance of branch 1 is equal to $1/R_1$, plus $j\omega C_1$, that is,

$$Y_1 = \frac{1}{R_1} + j\omega C_1 \tag{7.13}$$

or

$$Z_1 = \frac{1}{Y_1} = \frac{1}{\dfrac{1}{R_1} + j\omega C_1} = \frac{R_1}{1 + j\omega R_1 C_1} \tag{7.14}$$

The total input impedance, Z_{in}, of the circuit is the series combination of branch 1 and the branch formed by the parallel combination of branches 2 and 3.

$$Z = Z_{in} = Z_1 + Z_{23}$$
$$= Z_1 + \frac{Z_2 Z_3}{Z_2 + Z_3}$$
$$= \frac{R_1}{(1 + j\omega C_1 R_1)} + \frac{(R_2 + j\omega L_2)\left(R_3 + \dfrac{1}{j\omega C_3}\right)}{(R_2 + R_3) + j\omega L_2 + \dfrac{1}{j\omega C_3}} \tag{7.15}$$

The expression can be simplified once the element values are known. For example, for $R_1 = R_2 = R_3 = 1\ \Omega$, $C_1 = 2F$, $C_2 = 3F$, $C_3 = 1F$, and $L_2 = 1h$, the impedance expression becomes:

$$Z(j\omega) = \frac{6(j\omega)^3 + 14(j\omega)^2 + 12(j\omega) + 2}{6(j\omega)^3 + 15(j\omega)^2 + 8(j\omega) + 1} = |Z(j\omega)|\angle \arg Z(j\omega) \qquad (7.16)$$

Since the expression is a ratio of polynomials in $j\omega$ rather than just ω, we have expressed this by indicating $Z(j\omega)$ instead of $Z(\omega)$. The impedance amplitude and phase for any value of ω can be obtained from Eq. 7.16. Then if the current phasor for any value of ω is $I = |I|\angle\theta$ and $Z(j\omega)$ $|Z(j\omega)|\angle \arg Z(j\omega)$, the voltage phasor $V = |V|\angle\psi$ is given by

$$V = Z(j\omega)I \qquad (7.17)$$

and

$$\begin{aligned} |V| &= |Z(j\omega)|\,|I| \\ \psi &= \theta + \arg Z(j\omega) \end{aligned} \qquad (7.18)$$

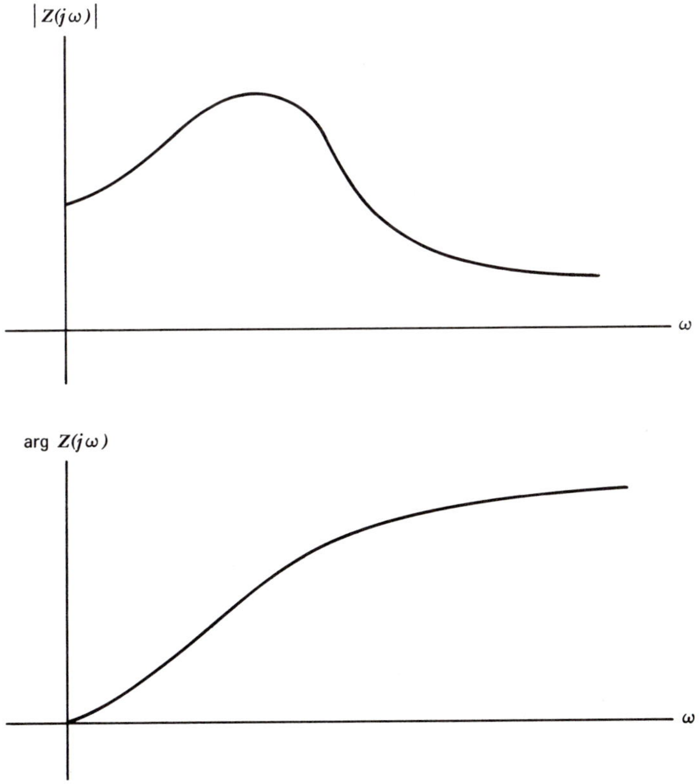

Fig. 7.4

The network impedance function is thus used either in the form of Eq. 7.16 or its amplitude and phase are expressed as functions of ω and used as in Eq. 7.18. Plots of such amplitude and phase functions are often used to show network functions as in Fig. 7.4.

Transfer Functions

A *transfer function* is a ratio expressed as a function of ω of the response (phasor) at one terminal-pair or port to the excitation (phasor) at a *different* terminal-pair. If the ratio is of voltage to current, it is a transfer impedance. It is a dimensionless ratio if both response and excitation are voltages or currents. Finally, it is called a transfer admittance if the response is a current and the excitation a voltage. These ratios as functions of ω must be determined in general by the solution of the equilibrium equations. For series-parallel circuits, it is quite easy to determine the transfer functions. The transfer function of a series-parallel circuit is determined in terms of the impedances of component circuits. We first show this for very simple circuits and then show that all series-parallel circuits are *equivalent* to these simple cases.

Let us first consider the series circuit as shown in Fig. 7.5 and determine the transfer function (voltage ratio):

$$T_V = \frac{V_2}{V_1} \tag{7.19}$$

For our simple circuit, the voltage V_2 is Z_2 times the current I_1, that is,

$$V_2 = Z_2 I_1 \tag{7.20}$$

The current I_1 in the circuit is the voltage V_1 divided by the circuit impedance that for the series circuit is $(Z_1 + Z_2)$. So

$$I_1 = \frac{V_1}{Z_1 + Z_2} \tag{7.21}$$

Substituting Eq. 7.21 into 7.20 and dividing by V_1, we have

$$\frac{V_2}{V_1} = \frac{Z_2}{Z_1 + Z_2} \tag{7.22}$$

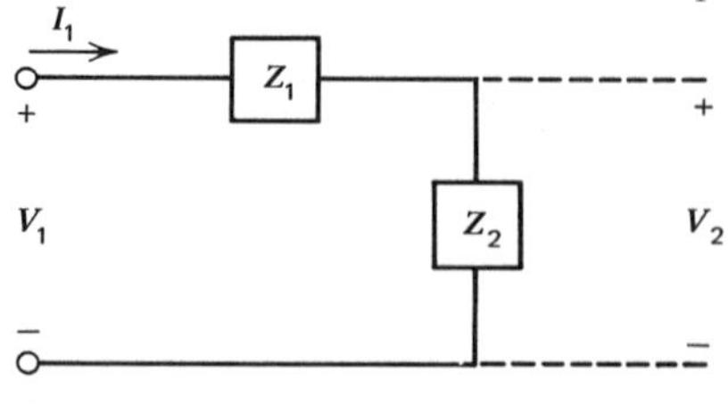

Fig. 7.5

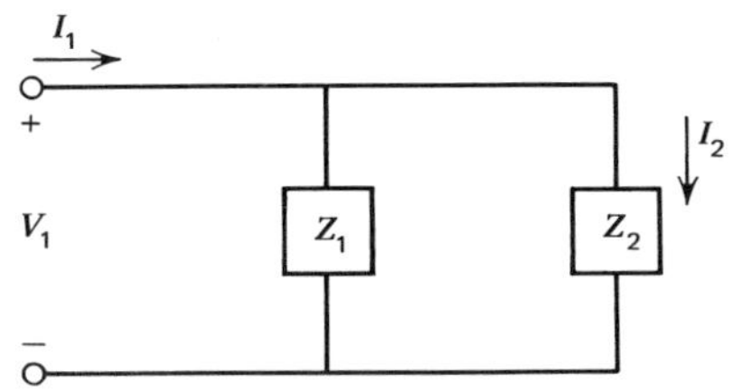

Fig. 7.6

The desired voltage ratio is thus obtained in terms of the component impedances. If these impedances are expressed as functions of ω, then so is the transfer function.

In a similar way, the transfer function as a current ratio is obtained for the circuit of Fig. 7.6. Here, we are interested in determining

$$T_I = \frac{I_2}{I_1} \tag{7.23}$$

The current I_2 is given by

$$I_2 = \frac{V_1}{Z_2} \tag{7.24}$$

and the voltage V_1 is I_1 times the total circuit impedance. Thus,

$$V_1 = I_1\left(\frac{Z_1Z_2}{Z_1+Z_2}\right) \tag{7.25}$$

Substituting into Eq. 7.24 and dividing by I_1,

$$\frac{I_2}{I_1} = \frac{Z_1}{Z_1+Z_2} \tag{7.26}$$

The two transfer functions obtained in Eqs. 7.22 and 7.26 are all that are needed to determine the transfer functions of any series-parallel circuit. All that is needed is to determine the impedance functions Z_1 and Z_2. The impedance function Z_2 is easily determined since it is the impedance of the circuit across or through which the response voltage or current is assumed. As shown in the next section, the rest of the circuit is *equivalent* to a *modified source* and a *single impedance* Z_1 to which Eqs. 7.22 and 7.26 may be applied. The equivalent circuit is therefore a very useful and important concept and is discussed in the next section.

The equivalent circuit concept is universally useful for both transient and steady-state analysis. While we will make immediate use of equivalent circuits for sinusoidal steady-state analysis, the concept is not restricted and can be used for transient analysis as well. For these reasons, our discussion in the next section is in terms of the time-domain, transient analysis where the response is given by zero-state response plus zero-input response.

7.3 THÉVENIN'S AND NORTON'S THEOREMS

In circuit analysis the objective of a problem is often to determine a *single* branch current through a given circuit element or a *single* voltage across an element. In problems of this kind it is generally not practicable to write a complete set of node or loop equations and solve a system of equations for this one current or voltage. It is then convenient to use two very important theorems on equivalent circuits, known as *Thévenin's and Norton's theorems.* We develop these theorems for complete circuit behavior as functions of time. The response is assumed to involve possible transients, and the initial state is allowed to be nonzero. The response is then known to be the sum of the zero-state and the zero-input responses.

Thévenin's Theorem

Suppose we have two parts of a circuit as shown in Fig. 7.7 connected by two wires. The part of the circuit designated A may include sources and may have a nonzero initial state. Part B is similarly an arbitrary network. The two parts are connected as shown by electrical connections only with current, $i(t)$, and voltage, $v(t)$, at the connection. There is no other coupling between parts A and B.

We want to replace circuit A by a simple but equivalent structure such that the terminal conditions for network B, voltage v and current i do not change. To do this we use the superposition principle that we derived earlier. The first part of the superposition principle states that the total response of a given circuit is equal to the zero-state response plus the zero-input response. The second part of the same principle states that the zero-state response of a circuit driven by several sources is the sum of the zero-state responses to each separate source with all other

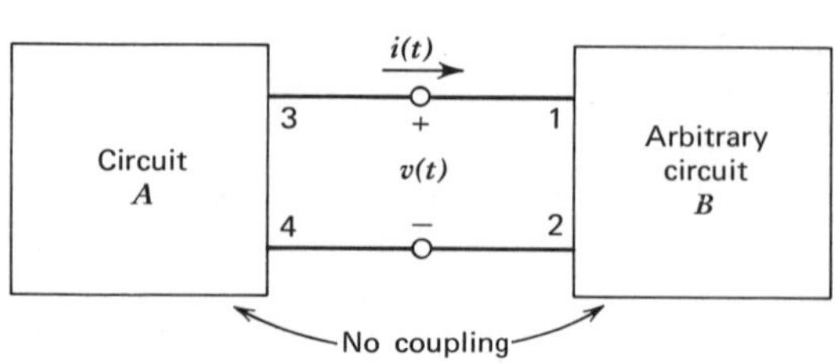

Fig. 7.7

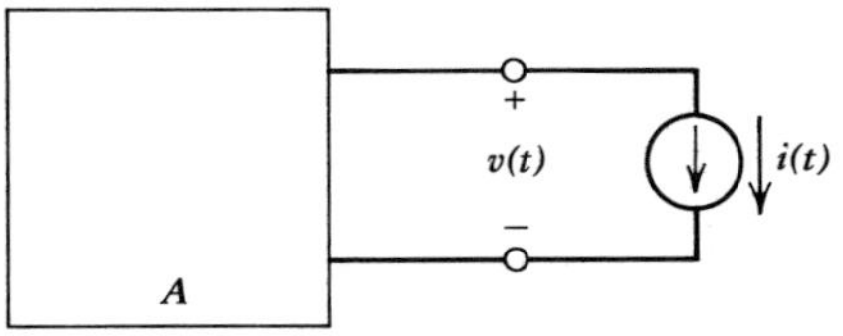

Fig. 7.8

sources set to zero. We use this principle of superposition to simplify our network.

In our representation of the circuit, the terminal conditions of the two parts were the current, $i(t)$, and the voltage, $v(t)$. If we replace the part of the circuit designated B with a current source that forces the *same* current $i(t)$ through the circuit A as shown in Fig. 7.8, then the voltage $v(t)$ must appear across the terminals of the circuit A. That is, the circuit A demonstrates the same voltage $v(t)$ across its terminals as long as the same current $i(t)$ flows through it.

Therefore, we can replace the circuit B temporarily by a current source producing current $i(t)$, connected to A. The response observed is the voltage $v(t)$. By the first part of the superposition theorem, the total voltage $v(t)$ is equal to the sum of the zero-state response and the zero-input response. And by the second part of the superposition principle, the zero-state response of the circuit A is equal to: (i) the zero-state response due to the source $i(t)$ with all other (internal) sources set to zero, plus (ii) the zero-state responses due to all internal sources with the external source, $i(t)$, set to zero. That is,

$$\begin{aligned} v(t) &= \text{zero-state response} \qquad\qquad\qquad\qquad + \text{zero-input response} \\ &= \begin{bmatrix} (\text{ZS response})_{i\neq 0} \\ \text{internal sources} = 0 \\ \text{IS} = 0 \end{bmatrix} + \begin{bmatrix} (\text{ZS response})_{i=0} \\ \text{int. sources} \neq 0 \\ \text{IS} = 0 \end{bmatrix} + \begin{bmatrix} (\text{ZI response})_{i=0} \\ \text{int. sources} = 0 \\ \text{IS} \neq 0 \end{bmatrix} \\ &= v_i(t) + v_a(t) + v_b(t) \end{aligned} \tag{7.37}$$

Note that setting $i(t) = 0$ is equivalent to an open circuit at the terminal of network A. Therefore, v_a, the zero-state response due to internal sources (with $i(t) = 0$), is the *open circuit voltage* due to only the internal sources of network A. And v_b, the zero-input response, is the *open-circuit voltage* due to only the initial state.

Now consider the circuit in Fig. 7.9 with circuit B as before but circuit A replaced by two sources v_a and v_b *in series* with the circuit A whose initial state and internal sources are all set equal to zero. The voltage across the terminals of the circuit A with initial state and internal sources zero is v_i since that is its zero-state response to the current $i(t)$ with all other

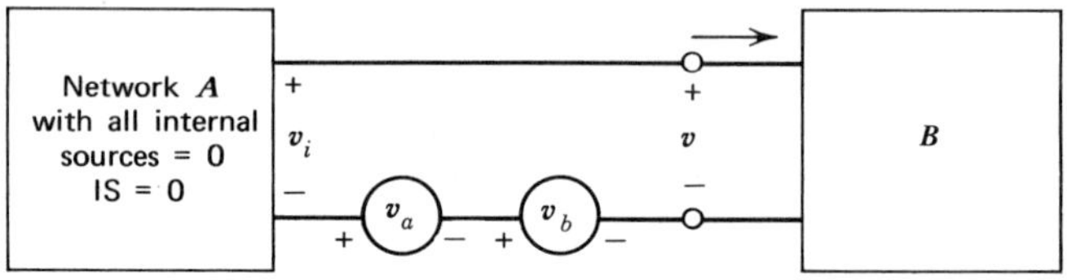

Fig. 7.9 Thévenin equivalent.

sources zero. Thus, the voltage across the terminals of B is $v_a + v_b + v_i = v(t)$ as before from Eq. 7.27.

Figure 7.9 represents a simplified equivalent of the original circuit A with the circuit B unchanged. This equivalent circuit is called the *Thévenin equivalent* given by setting all sources and initial state in the circuit A to zero and placing two voltage sources in series with it. The voltage sources are equal to the voltages v_a and v_b. The voltage v_a is obtained at the *open-circuited terminals of the original circuit A* due to the internal sources under zero-initial state. The voltage v_b is the open-circuit voltage of A with internal sources zero and nonzero initial state.

The voltage v_b, due to the initial state and with zero excitation, is a transient voltage that vanishes as time increases indefinitely. The voltage v_a is due to excitation sources and thus contains both transient and driven parts. For the sinusoidal steady-state case, v_b is zero. In this case, v_a should contain only the steady-state or driven response and is represented by the phasor value. Also, the circuit A is represented by its impedance in the Thévenin equivalent.

Note that the Thévenin equivalent replaces a given circuit A by its simplified equivalent such that the current and voltage at its terminals would be unchanged for connection of A to any other circuit. The internal circuit of A, of course, is changed and the signals *inside* of A cannot be given by the equivalent circuit. Only signals external to A are unaltered by the Thévenin equivalent.

Example 7.3.1 Determine the Thévenin equivalent for the circuits in Fig. 7.10.

For (a), only v_a need be determined since the state is always zero. The open circuit voltage v_a is determined at the open terminals aa':

$$v_a = v(t)$$

The circuit with the voltage source set to zero gives only the resistor R_2. Thévenin equivalent to Fig. 7.10a shown in Fig. 7.11a is simply the source $v(t)$ in series with R_2.

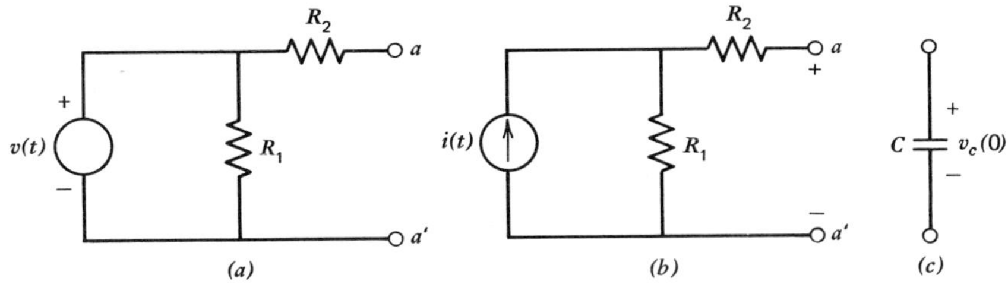

Fig. 7.10

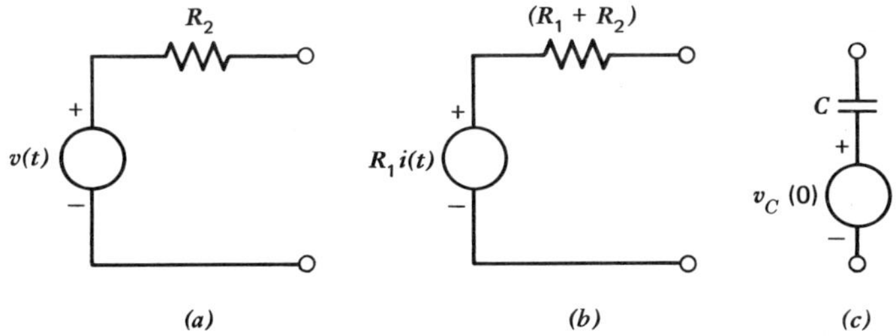

Fig. 7.11

For the circuit in Fig. 7.10*b*, the open-circuit voltage at *aa'* is

$$v_a = R_1 i(t)$$

the circuit with the source set to zero is R_1 and R_2 in series. The Thévenin equivalent is shown in Fig. 7.11*b*.

For the capacitor in Fig. 7.10*c*, the initial voltage is $v_c(0)$ and there is no other source. The equivalent circuit is a source $v_b = v_c(0)$ and an uncharged (zero-state) capacitor C in series as in Fig. 7.11*c*.

Example 7.3.2 For sinusoidal steady-state, find Thévenin's equivalent for the circuits in Fig. 7.12 where only phasor signals are shown.

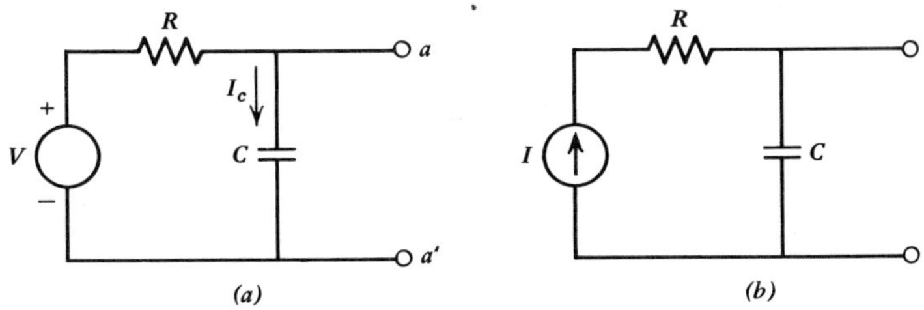

Fig. 7.12

For the circuit in Fig. 7.12a, the open-circuit voltage (phasor), V_a, at the terminals aa' is equal to the impedance of the capacitor times the current through it, that is,

$$V_a = \frac{1}{j\omega C} I_C$$

But, the current I_C is V divided by the total series impedance,

$$I_C = \frac{V}{R + \dfrac{1}{j\omega C}}$$

and

$$V_a = \frac{V}{1 + j\omega RC}$$

The Thévenin equivalent is shown in Fig. 7.13a where the series impedance is that of R and C in parallel.

The circuit in Fig. 7.12b gives the open-circuit voltage at aa'.

$$V_a = \frac{1}{j\omega C} I$$

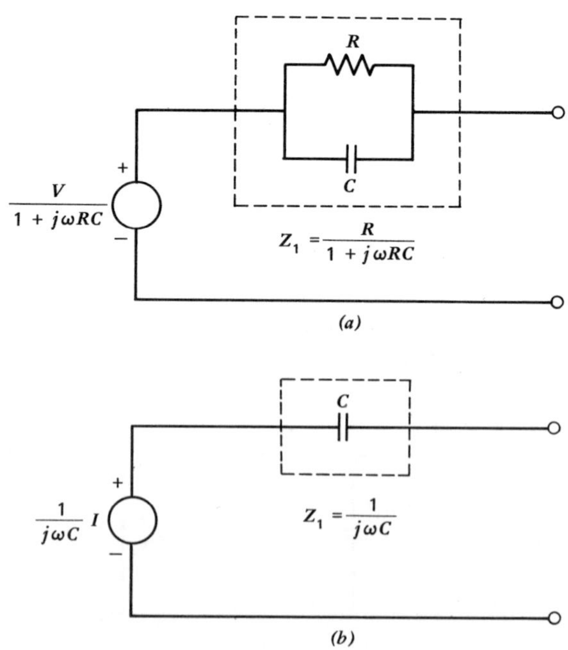

Fig. 7.13

The circuit with the source zero is just the capacitor and the equivalent is shown in Fig. 7.13*b*.

Norton's Theorem

Thévenin's theorem for equivalent circuits relied on obtaining the voltage response of the circuit A in Fig. 7.7 with the current acting as the independent excitation. Alternatively, we could obtain the current response of A with the voltage $v(t)$ acting as the source as shown in Fig. 7.14.

The current response of A is given by the sum of the zero-state response and the zero-input response. The zero-input response i_b is obtained by setting all sources to zero, that is, the external and internal sources of A set to zero. This is the current with the terminals of A short-circuited and all internal sources of A set to zero. The zero-state current is the zero-state current i_a due to the internal sources with the external source $v(t)$ set to zero plus the zero-state current i_v due to $v(t)$ with internal sources set to zero. That is,

$$i(t) = i_a + i_b + i_v \tag{7.28}$$

and the equivalent circuit is obtained by setting the initial state and all internal sources to zero in A with the current sources i_a and i_b in parallel. This is shown in Fig. 7.15. From the figure, the current i is again the sum of i_a, i_b, and i_v. The equivalent circuit in this case is called the Norton equivalent.

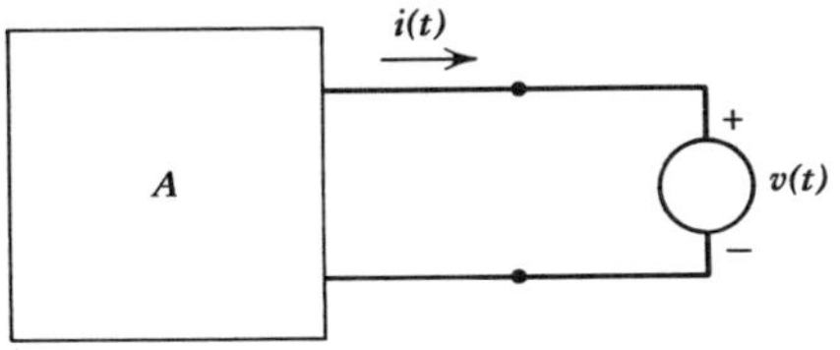

Fig. 7.14

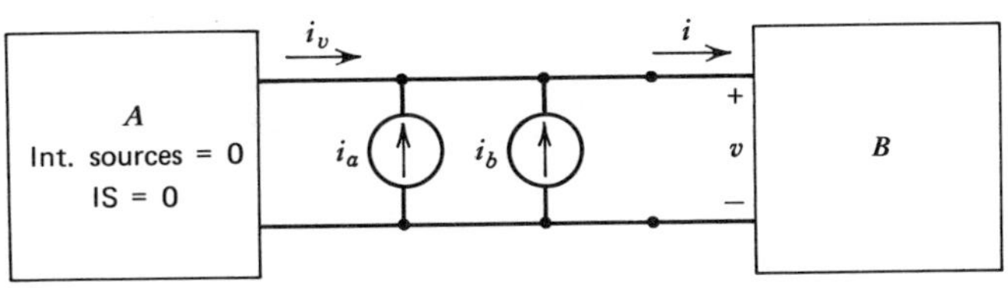

Fig. 7.15 Norton equivalent.

The Norton equivalent is the circuit A with initial state and internal sources set to zero with current sources in parallel. The current sources are the currents *in the short-circuit*, when the terminals of the original circuit A are *short-circuited*, due to the internal sources with initial state zero (zero-state) and due to the initial state with all internal sources zero (zero-input).

Example 7.3.3 Determine the Norton equivalents of the circuits in Fig. 7.16.

The equivalents are shown in Fig. 7.17 for the corresponding circuits.

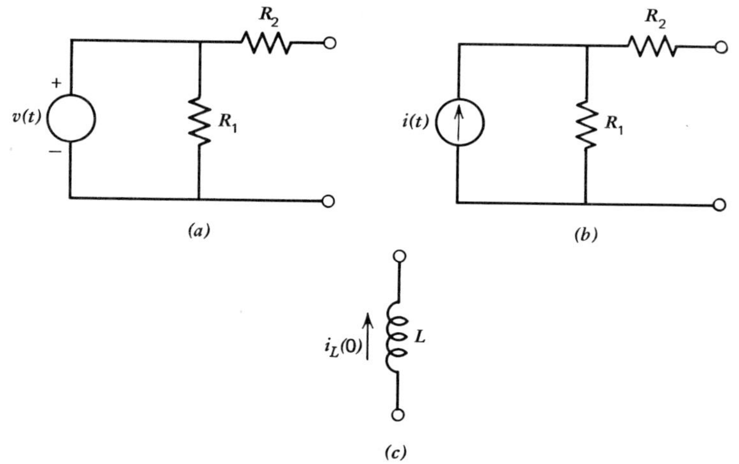

Fig. 7.16

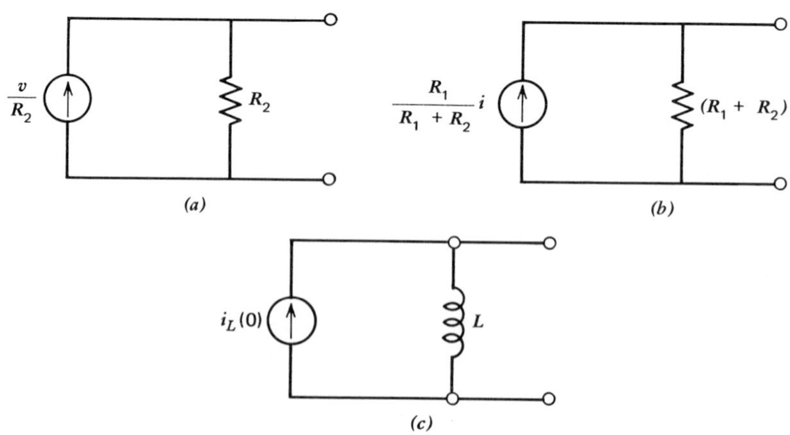

Fig. 7.17

For (a), the short-circuit current is $i_a = v/R_2$. For (b), the short-circuit current is $i_a = (R_1/(R_1+R_2))i$. For (c), the current is $i_b = i_L(0)$. Observe that the circuits are now placed in parallel with the sources.

Example 7.3.4 The Norton equivalents for the sinusoidal steady-state case are given in Fig. 7.18 for the circuits in Fig. 7.12. The short-circuit current phasors are V/R and I in the two cases.

Use of Thévenin's and Norton's equivalent circuits are of considerable help in simplifying analysis. For example, we saw how a charged capacitor is equivalent to a voltage source in series with an uncharged capacitor. Similarly, an inductor is equivalent to a current source in parallel with a zero-state inductor. Thus, all circuits may be treated as zero-state with sources accounting for the nonzero state.

For sinusoidal steady-state analysis, the equivalent circuit approach is even more useful. Circuits are replaced by a voltage or a current source with an appropriate impedance in series or in parallel with the source. The determination of the response across a *single* impedance can be now carried out by replacing the rest of the circuit by its equivalent. All

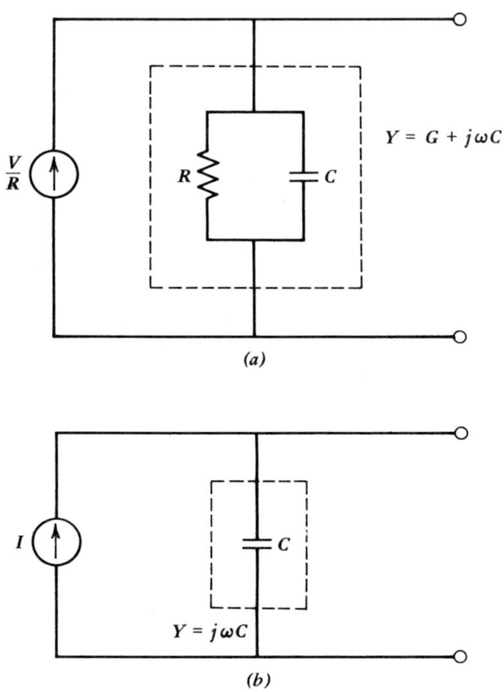

Fig. 7.18

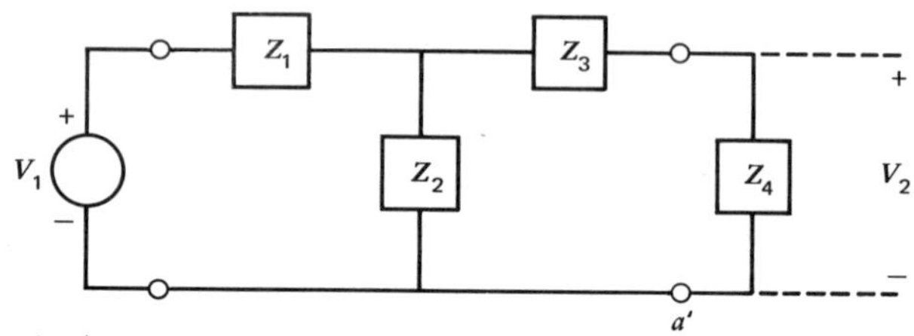

Fig. 7.19

circuits can thus be handled in a very simple way especially the series-parallel circuit, using Thévenin's and Norton's equivalents.

We now return to the sinusoidal steady-state analysis. All signals are again phasor signals and circuits are represented often by impedances.

As an example of the application of Thévenin's theorem, let us determine V_2/V_1 for the circuit of Fig. 7.19.

We will first replace the circuit to the left of aa' in Fig. 7.10, which is Thévenin's equivalent.

The equivalent circuit together with the rest of the network is shown in Fig. 7.20.

In the figure,

$$Z_e = Z_3 + \frac{Z_1 Z_2}{Z_1 + Z_2} \tag{7.29}$$

and the voltage (open circuit) at aa' is

$$V_a = \frac{V_1}{Z_1 + Z_2} \cdot Z_2 \tag{7.30}$$

Then

$$\frac{V_2}{V_a} = \frac{Z_4}{Z_e + Z_4} \tag{7.31}$$

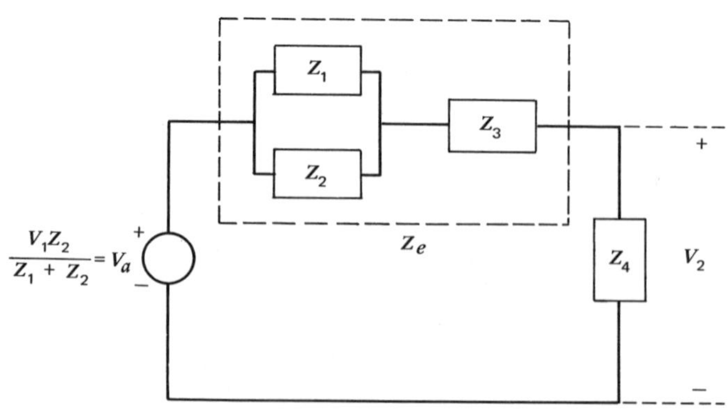

Fig. 7.20

and

$$\frac{V_2}{V_1} = \left(\frac{Z_4}{Z_4 + Z_3 + \dfrac{Z_1 Z_2}{Z_1 + Z_2}} \right) \left(\frac{Z_2}{Z_1 + Z_2} \right) \tag{7.32}$$

We are thus able to write down network functions for series-parallel networks in relatively simple manner.

The above illustration shows how Thévenin's and Norton's theorems can be used to handle series-parallel circuits. If the circuits are more complicated series-parallel circuits, repeated applications of the equivalent theorems may be used to simplify the circuits.

7.4 FREQUENCY RESPONSE AND FILTERING

Network or system functions (either transfer or driving-point) describe the behavior of the circuit in the sinusoidal steady-state case for each and every frequency of excitation. These functions expressed explicitly as a function of ω are derived either by the solution of the appropriate phasor equilibrium equations or by the use of the series-parallel methods. Thévenin and Norton equivalents can often be used to advantage in deriving the network functions.

The network or system function explicitly represents the response-excitation relationship for a circuit without specific and detailed reference to the internal circuit. Only the port or terminal-pair relationships are specified. The advantage is that the signal processing properties of the circuit are explicitly stated as a function of the frequency at the desired terminal-pairs and the unnecessary details of the circuit eliminated. We are able to state the response precisely for any given excitation once we know the network function. For example, consider the transfer function $T_V(\omega) = V_2/V_1$. The response V_2 is obtained from the given excitation V_1 at frequency ω by

$$V_2 = T_V(\omega) V_1 \tag{7.33}$$

The amplitude and phase of V_2 are obtained from Eq. 7.33. More explicitly, we can represent the amplitude and phase of V_2 in terms of the amplitudes and phases of T_V and V_1. Let

$$T_V(\omega) = |T_V(\omega)| \angle \arg T_V(\omega)$$

and (7.34)

$$V_1 = |V_1| \angle \arg V_1$$

where, arg T_V and arg V_1 represent phases of the respective functions.

Then

$$\begin{aligned} |V_2| &= |T_V(\omega)|\,|V_1| \\ \arg V_2 &= \arg T_V(\omega) + \arg V_1 \end{aligned} \tag{7.35}$$

Equation 7.35 conveniently gives the response amplitude and phase. It is, therefore, not uncommon that the transfer function as a function of ω is also expressed in terms of its *amplitude function* and *phase function* both in terms of ω. This is especially convenient when the signal processing properties of interest are specified in amplitude or phase. When the network function is represented graphically, the complex function necessarily must be represented by its amplitude and phase.

The system or network function expressed as a function of frequency is also called the *frequency response* of the circuit since it represents the response as a function of frequency to a unit excitation. An example of frequency response is shown in Fig. 7.21. The (unit excitation) frequency response concept allows one to examine the relative frequency dependent processing performed by the circuit.

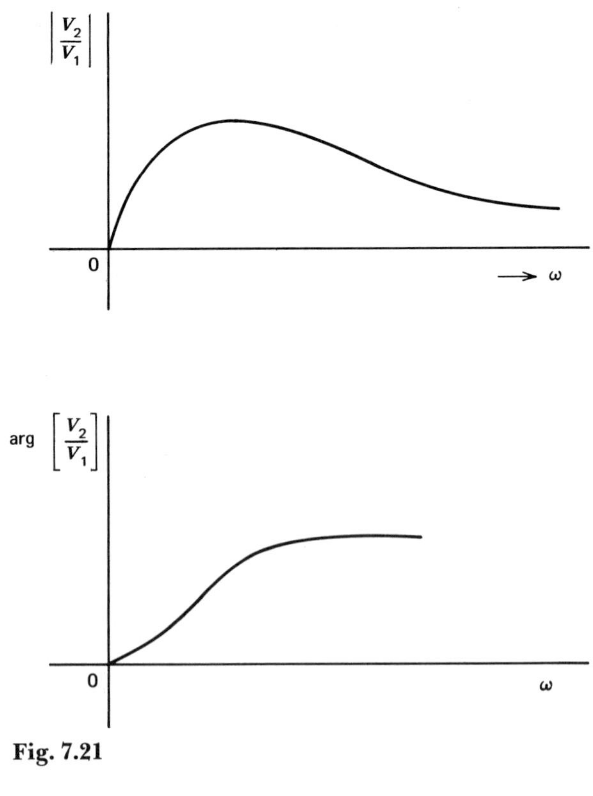

Fig. 7.21

The frequency response also gives a visual insight into the signal processing properties of a given circuit. If the magnitude of the frequency response is high at certain frequencies, it tells us that these frequencies are favored over those where the magnitude is small. A small magnitude of the system function implies that those frequency signals are eliminated or stopped from appearing at the output or response terminals. High magnitude implies that those frequencies are allowed to pass through and appear at the output as strong signals. Thus, a signal processing network of Fig. 7.21 allows the middle frequencies to pass through the circuit but the low and high frequencies are attenuated or eliminated.

An important class of signal processing circuits has the function of allowing certain frequency signals to pass through while stopping or eliminating other frequency signals from passing through. These circuits are known as *filters*. They keep out or stop certain frequencies, allowing others to pass through the filter (circuit).

Filtering of signals is carried out in many of our everyday appliances such as radio, TV, long distance telephone, etc. The incoming signals are combinations of many different messages out of which we must select the right one by appropriate filtering. In a crude sense, a message signal can be represented as a sum of several sine waves of different frequencies in a given range of frequencies. For example, a good deal of the energy in a voice signal is concentrated in the 0–3 kH. A music signal may have its energy concentrated in 0–10 kH, and so on. This fact is used to send many message signals simultaneously as described below for the case of telephone trunk lines.

Let us consider the voice signal with energy concentrated in a 3 kH band of frequencies. It is possible to process the signal without changing it so that the energy is translated from a 0–3 kH to any other band, say 3–6 kH, 6–9 kH, etc. We can thus stack several messages together as shown in Fig. 7.22. We have four different voice messages stacked from 0–12 kH band. We can send this composite signal over a telephone line

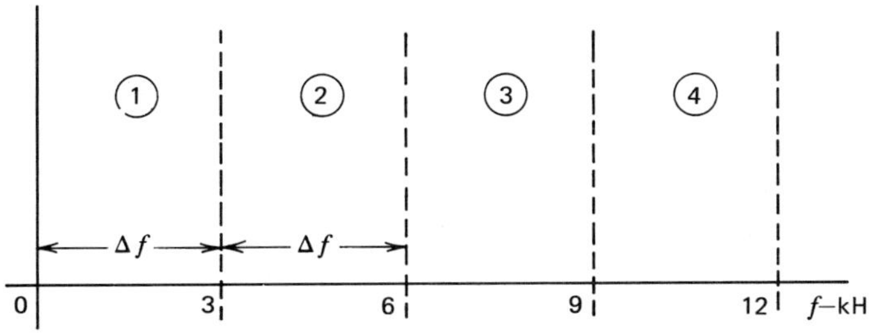

Fig. 7.22

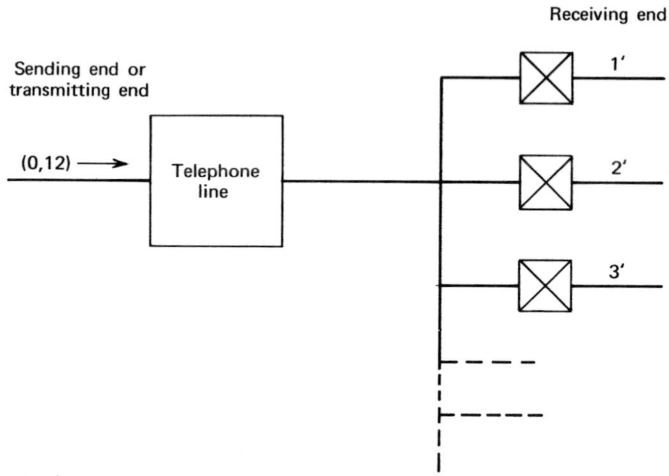

Fig. 7.23

but at the receiving end we need to separate the different messages. This is done by filtering as shown in Fig. 7.23.

The first box is a filter that allows only frequencies in 0–3 kH band to pass through. The second allows only 3–6 kH band to pass and so on. The filters are thus used to separate out the different messages at the receiving end. One telephone line can be used to handle many messages simultaneously only because signal processing circuits can be built to serve as filters.

Filters play an important role as signal processing networks, and we have defined filters as networks that filter out certain frequencies from the incoming signal. That is, they allow only certain filtered frequencies in the incoming signal to pass through. We describe below the performance characteristics of ideal filters and then see how we can achieve these in practice. We will consider only the magnitude of the ratio of the output to input signal since the magnitude function will determine which frequencies pass through and which do not.

Figure 7.24 shows the magnitude function $|T|$, representing an ideal filter. When this function has a value zero for any frequency, a signal at that frequency is eliminated and does not appear at the output or response terminals. When the function has value unity, the corresponding frequency is allowed to pass through the circuit unaltered in magnitude. Figure 7.24 shows that $|T|$ has a value unity for frequencies between ω_1 and ω_2. The value of $|T|$ is zero for frequencies that are outside this *pass-band*, that is, the range of frequencies the filter allows to pass through undistorted in amplitude. Those frequencies that are rejected constitute

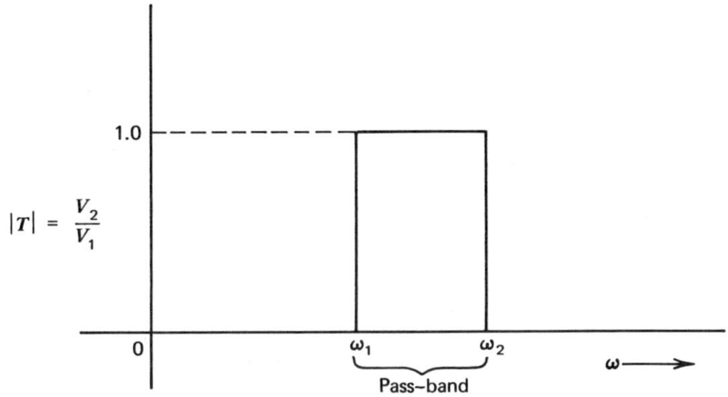

Fig. 7.24

the *stop-band.* If the filter passes a band of frequencies between ω_1 and ω_2, it is called the *band-pass filter.*

There are two other types of filters that are frequently used; they are (1) *low-pass filters* and (2) *high-pass filters.* In the case of a low-pass filter, the filter allows the signals of frequency from 0 up to some value ω_C to pass through while rejecting all other higher frequencies (Fig. 7.25). In the case of a high-pass filter, it allows only frequencies higher than a certain frequency ω_C to pass through unattenuated (Fig. 7.26). In other words, a low-pass filter passes only the low frequencies, whereas the high-pass filter passes only the high frequencies.

The performance measure of a filter is given by the range or the band of frequencies that are allowed to pass through unattenuated. This is measured in terms of the *bandwidth* of a filter. The bandwidth is defined by the range of frequencies that are allowed to pass through the filter.

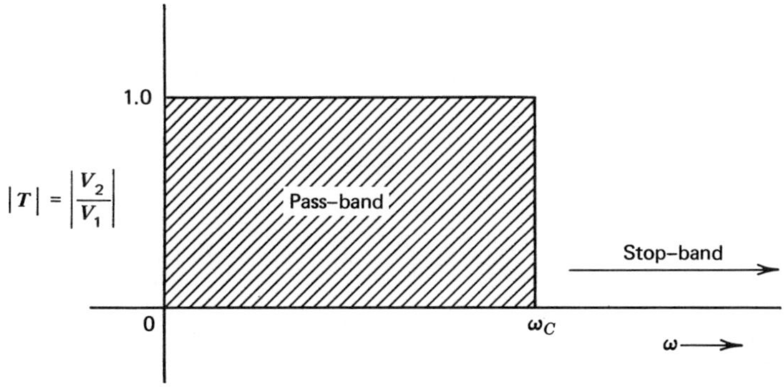

Fig. 7.25

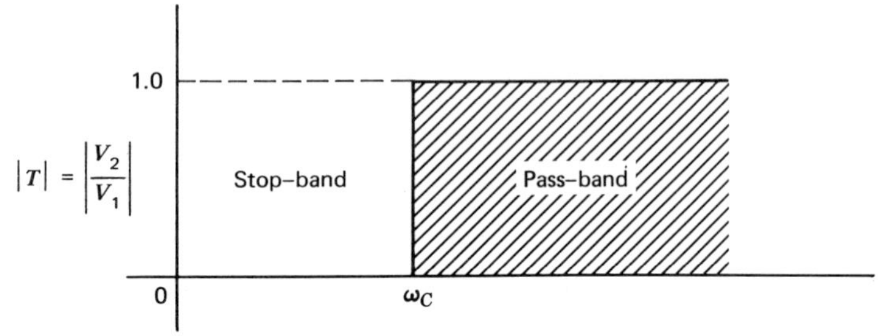

Fig. 7.26 High-pass filter.

The bandwidth or pass-band for the low-pass filter is ω_C. For the high-pass, we cannot really define the bandwidth since it is infinite, but can define the stopband.

The ideal curves discussed above are sharp or discontinuous and cannot be obtained in practice. We try to approximate as closely as possible these ideal characteristics by physical circuits. We now discuss some simple circuits that approximate these ideal filter characteristics.

We refer to the excitations as the *input* signals and the responses as the *output* signals as natural terms to represent the signals that are processed by the circuit. The simplest approximations to ideal filters are obtained by the choice of an appropriate frequency dependent element. Consider the general circuit of Fig. 7.27. The voltage transfer function magnitude is

$$|T| = \frac{|Z_2|}{|Z_1 + Z_2|} \tag{7.36}$$

If we let $Z_1 = R$ and Z_2 frequency dependent such that its magnitude is large at low frequencies and small at high frequencies, then Eq. 7.36 would have a large magnitude at low frequencies and a small magnitude at high frequencies. An approximate low-pass filter is thus obtained. The simplest such impedance Z_2 is that of a capacitor. Similarly, it should be observed that if the roles of Z_1 and Z_2 were reversed, the result is a high-pass filter. Here $Z_2 = R$ and $Z_1 = 1/j\,\omega C$. At high frequencies, $|T|$ is large

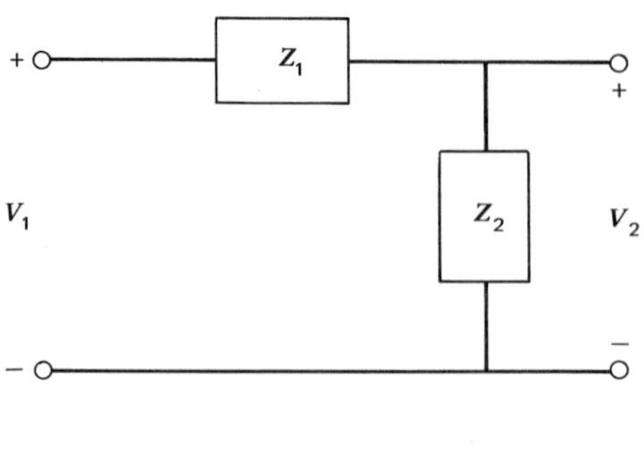

Fig. 7.27

and at low frequencies when $|Z_1|$ is very large, the value of $|T|$ is small. Let us examine these circuits in detail.

Low-Pass Filter

For the circuit of Fig. 7.28, let us derive the transfer function.

$$\frac{V_2}{V_1} = \frac{\dfrac{1}{j\omega C}}{R + \dfrac{1}{j\omega C}} = \frac{1}{1 + j\omega RC} \tag{7.37}$$

$$\left|\frac{V_2}{V_1}\right| = \frac{1}{\sqrt{1 + \omega^2 R^2 C^2}} \tag{7.38}$$

We observe, when $\omega = 0$,

$$\left|\frac{V_2}{V_1}\right| = 1$$

When ω becomes very large, the denominator becomes very large and $|V_2/V_1|$ becomes very small. Plots of $|V_2/V_1|$ as functions of ω are shown in Fig. 7.29 for different values of RC.

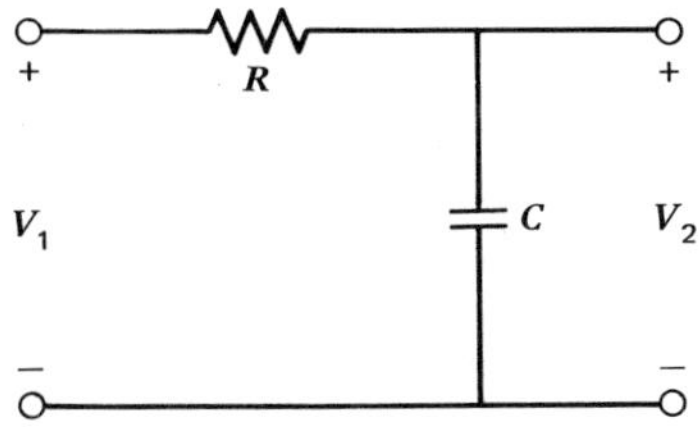

Fig. 7.28

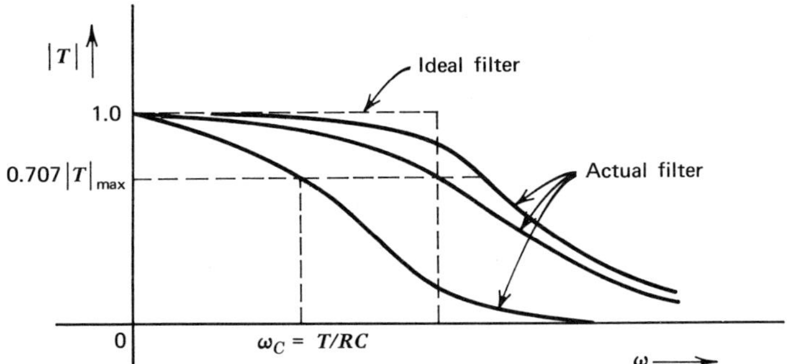

Fig. 7.29

The curves in Fig. 7.29 are crude approximations to a low-pass ideal filter curve that is shown dotted. Three curves for the approximate filter are drawn for various values of R and C. In order to arrive at the measure of bandwidth for such practical approximations, we use the following convention. The *cut-off frequency* is defined in terms of the power that would be delivered to a resistive load of one ohm at the output end. We define the cut-off frequency to be where *the power is one half the value of the maximum* power that can be delivered to the resistor.

The power that can be delivered by V_2 to a resistor of one ohm is proportional to the value of $|V_2|^2$. For comparison, we keep V_1 constant in magnitude. Thus, the power to the resistor is proportional to $|T|^2$. The maximum power is proportional to $|T|^2_{\max}$ and the cut-off frequency occurs when

$$|T(\omega_C)|^2 = \tfrac{1}{2}|T|^2_{\max}$$

or

$$|T(\omega_C)| = \tfrac{1}{\sqrt{2}}|T|_{\max} = 0.707|T|_{\max} \tag{7.39}$$

Thus, the cut-off frequency ω_C is where the transfer function magnitude is 0.707 its maximum value.

In our case, we have $|T|$ maximum at zero frequency where its value is unity. The frequency where power is one half is when $|T|$ is 0.707 of the maximum value. In Eq. 7.38, this occurs when $\omega_C{}^2R^2C^2 = 1$ or $\omega_C = 1/RC$. RC is the time constant of the series RC circuit. Therefore, *the bandwidth of a low-pass filter developed from a single RC combination is equal to the reciprocal of the time constant of the RC circuit.*

Next, let us examine a high-pass filter developed from the same two elements, a resistor and a capacitor. Here, the position of the capacitor and the resistor is interchanged.

High-Pass Filter

The transfer function for the circuit of Fig. 7.30 is

$$T = \frac{V_2}{V_1} = \frac{R}{R + \dfrac{1}{j\omega C}} = \frac{j\omega RC}{1 + j\omega RC} \tag{7.40}$$

Hence,

$$|T| = \frac{\omega RC}{\sqrt{1 + \omega^2R^2C^2}} \tag{7.41}$$

For $\omega = 0$, the numerator is zero and so $|T|$ is zero. When ω is very large,

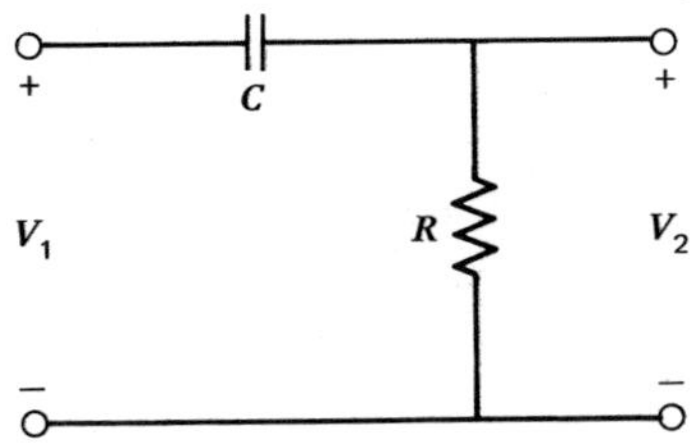

Fig. 7.30

$\omega^2R^2C^2$ becomes dominant with respect to one and

$$\sqrt{1+\omega^2R^2C^2} \simeq \sqrt{\omega^2R^2C^2} = \omega RC \tag{7.42}$$

So, for very large ω,

$$|T| = \frac{\omega RC}{\sqrt{1+\omega^2R^2C^2}} \simeq 1 \tag{7.43}$$

Finally, the maximum value occurs as $\omega \to \infty$ where

$$|T|_{\max} = 1 \tag{7.44}$$

and $0.707|T|_{\max}$ occurs again when $\omega_C{}^2R^2C^2 = 1$ or $\omega_C = 1/RC$.

Figure 7.31 shows the plot for the transfer function magnitude of Eq. 7.41. The curve represents an approximation to the ideal high-pass filter. In order to change the value of ω_C, we change the value of the time constant (RC).

All of the above are simple forms of filters. In order to achieve sharper cut-off characteristics, we must have elaborate circuits. A discussion of such circuits is beyond the scope of this book.

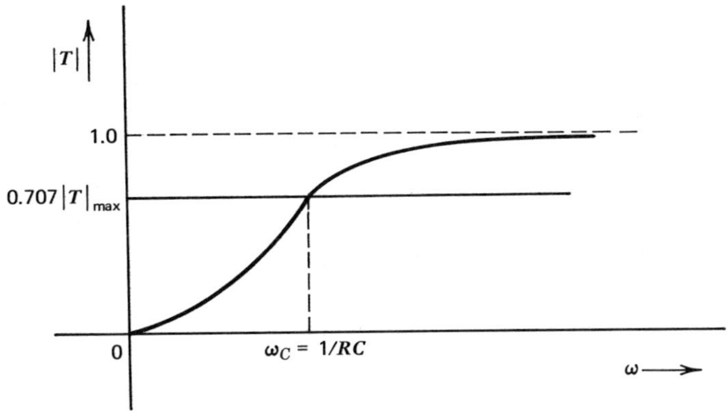

Fig. 7.31

Band-Pass Filter

The transfer function for a band-pass characteristic must have large magnitude at some intermediate frequency and low magnitudes at the low and the high frequency ends. If we can develop an impedance that is very small at the low frequency end and the high frequency end, we can achieve a band-pass characteristic for the current through it. The phenomenon known as resonance in a simple series *RLC* circuit of Fig. 7.32 gives us one such possibility.

For the circuit of Fig. 7.32, the impedance of the circuit is

$$Z = R + j\left(\omega L - \frac{1}{\omega C}\right) \tag{7.45}$$

The reactive impedance is high for high frequencies as well as low frequencies and low for some intermediate frequencies. The transfer function is

$$T = \frac{V_2}{V_1} = \frac{R}{R + j\left(\omega L - \frac{1}{\omega C}\right)} \tag{7.46}$$

and its magnitude is given by

$$|T| = \left|\frac{V_2}{V_1}\right| = \frac{R}{\sqrt{R^2 + \left(\omega L - \frac{1}{\omega C}\right)^2}} \tag{7.47}$$

Let us examine the magnitude of the transfer function at different values of ω. The transfer function magnitude is proportional to the current magnitude that depends on the magnitude of the impedance, $|Z|$. Near $\omega = 0$ and $\omega \to \infty$, $|Z|$ is infinitely large and the current and $|T|$ are very small or zero.

So, both the low and the high ends of the frequency spectrum give small values for the transfer function. At an intermediate value of frequency, we may expect to obtain a maximum value for the transfer function. The maximum occurs when the frequency dependent term in

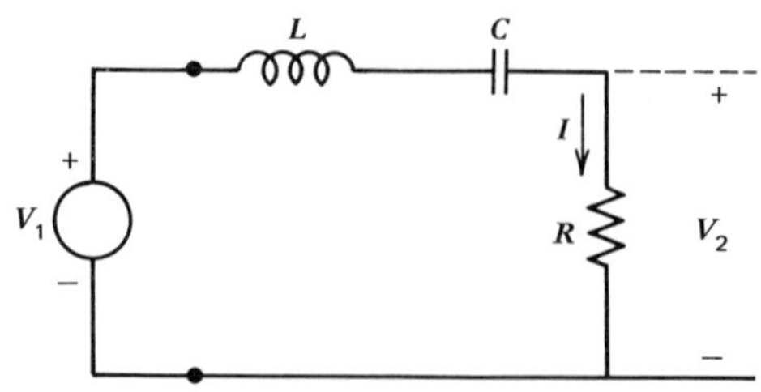

Fig. 7.32

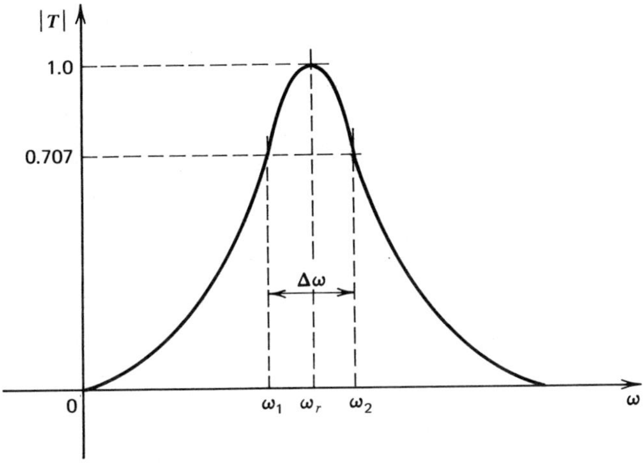

Fig. 7.33

the denominator of Eq. 7.47 is zero. That is

$$|T|_{\max} = \frac{R}{\sqrt{R^2 + \left(\omega_0 L - \dfrac{1}{\omega_0 C}\right)^2}} = 1 \tag{7.48}$$

which occurs when

$$\omega_0 L - \frac{1}{\omega_0 C} = 0$$

or

$$\omega_0 = \frac{1}{\sqrt{LC}} \tag{7.49}$$

This value of ω_0 gives the maximum value of the transfer function magnitude. At that frequency, we also have $Z = R$ and the phenomenon is known as *resonance*. The frequency ω_0 is called the *resonant frequency*. Note the impedance of the series *RLC* circuit is purely resistive at resonance. A plot of the transfer function magnitude is shown in Fig. 7.33. The shape of the curve resembles a band-pass characteristic. We examine in detail the properties of resonance in the next section. The resonant circuits play a key role in band-pass filters and a thorough study of these circuits is very important in the design of circuits.

7.5 RESONANT CIRCUITS

The resonance phenomenon provides the frequency selective behavior necessary in band-pass filters. The series *RLC* circuit used in the last section provided a transfer function magnitude approximating the ideal band-pass filter response. The magnitude of the transfer function was seen to be unity at the resonant frequency and approached zero at the low and the high frequency ends. The sharpness or speed with which the magnitude function approaches zero determines the nature of the band-pass filter.

The simple *RC* low-pass and high-pass filter characteristics were governed completely by the *RC* time constants. The center frequency of the simple *RLC* band-pass filter is the resonant frequency where the magnitude of the transfer function is unity. The resonant frequency occurs at $\omega_0 = 1/\sqrt{LC}$. The bandwidth of the filter is governed by other parameters of the resonant circuit. In this section we discuss the simple resonant circuits and develop expressions for the bandwidth and certain figures of merit associated with the filters. We study the series *RLC* circuit first, and for simplicity consider the current behavior in it. The transfer function is the current times the resistor and except for a proportionality constant, the current represents the transfer function. The resonance phenomenon in a parallel *RLC* circuit will also be discussed briefly at the end of this section.

For the series *RLC* circuit of Fig. 7.32, the impedance is given by

$$Z = R + j\left(\omega L - \frac{1}{\omega C}\right) \tag{7.50}$$

The voltage V_1 and the current I are related by

$$V_1 = ZI \tag{7.51}$$

The magnitude of the current in the circuit is given by

$$|I| = \frac{|V_1|}{\sqrt{R^2 + \left(\omega L - \frac{1}{\omega C}\right)^2}} \tag{7.52}$$

and, the maximum value of $|I|$ occurs at the resonant frequency $\omega_0 = 1/\sqrt{LC}$ where

$$|I|_{\max} = \frac{|V_1|}{R} \tag{7.53}$$

At the frequency ω_0, the impedance $Z(j\omega_0) = R$, the voltage and current are in phase with each other, the current magnitude is a maximum and the circuit is said to be in *resonance*. The frequency ω_0 is called the resonant frequency. The current magnitude plotted as a function of ω is shown in Fig. 7.34.

The cut-off frequencies ω_1 and ω_2 occur, by definition, at those points

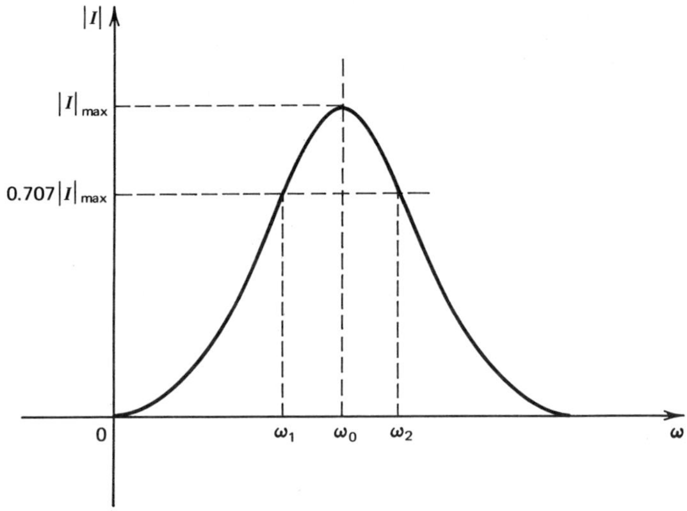

Fig. 7.34 Magnitude curve for series resonant circuit.

where the magnitude of the current phasor equals 0.707 of the maximum. We next proceed to develop expressions for ω_1 and ω_2 and determine the bandwidth $(\omega_2 - \omega_1)$.

We know that

$$|I| = \frac{|V_1|}{\sqrt{R^2 + \left(\omega L - \frac{1}{\omega C}\right)^2}} \tag{7.54}$$

At the cut-off frequencies,

$$|I| = 0.707 \frac{|V_1|}{R} = \frac{|V_1|}{\sqrt{2R^2}} \tag{7.55}$$

From Eq. 7.54, the frequency at which $|I| = |V_1|/\sqrt{2\,R^2}$ is given by the equality

$$R^2 = \left(\omega L - \frac{1}{\omega C}\right)^2$$

that is,

$$\left(\omega L - \frac{1}{\omega C}\right) = \pm R \tag{7.56}$$

The frequencies that satisfy this equation are the cut-off frequencies. Simplifying, we obtain the following two equations:

$$LC\omega^2 + RC\omega - 1 = 0 \tag{7.57}$$

$$LC\omega^2 - RC\omega - 1 = 0 \tag{7.58}$$

When we consider the first of these two equations, the two possible solutions of the quadratic equation are

$$\omega = \frac{-RC \pm \sqrt{(RC)^2 + 4LC}}{2LC} \tag{7.59}$$

In a physical system, ω must be positive. We do not consider the solution with a negative sign before the radical (which gives a negative value for ω). Hence, we have one of the cut-off frequencies as

$$\omega_1 = \frac{-RC + \sqrt{(RC)^2 + 4LC}}{2LC} \tag{7.60}$$

The second quadratic equation is

$$LC\omega^2 - RC\omega - 1 = 0$$

The values of ω satisfying this quadratic are

$$\omega = \frac{RC \pm \sqrt{(RC)^2 + 4LC}}{2LC} \tag{7.61}$$

Since ω is required to be positive in a physical system, the second cut-off frequency is given by

$$\omega_2 = \frac{RC + \sqrt{(RC)^2 + 4LC}}{2LC} \tag{7.62}$$

The difference between the two cut-off frequencies gives the bandwidth $\Delta\omega$ for the series RLC circuit. That is,

$$\begin{aligned} \Delta\omega &= \omega_2 - \omega_1 \\ &= \frac{2RC}{2LC} = \frac{R}{L} \end{aligned} \tag{7.63}$$

As a further point of interest, we note that the product of ω_1 and ω_2 yields the following expression:

$$\omega_1\omega_2 = \frac{(RC)^2 + 4LC - (RC)^2}{4L^2C^2} \tag{7.64}$$

that is,

$$\omega_1\omega_2 = \frac{1}{LC} = \omega_0{}^2$$

Hence, the resonant frequency ω_0 is the geometric mean of the two cut-off frequencies ω_1 and ω_2.

We now develop certain figures of merit or criteria to determine which circuit behaves as a better resonant circuit or band-pass filter. A series *RLC* circuit approaches an ideal resonant or oscillator circuit when the resistive part tends to zero. The bandwidth would then approach zero and, as a consequence, only the resonant frequency would be passed through and nothing else. Obviously, the resistive part in relation to the reactive part determines the quality of the circuit.

If the resistance is very small, the energy dissipated in the resistor would be negligible compared to the energy stored in the reactive part. A figure of merit can thus be defined in terms of energy as proportional to *the ratio of the energy stored in the reactive elements at any instant at resonance to energy dissipated in the resistor in one cycle.* This figure of merit, defined as the *quality factor* or simply, Q, is

$$Q = 2\pi \times \frac{\text{Energy stored at resonance}}{\text{Energy dissipated per cycle}} \tag{7.65}$$

For a series resonant circuit we can derive the figure of merit, Q, as follows. Let the current be given by

$$i(t) = |I| \cos \omega t$$

The stored energy in the inductor is

$$W_L(t) = \tfrac{1}{2} L i^2$$

or

$$W_L(t) = \tfrac{1}{2} L |I|^2 \cos^2 \omega t \tag{7.66}$$

Since the same current $i(t)$ flows through the capacitor, the voltage across the capacitor has a magnitude $|I|/\omega C$ and lags the current by 90°. Therefore,

$$v_C(t) = \frac{|I|}{\omega C} \cos\left(\omega t - \frac{\pi}{2}\right) = \frac{|I|}{\omega C} \sin \omega t \tag{7.67}$$

The energy stored in the capacitor is

$$W_C(t) = \tfrac{1}{2} C V_C^2$$

or

$$W_C(t) = \tfrac{1}{2} \frac{|I|^2}{\omega^2 C} \sin^2 \omega t \tag{7.68}$$

The total energy stored at any instant is given by

$$W = W_L + W_C$$

or

$$W = \tfrac{1}{2} L |I|^2 \left(\cos^2 \omega t + \frac{1}{\omega^2 LC} \sin^2 \omega t\right) \tag{7.69}$$

At resonant frequency where $\omega_0^2 = 1/LC$, the total energy is

$$W_r = \tfrac{1}{2}L|I|^2 \tag{7.70}$$

Equations 7.69 and 7.70 show that the total energy stored at any instant is a function of time, in general; but, at the resonant frequency it is a constant.

The energy dissipated in a cycle is the product of the average power and the period,

$$\begin{aligned} W_d &= P_{\text{av}} \times \text{period} \\ &= I_{\text{rms}}^2 R \times T \end{aligned} \tag{7.71}$$

or

$$W_d = \frac{|I|^2 R}{2f_0} \tag{7.72}$$

where

$$T = \frac{1}{f_0} = \frac{2\pi}{\omega_0} \tag{7.73}$$

and

$$I_{\text{rms}} = \frac{|I|}{\sqrt{2}} \tag{7.74}$$

The ratio of the energy stored at resonance to the energy dissipated per cycle multiplied by 2π gives the figure of merit Q. Thus

$$\begin{aligned} Q &= 2\pi \frac{\frac{1}{2}L|I|^2}{\frac{1}{2}\dfrac{R|I|^2}{f_0}} \\ &= \frac{\omega_0 L}{R} \end{aligned} \tag{7.75}$$

We see from this expression that as we reduce the resistance, the figure of merit or Q of the circuit increases. A smaller energy loss leads to a better sharpness of the curve.

The sharpness of the resonance curve is expressed in terms of *selectivity*, which is defined as the ratio of the resonant-frequency to the bandwidth,

$$\text{Selectivity} = S = \frac{\omega_0}{\Delta\omega} \tag{7.76}$$

But $\Delta\omega = R/L$ by Eq. 7.63; so

$$S = \frac{\omega_0}{R/L} = \frac{\omega_0 L}{R} = Q \tag{7.77}$$

Thus for a series R/C circuit, the selectivity is equal to the Q of the circuit.

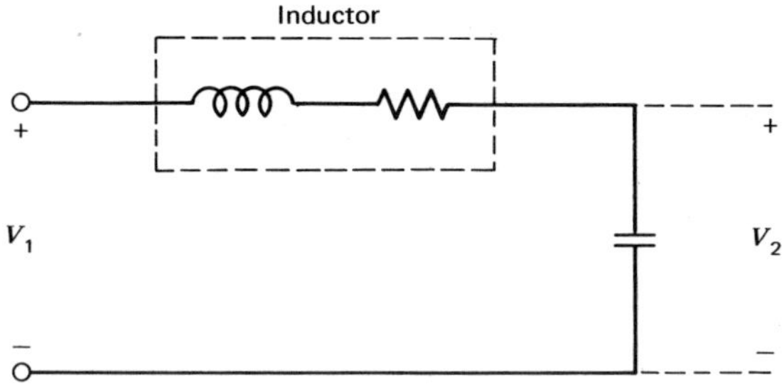

Fig. 7.35

A highly selective filter thus requires making R as small as possible. In the limit as $R \to 0$, only the frequency ω_0 gives a large current magnitude in the circuit. The resistance cannot ever be made exactly zero since the inductor coil always has some small but finite resistance. Thus a high selectivity is obtained by a capacitor in series with a *good* inductor having very small resistance. Since the resistance is now not directly available for the output voltage V_2, the output may be taken across the capacitor as shown in Fig. 7.35. The magnitude of the voltage V_2 is given by

$$|V_2| = \frac{1}{\omega C}|I| \tag{7.78}$$

which in the neighborhood of resonance behaves as $|I|$ and gives the desired highly selective characteristic.

Resonant circuits can also be obtained by a parallel combination of an inductor and a capacitor as shown in Fig. 7.36. Practical inductors always have some resistance that is included as a resistance in series with the inductance as shown.

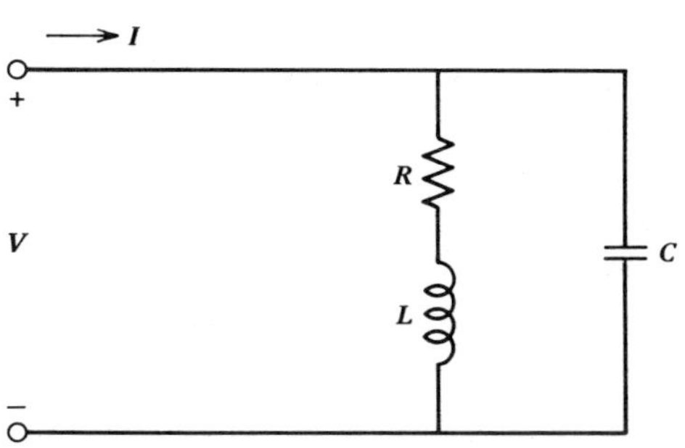

Fig. 7.36 Parallel resonant circuit.

The admittance of the circuit is given by

$$Y = j\omega C + \frac{1}{R + j\omega L} \tag{7.79}$$

$$= \frac{R}{R^2 + \omega^2 L^2} + \frac{-j\omega L}{R^2 + \omega^2 L^2} + j\omega C \tag{7.80}$$

The voltage current relationship in terms of the admittance function is

$$I = YV \tag{7.81}$$

At resonance the admittance is real and the current and voltage are in phase. This occurs when

$$\omega C - \frac{\omega L}{R^2 + \omega^2 L^2} = 0 \tag{7.82}$$

that is, when $\omega = 0$ or when

$$\omega = \frac{1}{\sqrt{LC}} \sqrt{(1 - R^2 C/L)} \tag{7.83}$$

When R is very small, $R^2 \rightarrow 0$ and

$$\omega_0 \simeq \frac{1}{\sqrt{LC}} \tag{7.84}$$

A nonzero resonant frequency thus occurs only when $R^2C/L < 1$, and it is equal to $1/\sqrt{LC}$ when R is very small. The magnitude of the voltage for a current excitation is proportional to the impedance magnitude. For small R, the impedance magnitude has a highly selective band-pass filter shape. The impedance magnitude approaches zero at the high frequency end. It is also very small at the low frequency end if R is very small. The impedance has a large magnitude at the middle frequences. The impedance magnitude may be plotted as a function of ω for specific R, L, and C, and the value of ω for which $|Z|$ is the largest obtained from the plot. The bandwidth can also be determined from the same plot.

A mathematical derivation of the frequency where $|Z|$ is the largest, and the determination of $\Delta\omega$ becomes quite unwieldy, in general. For the special case, when R is small compared to ωL in the passband of interest (i.e., high Q) we can obtain approximate values of the peak frequency and the bandwidth. In this case, $R^2 + \omega^2 L^2 \simeq \omega^2 L^2$ in Eq. 7.80, and

$$|Z| = \frac{1}{|Y|} \approx \frac{1}{\sqrt{\left(\frac{R}{\omega^2 L^2}\right)^2 + \left(\omega C - \frac{1}{\omega L}\right)^2}} \tag{7.85}$$

Assuming that the selectivity of the circuit is high, ω may be assumed approximately constant in the passband. Then $|Z|$ is approximately maximum when

$$\omega C - \frac{1}{\omega L} = 0,$$

that is, when

$$\omega = \omega_0 = \frac{1}{\sqrt{LC}} \tag{7.86}$$

Thus the peak in the response curve occurs in the vicinity of $\omega = 1/\sqrt{LC}$ that is also in the vicinity of resonance. Observe also from Eqs. 7.85 and 7.54 that the denominator expressions are similar except that R in Eq. 7.54 is replaced by R/ω^2L^2, L by C, and C by L in Eq. 7.85. However, for our approximations $R/\omega^2L^2 = R/\omega_0{}^2L^2$ is a constant where ω_0 is the resonant frequency. The bandwidth for Eq. 7.85 is then determined the same way as for Eq. 7.54 except that in the expression of Eq. 7.63, we now let $R = R/\omega_0{}^2L^2$ and interchange L and C. Thus the bandwidth in our case is

$$\begin{aligned}\Delta\omega &= \frac{R/\omega_0{}^2L^2}{C} = \frac{R}{\omega_0{}^2L^2C}\\ &= \frac{R}{L}\end{aligned} \tag{7.87}$$

since $\omega_0{}^2 = 1/LC$.

The approximate selectivity then is given by the same expression as before,

$$S = \frac{\omega_0}{\Delta\omega} = \frac{\omega_0 L}{R} = Q \tag{7.88}$$

Figure 7.37 shows the plot of $|Z|$ versus ω for the parallel resonant

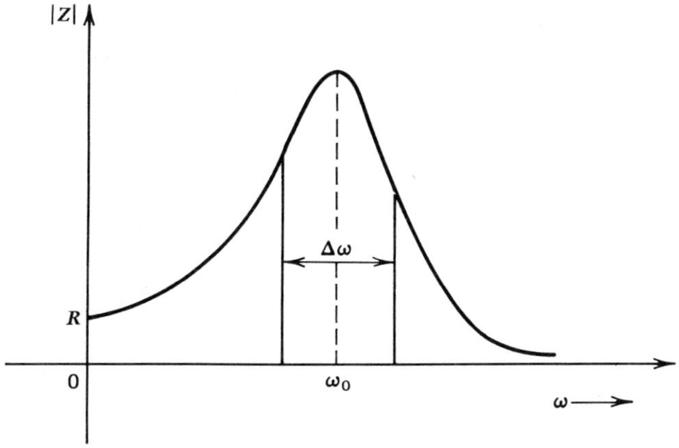

Fig. 7.37 Frequency response of a parallel resonant circuit.

circuit. The impedance approaches zero as $\omega \to \infty$ since the capacitor impedance ($1/j\omega C \to 0$). At $\omega = 0$, however, the capacitor behaves like an open-circuit ($1/j\omega C = \infty$), the inductor like a short-circuit ($j\omega L = 0$), and the impedance is that of the resistor R. The value of R is small but not zero, and thus the impedance at $\omega = 0$ is R but not zero.

Our approximations have assumed high selectivity and $R \ll \omega L$. The two conditions are equivalent from Eq. 7.88. In order for our approximations to be justified, the resulting value of S should be no less than, say, ten. Thus, for high values of S the parallel resonant circuit has the same expressions for the peak frequency, bandwidth, and selectivity as for the series resonant circuit.

7.6 DIGITAL COMPUTATION: AMPLITUDE AND PHASE RESPONSE OF THE TRANSFER FUNCTION

The linear algebraic formulation of the set of phasor equations for a certain circuit relates the circuit excitations with the multivariable phasor responses. The solution to this set of equations then expresses the responses in terms of the external excitations. However, we are usually interested in the behavior of a single phasor in response to a single source. It is then desirable to find the system function relating the particular response-excitation pair.

The system or network function is a general term defining the response-excitation ratio. It includes impedances, voltage ratios, current ratios, and so on. A general symbol $H(j\omega)$ is commonly used to denote a system function. Thus we first derive the system or network function $H(j\omega)$ for the particular response-excitation pair, then the frequency could be varied and a set of magnitude versus frequency or phase versus frequency data points obtained by using complex arithmetic operations.

For example, let us consider the circuit used in Section 6.7. This is again shown in Fig. 7.38 for convenience. The node equations describing

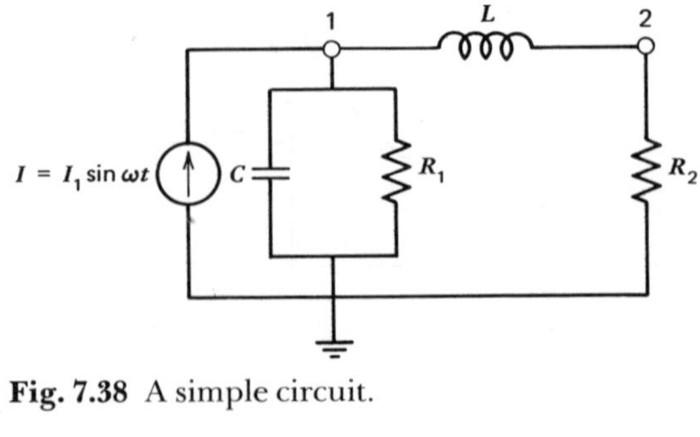

Fig. 7.38 A simple circuit.

this circuit are obtained in Section 6.7 as the following:

$$\begin{aligned} V_1\left(\frac{1}{R_1}+\frac{1}{j\omega L}+j\omega C\right)-V_2\left(\frac{1}{j\omega L}\right)&=I_1 \\ V_1\left(-\frac{1}{j\omega L}\right)+V_2\left(\frac{1}{R_2}+\frac{1}{j\omega L}\right)&=0 \end{aligned} \tag{7.89}$$

If we are only interested in the phasor response V_2, then it can be shown that the transfer function V_2/I_1 is given as

$$\frac{V_2}{I_1}=R_2\cdot\frac{\dfrac{1}{\dfrac{1}{R_1}+j\omega C}}{j\omega L+R_2+\dfrac{1}{\dfrac{1}{R_1}+j\omega C}} \tag{7.90}$$

If we set $s=j\omega$, then we have, after simplifying Eq. 7.90, the following expression:

$$\frac{V_2}{I_1}=\frac{R_1R_2}{s^2R_1LC+s(L+R_1R_2C)+R_1+R_2} \tag{7.91}$$

Thus, the steady-state response V_2 with respect to a reference phasor $I_1=|I_1|e^{j\theta}$ can be obtained by substituting a specific $j\omega$ for s.

In general, the system function, $H(s)$, relating a certain excitation-response pair can usually be written as the ratio of two polynomials in $j\omega$, or s. Let the denominator polynomial be denoted as $p(s)$ and the numerator one be denoted as $q(s)$. Then we have

$$H(s)=\frac{q(s)}{p(s)}=\frac{b_0+b_1s+b_2s^2+\cdots+b_ms^m}{a_0+a_1s+a_2s^2+\cdots+a_ns^n} \tag{7.92}$$

where the degree of $q(s)$, m, is generally less than the degree of $p(s)$, n. Thus the magnitude or phase of $H(s)$ can be evaluated by first finding the values of $p(s)$ and $q(s)$ at a particular frequency $s=j\omega$. Since $p(s)$ and $q(s)$ are to be computed many times for different values of s, it is important to do this as efficiently as possible. Straightforward term-by-term evaluation of $p(s)$ or $q(s)$ is inefficient, particularly for large values of n or m. For example, in the evaluation of $q(s)$, if each factor s^k is computed by $k-1$ repeated multiplications of s, then $m(m-1)/2$ multiplications and m additions are required to evaluate $q(s)$. If each factor is calculated by successive multiplications, ss^{k-1}, then $2m-1$ multiplications and m additions are required. However, if the *nested* form of $q(s)$

$$q(s)=b_0+s(b_1+s(b_2+\cdots+s(b_{m-1}+s(b_m)))) \tag{7.93}$$

is used, then only m multiplications and m additions are required per evaluation. This procedure of evaluating a polynomial is called *Horner's rule*. We have the following recursion formula for evaluating $q(s)$ for any specific s.

For an mth degree polynomial

$$q(s) = b_0 + b_1 s + b_2 s^2 + \cdots + b_m s^m \tag{7.94}$$

if we set

$$\begin{aligned} t_m &= b_m \\ t_{m-k} &= b_{m-k} + s t_{m-k+1} \qquad k = 1, 2, \cdots, m \end{aligned} \tag{7.95}$$

then $t_0 = q(s)$ for any specific value s.

After $p(s)$ and $q(s)$ are evaluated by using Horner's rule for any $s = j\omega$, then $H(s)$ is obtained easily by complex division. Let the result be expressed as

$$H(s) = R + jI \tag{7.96}$$

then the magnitude and phase response of $H(s)$ are simply obtained by the following formula:

$$\begin{aligned} |H(s)| &= \sqrt{R^2 + I^2} \\ \angle H(s) &= \tan^{-1}\frac{I}{R} \text{ radians} \\ &= \tan^{-1}\frac{I}{R} \times 57.29578 \text{ degrees} \end{aligned} \tag{7.97}$$

where $|H(s)|$ is the magnitude of $H(s)$ and $\angle H(s)$ is the phase of the resultant phasor $H(s)$.

Computer Program

To compute $H(s)$, we first store the coefficients of the denominator and numerator polynomials $q(s)$ and $p(s)$ in the vector arrays $Q(I)$ and $P(J)$ where $I = 1, 2, \cdots, m+1$ and $J = 1, 2, \cdots, n+1$. Then for a specific s, Horner's rule is used to compute $q(s)$ and $p(s)$. With $H(s)$ obtained, the magnitude and phase of $H(s)$ are easily obtained by using the formulas in Eq. 7.97.

A flowchart to evaluate the magnitude and phase response of $H(s)$ based on Horner's rule is given in Fig. 7.39. Note the following comment:

1. Since it is possible that $p(s)$ at a certain s is equal to zero, it is necessary to check the magnitude of $p(s)$ before the division of $q(s)/p(s)$ takes place. If $p(s) = 0$, then a message "$H(s)$ is singular" will be printed and $H(s)$ is set as $0 + j0$. The evaluation of $H(s)$ could be best accomplished by

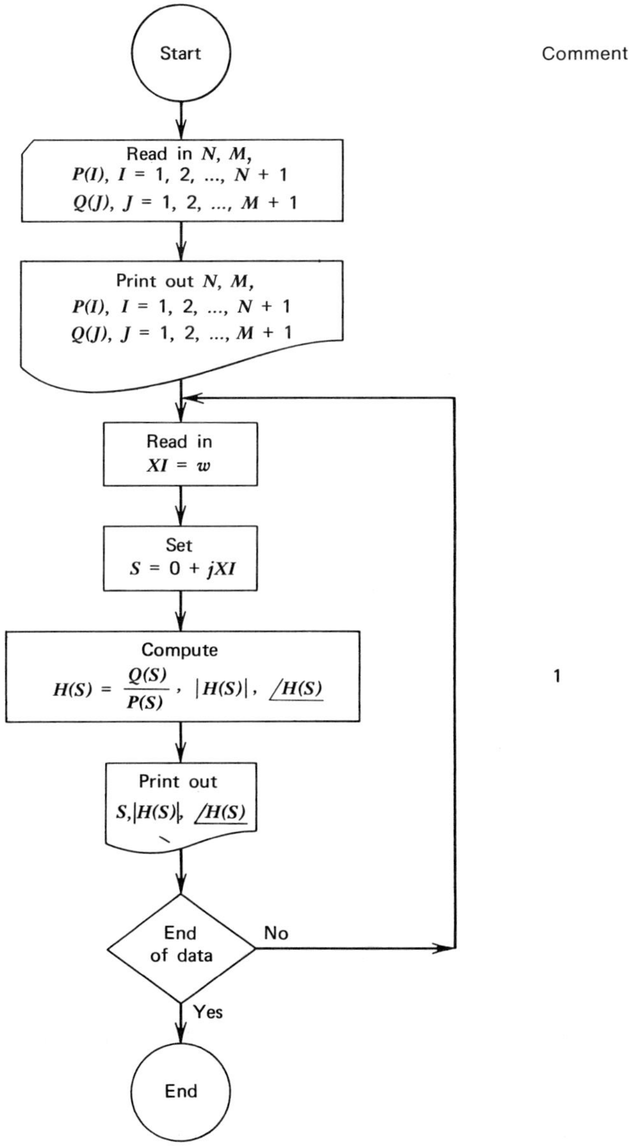

Fig. 7.39 Frequency response program.

using a subroutine AMPHA(P,Q,S,N,M,H) where P,Q are the denominator and numerator polynomials given, N and M are the respective degrees of the given polynomials. H is the complex result of $H(S)$ evaluated at $S = j\omega$ and is expressed as $|H|\angle H(S)$ with $|H|$ stored in the real part of H and $\angle H(S)$ stored in the imaginary part of H.

Example 7.6.1 Find the amplitude and phase response of the node voltage V_2 with respect to the current excitation I in the circuit of Fig. 7.38. The frequency range is from $\omega = 1$ rad/sec to $\omega = 20$ rad/sec with increment of 1 rad/sec. The circuit parameters are set as $R_1 = R_2 = L = C = 1$. Hence the given $H(s)$ is

$$H(s) = \frac{1}{s^2 + 2s + 2}$$

Therefore we have

$$q(s) = 1 \qquad m = 0$$
$$p(s) = 2 + 2s + s^2 \qquad n = 2$$

Solution. The FORTRAN program, the subroutine AMPHA, the input data listing and the computer output are shown in Figs. 7.40 to 7.43. The

```
  DIMENSION P(21),Q(20)
1 FORMAT(2I5)
2 FORMAT(F10.5)
3 FORMAT(10X,2I5)
4 FORMAT(10X,E14.5)
5 FORMAT(10X,2E14.5,2E14.5)
  COMPLEX S,H
  READ 1, N,M
  PRINT 3, N,M
  NI=N+1
  MJ=M+1
  READ 2, (P(I), I=1,NI)
  READ 2, (Q(J), J=1,MJ)
  PRINT 4, (P(I), I=1,NI)
  PRINT 4, (Q(J), J=1,MJ)
  XR=0.
6 READ 2, XI
  S=CMPLX(XR,XI)
  CALL AMPHA(P,Q,S,N,M,H)
  PRINT 5, S,H
  GO TO 6
  END
```

Fig. 7.40 The main program for amplitude-phase response.

parameters are the following:

P Vector storage for the $P(S)$ polynomial of $H(S) = Q(S)/P(S)$
Q Vector storage for the $Q(S)$ polynomial of $H(S) = Q(S)/P(S)$
S Imaginary frequency variable

H	Storage for $H(S) = \lvert H(S)\rvert \angle H(S)$
NI	$N+1$
MJ	$M+1$
XR	Real part of $S=0$
XI	Imaginary part of *S*
KN	$N+1$
PN	Value of $P(S)$ for a specific *S*
KM	$M+1$
QM	Value of $Q(S)$ for a specific *S*
HM	Magnitude of $H(S)$
HP	Phase of $H(S)$

```
      SUBROUTINE AMPHA(P,Q,S,N,M,H)
   10 FORMAT(17H H(S) IS SINGULAR)
      DIMENSICN P(21),Q(20)
      COMPLEX S,H,PN,QM
      KN=N+1
      PN=P(KN)
      CO 1 I=1,N
        PN=S*PN+P(KN-I)
    1 CONTINUE
      KM=M+1
      QM=Q(KM)
      IF(KM-1) 4,4,2
    2 CO 3 J=1,M
        QM=S*QM+Q(KM-J)
    3 CCNTINUE
    4 IF(CABS(PN)-0.C) 5,6,5
    6 PRINT 1C
      H=(0.C,C.C)
    5 H=QM/PN
      HM=CABS(H)
      HP=ATAN(AIMAG(H)/REAL(H))*57.2957795
      IF(REAL(H)) 7,8,8
    7 HP=HP+180.00
    8 H=CMPLX(HM,HP)
      RETURN
      END
```

Fig. 7.41 The subroutine for amplitude-phase computation.

```
          n     m
          2     0

P(1)      0.20000E 01
P(2)      0.20000E 01
P(3)      0.10000E 01
Q(1)      0.10000E 01
```

Fig. 7.42 Input data listing.

Re(S)	Im(S)	|H(S)|	∠H(S)
0.0	0.10000E 01	0.44721E 00	-0.63435E 02
0.0	0.20000E 01	0.22361E 00	0.24343E 03
0.0	0.30000E 01	0.10847E 00	0.22060E 03
0.0	0.40000E 01	0.62017E-01	0.20974E 03
0.0	0.50000E 01	0.39873E-01	0.20350E 03
0.0	0.60000E 01	0.27735E-01	0.19944E 03
0.0	0.70000E 01	0.20391E-01	0.19659E 03
0.0	0.80000E 01	0.15617E-01	0.19447E 03
0.0	0.90000E 01	0.12342E-01	0.19284E 03
0.0	0.10000E 02	0.99980E-02	0.19153E 03
0.0	0.11000E 02	0.82633E-02	0.19047E 03
0.0	0.12000E 02	0.69438E-02	0.18959E 03
0.0	0.13000E 02	0.59167E-02	0.18885E 03
0.0	0.14000E 02	0.51018E-02	0.18821E 03
0.0	0.15000E 02	0.44443E-02	0.18766E 03
0.0	0.16000E 02	0.39061E-02	0.18718E 03
0.0	0.17000E 02	0.34601E-02	0.18676E 03
0.0	0.18000E 02	0.30864E-02	0.18638E 03
0.0	0.19000E 02	0.27700E-02	0.18604E 03
0.0	0.20000E 02	0.25000E-02	0.18574E 03

Fig. 7.43 Computer output.

PROBLEMS

7-1. For the circuit of Fig. P7-1 the excitation and responses are all expressed as phasors.

(i) Find the phasor ratios V_{L2}/V, V_C/V, I_{L1}/V.

(ii) Find V_{L2}, V_C, I_{L1} for $V = V\angle 0$. What happens for the cases $w = 0$ and $w = \infty$?

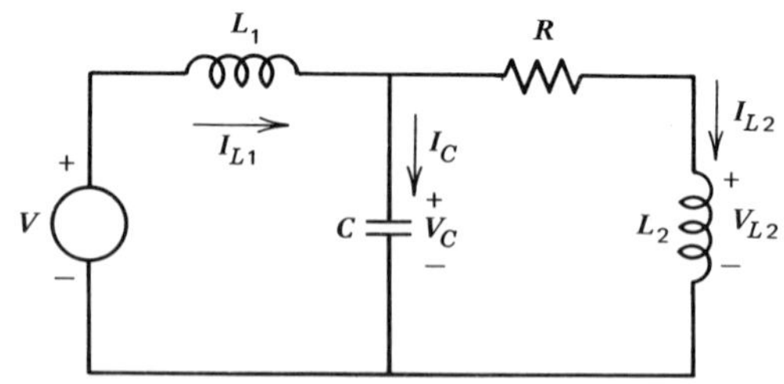

Fig. P7-1

7-2. Thévenin's theorem is often used to find the steady-state response of a specific portion of the network. The procedure is as follows. First, this specific part is taken *away* from the network and then the voltage across the disconnected terminals is mea-

sured; this is the equivalent source V_e; next, the excitations and initial states are all eliminated and the driving-point impedance of the disconnected terminals is obtained, this is the equivalent impedance Z_e; finally the series connection of the equivalent source and the equivalent impedance is added to the specific element for further calculations.
For the circuit of Fig. P7-1, find the equivalent source and impedance V_e and Z_e for the response of
(i) R
(ii) L_1
(iii) C
(iv) L_2.

7-2a. Solve 7-1(i) using the results of 7-2. Are the results identical?

7-3. For the circuit of Fig. P7-3, find V_L using Thévenin's theorem.

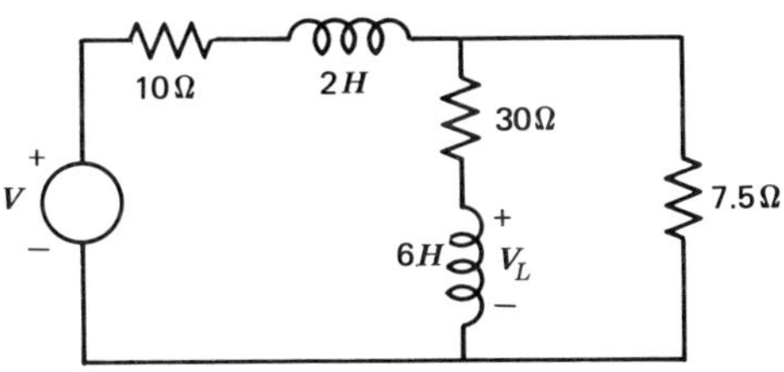

Fig. P7-3

7-4. Repeat 7-2 if the voltage source V of Fig. P7-1 is changed to current source I pointing to the positive terminal of V.

7-5. Repeat 7-3 for a current source I.

7-6. Repeat 7-3 using Norton's theorem.

7-7. Repeat 7-5 using Norton's theorem.

7-7a. Solve 6-13 using Thévenin's theorem.

7-7b. Solve 6-14 using Thévenin's theorem.

7-8. The Norton's and Thevenin's theorems are in fact related by the following equation:

$$I_e = V_e/Z_e$$

where I_e is the equivalent current source to be connected in parallel with Z_e. This equation can be used to find I_e if V_E and Z_e are known.

(i) For the circuit of Fig. P7-8, find Z_e and V_e for the elements R_3 and R_4.

(ii) Repeat (i) using Norton's theorem. Is it possible to obtain I_e by directly shorting the terminals of R_4?

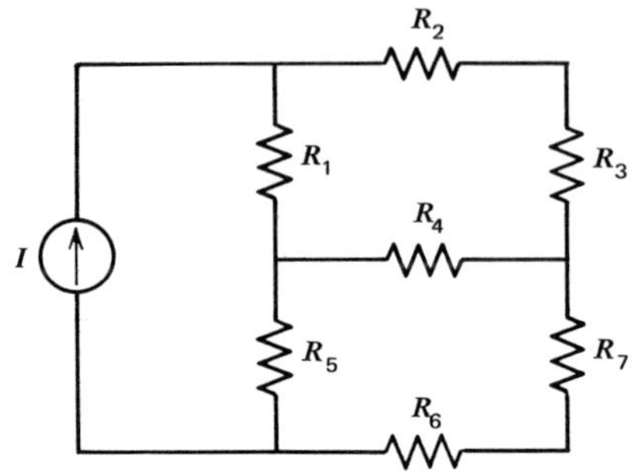

Fig. P7-8

7-8a. Find the transfer function V_0/V_i for the circuit shown in Fig. P7.8*a* using Thévenin's theorem. The circuit parameters are such that $R/L = 1/RC = a$. Discuss the case when $R = \infty\Omega$.

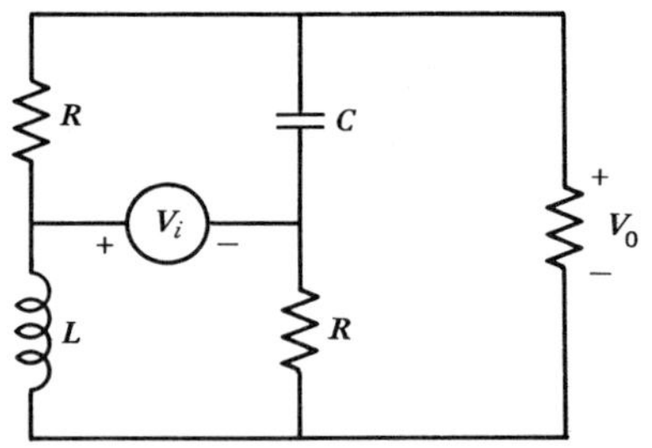

Fig. P7-8*a*

7-9. For the circuit of Fig. P6-17, the voltage across R is the desired response.
(i) Find Z_e and V_e using Thévenin's theorem.
(ii) Find V_R for $V_1 = 2\angle 0$ and $R = \frac{1}{4}\Omega$, 10ω, $\infty\Omega$.

7-10. For the circuits of Fig. P7-10, the impedances and admittances are related by the following equations:

$$Z_a = \frac{Y_A}{Y_A Y_B + Y_B Y_C + Y_C Y_A}$$

$$Z_b = \frac{Y_B}{Y_A Y_B + Y_B Y_C + Y_C Y_A}$$

$$Z_c = \frac{Y_C}{Y_A Y_B + Y_B Y_C + Y_C Y_A}$$

(i) If the response at 2-2′ is desired, find the Thévenin V_e and Z_e for the two circuits of Fig. P7-10 with a current source at the terminal pair 1-1′.

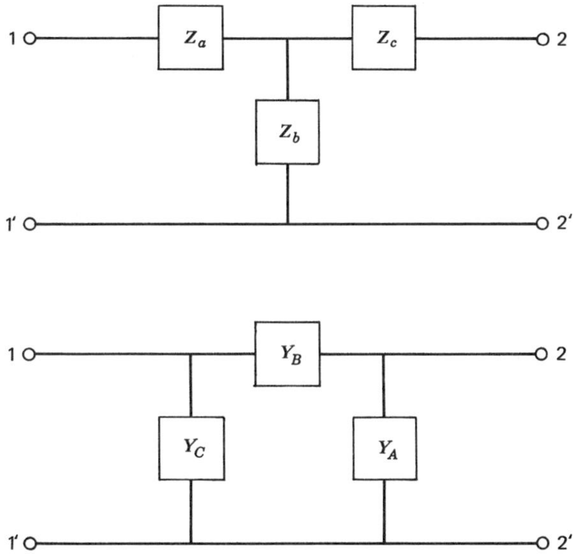

Fig. P7-10

(ii) Repeat (i) for a voltage source at 1-1′.

(iii) Repeat (i) and (ii) if the response at 1-1′ is desired.

(iv) Using the identities given relating Z_a, Z_b, and Z_c with Y_A, Y_B, and Y_C, show that these two circuits are equivalent. In other words, the Z_e and V_e are identical for same excitations at either 1-1′ or 2-2′ terminals of these two circuits.

(v) Using the identities given derive another set of expressions such that Y_A, Y_B, Y_C could be calculated in terms of Z_a, Z_b and Z_c.

The identities relating Y_A, Y_B, Y_C and Z_a, Z_b, Z_c are the so-called Δ-Y transformations.

7-10a. Find the transfer function V_2/I_1 of the circuit of Fig. P7-10a. How would you classify it as a filter?

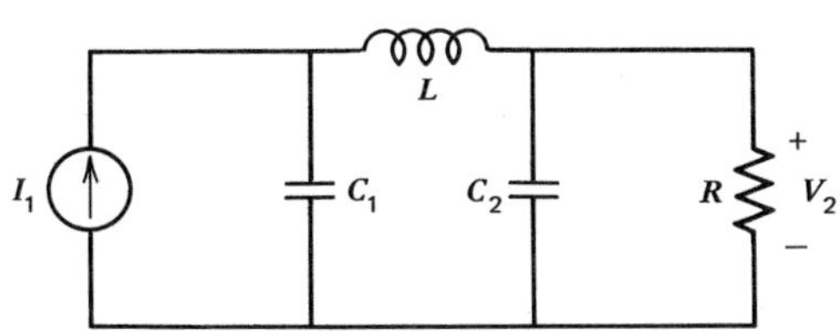

Fig. P7-10*a*

7-10b. Repeat 7-10a for the circuit shown in Fig. P7-10*b*.

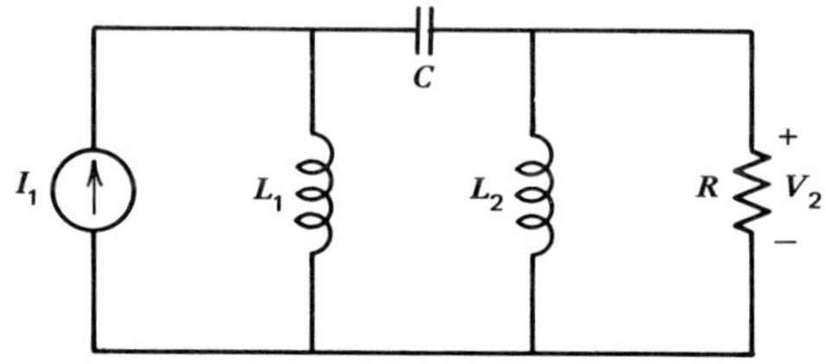

Fig. P7-10*b*

7-10c. Repeat 7-10a for the circuit shown in Fig. P7-10*c*.

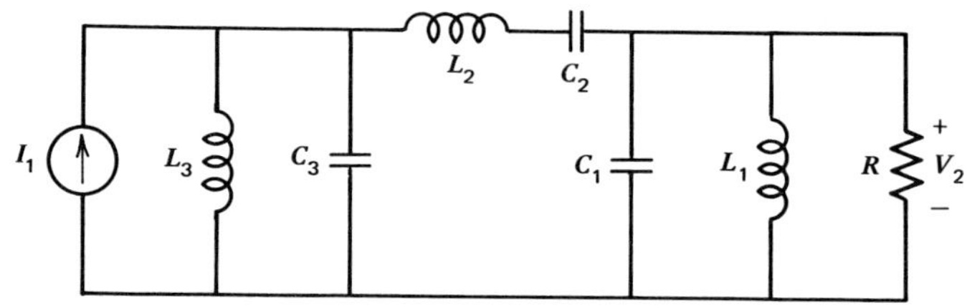

Fig. P7-10*c*

7-10d. Find the driving-point transfer function I_1/V_1 of the circuit shown in Fig. P7-10*d* as a function of L_1 and L_2.

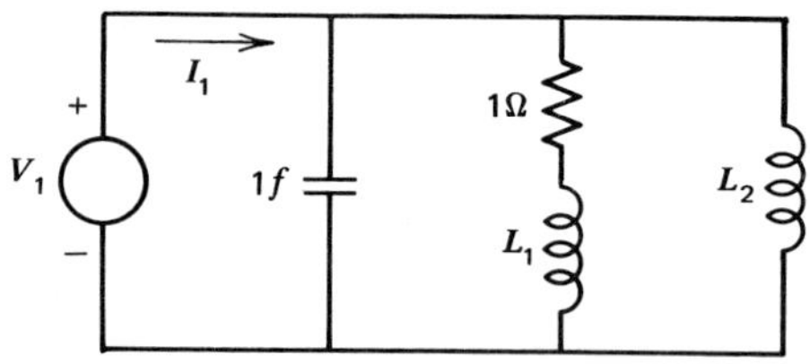

Fig. P7-10*d*

7-10e. For the circuit of Fig. P7-10*d*, let $L_2 = 2H$ and $\omega = 1$ rad./sec. Determine the values of L_1 which will cause resonance for I_1.

7-10f. Repeat 7-10e for $w = 1/2$ rad./sec.

7-10g. Repeat 7-10e if $L_1 = 2H$ and L_2 is a variable for $w = 1$ rad./sec.

7-11. For the circuit of Fig. P7-11:

(i) Find the transfer functions $H_L = V_L/V$, and $H_C = V_C/V$.

(ii) Find the frequencies $w_{\max}$ such that $|H_L|$ and $|H_C|$ attain their maxima.

(iii) State the condition that must be satisfied to achieve maxima for V_L and V_C. What happens if $R = 0$?

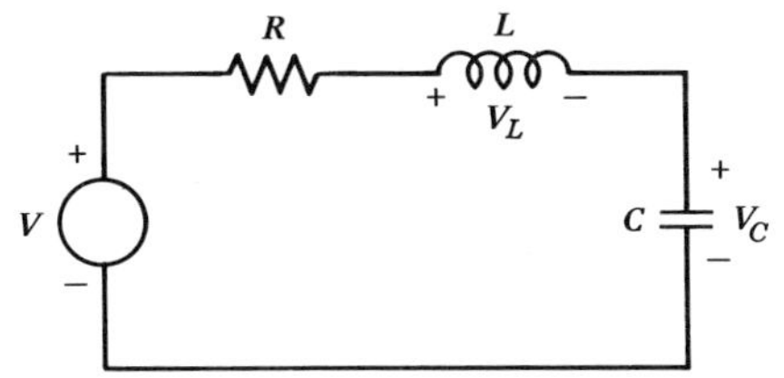

Fig. P7-11

7-12. For the amplitude responses of V_L and V_C of problem 7-11, let $R = 1\Omega, C = 0.1F, L = 5H$. Find
  (i) the cut-off frequencies,
  (ii) the bandwidths,
  (iii) the selectivities.

7-13. Find the magnitude of V_L and V_C of 7-12 at $w = w_{\max}$. Are they larger than the input voltage V? Why?

7-14. For the circuit of Fig. P7-14:
  (i) Find the transfer impedance $Z_{21} = V_2/I_1$. Determine whether this circuit is a low-pass or high-pass filter.
  (ii) Determine $w_{\max}$ and the pass-band for Z_{21}.

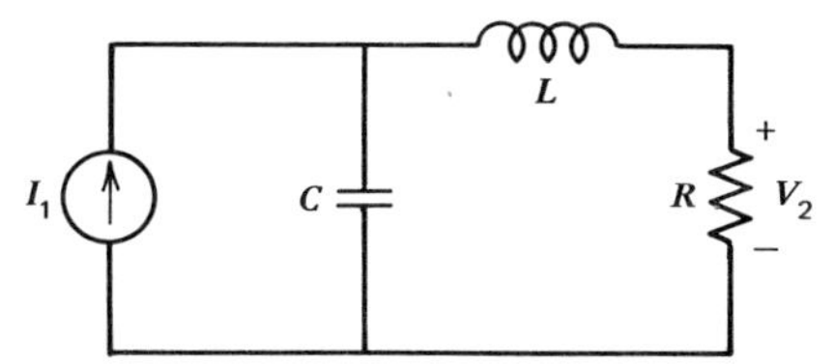

Fig. P7-14

7-15. For the circuit of Fig. P7-15:
  (i) Determine whether it is a high-pass or low-pass filter by finding the transfer impedance $Z_{21} = V_2/I_1$.
  (ii) Determine $w_{\max}$ and the pass-band or stop-band for Z_{21}.

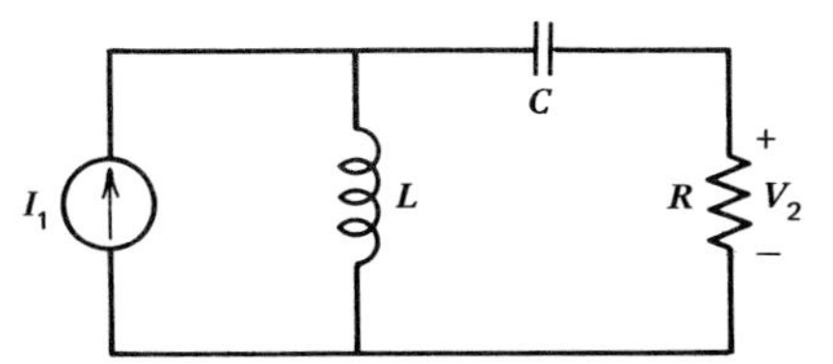

Fig. P7-15

7-16. For the three stage RC low-pass filter shown in Fig. P7-16:
  (i) Find the transfer function $H = V_o/V_i$.
  (ii) Repeat (i) if only two stages of RC circuit is used.
  (iii) Determine the cut-off frequencies of the transfer functions

of (i) and (ii). Is it true that the cut-off frequency will increase as the number of stages of RC circuit decreases?

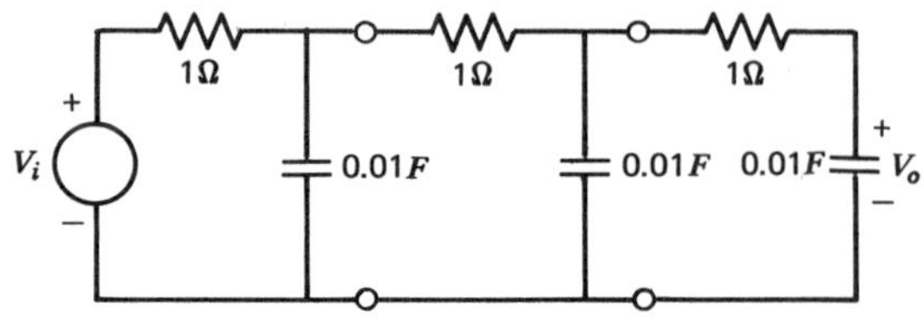

Fig. P7-16

7-17. It is possible for the driving point impedance of a general network to be purely resistive. Consider the circuit of Fig. P7-17.

(i) Find the driving-point impedance $Z = V/I$.

(ii) Draw phasor diagrams for I, I_C, and I_L.

(iii) Repeat (ii) for V, V_C, V_{R1}, V_L and V_{R2}.

(iv) Find the restrictions on R_1, R_2, L, and C so that Z is a pure resistance.

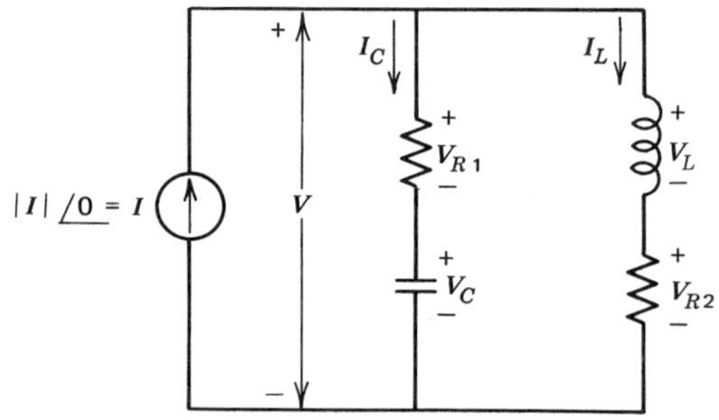

Fig. P7-17

7-17a. The Bode plots of the transfer function $H(s)$ are defined as plots of

$$20 \log_{10} |H(jw)|_{\text{in decibel } (db)} \text{ vs. } \log_{10} w$$

and

$$\angle H(jw)_{\text{in degrees}} \text{ vs. } \log_{10} w$$

for positive w's. This practice will allow greater ranges for both the frequency and the magnitude of the frequency responses of $H(s)$. Find approximate Bode plots for the frequency range of $1 \leqslant \omega \leqslant 1000$ for each of the following transfer functions by finding first the magnitude and phase responses at a discrete set of frequencies and then connecting them with a smooth curve. How would you classify them as filters?

(i) $H(s) = 10$
(ii) $H(s) = 1/s$
(iii) $H(s) = 10 + s$
(iv) $H(s) = 1/(10 + s)$

7-17b. The transfer function $H(s)$ is sometimes given as the ratio of two polynomials in factored form

$$H(s) = \frac{p(s)}{q(s)} = \frac{K_q(s - z_0)(s - z_1) \cdots (s - z_m)}{K_p(s - p_0)(s - z_1) \cdots (s - z_n)}$$

then the Bode plots can be expressed as

$$\begin{aligned} 20 \log_{10}|H(jw)|_{\text{in } db} = 20 \log_{10}|K_q| &+ 20 \log_{10}|jw - z_0| + \cdots \\ &+ 20 \log_{10}|jw - z_m| \\ - 20 \log_{10}|K_p| &- 20 \log_{10}|jw - p_0| - \cdots \\ &- 20 \log_{10}|jw - p_n| \end{aligned}$$

and

$$\begin{aligned} \angle H(jw) = {} & \angle K_q + \angle(jw - z_0) + \angle(jw - z_1) + \cdots + \angle(jw - z_m) \\ & - \angle K_p - \angle(jw - p_0) - \angle(jw - p_1) - \cdots - \angle(jw - p_n) \end{aligned}$$

Repeat 7-17a for each of the following transfer functions:

(i)

$$H(s) = \frac{1}{s(s+1)}$$

(ii)

$$H(s) = \frac{s}{(s+1)(s+2)}$$

(iii)

$$H(s) = \frac{(s+1)(s+3)}{(s+2)(s+4)}$$

(iv)

$$H(s) = \frac{1}{s^2(s+1)}$$

7-17c. Using the subroutine AMPHA, find the amplitude and phase responses of the transfer function $H(s) = V_L/V$ of the circuit of Fig. P7-3. Choose an appropriate frequency range.

7-18. Find the magnitude and phase responses of the transfer ratio $H = V_0/V_i$ of the circuit of Fig. P7-16. Use the subroutine AMPHA for a frequency range of $w = 0$ to $w = 100$ rad/s with step size $\Delta w = 1$ rad/sec.

7-18a. Repeat 7-18 for the driving-point impedance Z of the circuit of Fig. P7-17 with $R_1 = R_2 = 1\Omega$ and $L = 1H$ and $C = 1F$.

7-18b. Find the amplitude and phase responses for each of the following transfer functions for a frequency range of $w = 0$ to $w = 100$ rad/sec with step size $\Delta w = 1$ rad/sec. using the subroutine AMPHA.

(i)

$$H(s) = \frac{1+2s+3s^2}{4+5s+6s^2+7s^3}$$

(ii)

$$H(s) = \frac{1+s^2}{4+6s+4s^2+s^3+s^4}$$

7-19. For the circuit of Fig. P7-19, form a set of loop equations and find the transfer ratio $H = V_2/V_1$.

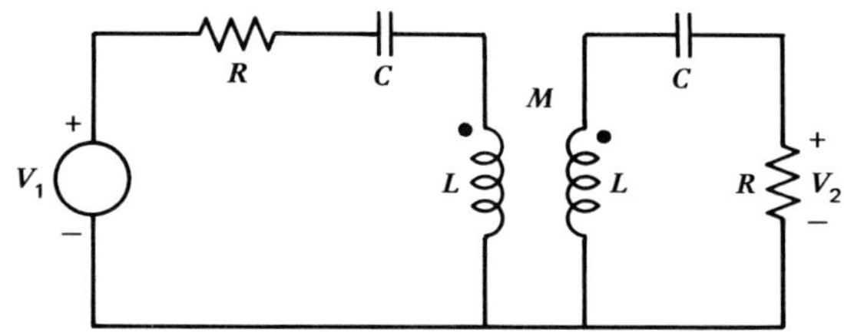

Fig. P7-19

7-20. For the circuit of Fig. P7-19, we have $C = 0.01F$, $L = 1H$, $R = 1\Omega$. For $w = 1$ to $w = 20$ rad/sec. with increment $\Delta w = 0.1$ rad/sec., find the phase and amplitude responses of the transfer ratio H found in 7-19. Use the following set of M values:

(i) $M = 0.01H$,
(ii) $M = 0.1H$,
(iii) $M = 0.5H$.

Determine whether this is a low-pass, high-pass, or band-pass filter. Find the bandwidth of the circuit if it is a bandpass filter.

7-20a. Modify the subroutine AMPHA so that its output could be used for Bode plots. Use this modified subroutine for the transfer functions given in 7-18b.

7-20b. Repeat 7-20a for the transfer functions given in 7-17b.

7-20c. Write a FORTRAN subroutine to implement the computation of data for the Bode plots. Specifically, the polynomials $p(s)$ and $q(s)$ are given in terms of the sets of constants

$$K_p, p_0, p_1, \ldots, p_n$$

and

$$K_q, z_0, z_1, \ldots, z_n$$

respectively. The outputs are required for a set of w's.

7-20d. Use the resultant subroutine of 7-20c to find the Bode plots for each of the transfer functions given in 7-17b. Compare with the results found in 7-17b.

CHAPTER 8
Fourier Analysis

We have discussed in the previous chapters how to find the responses of a linear circuit subject to various excitations. The excitation waveforms are primarily restricted to either step functions or exponential functions. However, there exist periodic or aperiodic functions that are neither step functions nor exponential functions and are used frequently as excitations to a circuit. For example, the voltages representing the voice of a person using a telephone or the voltages generated from the pick-up head of a record player are all of this type. So the natural question is: Do we have to develop a particular method for finding the response of a circuit each time when there is a new excitation function?

Fortunately the answer is no. Otherwise it would be meaningless to devote our previous chapters in discussing the circuit responses subject only to those two types of signals. And it would certainly need thousands of volumes of books just to list the responses of a simple circuit due to various excitation functions. The reasons for the answer are: first, the superposition principle for linear circuits allows us to construct the total response due to several excitation sources by adding the responses due to each source alone; second, *Fourier analysis*, which we will study in this chapter, allows us to *decompose* any arbitrary excitation function into *components*, and it is easier to find the responses of the circuit due to these components. In fact, these components are exponential functions and the methods used in the previous chapters can all be used to determine the circuit response due to these components. We will discuss in this chapter the methods for decomposing an arbitrary excitation function that fall under the classification of *Fourier analysis*.

8.1 FOURIER SERIES FOR PERIODIC FUNCTIONS

One of the most common classes of signals encountered are periodic signals. If T is the period of the signals, then

$$s(t) = s(t \pm nT) \qquad n = 0, 1, 2, \ldots \tag{8.1}$$

In addition to being periodic, if $s(t)$ has only a finite number of discontinuities in any finite period, and if the integral

$$\int_{\alpha}^{\alpha+T} |s(t)|\, dt$$

is finite (where α is an arbitrary constant), it can be shown that $s(t)$ can be expanded into an infinite trigonometric series,

$$s(t) = \frac{a_0}{2} + a_1 \cos \omega t + a_2 \cos 2\omega t + \cdots + b_1 \sin \omega t + b_2 \sin 2\omega t + \cdots \tag{8.2}$$

where $\omega = 2\pi/T$. This trigonometric series is generally referred to as the *Fourier series*. In compact form, the Fourier series is

$$s(t) = \frac{a_0}{2} + \sum_{n=1}^{\infty} (a_n \cos n\omega t + b_n \sin n\omega t) \tag{8.3}$$

Let us look at an example to see what Eq. 8.3 really means. Suppose we have a fully rectified sine wave shown in Fig. 8.1. As we observe, the period is $T = \pi$. For this signal, there are no discontinuities in any finite period. Also we see that the integral

$$\int_{\alpha}^{\alpha+T} |s(t)|\, dt$$

is finite for any α. Thus $s(t)$ can be expanded into an infinite series. Without going into the details of *how* the results are calculated, (which we will

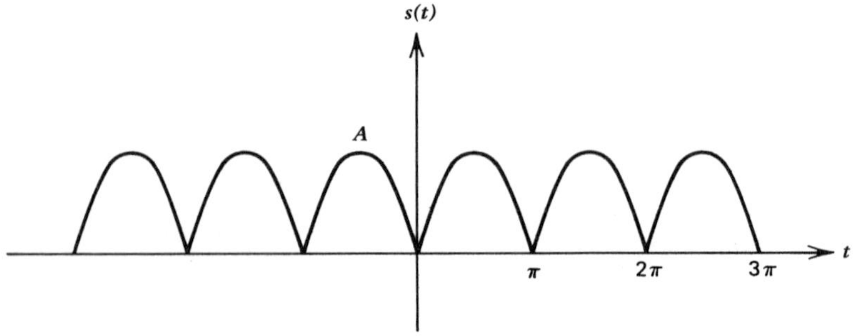

Fig. 8.1 A rectified sine wave.

supply later) the Fourier coefficients of the rectified sine wave are:

$$b_n = 0 \qquad n = 0, 1, 2, \ldots$$

$$a_0 = \frac{4A}{\pi}$$

$$a_n = \frac{1}{1-4n^2}\frac{4A}{\pi} \qquad n = 0, 1, 2, \ldots$$

For simplicity, let $A = \pi$, hence the Fourier series of the rectified sine wave can be written as

$$s(t) = 2\left(1 + \sum_{n=1}^{\infty} \frac{2}{1-4n^2} \cos 2nt\right) \tag{8.4}$$

Consider the implications of the Fourier series expansion using this example. The left-hand side of Eq. 8.4 is the rectified sine wave shown in Fig. 8.1. On the right-hand side is an infinite series of a sinusoidal function (a special form of exponential function).

The equality sign tells us that the sum of this infinite series is equal to the rectified sine wave for every value of t. Hence if we plot the various sinusoidal functions on the right-hand side and then add them together for each t, then the right-hand side is approaching the left-hand side as more terms are added up on the right-hand side. Let us look at a particular instant of time, say $t = \pi/2$. Then we have the following results for different partial sums on the right-hand side of Eq. 8.4.

No. of Function Added	Results of Partial Sum
1	2
2	3.3333
3	3.0667
4	3.1809
5	3.1175
6	3.1579
7	3.1299
8	3.1505
9	3.1349
10	3.1473
11	3.1373
12	3.1443

Note that the exact value for $s(t)$ on the left-hand side of Eq. 8.4 is π or 3.1416, hence we see the partial sum is indeed approaching π as more terms are included in the partial sum. We conclude that the Fourier series expansion of any periodic function is an approximation to the function in terms of a summation of sinusoids.

We see from Eq. 8.3 that once $s(t)$ is expanded in a Fourier series, we can describe $s(t)$ completely in terms of the coefficients of its harmonic terms, $a_0, a_1, a_2, \ldots, b_1, b_2, \ldots$. These coefficients constitute a *frequency domain* description of the signal. In other words, we are looking at the signal $s(t)$ from the frequency point of view. Normally for a signal $s(t)$, if the value of $s(t)$ for each t is defined, then we are sure that $s(t)$ is completely defined. Now if the coefficients a_i, b_i are defined for all i then $s(t)$ is also completely defined. Our task now is to find the equations for the coefficients a_i, b_i in terms of the given signal $s(t)$. It can be shown that if $s(t)$ is expandable into Fourier series, then we have

$$a_0 = \frac{2}{T}\int_{\alpha}^{\alpha+T} s(t)\,dt \tag{8.5}$$

$$a_n = \frac{2}{T}\int_{\alpha}^{\alpha+T} s(t)\cos n\omega t\,dt \qquad n = 1, 2, \ldots \tag{8.6}$$

$$b_n = \frac{2}{T}\int_{\alpha}^{\alpha+T} s(t)\sin n\omega t\,dt \qquad n = 1, 2, \ldots \tag{8.7}$$

where α is any real constant. Furthermore, when $s(t)$ contains a jump discontinuity, say at t_0, then $s(t)$ is not defined at $t = t_0$; however, the Fourier series expansion of $s(t)$, or the right-hand side of Eq. 8.3, does have a fixed value of

$$\tfrac{1}{2}(s(t_0+) + s(t_0-)) \tag{8.8}$$

At any point t_1 that $s(t)$ is differentiable (thus naturally continuous) then both sides of Eq. 8.3 are indeed equal.*

As an example, let us return to our previous example. We wish to find the Fourier coefficients of the fully rectified sine wave in Fig. 8.1. As we observe, the period is $T = \pi$ so that the fundamental frequency $\omega = 2\pi/T = 2$. The signal is given as

$$s(t) = A|\sin t| \tag{8.9}$$

Let us take $\alpha = 0$ and so we have

$$a_0 = \frac{2A}{\pi}\int_0^{\pi} \sin t\,dt = \frac{4A}{\pi} \tag{8.10}$$

$$a_n = \frac{2}{\pi}\int_0^{\pi} s(t)\cos 2nt\,dt$$

*For a proof see H. F. Davis, *Fourier Series and Orthogonal Functions*, Allyn & Bacon, Boston, 1963, pp. 92–95.

$$= \frac{1}{1-4n^2}\frac{4A}{\pi} \tag{8.11}$$

$$b_n = \frac{2}{\pi}\int_0^{\pi} s(t)\sin 2nt\,dt = 0 \tag{8.12}$$

Thus the Fourier series of the rectified sine wave is

$$s(t) = \frac{4A}{\pi}\left(1+\sum_{n=1}^{\infty}\frac{2}{1-4n^2}\cos 2nt\right) \tag{8.13}$$

which simplifies to Eq. 8.4 if we set $A = \pi$.

8.2 EVALUATION OF FOURIER COEFFICIENTS

Although the general formula for finding the Fourier coefficients in Eqs. 8.5 to 8.7 is applicable for any periodic function $s(t)$, it is possible to simplify the operations if $s(t)$ is of a certain type. We will discuss in this section some methods to simplify the evaluation of Fourier coefficients. In addition, we will also consider Fourier series in exponential form.

First, let us examine how the evaluation of coefficients is simplified by symmetry considerations. From Eqs. 8.5 to 8.7, let us take $\alpha = -T/2$ and represent the integrals as the sum of two separate parts, that is,

$$\begin{aligned} a_n &= \frac{2}{T}\left[\int_0^{T/2} s(t)\cos n\omega t\,dt + \int_{-T/2}^{0} s(t)\cos n\omega t\,dt\right] \\ b_n &= \frac{2}{T}\left[\int_0^{T/2} s(t)\sin n\omega t\,dt + \int_{-T/2}^{0} s(t)\sin n\omega t\,dt\right] \end{aligned} \tag{8.14}$$

Since the variable (t) in the above integrals is a dummy variable, let us substitute $x = t$ in the integrals with limits $(0; T/2)$ and let $x = -t$ in the integrals with limits $(-T/2; 0)$. Then we have

$$\begin{aligned} a_n &= \frac{2}{T}\int_0^{T/2} [s(x)+s(-x)]\cos n\omega x\,dx \\ b_n &= \frac{2}{T}\int_0^{T/2} [s(x)-s(-x)]\sin n\omega x\,dx \end{aligned} \tag{8.15}$$

Suppose now the function is odd, that is, $s(x) = -s(-x)$, then we see that $a_n = 0$ for all n, and

$$b_n = \frac{4}{T}\int_0^{T/2} s(x)\sin n\omega x\,dx \tag{8.16}$$

This implies that, if a function is odd, its Fourier series will contain only sine terms. On the other hand, suppose the function is even, that is, $s(x) = s(-x)$, then $b_n = 0$ and

$$a_n = \frac{4}{T}\int_0^{T/2} s(x) \cos n\omega x\, dx \tag{8.17}$$

Consequently the Fourier series of an even function will contain only cosine terms.

Suppose next, the function $s(t)$ obeys the condition

$$s\left(t \pm \frac{T}{2}\right) = -s(t) \tag{8.18}$$

as given by the example in Fig. 8.2. Then we can show that $s(t)$ contains only odd harmonic terms, that is,

$$a_n = b_n = 0, \qquad n \text{ even}$$

and

$$\begin{aligned} a_n &= \frac{4}{T}\int_0^{T/2} s(t) \cos n\omega t\, dt \\ b_n &= \frac{4}{T}\int_0^{T/2} s(t) \sin n\omega t\, dt \qquad n \text{ odd} \end{aligned} \tag{8.19}$$

Let us now look at some examples. The signal of a full-wave rectified sine wave is a good example of even functions; hence the result of Eq. 8.13 at the end of the last section contains only cosine terms. For an odd function, let us consider a square wave shown in Fig. 8.3. Applying Eqns.

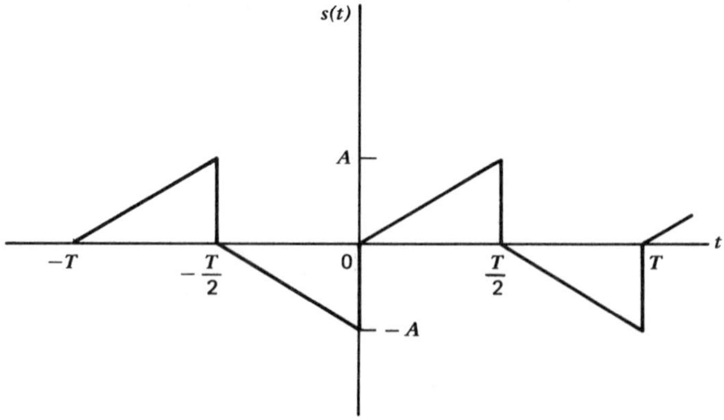

Fig. 8.2 Signal that satisfies Eq. 8.18.

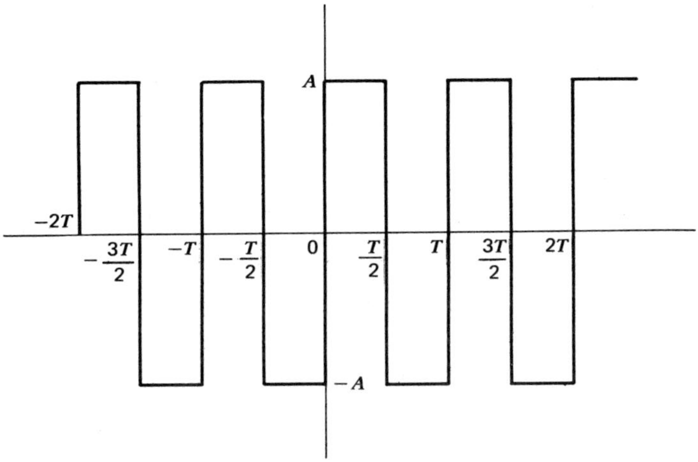

Fig. 8.3 An odd square wave.

8.5 to 8.7, we have

$$a_0 = \frac{2}{T}\int_0^{T/2} dt - \frac{2}{T}\int_{T/2}^{T} dt = 0$$

$$a_n = \frac{2}{T}\int_0^{T/2} \cos n\frac{2\pi}{T}t\,dt - \frac{2}{T}\int_{T/2}^{T} \cos n\frac{2\pi}{T}t\,dt$$

$$= 0$$

$$b_n = \frac{2}{T}\int_0^{T/2} \sin n\frac{2\pi}{T}t\,dt - \frac{2}{T}\int_{T/2}^{T} \sin n\frac{2\pi}{T}t\,dt$$

$$= \frac{4}{T}\int_0^{T/2} \sin n\frac{2\pi}{T}t\,dt = \frac{4}{T}(\cos n\pi - 1) \tag{8.20}$$

Now let us consider the *complex exponential form* of a Fourier series. That is, if we express cos $n\omega t$ and sin $n\omega t$ in terms of complex exponentials, the Fourier series in Eqs. 8.5 to 8.7 can be written as

$$s(t) = \frac{a_0}{2} + \sum_{n=1}^{\infty}\left[a_n\left(\frac{e^{jn\omega t} + e^{-jn\omega t}}{2}\right) + b_n\left(\frac{e^{jn\omega t} - e^{-jn\omega t}}{2j}\right)\right]$$

$$= \frac{a_0}{2} + \sum_{n=1}^{\infty}\left[\left(\frac{a_n - jb_n}{2}\right)e^{jn\omega t} + \left(\frac{a_n + jb_n}{2}\right)e^{-jn\omega t}\right] \tag{8.21}$$

If we define

$$\beta_n = \frac{a_n - jb_n}{2} \qquad \beta_{-n} = \frac{a_n + jb_n}{2} \qquad \beta_0 = \frac{a_0}{2}$$

then the complex form of the Fourier series is

$$s(t) = \beta_0 + \sum_{n=1}^{\infty} (\beta_n e^{jn\omega t} + \beta_{-n} e^{-jn\omega t})$$

$$= \sum_{n=-\infty}^{\infty} \beta_n e^{jn\omega t} \tag{8.22}$$

We can readily express the coefficient β_n as a function of $s(t)$ since

$$\beta_n = \frac{a_n - jb_n}{2}$$

$$= \frac{1}{T} \int_0^T s(t)\,(\cos n\omega t - j \sin n\omega t)\; dt$$

$$= \frac{1}{T} \int_0^T s(t) e^{-jn\omega t}\, dt \tag{8.23}$$

Observe that β_n is usually complex and can be represented as

$$\beta_n = Re\beta_n + jIm\beta_n \tag{8.24}$$

where we have

$$Re\beta_n = \frac{a_n}{2} = \frac{1}{T} \int_0^T s(t) \cos n\omega t\; dt \tag{8.25}$$

$$Im\beta_n = \frac{-b_n}{2} = -\frac{1}{T} \int_0^T s(t) \sin n\omega t\; dt \tag{8.26}$$

We also see from Eq. 8.22 that $s(t)$ is expressed as an infinite sum of phasors, hence the *amplitude spectrum* of the Fourier series is defined as the magnitude of the phasor

$$|\beta_n| = (Re^2\beta_n + Im^2\beta_n)^{1/2} \tag{8.27}$$

as a function of n or $n\omega$.

And the *phase spectrum* is defined as the phase of the phasor

$$\phi_n = \tan^{-1}\left(\frac{Im\beta_n}{Re\beta_n}\right) \tag{8.28}$$

as a function of n or $n\omega$.

It is easily seen that the amplitude spectrum is an even function and the phase spectrum is an odd function of n.

The amplitude spectrum provides us with valuable insight as to where to truncate the infinite series and still maintain a good approximation to the original waveform. As an example, let us again consider the square wave in Fig. 8.3. From Eq. 8.23 we obtain the coefficients of the complex Fourier series as

$$\beta_n = \frac{1}{T}\int_0^{T/2} Ae^{-jn\omega t}\,dt - \frac{1}{T}\int_{T/2}^{T} Ae^{-jn\omega t}\,dt$$

$$= \frac{A}{jn\omega T}\left(1 - 2e^{-jn\omega(T/2)} + e^{-jn\omega t}\right) \tag{8.29}$$

Since $n\omega T = n2\pi$, β_n can be simplified to

$$\beta_n = \frac{A}{jn\omega T}\left(1 - 2e^{-jn\pi} + e^{-j2n\pi}\right) \tag{8.30}$$

Simplifying β_n one step further, we obtain

$$\beta_n = \frac{1}{T}\cdot\frac{4A}{jn\omega} \qquad n \text{ odd}$$

$$= 0 \qquad n \text{ even} \tag{8.31}$$

The discrete spectra of the square wave is shown in Fig. 8.4.

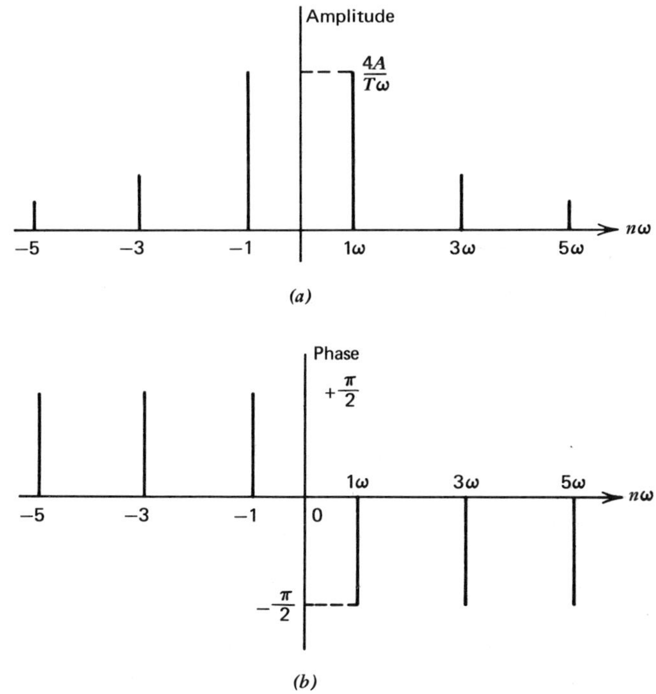

Fig. 8.4 Discrete spectra of square wave. (*a*) Amplitude. (*b*) Phase.

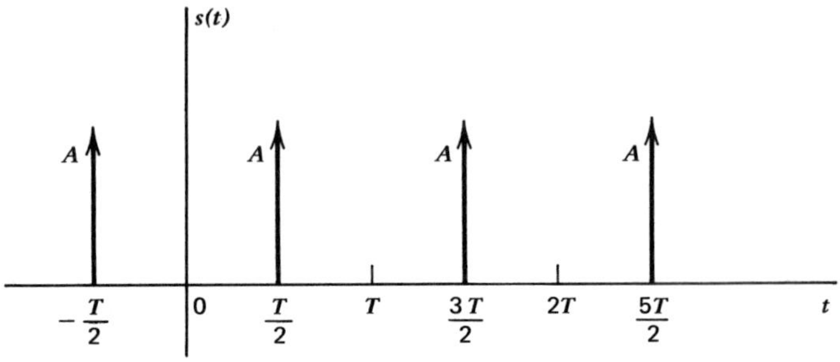

Fig. 8.5 A train of impulses.

For functions which are made up of *straight-line* components only, we can make use of a basic property of impulse functions to simplify the calculation of complex Fourier coefficients. The square wave we just discussed is of such type of functions. The method is based upon the relation

$$\int_{-\infty}^{\infty} f(t)\delta(t-T_1)\,dt = f(T_1) \tag{8.32}$$

Let us first use this equation to evaluate the complex Fourier coefficients for the periodic impulse train as is shown in Fig. 8.5. Using Eq. 8.32 with $f(t) = e^{-jn\omega t}$, we have

$$\beta_n = \frac{A}{T}\int_0^T \delta\left(t-\frac{T}{2}\right) e^{-jn\omega t}\,dt = \frac{A}{T}e^{-(jn\omega(T/2))} \tag{8.33}$$

We see that the complex Fourier coefficients for impulse functions are obtained by simply substituting the time at which the impulses occur into the expression, $e^{-jn\omega t}$.

In the evaluation of Fourier coefficients, we must remember that the limits for the β_n integral are taken over *one period* only, that is, we consider only a single period of the signal in the analysis. As an example, consider again the square wave in Fig. 8.3. To evaluate β_n, we consider only a single period of the square wave, say, from $t = 0$ to $t = T$, as shown in Fig. 8.6*a*. Since the square wave is not made up of impulses, let us differentiate the single period of the square wave to give $s'(t)$ as shown in Figure 8.6*b*. We can now evaluate the complex Fourier coefficients for the derivative $s'(t)$, which clearly is made up of impulses alone. Analytically, if $s(t)$ is given as

$$s(t) = \sum_{n=-\infty}^{\infty} \beta_n e^{jn\omega t} \tag{8.34}$$

then the derivative of $s(t)$ is

$$s'(t) = \sum_{n=-\infty}^{\infty} jn\omega\beta_n e^{jn\omega t} \tag{8.35}$$

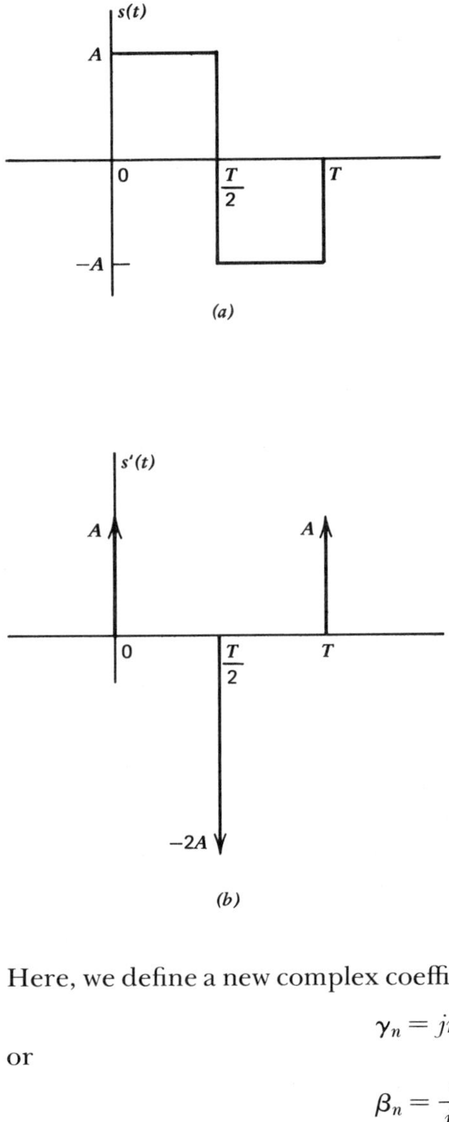

Fig. 8.6 (*a*) Square wave over period $[0, T]$. (*b*) Derivative of square wave over period $[0, T]$.

Here, we define a new complex coefficient

$$\gamma_n = jn\omega\beta_n$$

or

$$\beta_n = \frac{\gamma_n}{jn\omega} \tag{8.36}$$

If the derivative $s'(t)$ is a function that consists of impulse components alone, then we simply evaluate γ_n first and then obtain β_n from Eq. 8.36. For example, the derivative of the square wave yields the impulse train in Fig. 8.6. In the interval $[0, T]$, the signal $s'(t)$ is given as

$$s'(t) = A\delta(t) - 2A\delta\left(t - \frac{T}{2}\right) + A\delta(t - T) \tag{8.37}$$

Then the complex coefficients are

$$\gamma_n = \frac{1}{T}\int_0^T s'(t)e^{-jn\omega t}\,dt$$

$$= \frac{A}{T}(1 - 2e^{-(jn\omega(T/2))} + e^{-jn\omega T}) \tag{8.38}$$

The Fourier coefficients of the square wave are

$$\beta_n = \frac{\gamma_n}{jn\omega}$$

$$= \frac{A}{jn\omega T}(1 - 2e^{-(jn\omega(T/2))} + e^{-jn\omega T}) \tag{8.39}$$

which checks with the solution obtained in the standard way in Eq. 8.29.

If the first derivative, $s'(t)$, does not contain impulses, then we must differentiate again to yield

$$s''(t) = \sum_{n=-\infty}^{\infty} \lambda_n e^{jn\omega t} \tag{8.40}$$

where

$$\lambda_n = jn\omega\gamma_n = (jn\omega)^2\beta_n$$

For the triangular pulse in Fig. 8.7, the second derivative over the period $[0, T]$ is

$$s''(t) = \frac{2A}{T}\left[\delta(t) - 2\delta\left(t - \frac{T}{2}\right) + \delta(t - T)\right] \tag{8.41}$$

The coefficients λ_n are now obtained as

$$\lambda_n = \frac{1}{T}\int_0^T s''(t)e^{-jn\omega t}\,dt$$

$$= \frac{2A}{T^2}(1 - 2e^{-(jn\omega(T/2))} + e^{-jn\omega T})$$

which simplifies to give

$$\lambda_n = \frac{8A}{T^2} \qquad n \text{ odd}$$

$$= 0 \qquad n \text{ even} \tag{8.42}$$

From λ_n we obtain

$$\beta_n = \frac{\lambda_n}{(j\omega n)^2}$$

$$= -\frac{2A}{n^2\pi^2} \qquad n \text{ odd}$$

$$= 0 \qquad n \text{ even} \tag{8.43}$$

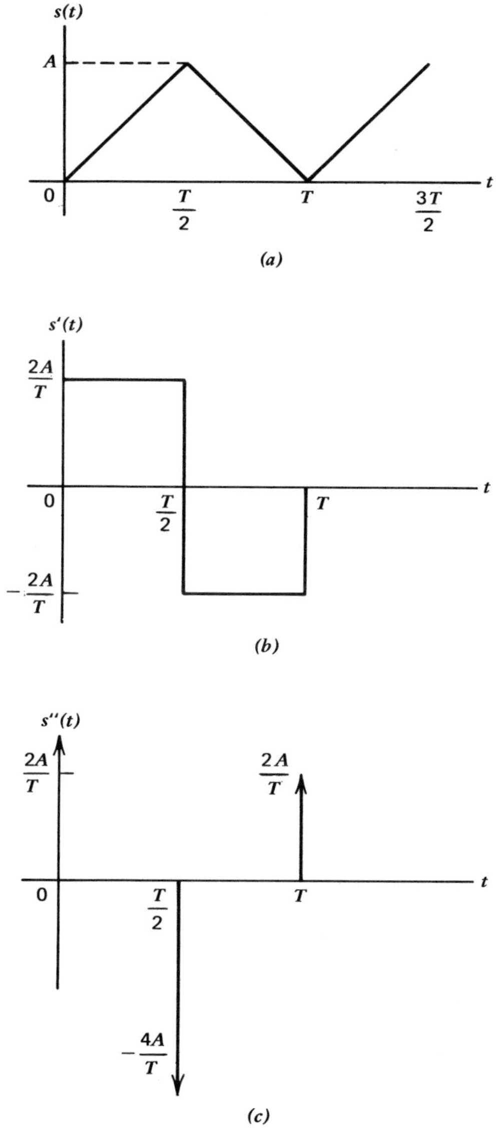

Fig. 8.7 The triangular wave and its derivatives.

A slight difficulty arises if the expression for $s'(t)$ contains an impulse in addition to other straight-line terms. Because of these straight-line terms we must differentiate once more. However, from this additional differentiation, we obtain the derivative of the impulse as well. This presents no difficulty, however, because we know from Chapter 1 that

$$\int_{-\infty}^{\infty} s(t)\delta'(t-T) = -s'(T) \tag{8.44}$$

so that

$$\int_{\infty}^{\infty} \delta'(t-T)e^{-jn\omega t}\,dt = jn\omega e^{-jn\omega T} \tag{8.45}$$

We can therefore tolerate doublets or even higher derivatives of impulses in the analysis. Consider the signal $s(t)$ given in Fig. 8.8*a*. Its

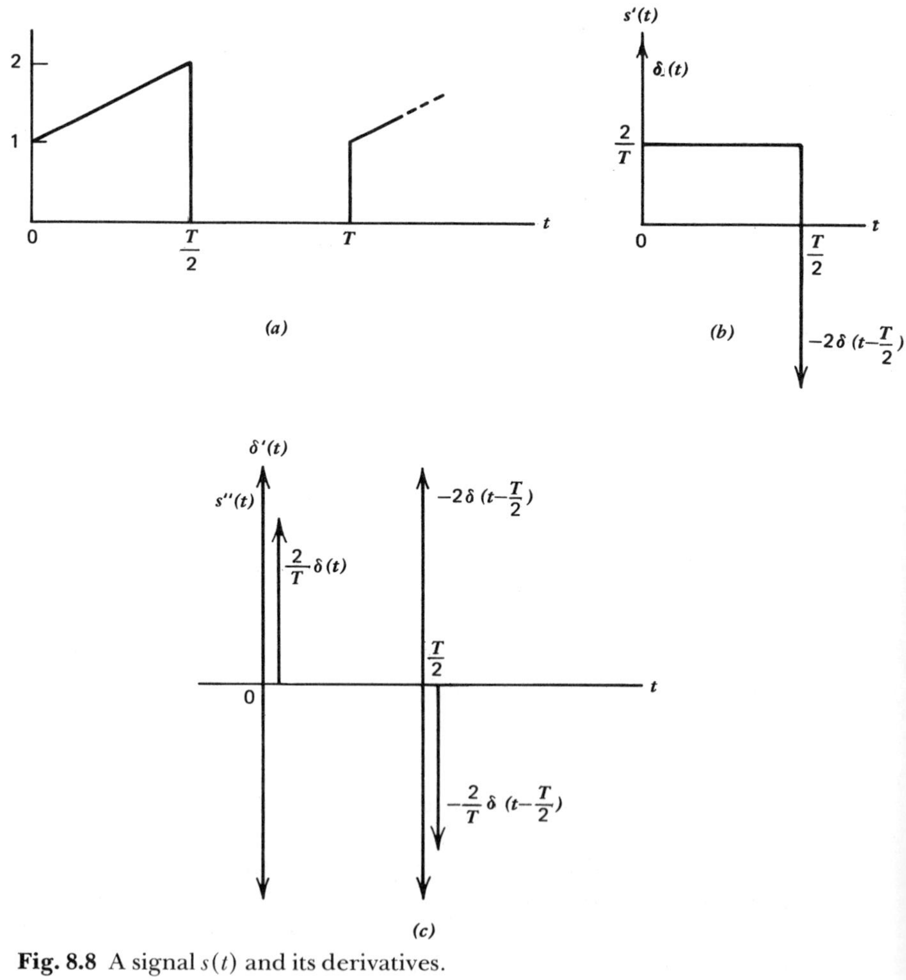

Fig. 8.8 A signal $s(t)$ and its derivatives.

derivative $s'(t)$, shown in Fig. 8.8*b* can be expressed as

$$s'(t) = \frac{2}{T}\left[u(t) - u\left(t - \frac{T}{2}\right)\right] + \delta(t) - 2\delta\left(t - \frac{T}{2}\right) \tag{8.46}$$

The second derivative $s''(t)$ consists of a pair of impulses and a pair of doublets as given by

$$s''(t) = \frac{2}{T}\left[\delta(t) - \delta\left(t - \frac{T}{2}\right)\right] + \delta'(t) - 2\delta'\left(t - \frac{T}{2}\right)$$

as shown in Fig. 8.8*c*. We therefore evaluate λ_n as

$$\begin{aligned}\lambda_n &= \frac{1}{T}\int_0^T s''(t)e^{-jn\omega t}\,dt \\ &= \frac{2}{T^2}\left(1 - e^{-(jn\omega(T/2))}\right) + \frac{j\omega n}{T}\left(1 - 2e^{-(jn\omega(T/2))}\right)\end{aligned} \tag{8.47}$$

The complex coefficients β_n are now obtained as

$$\begin{aligned}\beta_n &= \frac{\lambda_n}{(j\omega n)^2} \\ &= \frac{2}{(j\omega nT)^2}\left(1 - e^{-(jn\omega(T/2))}\right) + \frac{1}{j\omega nT}\left(1 - 2e^{-(jn\omega(T/2))}\right)\end{aligned}$$

Simplifying, we have

$$\begin{aligned}\beta_n &= -\frac{1}{n^2\pi^2} + \frac{3}{j2\pi n} \qquad n \text{ odd} \\ &= -\frac{1}{j2\pi n} \qquad n \text{ even}\end{aligned} \tag{8.48}$$

In conclusion, it must be pointed out that the method of using impulses to evaluate Fourier coefficients does not give the d-c coefficient, $a_0/2$ or β_0. We obtain this coefficient through standard methods as given by Eq. 8.5.

8.3 THE FOURIER INTEGRAL

In the previous section we discussed methods for decomposing a periodic signal $s(t)$ with period T into a sum of elementary functions. However, in many cases the excitation signal is aperiodic, such as a single unrepeated pulse. Functions of this sort cannot be handled directly through the use of Fourier series. However, by investigating the limit (if any) that is approached by a Fourier series as the period of the given function becomes infinite, a suitable representation for aperiodic functions can be obtained. Recall the complex coefficient β_n for periodic signals is defined as

$$\beta_n = \frac{1}{T}\int_{\alpha}^{\alpha+T} s(t)\, e^{-jn\omega t}\, dt \qquad n = 0, \pm 1, \pm 2, \ldots \tag{8.49}$$

where $\omega = 2\pi/T$. And the signal $s(t)$ is then represented as

$$s(t) = \sum_{n=-\infty}^{\infty} \beta_n e^{jn\omega t} \tag{8.50}$$

Specifically, if we take $\alpha = -T/2$, then Eq. 8.49 can be rewritten as

$$\beta_n = \frac{1}{T}\int_{-T/2}^{T/2} s(t) e^{-jn\omega t}\, dt \qquad n = 0, \pm 1, \pm 2, \ldots \tag{8.51}$$

Observe that the amplitude and phase spectra of β_n as a function of $n\omega$ are discrete lines spaced at an interval

$$\Delta\omega_n = (n+1)\omega - (n\omega) = \omega = \frac{2\pi}{T} \tag{8.52}$$

An initial investigation of Eqs. 8.51 and 8.52 tells us that as T becomes larger and larger, the magnitude of β_n and the spacing between adjacent spectrum lines are getting smaller and smaller. Let us look at an example for best illustration. Consider a train of pulses with finite width $2p$ and period T, as shown in Fig. 8.9. By applying Eq. 8.51, we obtain

$$\begin{aligned}\beta_n &= \frac{1}{T}\int_{-T/2}^{T/2} s(t) e^{-jn\omega t}\, dt \\ &= \frac{1}{T}\int_{-p}^{p} e^{-jn\omega t}\, dt = \frac{1}{T}\left(e^{-jn\omega p} - e^{jn\omega p}\right)\left(\frac{1}{-jn\omega}\right) \\ &= \frac{p}{T}\frac{\sin(n\omega p)}{n\omega p}\end{aligned} \tag{8.53}$$

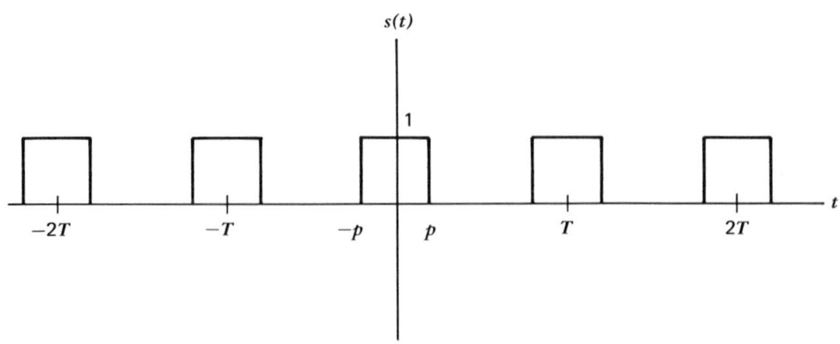

Fig. 8.9 A train of rectangular pulses.

For convenience let $p = 1$. So Eq. 8.53 is simplified to

$$\beta_n = \frac{1}{T}\frac{\sin(n\omega)}{n\omega} \tag{8.54}$$

The amplitude spectrum of β_n as a function of $n\omega$ is shown in Fig. 8.10 for $T = 4$ and 8. We see that as T increases, the amplitude of β_n decreases and the spacing between adjacent lines is getting smaller. As T goes to infinity we have a continuous spectrum with infinitesimal magnitude. This suggests that the summation in Eq. 8.50 is approaching an integral. We can also see from Eq. 8.51 that $\beta_n T$ is finite if the limit of the integral

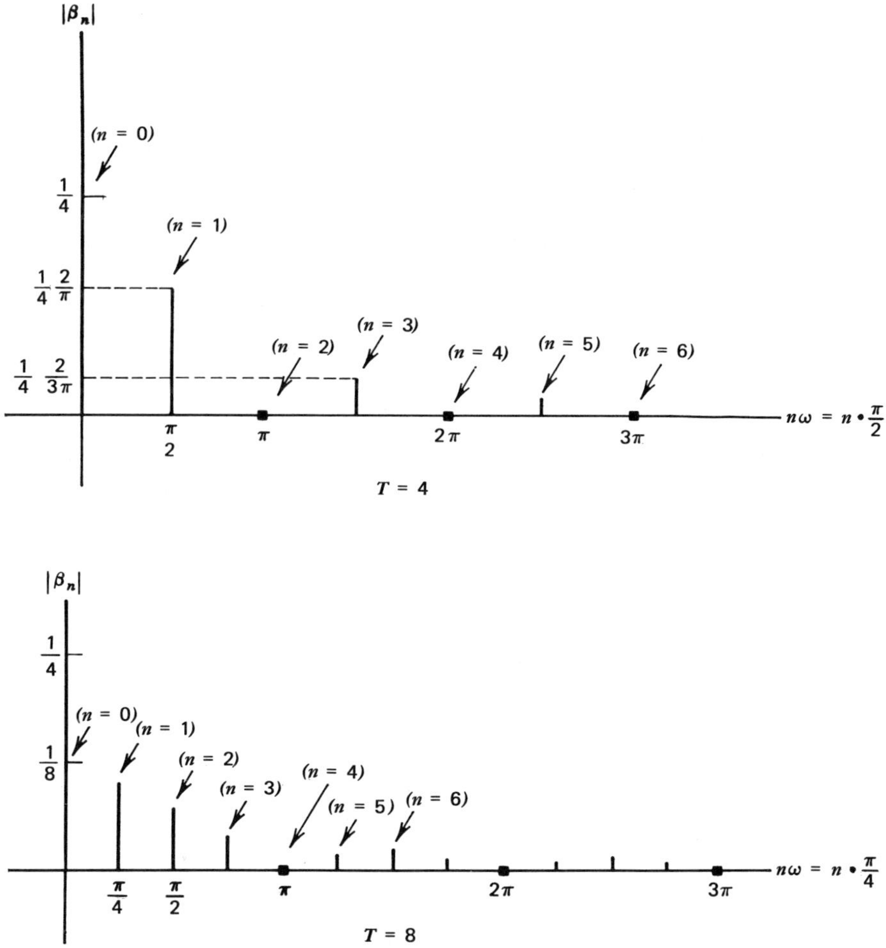

Fig. 8.10 The amplitude spectra of a rectangular pulse train.

$$\lim_{T\to\infty}\int_{-T/2}^{T/2} s(t)e^{-jn\omega t}\,dt \tag{8.55}$$

does exist. Hence we could define

$$S(\omega_n) = T\beta_n = \int_{-\infty}^{\infty} s(t)e^{-j\omega_n t}\,dt \tag{8.56}$$

where $\omega_n = n\omega$ as a new coefficient that is called the *Fourier transform* of the given function $s(t)$ if Eq. 8.55 does exist. Substituting Eq. 8.56 into Eq. 8.50, we have

$$\begin{aligned} s(t) &= \sum_{n=-\infty}^{\infty} T\beta_n e^{jn\omega t}\frac{1}{T} \\ &= \frac{1}{2\pi}\sum_{n=-\infty}^{\infty} S(\omega_n)e^{j\omega_n t}\frac{2\pi}{T} \\ &= \frac{1}{2\pi}\sum_{n=-\infty}^{\infty} (S(\omega_n)e^{j\omega_n t})\Delta\omega_n \end{aligned} \tag{8.57}$$

As $T \to \infty$, we see $\Delta\omega_n \to 0$ and Eq. 8.57 approaches the integral

$$s(t) = \frac{1}{2\pi}\int_{-\infty}^{\infty} S(\omega_n)e^{j\omega_n t}\,d\omega_n \tag{8.58}$$

For simplicity, let $\omega_n = 2\pi f$, then we have $d\omega_n = 2\pi\, df$. Hence Eqs. 8.56 and 8.58 are simplified to

$$S(f) = \int_{-\infty}^{\infty} s(t)e^{-j2\pi ft}\,dt \tag{8.59}$$

$$s(t) = \int_{-\infty}^{\infty} S(f)e^{j2\pi ft}\,df \tag{8.60}$$

Equation 8.60 tells us that the aperiodic function $s(t)$ is now decomposed into an infinite number of elementary functions that are "spread" continuously on the spectrum plane.

Equations 8.59 and 8.60 are sometimes called the *Fourier transform pair.* If we let script $\mathscr{F}$ denote the operation of Fourier transformation and $\mathscr{F}^{-1}$ denote inverse transformation, then

$$\begin{aligned} S(f) &= \mathscr{F}\cdot s(t) \\ s(t) &= \mathscr{F}^{-1}\cdot S(f) \end{aligned} \tag{8.61}$$

In general, the Fourier transform $S(f)$ is complex and can be denoted as

$$S(f) = \mathrm{Re}\,S(f) + j\,\mathrm{Im}\,S(f) \tag{8.62}$$

The real part of $S(f)$ is obtained through the formula

$$\operatorname{Re} S(f) = S(f) + S(-f)$$

$$= 2\int_{-\infty}^{\infty} s(t) \cos 2\pi f t\, dt$$

and the imaginary part through

$$\operatorname{Im} S(f) = j[S(f) - S(-f)]$$

$$= 2\int_{-\infty}^{\infty} s(t) \sin 2\pi f t\, dt \qquad (8.64)$$

The amplitude spectrum of $S(f)$ is defined as

$$A(f) = [\operatorname{Re} S(f)^2 + \operatorname{Im} S(f)^2]^{1/2} \qquad (8.65)$$

and the phase spectrum is

$$\phi(f) = \arctan\left[\frac{\operatorname{Im} S(f)}{\operatorname{Re} S(f)}\right] \qquad (8.66)$$

Let us examine some examples.

Example 8.3.1 The first example is $s(t) = A\delta(t - t_0)$, we have

$$S(f) = \int_{-\infty}^{\infty} A\delta(t - t_0) e^{-j2\pi f t}\, dt$$

$$= Ae^{-j2\pi f t_0} \qquad (8.67)$$

Its amplitude spectrum is

$$A(f) = A \qquad (8.68)$$

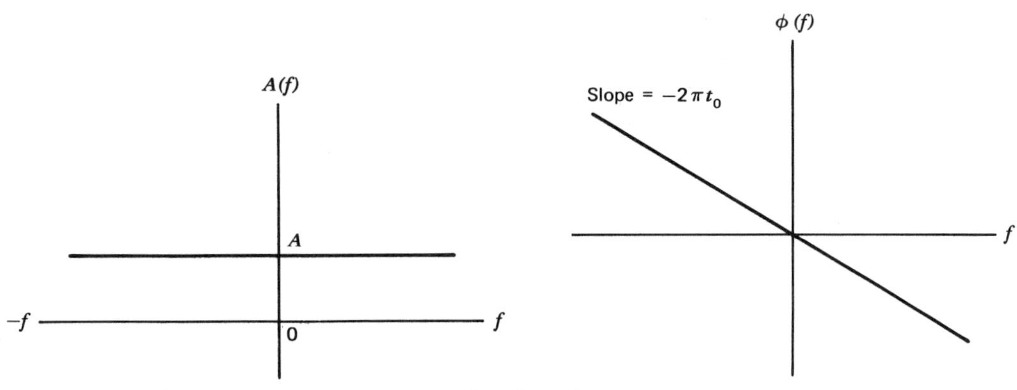

Fig. 8.11 Amplitude and phase spectrum of $A\delta(t - t_0)$.

while its phase spectrum is

$$\phi(f) = -2\pi f t_0 \tag{8.69}$$

as shown in Fig. 8.11.

Example 8.3.2 Next let us consider again the isolated rectangular pulse shown in Fig. 8.12 which is the limiting case of the pulse train shown in Fig. 8.9. Formally, it is defined as

$$s(t) = \begin{cases} 1 & |t| \leq p \\ 0 & |t| > p \end{cases} \tag{8.70}$$

The Fourier transform of $s(t)$ is easily found by using Eq. 8.59 and we have

$$S(f) = \int_{-p}^{p} e^{-j2\pi f t}\, dt$$

$$= 2p \frac{\sin(2\pi f p)}{2\pi f p} \tag{8.71}$$

The plot of Eq. 8.71 is shown in Fig. 8.13.

Let us examine the implications of these two examples. We see from the first example that the Fourier transform of a delta function has constant amplitude. In other words, if we regard the amplitude spectrum as the magnitude versus frequency plot of the "decomposed" elementary functions, then for a function with an extremely short duration in the time domain, the "decomposed" ones are spread uniformly over the frequency domain. Suppose that this signal is transmitted through some filter, then we would need an infinite bandwidth filter for its transmis-

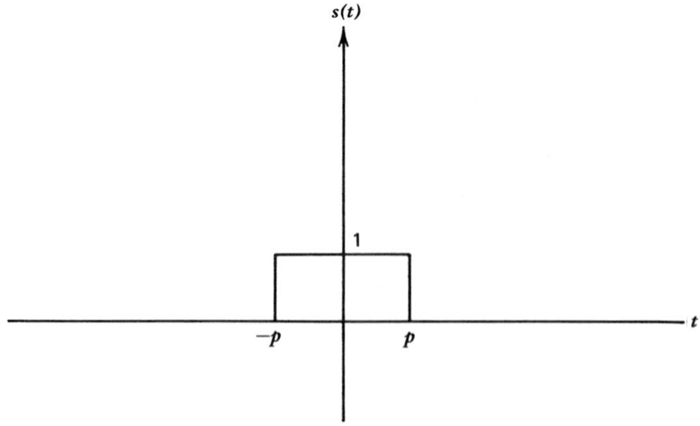

Fig. 8.12 An isolated rectangular pulse.

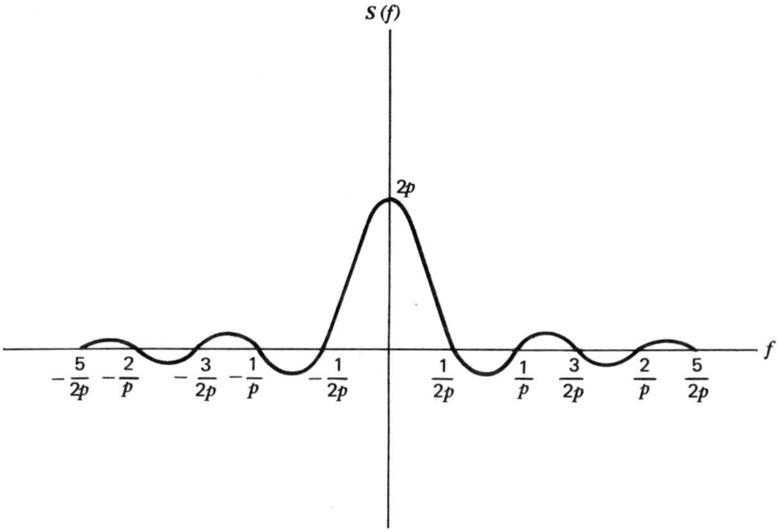

Fig. 8.13 The Fourier transform of the rectangular pulse.

sion. We can also say that to reconstruct the delta function, we need all its frequency components. This is possible only if we had an infinite bandwidth filter.

In practical cases we usually have signals ϕ with finite time duration such as the pulse in Example 8.3.2. We see from this example that the frequency components of this pulse are primarily concentrated in the frequency plane between $-1/2p \leq f \leq 1/2p$. If p decreases then this interval increases and the maximum magnitude decreases. Hence the shorter the time duration, the larger is the spread of its frequency components in the frequency domain.

These observations are important in that they illustrate why in pulse transmission, narrow pulses could only be transmitted through filters with large bandwidths; whereas pulses with longer time durations can be transmitted with relatively narrow bandwidth filters.

8.4 PROPERTIES OF FOURIER TRANSFORMS

We will now consider some important properties of Fourier transforms.

Linearity

The linearity property of Fourier transforms states that the Fourier transform of a sum of two signals is the sum of their individual Fourier transforms, that is,

$$\mathscr{F}[c_1 s_1(t) + c_2 s_2(t)] = c_1 S_1(f) + c_2 S_2(f) \tag{8.72}$$

This follows directly from the definition in Eq. 8.59.

Differentiation

This property states that the Fourier transform of the derivative of a signal is $j2\pi f$ times the Fourier transform of the signal itself,

$$\mathscr{F} \cdot s'(t) = j2\pi f S(f)$$

or more generally,

$$\mathscr{F} \cdot s^{(n)}(t) = (j2\pi f)^n S(f) \tag{8.74}$$

The proof is obtained by taking the derivative of both sides of the inverse transform definition,

$$\begin{aligned} s'(t) &= \frac{d}{dt}\int_{-\infty}^{\infty} S(f)e^{j2\pi ft}\,df \\ &= \int_{-\infty}^{\infty} (j2\pi f)S(f)e^{j2\pi ft}\,df \end{aligned} \tag{8.75}$$

Similarly, it is easily shown that the transform of the integral of $s(t)$ is

$$\mathscr{F}\left[\int_{-\infty}^{t} s(\tau)d\tau\right] = \frac{1}{j2\pi f}S(f) \tag{8.76}$$

Consider the following example

$$s(t) = e^{-at}u(t) \tag{8.77}$$

Its Fourier transform is

$$\begin{aligned} S(f) &= \int_{-\infty}^{\infty} e^{-at}u(t)e^{-j2\pi ft}\,dt \\ &= \int_{0}^{\infty} e^{-at}e^{-j2\pi ft}dt = \frac{1}{a+j2\pi f} \end{aligned} \tag{8.78}$$

The derivative of $s(t)$ is

$$s'(t) = \delta(t) - ae^{-at}u(t) \tag{8.79}$$

Its Fourier transform is

$$\begin{aligned} \mathscr{F}[s'(t)] &= 1 - \frac{a}{a+j2\pi f} = \frac{j2\pi f}{a+j2\pi f} \\ &= j2\pi f S(f) \end{aligned} \tag{8.80}$$

Symmetry

The symmetry property of Fourier transforms states that if

$$\mathscr{F} \cdot x(t) = X(f) \tag{8.81}$$

then

$$\mathscr{F} \cdot X(t) = x(-f) \tag{8.82}$$

This property follows directly from the symmetrical nature of the Fourier transform pair in Eqs. 8.59 and 8.60.

Example 8.4.1 Consider the Fourier transform of the unit impulse. We have

$$\begin{aligned} \mathscr{F} \cdot \delta(t) &= \int_{-\infty}^{\infty} \delta(t) e^{-j2\pi ft}\, dt \\ &= 1 \end{aligned} \tag{8.83}$$

From the symmetry property we can show that

$$\mathscr{F} \cdot 1 = \delta(f) \tag{8.84}$$

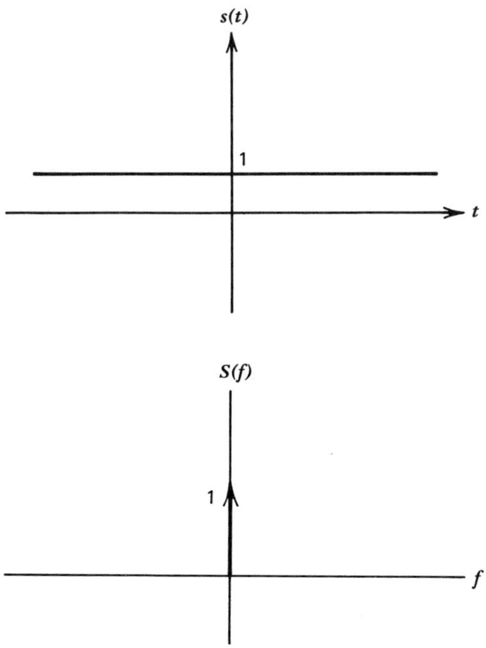

Fig. 8.14 Fourier transform of $s(t) = 1$.

as shown in Fig. 8.14. The above example is also an extreme example of the time-duration and bandwidth reciprocity relationship. It says that zero time duration, $\delta(t)$, gives rise to infinite bandwidth in the frequency domain; while zero bandwidth, $\delta(f)$ corresponds to infinite time duration.

Scale Change

The scale change property describes the time duration and bandwidth reciprocity relationship. It states that

$$\mathcal{F}\left[s\left(\frac{t}{a}\right)\right] = |a|S(af) \tag{8.85}$$

We prove this property most easily through the inverse transform,

$$\mathcal{F}^{-1}[|a|S(af)] = |a| \int_{-\infty}^{\infty} S(af)e^{-j2\pi ft}df \tag{8.86}$$

Let

$$f' = af$$

then

$$\mathcal{F}^{-1}[|a|S(f')] = |a| \int_{-\infty}^{\infty} S(f')e^{-j2\pi f'(t/a)} \frac{df'}{a}$$

$$= \mathcal{F}\left[s\left(\frac{t}{a}\right)\right] \tag{8.87}$$

As an example, consider

$$\mathcal{F}[e^{-at}u(t)] = \frac{1}{j2\pi f + a} \tag{8.88}$$

then

$$\mathcal{F}[e^{-t}u(t)] = \frac{|a|}{j2\pi af + a}$$

$$= \frac{1}{j2\pi f + 1} \tag{8.89}$$

if $a > 0$.

Folding

The Folding property states that

$$\mathcal{F}[s(-t)] = S(-f) \tag{8.90}$$

The proof follows directly from the definition of the Fourier transform. An example is

$$\mathcal{F}[e^{t}u(-t)] = \frac{1}{a - j2\pi f} \tag{8.91}$$

Delay

If a signal is delayed by an amount t_0 in the time domain, the corresponding effect in the frequency domain is to multiply the transform of the undelayed signal by $e^{-j2\pi f t_0}$, that is,

$$\mathcal{F}[s(t-t_0)] = e^{-j2\pi f t_0} S(f) \tag{8.92}$$

For example,

$$\mathcal{F}[e^{-a(t-t_0)}u(t-t_0)] = \frac{e^{-j2\pi f t_0}}{a + j2\pi f} \tag{8.93}$$

Modulation

The *modulation* or *frequency shift* property of Fourier transforms states that if a Fourier transform is shifted in frequency by an amount f_0, the corresponding effect in time is described by multiplying the original signal by $e^{j2\pi f_0 t}$, that is,

$$\mathcal{F}^{-1}[S(f-f_0)] = e^{j2\pi f_0 t} s(t) \tag{8.94}$$

Example 8.4.2 Given $S(f)$ in Fig. 8.15*a*, let us find the inverse transform of $S_1(f)$ in Fig. 8.15*b* in terms of $s(t) = \mathcal{F}^{-1}S(f)$. We know that

$$S_1(f) = S(f-f_0) + S(f+f_0) \tag{8.95}$$

Then

$$\begin{aligned} \mathcal{F}^{-1}S_1(f) &= e^{j2\pi f_0 t} s(t) + e^{-j2\pi f_0 t} s(t) \\ &= 2s(t)\cos 2\pi f_0 t \end{aligned} \tag{8.96}$$

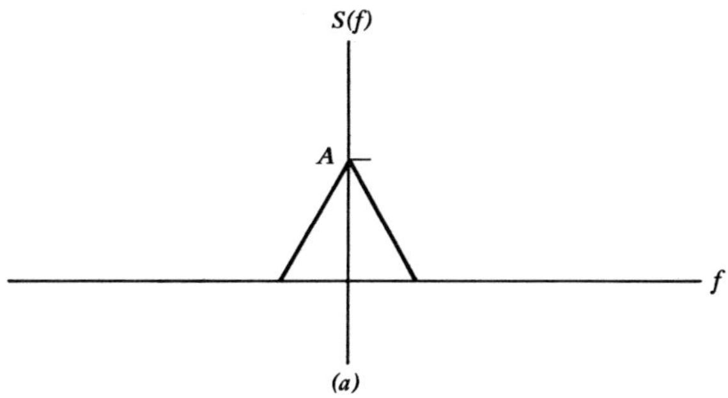

Fig. 8.15*a* Demonstration of amplitude modulation.

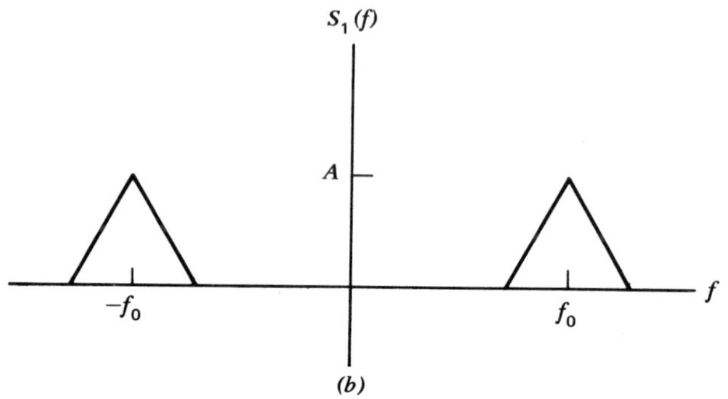

Fig. 8.15b Demonstration of amplitude modulation.

Thus we see that multiplying a signal by a cosine or sine wave in the time domain corresponds to shifting its spectrum by an amount $\pm f_0$. In transmission terminology, f_0 is the *carrier frequency* and the process of multiplying $s(t)$ by $\cos 2\pi f_0 t$ is called *amplitude modulation.*

Parseval's Theorem

An important theorem that relates energy in the time and frequency domains is Parseval's theorem that states that,

$$\int_{-\infty}^{\infty} s_1(t)s_2(t)\,dt = \int_{-\infty}^{\infty} S_1(f)S_2(-f)\,dt \tag{8.97}$$

The proof is obtained very simply as follows:

$$\begin{aligned}\int_{-\infty}^{\infty} s_1(t)s_2(t)\,dt &= \int_{-\infty}^{\infty} s_2(t)\,dt \int_{-\infty}^{\infty} S_1(f)e^{j2\pi ft}df \\ &= \int_{-\infty}^{\infty} S_1(f)\,df \int_{-\infty}^{\infty} s_2(t)e^{j2\pi ft}dt \\ &= \int_{-\infty}^{\infty} S_1(f)S_2(-f)\,df \end{aligned} \tag{8.98}$$

In particular, when $s_1(t) = s_2(t)$ we have a corollary of Parseval's theorem known as *Plancheral's* theorem,

$$\int_{-\infty}^{\infty} |s(t)|^2dt = \int_{-\infty}^{\infty} |S(f)|^2df \tag{8.99}$$

If $s(t)$ is equal to the current through, or the voltage across, a one ohm resistor, then the total energy is

$$\int_{-\infty}^{\infty} s^2(t)\,dt \tag{8.100}$$

We see from Eq. (8.99) that the total energy is also equal to the area under the curve of $|S(f)|^2$. Thus $|S(f)|^2$ is sometimes called an *energy density* or *energy spectrum*.

8.5 CONVOLUTION II

During the course of our discussion of time-domain theory, we have developed the fact that the response of a circuit to an arbitrary excitation is given by the convolution integral. This integral involves the impulse response and the arbitrary excitation. Specifically, we have

$$g(t) = \int_0^t f(\tau) \cdot h(t-\tau)\,d\tau \tag{8.101}$$

where $g(t)$ is the response function, $f(t)$ is the excitation and $h(t)$ is the response of the circuit to a unit impulse. Furthermore, for most of the practical cases, the excitation is applied only after t is greater than zero and the response of the circuit before the impulse is applied is also zero. Hence we could have

$$f(t) = g(t) = 0 \qquad t < 0 \tag{8.102}$$

Thus we can rewrite Eq. 8.101 as

$$g(t) = \int_{-\infty}^{0} f(\tau)h(t-\tau)\,d\tau + \int_0^t f(\tau)h(t-\tau)\,d\tau + \int_t^{\infty} f(\tau)h(t-\tau)\,d\tau \tag{8.103}$$

since the first and third terms on the right-hand side of Eq. 8.103 are identically zero. Hence, finally we have

$$g(t) = \int_{-\infty}^{\infty} f(\tau)h(t-\tau)\,d\tau \tag{8.104}$$

A natural question at this point: is there some sort of relationship between the Fourier transform of $g(t)$ and that of $f(t)$? The answer is yes, and we can find it out by applying Eq. 8.104 directly to the integral

$$G(f) = \int_{-\infty}^{\infty} g(t)e^{-j2\pi ft}\,dt \tag{8.105}$$

Thus we have

$$G(f) = \int_{-\infty}^{\infty} \left[\int_{-\infty}^{\infty} f(\tau)h(t-\tau)\,d\tau \right] e^{-j2\pi ft}\,dt \tag{8.106}$$

After interchanging the order of integration, we get

$$G(f) = \int_{-\infty}^{\infty} f(\tau)\left[\int_{-\infty}^{\infty} h(t-\tau)e^{-j2\pi ft}dt\right]d\tau \tag{8.107}$$

Now we have

$$\int_{-\infty}^{\infty} h(t-\tau)e^{-j2\pi ft}dt = H(f)e^{-j2\pi f\tau} \tag{8.108}$$

because of the property expressed in Eq. 8.92. Hence, Eq. 8.107 becomes

$$\begin{aligned} G(f) &= \int_{-\infty}^{\infty} f(\tau)H(f)e^{-j2\pi f\tau}d\tau \\ &= H(f)\cdot F(f) \end{aligned} \tag{8.109}$$

So we see that *convolution in the time domain corresponds to multiplication in the frequency domain.* In other words, if the transforms of $h(t)$ and $f(t)$, or $H(f)$ and $F(f)$, are known, then the response transform $G(f)$ can be found by multiplying $H(f)$ and $F(f)$. Once Eq. 8.109 is known for every f in the continuous frequency domain, then $g(t)$ can be obtained by using the inverse Fourier transform. However, the inversion process is not very easy in practice. In many cases, only by the use of a digital computer can we find the inverse transform. In fact, for many nontrivial functions $s(t)$, their Fourier transforms can rarely be found in analytical form. We usually satisfy ourselves by having an approximation of $S(f)$ by using a computer. This will be discussed further in sections five and six later.

As an illustration, consider a simple RC circuit shown in Fig. 8.16. The impulse response can be found as

$$h(t) = \frac{1}{RC}e^{-t/RC} \qquad t > 0 \tag{8.110}$$

Suppose we want to find the circuit response $g(t)$ due to a square pulse of height and width in 1 unit. By applying the convolution integral, we have then

$$g(t) = \int_{0}^{t} f(\tau)\frac{1}{RC}e^{-\frac{(t-\tau)}{RC}}d\tau \tag{8.111}$$

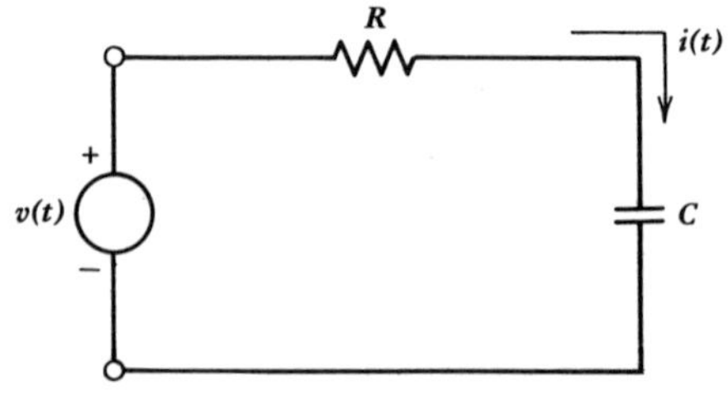

Fig. 8.16 An RC circuit.

We have

$$f(t) = \begin{cases} 1 & 0 \leqslant t \leqslant 1 \\ 0 & \text{otherwise} \end{cases} \tag{8.112}$$

Hence from Eq. 8.111 we have

$$g(t) = \begin{cases} \displaystyle\int_0^t \frac{1}{RC} e^{-\frac{(t-\tau)}{RC}} dt & 0 \leqslant t \leqslant 1 \\ \\ \displaystyle\int_0^1 \frac{1}{RC} e^{-\frac{(t-\tau)}{RC}} dt & t \geqslant 1 \end{cases} \tag{8.113}$$

that simplifies to

$$g(t) = \begin{cases} 1 - e^{-\frac{t}{RC}} & t \leqslant 1 \\ e^{-\frac{t}{RC}} (e^{\frac{1}{RC}} - 1) & t \geqslant 1 \end{cases} \tag{8.114}$$

If we apply Fourier analysis to find the response, then we have

$$\begin{aligned} H(f) &= \int_{-\infty}^{\infty} \frac{1}{RC} e^{-\frac{t}{RC}} \; e^{-j2\pi ft} dt \\ &= \frac{1}{1 + j2\pi fRC} \end{aligned} \tag{8.115}$$

$$\begin{aligned} F(f) &= \int_{-\infty}^{\infty} f(t) e^{-j2\pi ft} dt \\ &= \int_0^1 e^{-j2\pi ft} dt \\ &= \frac{1}{j2\pi f} [1 - e^{-j2\pi f}] \end{aligned} \tag{8.116}$$

Finally we have

$$\begin{aligned} G(f) &= H(f)F(f) \\ &= \frac{1}{1 + j2\pi fRC} \cdot \frac{1}{j2\pi f} [1 - e^{-j2\pi f}] \end{aligned} \tag{8.117}$$

Equation 8.117 can be verified by applying Fourier transform directly to Eq. 8.114. However, to find $g(t)$ by finding the inverse of Eq. 8.117 is much more difficult.

8.6 DISCRETE FOURIER TRANSFORM

The Fourier transform for a time signal $s(t)$ is defined in section 8.3 as

$$S(f) = \int_{-\infty}^{\infty} s(t)\, e^{-j2\pi ft}\, dt \tag{8.118}$$

Then the inverse relationship is

$$s(t) = \int_{-\infty}^{\infty} S(f)\, e^{j2\pi ft}\, df \tag{8.119}$$

It is assumed that $s(t)$ exists for all t so that its Fourier transform $S(f)$ is defined by Eq. 8.118. However, if we only know the signal $s(t)$ at a set of equally spaced time points $t = i\Delta t_1$ where $i = 0, \pm 1, \pm 2, \cdots$ and Δt_1 is the length of the interval, then Eq. 8.119 can be written as

$$s(i\Delta t_1) = \int_{-\infty}^{\infty} S(f)\, e^{j2\pi i f \Delta t_1}\, df \tag{8.120}$$

If we divide the range of integration into segments each with length F, then we could rewrite Eq. 8.120 as

$$s(i\Delta t_1) = \sum_{k=-\infty}^{\infty} \int_{kF}^{(k+1)F} S(f)\, e^{j2\pi i f/F}\, df \tag{8.121}$$

where $F = 1/\Delta t_1$. Now the exponential function $e^{j2\pi i f/F}$ is a periodic function of f with period F or, in other words, if we have $f' = f + kF$, $k = 0, \pm 1, \pm 2, \cdots$ then

$$\begin{aligned} e^{j2\pi i f'/F} = e^{j2\pi i(f+kF)/F} &= e^{j2\pi i f/F} \cdot e^{j2\pi ik} \\ &= e^{j2\pi i f/F} \cdot 1 \end{aligned}$$

(since $e^{j2\pi ik} = 1$ for any integer product ik).

Thus we have a periodic exponential function of f with period F. Let the variable of integration be changed to

$$f = f_1 + kF$$

then Eq. 8.121 can be rewritten as

$$\begin{aligned} s(i\Delta t_1) &= \sum_{k=-\infty}^{\infty} \int_0^F S(f_1 + kF)\, e^{j2\pi i(f_1+kF)/F}\, df_1 \\ &= \int_0^F \sum_{k=-\infty}^{\infty} S(f_1 + kF)\, e^{j2\pi i f_1/F}\, df_1 \\ &= \int_0^F S_p(f_1)\, e^{j2\pi i f_1/F}\, df_1 \end{aligned} \tag{8.122}$$

where

$$S_p(f_1) = \sum_{k=-\infty}^{\infty} S(f_1 + kF) \tag{8.123}$$

is a periodic function formed by the superposition of the generally non-periodic function shifted by all multiples of a fundamental period F.

Let us look at a graphical illustration of Eq. 8.123. In Fig. 8.17, $S(f_1)$ is the given Fourier transform of the time signal $s(t)$, then $S_p(f_1)$ is simply the summation formed by adding the shifted $S(f_1)$ according to Eq. 8.123. Also notice that if we take F sufficiently large (or Δt_1 sufficiently small) such that the magnitude of $S(f_1)$ is negligible for $|f_1| \geqslant F/2$, then $S(f_1)$ could be approximated by $S_p(f_1)$ within the range $-F/2 \leqslant f \leqslant F/2$. This observation is important as we will see later that practically we can only find $S_p(f_1)$ at a discrete set of frequency points for a given $s(t)$.

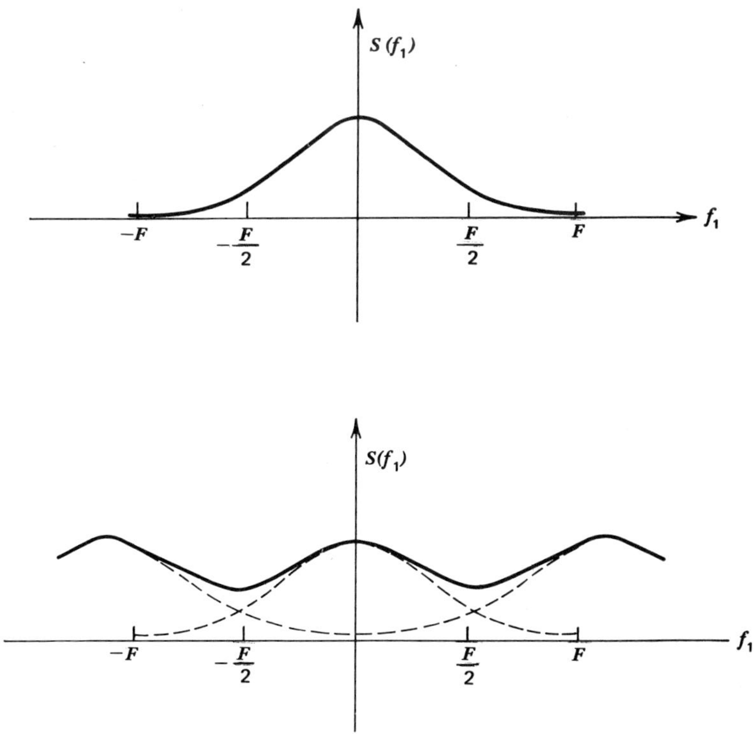

Fig. 8.17 Relationship between $S(f_1)$ and $S_p(f_1)$.

Now since $S_p(f_1)$ is a periodic function of f_1 it has a Fourier series expansion. Hence we have

$$S_p(f_1) = \sum_{n=-\infty}^{\infty} \beta_n e^{j2\pi n f_1/F} \tag{8.124}$$

where

$$\beta_n = \frac{1}{F}\int_0^F S_p(f_1)\, e^{-j2\pi n f_1/F}\, df_1 \tag{8.125}$$

Thus we see from Eq. 8.122 and Eq. 8.125 that we have

$$\beta_n = \frac{1}{F} s(-n\Delta t) \tag{8.126}$$

Substituting Eq. 8.126 into Eq. 8.124, we have

$$\begin{aligned} S_p(f_1) &= \frac{1}{F}\sum_{n=-\infty}^{\infty} s(-n\Delta t_1) e^{j2\pi n f_1/F} \\ &= \frac{1}{F}\sum_{n_1=-\infty}^{\infty} s(n_1\Delta t_1) e^{-j2\pi n_1 f_1/F} \end{aligned} \tag{8.127}$$

where we have substituted $-n$ with n_1. In this relationship between $S_p(f_1)$ and $s(n_1\Delta t_1)$ the usual roles of time and frequency are interchanged; that is, a periodic, continuous function of frequency corresponds to a sequence of time coefficients. The inverse relationship of Eq. 8.127 is

$$s(n_1\Delta t) = F\beta_n = \int_0^F S_p(f_1) e^{j2\pi n_1 f_1/F} df_1 \tag{8.128}$$

Thus, we have defined a *Discrete Fourier Transform* pair for the *discrete* time sequence $s(n_1\Delta t_1)$ in Eq. 8.127 and Eq. 8.128. So for a given time sequence $s(n_1\Delta t)$, $n_1 = 0, \pm 1, \pm 2, \ldots$ we can find a periodic function $S_p(f_1)$ by using Eq. 8.127. The inverse relation in Eq. 8.127 can then be used to reconstruct the time sequence if we know the periodic function $S_p(f_1)$.

Now Eq. 8.127 is an *infinite* discrete Fourier function. However, if we consider the values of $S_p(f_1)$ at N equally spaced points between 0 and F (after all, $S_p(f_1)$ is a periodic function with period F'), it is sufficient to describe $S_p(f_1)$ if only the values of $S_p(f_1)$ in $0 \leqslant f_1 \leqslant F$ are given; that is, sample $S_p(f_1)$ at intervals $\Delta f_1 = F/N = 1/(N\Delta t_1)$, we obtain

$$\begin{aligned} S_p(k\Delta f_1) &= \frac{1}{F}\sum_{n_1=-\infty}^{\infty} s(n_1\Delta t_1) e^{-j2\pi n_1 k\Delta f_1/F} \\ &= \frac{1}{F}\sum_{n=-\infty}^{\infty} s(n_1\Delta t_1) e^{-j2\pi n_1 k/N} \qquad k = 0, 1, 2, \cdots, N-1 \end{aligned} \tag{8.129}$$

Again $e^{-j2\pi n_1 k/N}$ is a periodic sequence of n_1 with period N. So that we have

$$s(n_1\Delta t_1 + \ell N\Delta t_1)e^{-j2\pi(n_1+\ell N)k/N} \\ = s(n_1\Delta t_1 + \ell N\Delta t_1)e^{-j2\pi n_1 k/N} \qquad \ell = 0, \pm 1, \pm 2, \cdots \quad (8.130)$$

From Eq. 8.130 we see for each n_1 we have ∞ terms in the summation of Eq. 8.129 with the same exponential function $e^{-j2\pi n_1 k/N}$. Furthermore we have only N distinct exponential functions $e^{-j2\pi n_1 k/N}$ with n_1 that range from 0 to $N-1$. Hence we could sum up Eq. 8.129 in N groups of infinite summations of the time sequence each multiplied by a distinct exponential function. Thus we have

$$S_p(k\Delta f_1) = \frac{1}{F}\sum_{n_1=0}^{N-1}\left\{\sum_{\ell=-\infty}^{\infty} s(n_1\Delta t_1 + \ell N\Delta t_1)\right\}e^{-j2\pi n_1 k/N}$$

$$= \frac{1}{F}\sum_{n_1=0}^{N-1} s_p(n_1\Delta t_1)e^{-j2\pi n_1 k/N} \qquad k = 0, 1, 2, \cdots, N-1 \quad (8.131)$$

$$s_p(n_1\Delta t_1) = \sum_{\ell=-\infty}^{\infty} s(n_1\Delta t_1 + \ell T) \quad (8.132)$$

is periodic of period $T = N\Delta t_1$. Figure 8.18 is a graphical illustration of Eq. 8.132 for a triangular pulse $s(t)$.

For practical case $s(t)$ is usually restricted to have nonzero values for $t > 0$, then from Eq. 8.132 we see that if we take $T = N\Delta t_1$ sufficiently large so that $s(t)$ is negligible for $t \geqslant T$, then we have

$$s_p(n_1\Delta t_1) \approx s(n_1\Delta t_1) \qquad 0 \leqslant n_1\Delta t_1 \leqslant T \quad (8.133)$$

Hence, if we take N samples of $s(t)$ as the approximations to $s_p(t)$, then from Eq. 8.131 we can find a set of N values for the frequency function $S_p(f_1)$ at evenly spaced points

$$f_1 = k\Delta f_1 = k\frac{1}{N\Delta t_1} = k\frac{F}{N} \qquad k = 0, 1, 2, \cdots, N-1 \quad (8.134)$$

If again F is large enough, then the sampled $S_p(k\Delta f_1)$ will be a good approximation to $S(f)$ at those sampling points within the range $[F/2, F/2]$.

Let us now look at one example. For the single triangular pulse $s(t)$ shown in Fig. 8.19, if we take $T = 1$, then the $s_p(t)$ generated from $s(t)$ in Fig. 8.19 by using Eq. 8.132 is identical to $s(t)$ in the interval $[0, T]$. The

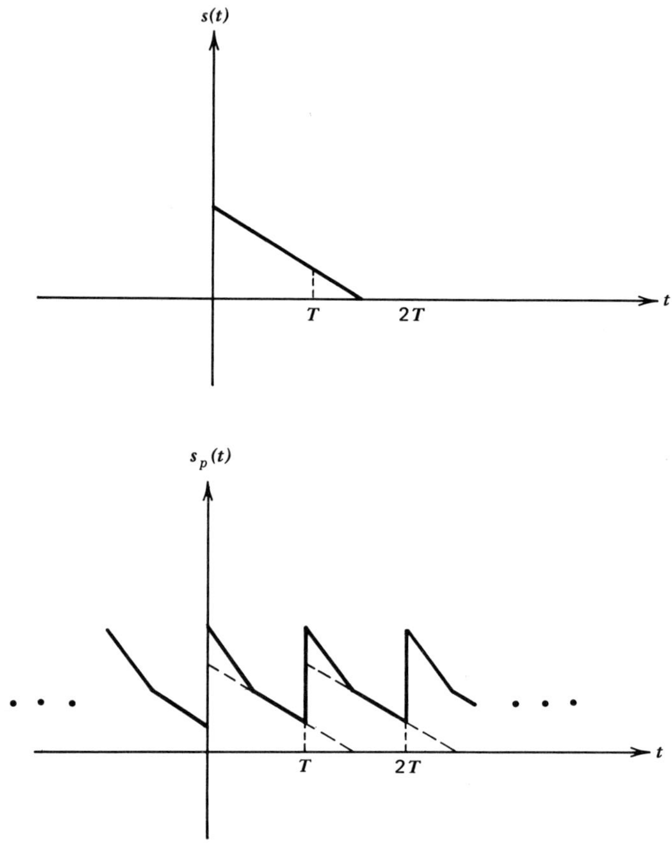

Fig. 8.18 The relationship between $s(t)$ and $s_p(t)$.

Fourier transform of the triangular pulse $s(t)$ is found to be

$$S(f) = \int_{-\infty}^{\infty} s(t)e^{-j2\pi ft}dt = \int_0^1 te^{-j2\pi ft}dt = \frac{j}{2\pi f} \qquad \text{for } f = \pm 1, \pm 2, \ldots$$

$$S(0) = \int_0^1 tdt = \frac{1}{2} \tag{8.135}$$

So for $f > 10$, the magnitude of $S(f)$ is negligible compared to the magnitude of $S(f)$ for $f < 10$. Hence let $F = 8$, then we have

$$\Delta t_1 = \frac{1}{F} = 0.125$$

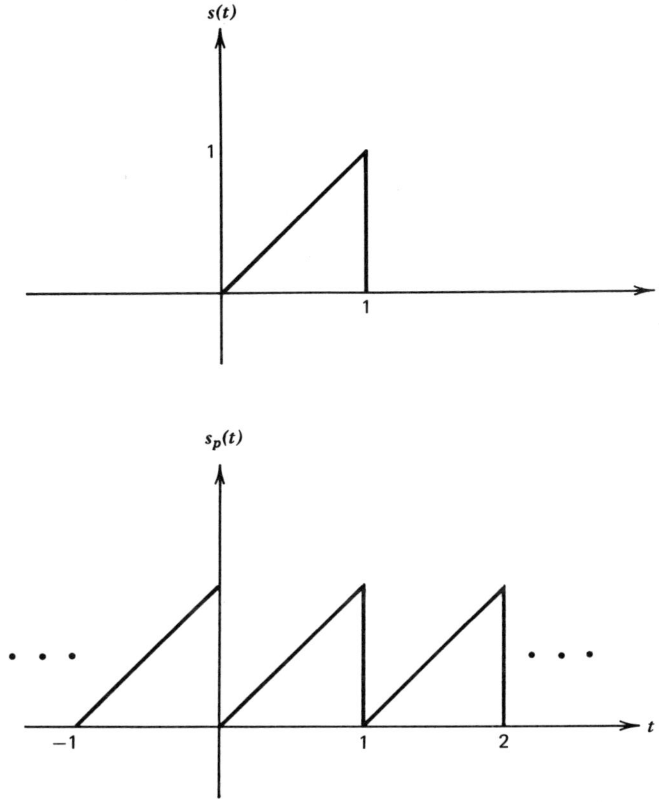

Fig. 8.19 A triangular pulse and the pulse train.

and the number of samples N and increment frequency Δf_1 are

$$N = \frac{T}{\Delta t_1} = \frac{1}{0.125} = 8 \qquad \Delta f_1 = \frac{F}{N} = 1$$

So we take $N = 8$ samples within the interval [0, 1] and we have

$$\begin{aligned}
s(0) &= 0 &&= s_p(0)\\
s(\Delta t_1) &= 0.125 &&= s_p(\Delta t_1)\\
s(2\Delta t_1) &= 0.25 &&= s_p(2\Delta t_1)\\
s(3\Delta t_1) &= 0.375 &&= s_p(3\Delta t_1)\\
s(4\Delta t_1) &= 0.50 &&= s_p(4\Delta t_1)\\
s(5\Delta t_1) &= 0.625 &&= s_p(5\Delta t_1)\\
s(6\Delta t_1) &= 0.75 &&= s_p(6\Delta t_1)\\
s(7\Delta t_1) &= 0.875 &&= s_p(7\Delta t_1)
\end{aligned}$$

Now we can apply Eq. 8.131 and have the following result for the first four values of $S_p(k\Delta f_1)$ with $k = 0, 1, 2, 3$.

$$S_p(0\Delta f_1) = S_p(0) = \tfrac{1}{8} \sum_{n_1=0}^{N-1} s_p(n_1\Delta t_1)\, e^{-j2\pi n_j 0/8} = 0.4375$$

$$\begin{aligned} S_p(\Delta f_1) &= S_p(1) = -0.0625 + j0.1509 \\ S_p(2\Delta f_1) &= S_p(2) = -0.0625 + j0.0625 \\ S_p(3\Delta f_1) &= S_p(3) = -0.0625 + j0.0259 \end{aligned}$$

The exact values of the nonperiodic $S(f)$ at $f = 0, 1, 2, 3$ are the following from Eq. 8.135.

$$\begin{aligned} S(0) &= 0.5 \\ S(1) &= j0.1592 \\ S(2) &= j0.0796 \\ S(3) &= j0.0531 \end{aligned}$$

For only 8 samples we see that $S_p(k\Delta f_1)$ is already a good approximation for the nonperiodic $S(f)$ within the interval $[0, F]$.

Now we are ready to define the *Finite Discrete Fourier Transform* or FDFT for a general discrete time sequence. Let us assume that we have a set of N evenly spaced time sequence $s(n_1\Delta t_1)$ with time intervals Δt_1 and $n_1 = 0, 1, \cdots, N-1$. Then we can define periodic time sequence as

$$s_p(n_1\Delta t_1) = \sum_{\ell=-\infty}^{\infty} s(n_1\Delta t_1 + \ell N\Delta t_1) \tag{8.136}$$

which is simply $s(n_1\Delta t_1)$ repeated at each period $T = N\Delta t_1$. In the interval $[0, N\Delta t_1]$ we have

$$s_p(n_1\Delta t_1) = s(n_1\Delta t_1) \qquad n_1 = 0, 1, \cdots, N-1 \tag{8.137}$$

Let $F = 1/\Delta t_1$. Now the FDFT of $s_p(n_1\Delta t_1)$ is defined as

$$S_p(k_1\Delta f_1) = \frac{1}{F} \sum_{n_1=0}^{N-1} s_p(n_1\Delta t_1)\, e^{-j2\pi n_1 k_1/N} \qquad k_1 = 0, 1, \cdots, N-1 \tag{8.138}$$

which can easily be shown as a periodic function with a period $N\Delta f_1 = F$ where $\Delta f_1 = 1/N\Delta t_1 = F/N$. For convenience in notation let us drop the subscript p and assume the given sequence $s(n_1\Delta t_1)$ to be periodic. Hence we define the FDFT of a periodic sequence $s(n_1\Delta t_1)$ with period $T = N\Delta t_1$ as

$$S(k_1\Delta f_1) = \frac{1}{F} \sum_{n_1=0}^{N-1} s(n_1\Delta t_1)\, e^{-j2\pi n_1 k_1/N} \qquad k_1 = 0, 1, \cdots, N-1 \tag{8.139}$$

which is again a periodic function with period $F = N\Delta f_1$. Further let us put $S_{k_1} \triangleq S(k_1\Delta f_1)$, $s_{n_1} \triangleq s(n_1\Delta t_1)$ and $\omega = e^{j2\pi/N}$ to simplify our notations. Then we have

$$S_{k_1} = \frac{1}{F}\sum_{n_1=0}^{N-1} s_{n_1}\omega^{-n_1k_1} \qquad k_1 = 0, 1, \cdots, N-1 \tag{8.140}$$

The inverse of Eq. 8.140 can be shown as

$$s_{n_2} = \frac{F}{N}\sum_{k_1=0}^{N-1} S_{k_1}\omega^{n_2k_1} \qquad n_2 = 0, 1, \cdots, N-1 \tag{8.141}$$

In fact if we substitute Eq. 8.140 into the above equation we have

$$\begin{aligned} s_{n_2} &= \frac{F}{N}\sum_{k_1=0}^{N-1}\frac{1}{F}\sum_{n_1=0}^{N-1} s_{n_1}e^{j2\pi k_1(n_2-n_1)/N} \\ &= \frac{1}{N}\sum_{n_1=0}^{N-1} s_{n_1}\sum_{k_1=0}^{N-1} e^{j2\pi k_1(n_2-n_1)/N}. \end{aligned} \tag{8.142}$$

Now for the sum on k_1, we have

$$\begin{aligned} \sum_{k_1=0}^{N-1} e^{j2\pi k_1(n_2-n_1)/N} &= 0 \qquad n_1 \neq n_2 \\ &= N \qquad n_1 = n_2 \end{aligned} \tag{8.143}$$

Hence we have

$$s_{n_2} = \frac{1}{N}\sum_{n_1=n_2}^{N-1} s_{n_1}N = s_{n_1} \tag{8.144}$$

The *Finite Discrete Fourier Series* or FDFS can similarly be defined as follows. For a periodic time sequence

$$s(n_1\Delta t_1) = s(n_1\Delta t_1 \pm \ell N\Delta t_1) \qquad n_1 = 0, 1, \cdots, N-1 \tag{8.145}$$

where ℓ is any integer and $T = N\Delta t_1$, then the FDFS of this sequence is

$$\alpha_{k_1} = \frac{1}{N}\sum_{n_1=0}^{N-1} s_{n_1}e^{-j2\pi n_1k_1/N} \qquad k_1 = 0, 1, \cdots, N-1 \tag{8.146}$$

A heuristic derivation of this definition is the following. First let $s(n_1\Delta t_1)$ be the value of a periodic function $s(t)$ at $t = n_1\Delta t_1$. Then we can approximate $s(t)$ over the interval $[0, T]$ as an impulse train. Thus we have

$$\begin{aligned} s(t) &= \int_0^T s(\tau)\delta(t-\tau)d\tau \\ &\approx \sum_{i=0}^{N-1} s(i\Delta t_1)\delta(t-i\Delta t_1)\Delta t_1 \end{aligned}$$

where $\Delta t_1 = T/N$. Then the coefficients of the Fourier series expansion of $s(t)$ is obtained as

$$\begin{aligned}
\beta_n &= \frac{1}{T}\int_0^T s(t)e^{-jn(2\pi/T)t}dt \\
&\approx \frac{\Delta t_1}{T}\int_0^T \sum_{i=0}^{N-1} s(i\Delta t_1)\delta(t-i\Delta t_1)e^{-jn(2\pi/T)t}dt \\
&= \frac{1}{N}\sum_{i=0}^{N-1} s(i\Delta t_1)e^{-jn(2\pi/T)i\Delta t_1} \\
&= \frac{1}{N}\sum_{i=0}^{N-1} s_i e^{-j2\pi in/N} \qquad n = 0, 1, 2, \cdots \\
&= \alpha_n
\end{aligned} \tag{8.147}$$

which is a periodic sequence with period N. Hence we could write

$$\alpha_n = \alpha_{n\pm N} = \alpha_{n\pm 2N} = \cdots$$

This leads to the definition of Eq. 8.146. Therefore we expect that α_{k_1} is a good approximation to β_{k_1} if N is sufficiently large. The time sequence s_{n_1} can easily be shown as equal to

$$s_{n_1} = \sum_{k_1=0}^{N-1} \alpha_{k_1}e^{-j2\pi n_1 k_1/N} \qquad n_1 = 0, 1, \cdots, N-1 \tag{8.148}$$

For another example, let us consider a periodic square pulse $s(t)$ as shown in Fig. 8.20. The Fourier series expansion of this signal is found as

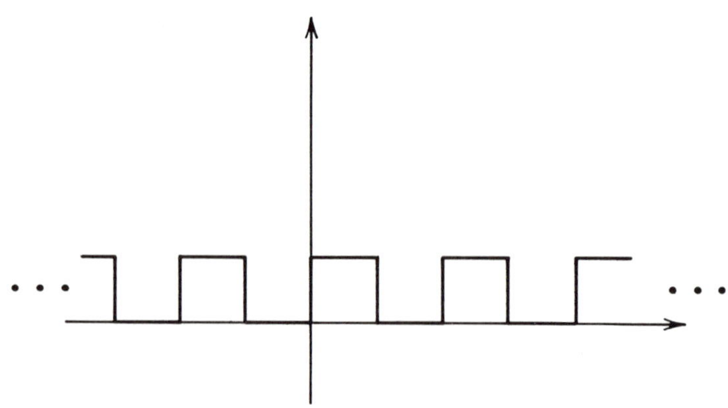

Fig. 8.20 A square pulse train.

$$\beta_k = \frac{1}{T}\int_0^T s(t)e^{-j2\pi kt/T}dt$$

$$= \frac{1}{2}\left[\int_0^1 e^{-j2\pi kt/2}dt\right]$$

$$= \frac{1}{2\pi k}\left[je^{-j\pi k} - j\right] \qquad k = \pm 1, \pm 2, \cdots$$

$$\beta_0 = \frac{1}{2}\int_0^1 dt = \frac{1}{2} \tag{8.149}$$

So the first four coefficients are found as

$$\beta_0 = \frac{1}{2} = 0.50$$

$$\beta_1 = \frac{1}{2\pi}\left[je^{-j\pi} - j\right] = -j0.3183$$

$$\beta_2 = \frac{1}{4\pi}\left[je^{-j2\pi} - j\right] = 0$$

$$\beta_3 = \frac{1}{6\pi}\left[je^{-j3\pi} - j\right] = -j0.1061$$

If we take sixteen samples from the given $s(t)$, then we have $N = 16$, and $\Delta t_1 = T/N = 2/16 = 0.125$. The samples values are:

$$s(i\Delta t_1) = 1 \qquad i = 0, 1, 2, \cdots, 7$$
$$= 0 \qquad i = 8, 9, \cdots, 15$$

For this set of sampled sequence we obtain 16 coefficients αk_1 with $k_1 = 0, 1, \cdots, 15$ using Eq. (8.147). The first four coefficients are

$$\alpha_0 = \frac{1}{16}\sum_{i=0}^{15} s(i\Delta t_1)\, e^{-j(2\pi/T)\cdot 0\cdot i\Delta t_1} = 0.50$$

$$\alpha_1 = \frac{1}{16}\sum_{i=0}^{15} s(i\Delta t_1)\, e^{-j(2\pi/2)\cdot 1\cdot i\Delta t_1} = 0.0625 - j0.2883$$

$$\alpha_2 = \frac{1}{16}\sum_{i=0}^{15} s(i\Delta t_1)\, e^{-j\pi\cdot 2\cdot i\Delta t_1} = 0$$

$$\alpha_3 = \frac{1}{16}\sum_{i=0}^{15} s(i\Delta t_1)\, e^{-j\pi\cdot 3\cdot i\Delta t_1} = 0.0625 - j0.0935$$

We see they are indeed quite close to the first four exact Fourier coefficients of the original $s(t)$.

The convolution for two sets of time sequences s_{n_1} and r_{n_1} each with N samples spaced at intervals $\Delta t_1 = 1/F$ can be defined similarly as the convolution of two time functions. Thus we have

$$u_m = s_{n_1} r_{n_1} \stackrel{\Delta}{=} \frac{1}{F} \sum_{n_1=0}^{N-1} s_{n_1} r_{m-n_1} \qquad m = 0, 1, \cdots, N-1 \tag{8.150}$$

Substituting the inverse FDFT for the time sequence s_{n_1} on the right-hand side of Eq. 8.150 we have

$$\begin{aligned} u_m &= \sum_{n_1=0}^{N-1} \left[\frac{1}{N} \sum_{k_1=0}^{N-1} S_{k_1} e^{j2\pi n_1 k_1/N} \right] r_{m-n_1} \\ &= \frac{1}{N} \sum_{k_1=0}^{N-1} S_{k_1} e^{j2\pi m k_1/N} \left[\sum_{n_1=0}^{N-1} r_{m-n_1} e^{-j2\pi(m-n_1)k_1/N} \right] \\ & \qquad m = 0, 1, \cdots, N-1 \end{aligned} \tag{8.151}$$

Let $m - n_1 = m_1$, then we have

$$\begin{aligned} u_m &= \frac{1}{N} \sum_{k_1=0}^{N-1} S_{k_1} \omega^{mk_1} \left[\sum_{m_1=m}^{m+1-N} r_{m_1} e^{-j2\pi m_1 k_1/N} \right] \\ &= \frac{1}{N} \sum_{k_1=0}^{N-1} F S_{k_1} \omega^{mk_1} \left[\frac{1}{F} \sum_{m_1=0}^{N-1} r_{m_1} \omega^{-m_1 k_1} \right] \\ &= \frac{F}{N} \sum_{k_1=0}^{N-1} S_{k_1} R_{k_1} \omega^{mk_1} \qquad m = 0, 1, 2, \cdots N \end{aligned} \tag{8.152}$$

by using the fact that r_{m_1} is a periodic sequence with period N. Here R_{k_1} is the FDFT of the time sequence r_{n_1}. Equation 8.152 also says that the FDFT of the convolution of s_{n_1} and r_{n_1} can be expressed as

$$U_{k_1} = S_{k_1} R_{k_1} \qquad k_1 = 0, 1, \cdots, N-1 \tag{8.153}$$

which is of course also periodic with period N.

This relation is very useful in finding the convolution of two time functions $s(t)$ and $r(t)$. We first sample $s(t)$ and $r(t)$ and find very good approximations S_{k_1} and R_{k_1} for the Fourier transforms $S(f)$ and $R(f)$ respectively by using FDFT. Then Eq. 8.152 can be used to find the discrete convolution u_m that should be very close to the exact convolution since U_{k_1} is a good approximation for $S(f)R(f)$.

Table 8.6.1 Main Properties of the Usual and Finite Discrete Fourier Transforms

Property	Usual One	FDFT or FDFS
Definition	$S(f) = \int_{-\infty}^{\infty} s(t)\, e^{-j2\pi ft}\, dt$	$S_{k_1} = \frac{1}{F} \sum_{n_1=0}^{N-1} s_{n_1} e^{-j2\pi n_1 k_1/N}$ $F = \frac{1}{\Delta t_1}, k_1 = 0, 1, \cdots, N-1$
Linearity	Linear	Linear
Inverse	$s(t) = \int_{-\infty}^{\infty} S(f)\, e^{j2\pi ft}\, dt$	$s_{n_1} = \frac{F}{N} \sum_{k_1=0}^{N-1} S_{k_1} e^{+j2\pi n_1 k_1/N}$ $\frac{F}{N} = \Delta f_1$ $n_1 = 0, 1, \cdots, N-1$
Convolution	$u(t) = \int_{-\infty}^{\infty} s(\tau) r(t-\tau)\, d\tau$ $= \int_{-\infty}^{\infty} S(f) R(f) e^{j2\pi ft} df$ $U(f) = S(f) R(f)$	$s_{n_1} r_{n_1} = u_m$ $= \frac{F}{N} \sum_{k_1=0}^{N-1} S_{k_1} R_{k_1} e^{j2\pi m k_1/N}$ $m = 0, 1, \cdots, N-1$ $U_{k_1} = S_{k_1} R_{k_1}$ $k_1 = 0, 1, \cdots, N-1$
Series Expansion	$\beta_k = \frac{1}{T} \int_0^T s(t)\, e^{-j2\pi kt/T}\, dt$ $k = 0, \pm 1, \pm 2, \cdots$ $s(t) = \sum_{k=0}^{\infty} \beta_k e^{j2\pi kt/T}$	$\alpha_{k_1} = \frac{1}{N} \sum_{i=0}^{N-1} s_i e^{-j2\pi i k_1/N}$ $k_1 = 0, 1, \cdots, N-1$ $s_i = \sum_{k_1=0}^{N-1} \alpha_{k_1} e^{j2\pi k_1/N}$ $i = 0, 1, \cdots, N-1$

We list in Table 8.6.1 the main properties of the usual Fourier transform and the similar ones for the FDFT or FDFS.

The usefulness of the FDFT or FDFS is further enhanced by the discovery of a very efficient algorithm to carry out the finite summation involved. This is the so-called Fast Fourier Transform or FFT. We will discuss this in the next section.

8.7 DIGITAL COMPUTATION VIII: FAST FOURIER TRANSFORM

Suppose that we are given a set of N evenly spaced sampled values of a continuous signal $s(t)$ within an interval $[0, T]$. So $s(n_1\Delta t_1)$ is known for $n_1 = 0, 1, \cdots, N-1$ and

$$\Delta t_1 = \frac{1}{F} = \frac{T}{N}$$

Now we can find the FDFT of the given set of $s(n_1\Delta t_1)$ by using the following formula, that is,

$$S_{k_1} = \frac{1}{F}\sum_{n_1=0}^{N-1} s(n_1\Delta t_1)e^{-j2\pi k_1 n_1/N} \qquad k_1 = 0, 1, \cdots, N-1 \quad (8.154)$$

We see that we need $N+1$ complex multiplications to find each S_{k_1}. So in total N^2+N multiplications are needed to find the complete set of S_{k_1}'s. This is time consuming if N is large, which is often the case since Eq. 8.154 is used to find approximations for the Fourier transform $S(f)$ of $s(t)$.

Fortunately due to the periodicity of the exponential function involved in Eq. 8.154 we could evaluate the whole set of S_{k_1}'s in approximately $\frac{1}{2}N\log_2 N$ complex multiplications if N is chosen as powers of 2. This is the so-called Fast Fourier Transform or FFT used to evaluate the FDFT of a time sequence. Clearly it can also be used to evaluate the FDFS of a periodic time sequence or the discrete convolution of two time sequences.

The principle of FFT can best be explained using some examples. So let $s(t)$ be given at only four discrete points $t = k\Delta t_1$ where $k = 0,1,2,3$ and $\Delta t_1 = T/4$. Hence $N = 4$ and Eq. 8.154 becomes

$$S_{k_1} = \frac{T}{4}\sum_{n_1=0}^{3} s(n_1\Delta t_1)e^{-j2\pi k_1 n_1/4} \qquad k = 0,1,2,3 \qquad (8.155)$$

For simplicity we denote $s(n_1\Delta t_1)$ as s_{n_1} and $e^{-j2\pi/4}$ as ω. Since $\omega^{k_1 n_1}$ is a periodic function, we have the following identities

$$\omega^{2\ell} = \begin{cases} -1 & \text{for } \ell \text{ odd} \\ 1 & \text{for } \ell \text{ even} \end{cases} \qquad (8.156)$$

Now Eq. 8.155 can be written for each k_1 as follows:

$$\begin{aligned} S_0 &= \frac{T}{4}[s_0 + s_1 + s_2 + s_3] \\ S_1 &= \frac{T}{4}[s_0 + s_1\omega + s_2\omega^2 + s_3\omega^3] \\ S_2 &= \frac{T}{4}[s_0 + s_1\omega^2 + s_2\omega^4 + s_3\omega^6] \\ S_3 &= \frac{T}{4}[s_0 + s_1\omega^3 + s_2\omega^6 + s_3\omega^9] \end{aligned} \qquad (8.157)$$

Regrouping the terms using the identities in Eq. 8.156, we have

$$\begin{aligned}
\frac{4S_0}{T} &= (s_0+s_2)+(s_1+s_3)\\
\frac{4S_1}{T} &= (s_0-s_2)+(s_1-s_3)\omega\\
\frac{4S_2}{T} &= (s_0+s_2)-(s_1+s_3)\\
\frac{4S_3}{T} &= (s_0-s_2)-(s_1-s_3)\omega
\end{aligned} \tag{8.158}$$

We see in Eq. 8.158 only one multiplication is needed, that is, the multiplication of (s_1-s_3) with ω. The sequence of computation could be illustrated by using the diagram in Fig. 8.21, where $\rightarrow^{1}\underset{\uparrow 1}{\oplus}$ is used to indicate the operation of addition, the numerals attached to the arrowhead are the signs of the quantities added. We also use $\rightarrow \underset{\omega}{\odot}$ to indicate a multiplication of the arrowed signal with a factor ω.

Thus starting from the given s_0, s_1, s_2, s_3, we first form the sequence of values

$$\begin{aligned}
s_0+s_2 &= s_{02}\\
s_0-s_2 &= \bar{s}_{02}\\
s_1+s_3 &= s_{13}\\
s_1-s_3 &= \bar{s}_{13}
\end{aligned} \tag{8.159}$$

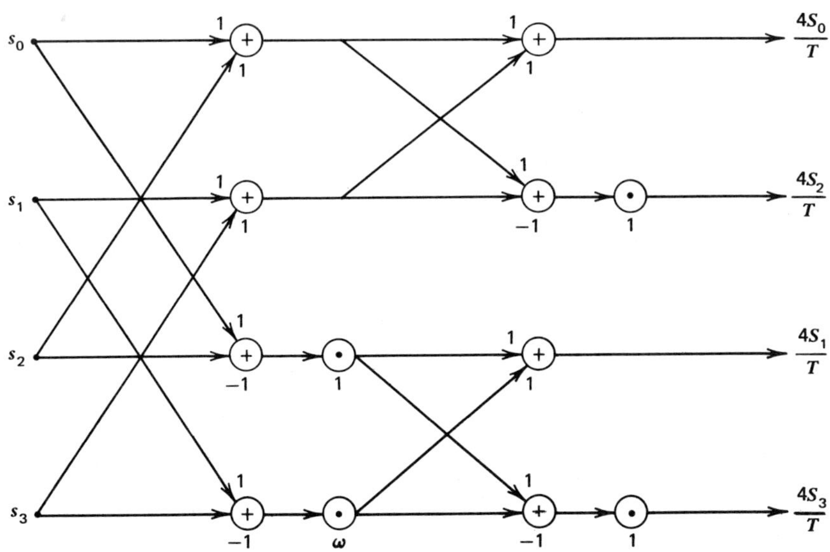

Fig. 8.21 Computation sequence of FDFT.

Next $\bar{s}_{13}$ is multiplied by ω and finally we have the FDFT in the sequence

$$
\begin{aligned}
s_{02}+s_{13} &= \frac{4}{T}S_0 \\
s_{02}-s_{13} &= \frac{4}{T}S_2 \\
\bar{s}_{02}+\omega\bar{s}_{13} &= \frac{4}{T}S_1 \\
\bar{s}_{02}-\omega\bar{s}_{13} &= \frac{4}{T}S_3
\end{aligned}
\tag{8.160}
$$

Suppose now the signal $s(t)$ is sampled at 8 points in the interval $[0, T]$. Therefore we have s_i with $i = 0, 1, \cdots, 7$, $N = 8$, and $\Delta t_1 = 1/F = T/N = T/8$. In this case let us denote $\omega = e^{-j2\pi/8}$ and we have the following identities:

$$
\omega^{4\ell} = \begin{cases} -1 & \text{for } \ell \text{ odd} \\ 1 & \text{for } \ell \text{ even} \end{cases}
\tag{8.161}
$$

With Eq. 8.161 we see that we only need three distinct values of the powers of ω—ω, ω^2, and ω^3—in order to evaluate the FDFT of the given set of s_i.

A result similar to Eq. 8.158 is given below:

$$
\begin{aligned}
\frac{8S_0}{T} &= [(s_0+s_4)+(s_2+s_6)]+[(s_1+s_5)+(s_3+s_7)] \\
\frac{8S_1}{T} &= [(s_0-s_4)+(s_2-s_6)\omega^2]+[(s_1-s_5)\omega+(s_3-s_7)\omega^3] \\
\frac{8S_2}{T} &= [(s_0+s_4)-(s_2+s_6)]+[(s_1+s_5)-(s_3+s_7)]\omega^2 \\
\frac{8S_3}{T} &= [(s_0-s_4)-(s_2-s_6)\omega^2]-[(s_1-s_5)\omega-(s_3-s_7)\omega^3]\omega^2 \\
\frac{8S_4}{T} &= [(s_0+s_4)+(s_2+s_6)]-[(s_1-s_5)+(s_3+s_7)] \\
\frac{8S_5}{T} &= [(s_0-s_4)+(s_2-s_6)\omega^2]-[(s_1-s_5)\omega+(s_3-s_7)\omega^3] \\
\frac{8S_6}{T} &= [(s_0+s_4)-(s_2-s_6)]-[(s_1+s_5)-(s_3+s_7)]\omega^2 \\
\frac{8S_7}{T} &= [(s_0-s_4)-(s_2-s_6)\omega^2]-[(s_1-s_5)\omega-(s_3-s_7)\omega^3]\omega^2
\end{aligned}
$$

And the sequence of computation is shown in Fig. 8.22. We see from this figure there are two distinct types of operations: one is purely addition and another is subtraction followed by multiplication. Let us denote the addition operation by A and the subtraction-multiplication operation by

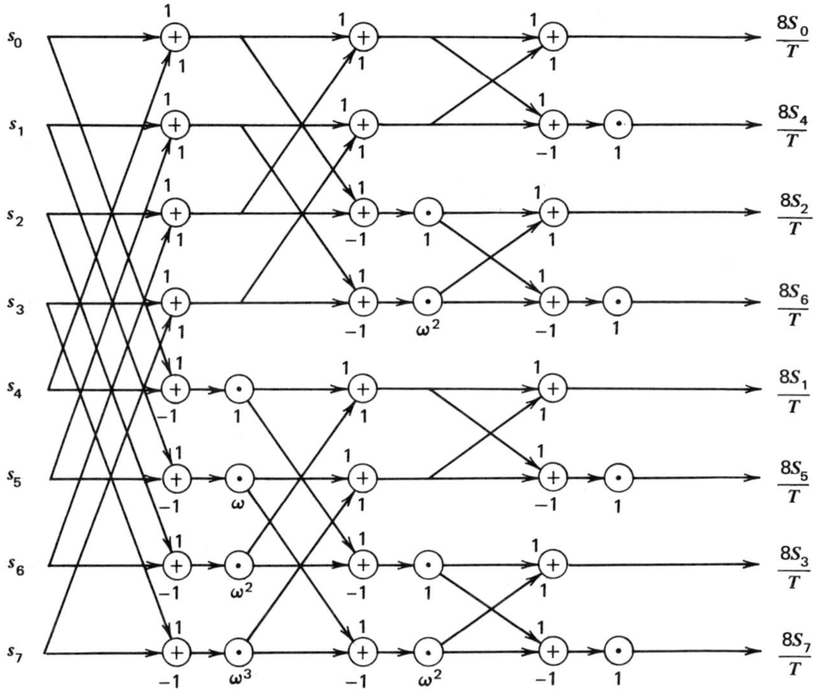

Fig. 8.22 An eight samples FFT computation sequence.

S-M. Then for this particular example we have the following stages of computation.

Stage 1: Two groups of computation are done with the first group purely *A* operations and the second group *S-M* operations.

Stage 2: Four groups of computations are done in the sequence of *A*, *S-M*, *A*, *S-M*.

Stage 3: Eight groups of computations are done in the sequence of *A*, *S-M*, *A*, *S-M*, *A*, *S-M*, *A*, *S-M*.

If we regard the multiplication by one as nontrivial, then for this example we need $1/2 \cdot 8 \log_2 8 = 12$ multiplications to find the FDFT of the given eight samples. We need also $8 \log_2 8 = 24$ additive operations.

For storage requirements we need, in this example, only eight registers for the storage of the original set of data and the intermediate result following each stage of computation.

The final results are in the so-called *bit-reversal* order. This can be explained using our eight samples example. Suppose that any data point

s_i is stored in a location addressed as i in binary digits. Hence we could replace the subscript of s_i with its equivalent binary digits. For example, s_3 would be denoted as s_{011} and s_5 is denoted as s_{101} since the binary representations of 3 and 5 are 011 and 101 respectively. Now from the results in Fig. 8.22 we have the following table indicating the contents of the eight registers before and after the computation.

Table 8.7.1 Register Contents Before and After the FFT Computation

Storage Location	Contents Before Computation	After FFT
000	s_{000}	$\frac{8}{T} S_{000}$
001	s_{001}	$\frac{8}{T} S_{100}$
010	s_{010}	$\frac{8}{T} S_{010}$
011	s_{011}	$\frac{8}{T} S_{110}$
100	s_{100}	$\frac{8}{T} S_{001}$
101	s_{101}	$\frac{8}{T} S_{101}$
110	s_{110}	$\frac{8}{T} S_{011}$
111	s_{111}	$\frac{8}{T} S_{111}$

In this table we have also replaced the subscript of S_i with its binary equivalent. We see that the final result in a certain register i is equal to $(8/T)S_{k_i}$ where the binary equivalent of k_i is expressed in reverse order of that i. Thus we see in register 001, the final result is $(8/T)S_{100}$ whose subscript 100 is just the reverse of 001. Hence we could easily restore the final results in its natural order using this bit-reversal information.

In general, for a set of $N = 2^m$ sampled points, we have m stages of computation. At the ith stage the N results from the previous stage are subdivided into 2^i groups with A and S-M operations alternatively executed. At the ith stage we also have $N2^{-i}$ distinct multipliers for the S-M operations. These are $1, \omega^{(2^i-1)\cdot 1}, \omega^{(2^i-1)\cdot 2}, \ldots, \omega^{(2^i-1)\cdot(N2^{-i}-1)}$ where $\omega = e^{-j2\pi/N}$. We also need about N storage registers for the computation. The total number of multiplications is at most equal to $(1/2)N \log_2 N$ or $(1/2)Nm$ and the number of additive operations is equal to Nm. The evaluation of FDFS or the finite discrete convolution can similarly be carried out using the FFT. The details of encoding the FFT into FORTRAN language is discussed in the actual computer program.

Computer Program

To implement the FFT for a given set of N evenly spaced points of a time sequence, we need the periodic exponential function w^i where $i = 0, 1, 2, \ldots, N-1$. Notice that since $\omega = e^{-j2\pi/N}$, we have

$$\omega^k = e^{-j(2\pi/N)k} = e^{-j(2\pi/N)k}$$

$$\omega^{k+(N/4)} = e^{-j(2\pi/N)(k+(N/4))} = -je^{-j(2\pi/N)(k)} \qquad 0 \leq k \leq \frac{N}{4} - 1$$

Hence only $1/4\,N$ distinct ω values need to be generated. These could be obtained by using the standard cosine and sine functions available in FORTRAN language. Once ω^i and s_j are available, then FFT could be carried out in $\log_2 N$ stages of computation as described previously. Finally bit-reversal for the index of the results are necessary to restore the final results in natural order.

A flowchart to evaluate the FFT of a given set of N data sequences is shown in Fig. 8.23. We have the following comments:

1. The ω^i generation and the FFT computation could be accomplished using a subroutine FFT(SCALE,S,N,M) where SCALE is the scale factor to be multiplied to the final results. S is started as the given set of data points and is ended as the FFT of the original data in bit-reversal order. N is the number of data points and M is equal to $\log_2 N$ or the number of bits to represent $0, 1, 2, \ldots, N-1$.
2. To restore the results from the FFT in natural order, a subroutine INDEX(I,M,IND) is used where I is an input integer and IND is the corresponding output integer with the following relation:

   $$(\mathrm{I}-1)_{\text{in bit}} = [(\mathrm{IND}-1)_{\text{in bit}}]_{\text{reversed}}$$

   In other words, the bit representation of IND-1 is the reverse of bit representation if $\mathrm{I}-1$. The value of I is restricted to integers equal to or greater than 1.

The subroutines are programmed to find the FFT of a set of $N = 2^M$ data points where M is at most equal to 10. If N is larger than 2^{10}, then the dimension statement in each subroutine has to be changed. Thus if

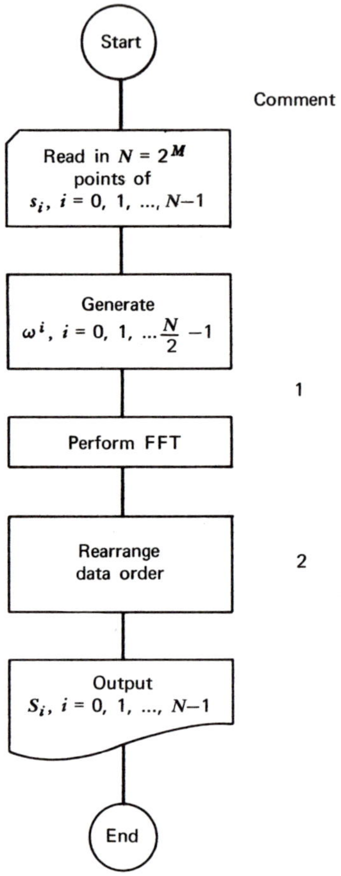

Fig. 8.23 A general flowchart for FDFT or FDES.

we have $N = 2^{11}$, then the dimension statement in INDEX is changed to

DIMENSION K(11)

and the dimension statement in FFT is changed to

DIMENSION S(2048), W(1024)

Example For the single ramp pulse in Fig. 8.19, find the FDFT of the sampled signal using 1024 samples.

Solution. For this problem we have $N = 1024$; hence $M = \log_2 1024 = 10$. From Table 8.6.1, we see the scale factor is $1/F$ or Δt_1 which is equal to $T/\mathrm{N} = 1/1024$. The given set of data points are generated by using a function $F(X) = X$.

```
      DIMENSION S(1024)
   10 FORMAT(10X,2E14.5)
      COMPLEX S
      F(X)=X
      T=1
      M=10
      N=2**M
      DT=T/N
      DO 1 I=1,N
        S(I)=F((I-1)*DT)
    1 CONTINUE
      SCALE=T/N
      CALL FFT(SCALE,S,N,M)
      N2=N/2
      DO 2 I=1,N2
        CALL INDEX(I,M,IND)
        PRINT 10, S(IND)
    2 CONTINUE
      END
```

Fig. 8.24 A FDFT program.

```
      SUBROUTINE INDEX(I,M,IND)
      DIMENSION K(10)
      I1=I-1
      DO 1 J=1,M
        IQ=I1/2
        K(J)=I1-2*IQ
        I1=IQ
    1 CONTINUE
      IND=K(1)
      DO 2 J=2,M
        IND=IND*2+K(J)
    2 CONTINUE
      IND=IND+1
      RETURN
      END
```

Fig. 8.25 The bit-reverse subroutine INDEX.

The FORTRAN program, the subroutines FFT and INDEX, and sample outputs are shown in Figs. 8.24 to 8.26. The parameters are the following:

S	N vector, original data set, also the result of FFT.
$F(X)$	Function to generate S.
T	Period or time duration of $F(X)$.
M	$\text{Log}_2 N$.
N	2^M, number of data points.
DT	T/N, interval length.
$SCALE$	Scale factor.
K	M vector, for bit-reverse purpose.
$W(I)$	Storage for $e^{-j(2\pi/n)(I-1)}$.

The computer output is given in Fig. 8.27.

```
      SUBROUTINE FFT(SCALE,S,N,M)
      DIMENSION S(1024), W(512)
      COMPLEX S,W,SAVE
    2 FORMAT(10X,2E14.5)
      N4=N/4
      ARG=(-1)*2*3.14159/N
      DO 1 I=1,N4
        X=(I-1)*ARG
        XR=COS(X)
        XI=SIN(X)
        W(I)=CMPLX(XR,XI)
        XR=-XR
        W(I+N4)=CMPLX(XI,XR)
    1 CONTINUE
      NL=N
      DO 10 L=1,M
        NL=NL/2
        L2=2**(L-1)
        DO 11 I=1,L2
          I1=I-1
          K1=2*NL*I1
          DO 12 J=1,NL
            KJ=J+K1
            SAVE=S(KJ)
            S(KJ)=SAVE+S(KJ+NL)
            S(KJ+NL)+(SAVE-S(KJ+NL))*W(1+(J-1)*L2)
   12     CONTINUE
   11   CONTINUE
   10 CONTINUE
      DO 13 K=1,N
        S(K)=SCALE*S(K)
   13 CONTINUE
      RETURN
      END
```

Fig. 8.26 The FFT subroutine.

i	S_i Real	S_i Imaginary
0	0.49951E 00	0.0
1	−0.48848E-03	0.15915E 00
2	−0.48838E-03	0.79576E-01
3	−0.48841E-03	0.53050E-01
4	−0.48833E-03	0.39787E-01
5	−0.48837E-03	0.31828E-01
6	−0.48835E-03	0.26523E-01
7	−0.48837E-03	0.22733E-01
8	−0.48831E-03	0.19890E-01
9	−0.48832E-03	0.17679E-01
10	−0.48833E-03	0.15910E-01
11	−0.48834E-03	0.14463E-01
12	−0.48831E-03	0.13257E-01
13	−0.48832E-03	0.12236E-01
14	−0.48833E-03	0.11361E-01
15	−0.48835E-03	0.10603E-01

Fig. 8.27 A sample computer output for S_i.

PROBLEMS

8-1. The Fourier series of a periodic function $s(t)$ is given in Eq. 8.2 as

$$s(t) = \tfrac{1}{2}a_0 + a_1 \cos wt + a_2 \cos 2wt + \cdots$$
$$+ b_1 \sin wt + b_2 \sin 2wt + \cdots$$

We note that the frequencies of the decomposed components are of integer multiples of the frequency of $s(t)$. To find the unknown coefficients a_i and b_i, we multiply both sides of Eq. 8.2 with a certain function, say $f(t)$, and then integrate both sides over an interval of time duration T. For example, let $f(t) = 1$ and the integration is taken over one period of $s(t) T = 2\pi/w$. We have

$$\int_0^T s(t)\,dt = \int_0^T \tfrac{1}{2}a_0\,dt + \int_0^T a_1 \cos wt\,dt + \cdots + \int_0^T b_1 \sin wt\,dt + \cdots$$

On the left-hand side, the integration depends on $s(t)$. On the right-hand side, let us look at a general term

$$\int_0^T a_k \cos kwt\,dt$$

which, upon substituting $w = 2\pi/T$, turns out to be

$$\int_0^T a_k \cos kwt\,dt = \int_0^T a_k \cos(2\pi kt/T)\,dt$$
$$= a_k(T/2\pi k) \sin(2\pi kt/T)\big|_0^T = 0 \qquad k \neq 0$$

Similarly we can show that

$$\int_0^T b_k \sin kwt\,dt = 0, \qquad k \neq 0$$

Hence we have

$$\int_0^T s(t)\,dt = \tfrac{1}{2}a_0 T$$

This gives us Eq. 8.5 for the unknown a_0. Now we are going to derive Eqs. 8.6 and 8.7.

(i) Show that

$$\int_{\alpha}^{\alpha+T} (\sin mwt \cos nwt)\, dt = 0 \qquad \text{for any } m, n.$$

$$\int_{\alpha}^{\alpha+T} (\sin mwt \sin nwt)\, dt$$

$$= \int_{\alpha}^{\alpha+T} (\cos mwt \cos nwt)\, dt = \begin{cases} 0 & n \neq m \\ \frac{1}{2}T & n = m \end{cases}$$

(Hint: Use the trigonometry identities

$$2 \sin A \cos B = \sin (A+B) + \sin (A-B)$$
$$2 \cos A \cos B = \cos (A+B) + \cos (A-B)$$
$$2 \sin A \sin B = \cos (A-B) - \cos (A+B)$$

(ii) Using the results of (i), derive Eq. 8.6 by multiplying both sides of Eq. 8.2 with cos kwt and integrating over an arbitrary period from $t = \alpha$ to $t = \alpha + T$.

(iii) Derive Eq. 8.7 by multiplying both sides of Eq. 8.2 with sin kwt and integrating over a full period.

8-2. For the periodical function $s(t) = |\cos t|$:

(i) Find the Fourier series expansion of $s(t)$.

(ii) Evaluate $s(1/2\pi)$ by adding the infinite series up to the first five terms. Compare with the exact $s(1/2\pi)$. (Using 5-digit mantissa chopped floating-point arithmetic.)

8-3. For the periodic signals of Fig. P8-3, find the Fourier series coefficients a_k and b_k.

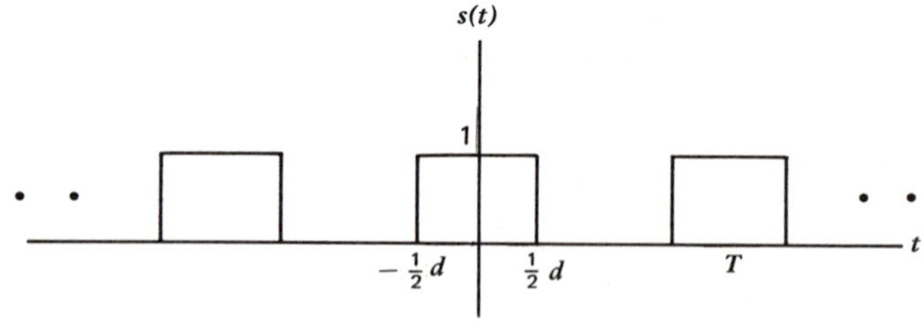

Fig. P8-3

8-4. Repeat 8-3 for the periodical signal shown in Fig. P8-4.

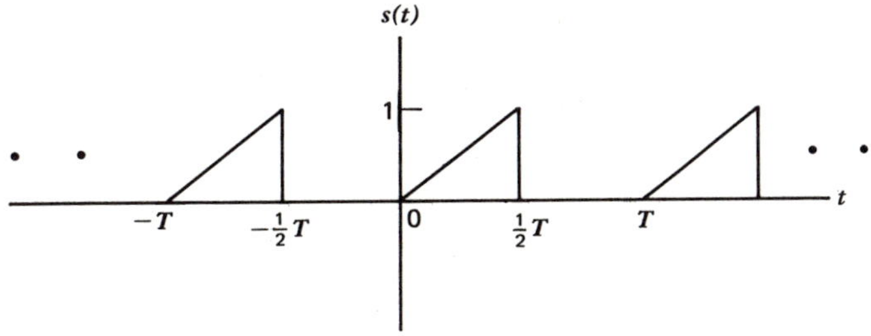

Fig. P8-4

8-4a. Repeat 8-3 for the waveforms shown for a period in Fig. P8-4*a*.

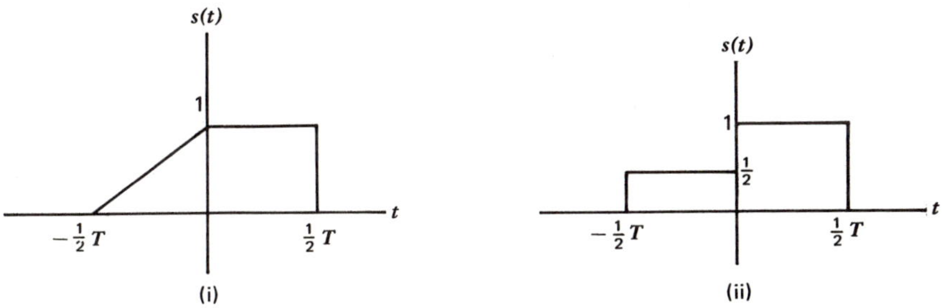

Fig. P8-4*a*

8-4b. Repeat 8-3 for the waveforms shown in Fig. P8-4*b* for a period.

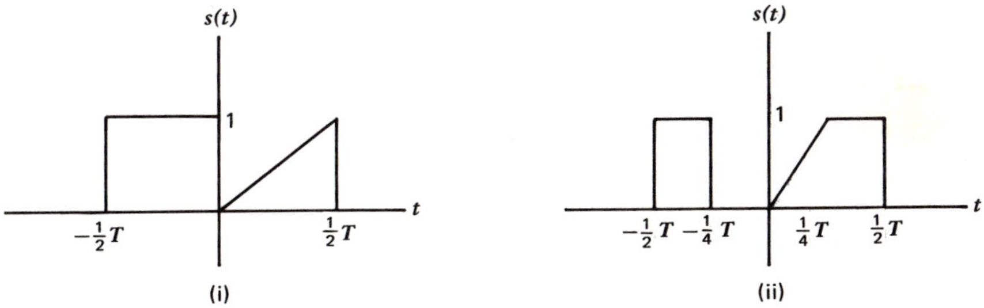

Fig. P8-4*b*

8-4c. Repeat 8-3 for the following waveforms shown for a period.

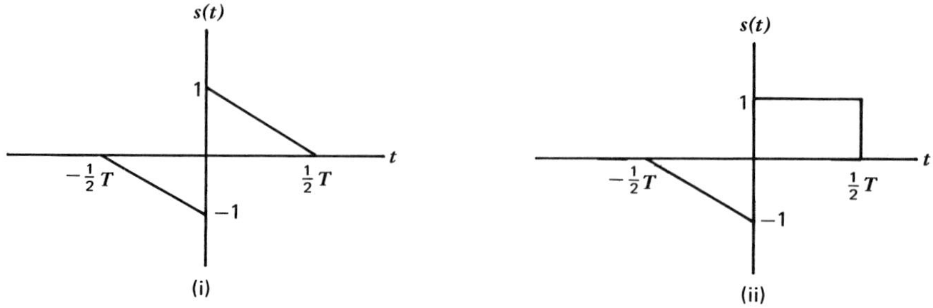

Fig. P8-4*c*

8-5. Find and plot the amplitude and phase spectra of the signal of Fig. P8-4.

8-6. For the signal of Fig. P8-4, determine the complex Fourier coefficients using the impulse function method.

8-7. Repeat 8-6 for the signal shown in Fig. P8-7.

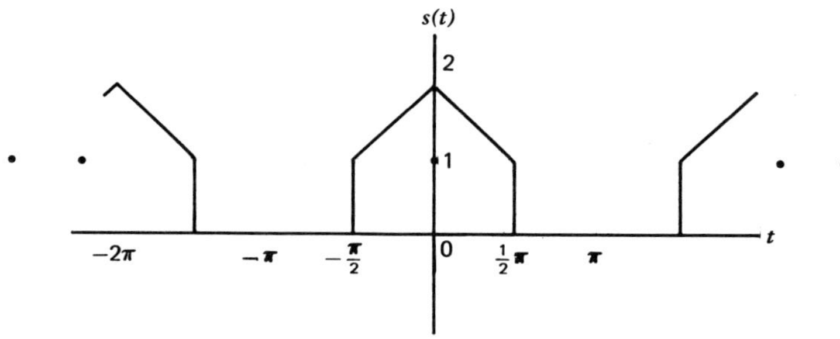

Fig. P8-7

8-7a. For the waveforms shown in Fig. P8-7*a* for a period, determine the complex Fourier coefficients and plot the magnitude and phase spectra.

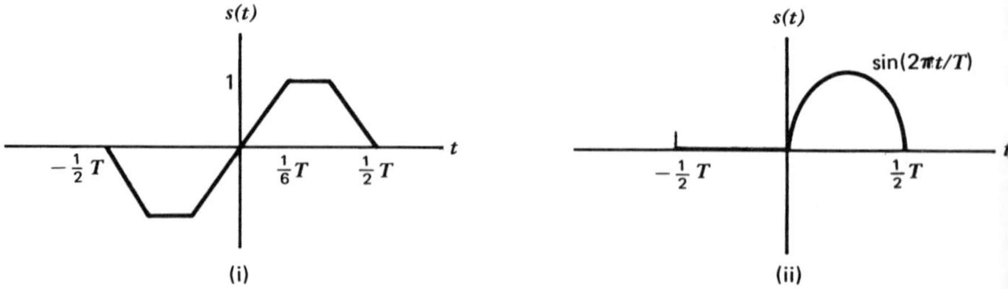

Fig. P8-7*a*

8-7b. Repeat 8-7a for the following signals shown for a period.

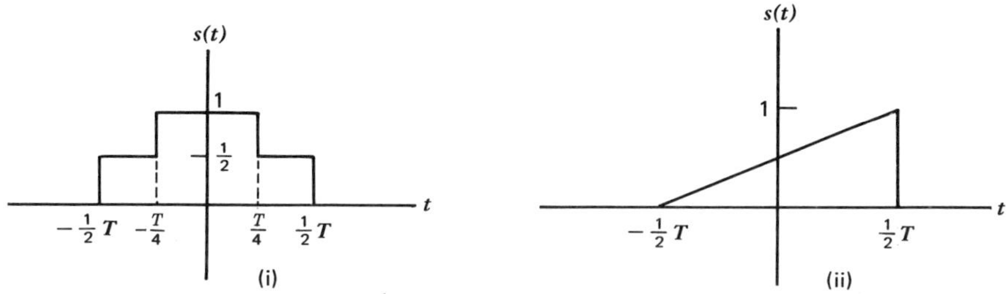

Fig. P8-7*b*

8-7c. Repeat 8-7a for the following waveforms shown for a period.

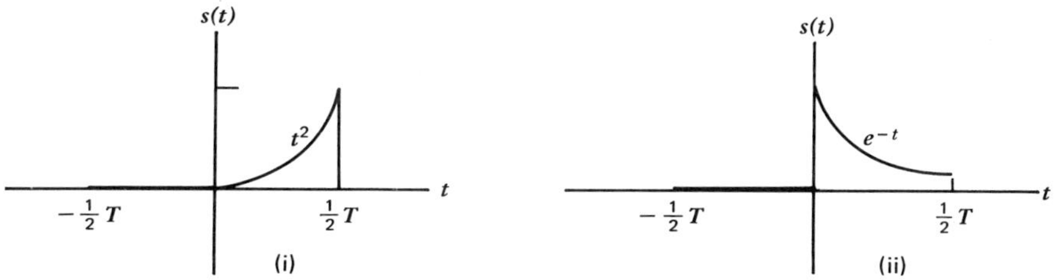

Fig. P8-7*c*

8-8. Find the Fourier transform for the functions shown in Fig. P8-8.

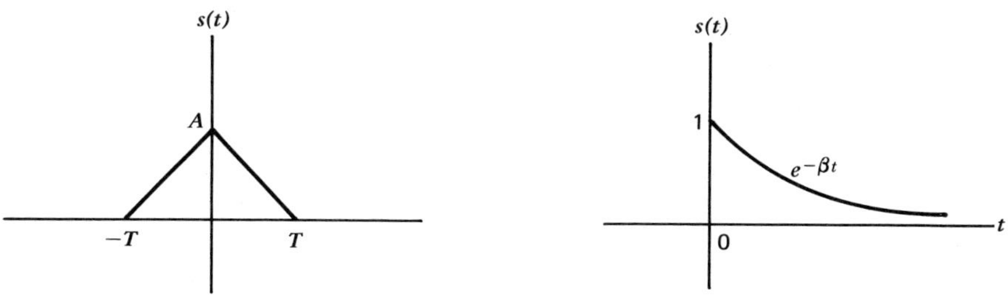

Fig. P8-8

8-9. Repeat 8-8 for the functions shown in Fig. P8-9.

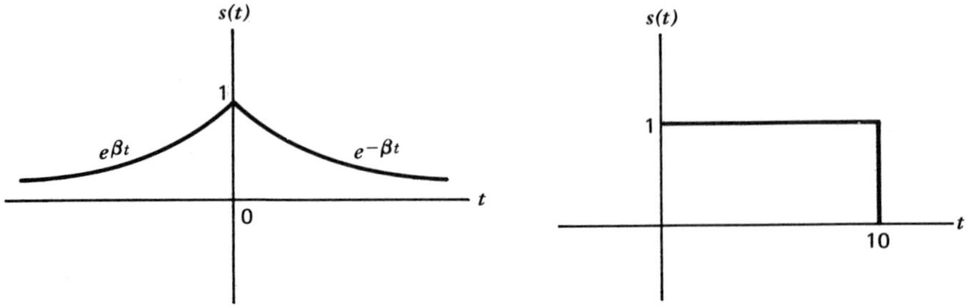

Fig. P8-9

8-9a. Find the Fourier transforms of the following functions:

(i) $s(t) = Ke^{-at} \sin w_0 t u(t), \qquad a > 0.$

(ii) $s(t) = K(t/t_0)\, e^{-t/t_0} u(t)$

8-9b. Repeat 8-9a for the following functions:

(i) $s(t) = Ke^{-(t/t_0)^2}$

(ii) $s(t) = Ke^{-a(t-t_0)} \sin w_0 t u(t), a > 0.$

8-9c. Find the Fourier transforms of the aperiodic functions shown in Fig. P8-9*c*.

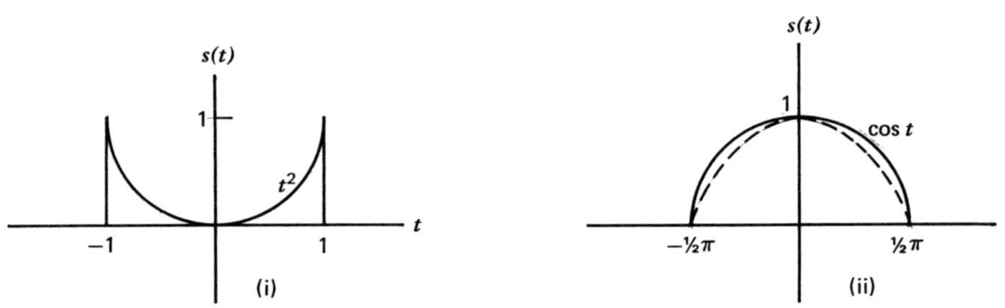

Fig. P8-9*c*

8-9d. Repeat 8-9c for the aperiodic functions shown in Fig. P8-9*d*.

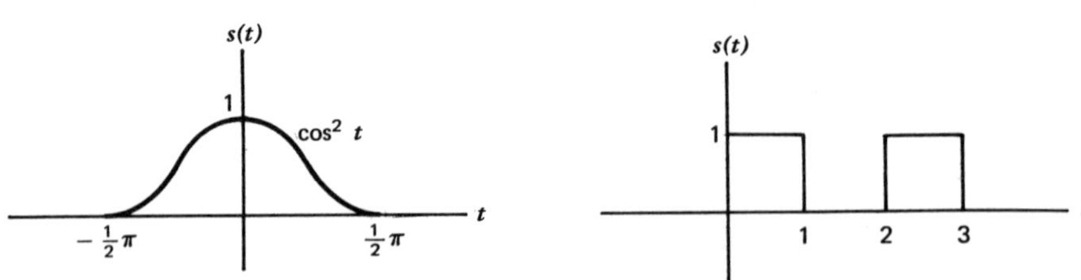

Fig. P8-9*d*

8-9e. Find the Fourier transform of the response $x(t)$ of the following equations:
(i) $x''(t)+4x'(t)+3x(t)=5e^{-t}\sin 2t$.
(ii) $x''(t)+6x'(t)+25x(t)=2\cos t\,[u(t)-u(t-1/2\pi)]$

8-9f. Repeat 8-9e using each of the signals of Fig. P8-9*c* as the right-hand side excitation.

8-9g. Repeat 8-9e using signals of Fig. P8-9*d* as excitations.

8-10. Find the Fourier transform for the following functions:
(i) $f(t)=A\delta(t)$
(ii) $f(t)=A\sin w_0 t$
(iii) $f(t)=\begin{cases}1-\dfrac{|t|}{T} & -T\leqslant t\leqslant T\\ 0 & |t|>T\end{cases}$

8-10a. Demonstrate the validity of Eq. 8-109 using the functions given in 4-8a.

8-10b. Repeat 8-10a using the functions shown in Fig. P4-9*a* for (i) $t_0=1$, (ii) $t_0=3$, (iii) $t_0=4$.

8-11. The impulse response transform of a certain network is given as

$$H(jw)=1/(1+jwT)$$

Sketch the amplitude and phase spectra of the system response due to the triangular excitation shown in Fig. P8-8. What kind of filter is this network?

8-12. Repeat 8-11 if the impulse response transform is given as

$$H(jw)=jwT/(1+jwT)$$

8-13. Find the Fourier transform of the function

$$s(t)=\begin{cases}e^{jwt} & -T\leqslant t\leqslant T\\ 0 & |t|>T\end{cases}$$

As $T\to\infty$, $f(t)$ becomes periodic in the infinite interval. What does $S(jw)$ become in this case? Is this reasonable?

8-14. For the complex form of the Fourier series

$$s(t)=\sum_{n=-\infty}^{\infty}\beta_n e^{jnwt}$$

where $s(t)$ is a periodic function with period $T=2\pi/w$, show that
(i) If $s(t)$ is real, then $\beta_n=\beta^*_{-n}$
(ii) If $s(t)$ is real and even, that is, $s(t)=s(-t)$, then β_n is real.
(iii) If $s(t)$ is real and odd, then β_n is imaginary.
(iv) If $s(t+1/2T)=s(t)$, then $\beta_n=0$ for odd n.
(v) If $s(t+1/2T)=-s(t)$, then $\beta_n=0$ for even n.

8-15. Find the time functions corresponding to the following Fourier transforms:
(i) $S(f) = \delta(2\pi f - 2\pi f_0)$
(ii) $S(f) = u(f+f_0) - u(f-f_0)$
(Hint: Use the symmetry property.)

8-16. Represent the function $f(t)$ defined by

$$f(t) = \begin{cases} 1 & |t| < 1 \\ 1/2 & |t| = 1 \\ 0 & |t| > 1 \end{cases}$$

in terms of Fourier integral, and thus show that

$$f(t) = (2/\pi) \int_0^\infty (\sin u \cos u/u)\, du$$

The integral is known as Dirichlet's discontinuous function.

8-16a. Demonstrate the validity of the Parseval's theorem, namely,

$$\int_{-\infty}^{\infty} v(t)\, i(t)\, dt = \int_{-\infty}^{\infty} V(f) I(-f)\, df$$

by actually evaluating both sides in the respective time and frequency domain for the following zero-state one-port elements subjected to the current excitation $i(t) = e^{-t}u(t)$.

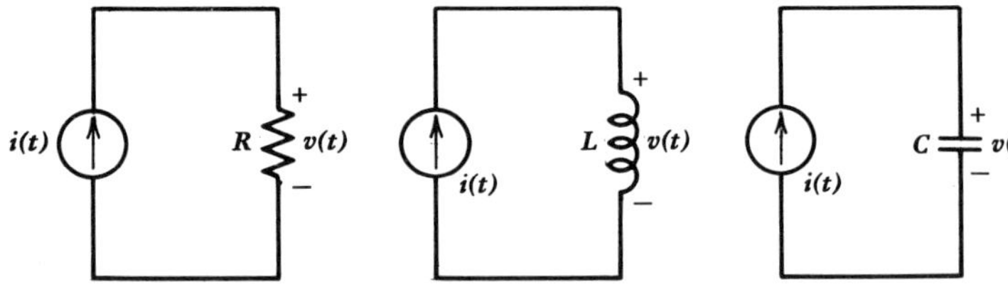

Fig. P8-16*a*

8-16b. Repeat 8-16a if the current excitation is replaced by a voltage excitation $v(t) = \sin t\,[u(t) - u(t-\pi)]$.

8-17. Using subroutine FFT with $N = 2^6, 2^8, 2^{10}$ for $0 \leqslant t \leqslant \pi$, solve problem 8-2(i) for the first ten coefficients. Compare with the exact results found in 8-2(i).

8-17a. Repeat 8-17 for the signals of 8-3 with $d = 1$, $T = \pi$.

8-17b. Repeat 8-17 for the signals of 8-4 with $T = \pi$.

8-18. Find $s(t)$ of 8-2 at $t = 0, \frac{1}{2}\pi, \pi$ by finding the FDFS inverse of the results of 8-17. Compare with the exact $s(t)$ at these time instants.

List the errors found for $N = 2^6, 2^8, 2^{10}$. Is it true that as N increases the errors will decrease?

8-18a. Repeat 8-18 for the signals of 8-3 with $d = 1$, $T = \pi$. Use the results of 8-17a.

8-18b. Repeat 8-18 for the signals of 8-4 with $T = \pi$ using the results of 8-17b.

8-19. For the exponential signal of Fig. P8-8, let $\beta = 1$. Find the FDFT of the signal by taking N samples of the signal for $0 \leq t \leq 10$ with $N = 2^6, 2^8, 2^{10}$. Compare with the exact answers at several discrete frequency points.

8-20. Repeat 8-19 for the rectangular pulse signal of Fig. P8-9.

8-21. Let the signal defined in 8-19 be the impulse response of a linear circuit C. Find the response of C due to an excitation of the rectangular pulse of Fig. P8-9 by discrete convolution using the results of 8-19 and 8-20. Compare with the exact answers at several discrete time instants. List the errors for $N = 2^6, 2^8, 2^{10}$.

8-21a. Find the FDFT of the functions $e(t)$ and $h(t)$ given in 4-8a(iii) with $T = 10$ and $N = 2^6$.

8-21b. Find the discrete convolution of $e(t)$ with $h(t)$ using the results of 8-21a. Compare with the exact results of 4-8a at some discrete time instants.

CHAPTER 9
The Laplace Transform

We have seen in the previous chapter how we can apply Fourier analysis to decompose most of the periodic or aperiodic functions into components in the frequency domain. The total response of the circuit due to any such function could then be obtained by summing up the responses due to these "decomposed" components. In many applications, the function to be decomposed is identically zero before some instant, usually $t = 0$. So the general Fourier transform pair, given by Eqs. 8.59 and 8.60, becomes

$$S(f) = \int_0^{\infty} s(t) e^{-j2\pi ft} dt \tag{9.1}$$

$$s(t) = \int_{-\infty}^{\infty} S(f) e^{j2\pi ft} df \tag{9.2}$$

However, an aperiodic function still has to satisfy certain conditions in order to be Fourier transformable. For example, one of the conditions is that the improper integral

$$\int_{-\infty}^{\infty} |s(t)| dt < \infty \tag{9.3}$$

exists. Thus, the unit step function $u(t)$ where

$$u(t) = \begin{cases} 0 & t < 0 \\ 1 & t \geqslant 0 \end{cases} \tag{9.4}$$

is not Fourier transformable. In fact, for this function we have

$$\begin{aligned} U(f) &= \int_0^\infty 1e^{-j2\pi ft}dt \\ &= \left.\frac{e^{-j2\pi ft}}{-j2\pi f}\right|_0^\infty \\ &= \left.\frac{\cos 2\pi ft - j\sin 2\pi ft}{-j2\pi f}\right|_0^\infty \end{aligned} \tag{9.5}$$

and this is completely meaningless as both the cosine and sine oscillate without limit as their arguments become infinite. The reason for such results is that the improper integral

$$\int_0^\infty 1 \cdot dt = t|_0^\infty$$

does not exist. Hence one remedy to handle this case and others like it is to multiply some *convergence factor* with the original function so that the resultant function satisfies the condition set in Eq. 9.3. It is found that $e^{-\sigma t}$ where σ is some positive constant is a good choice of a convergence factor. Thus for a function $f_0(t)$ that is identically zero before $t = 0$, we define

$$f(t) = \begin{cases} 0 & t < 0 \\ e^{-\sigma t} f_0(t) & t \geqslant 0 \end{cases} \tag{9.6}$$

The σ is chosen such that $f(t)$ is Fourier transformable. The Fourier transform of $f(t)$ is

$$\begin{aligned} F_0(f) &= \int_0^\infty f(t)e^{-j2\pi ft}dt \\ &= \int_0^\infty f_0(t)e^{-(\sigma+j2\pi f)t}dt \end{aligned} \tag{9.7}$$

And the inverse transform is

$$f(t) = e^{-\sigma t} f_0(t) = \int_{-\infty}^\infty F_0(f)e^{j2\pi ft}df \tag{9.8}$$

We could of course recover $f_0(t)$ by multiplying both sides of Eq. 9.8 by $e^{\sigma t}$. This gives us

$$f_0(t) = \int_{-\infty}^\infty F_0(f)e^{(\sigma+j2\pi f)t}df \tag{9.9}$$

For an example, if we have $f_0(t) = u(t)$, then $f(t) = e^{-\sigma t} f_0(t)$ will have a Fourier transform that is

$$F(f) = \int_0^\infty e^{-(\sigma + j2\pi f)t}\, dt$$

$$= \frac{e^{-(\sigma + j2\pi f)t}}{-(\sigma + j2\pi f)}\Bigg|_0^\infty = \frac{1}{\sigma + j2\pi f} \tag{9.10}$$

for any $\sigma > 0$.

For simplicity let $s = \sigma + j2\pi f$, then we have $ds = j2\pi\, df$, then Eqs. 9.7 and 9.9 become

$$F_0(s) = \int_0^\infty f_0(t)\, e^{-st}\, dt \tag{9.11}$$

$$f_0(t) = \frac{1}{j2\pi} \int_{\sigma - j\infty}^{\sigma + j\infty} F_0(s)\, e^{st}\, ds \tag{9.12}$$

for some appropriate constant σ. Equation 9.11 is now called the *Laplace transform* of a given function $f_0(t)$. Equation 9.12 is the *inverse Laplace transform.* We should again emphasize here that to get the Laplace transform of a given function $f_0(t)$ is easier than the inversion process of finding $f_0(t)$ from its given Laplace transform $F_0(s)$. We will discuss in this chapter the properties of the Laplace transform and its applications.

9.1 THE LAPLACE TRANSFORM AND ITS PROPERTIES

We have seen how the necessity of having a larger class of functions to be Fourier transformable leads us to the Laplace transform defined in Eqs. 9.11 and 9.12. Formally, the Laplace transform of a function of time $f(t)$ is defined as

$$\mathbf{L}[f(t)] = F(s) = \int_{0-}^\infty f(t)\, e^{-st}\, dt \tag{9.13}$$

where s is the complex frequency variable,

$$s = \sigma + j\omega$$

This definition of the Laplace transform is different from the definition given in most standard texts*, in that the lower limit of integration is $t = 0-$ instead of $t = 0+$. We thus take into account the possibility that

*M. E. Van Valkenburg, *Network Analysis*, Prentice-Hall, Englewood Cliffs, N.J., 1956.

$f(t)$ may be an impulse or one of its higher derivatives. It is clear that $\mathbf{L}[\delta(t)] = 0$ for the $0+$ definition; whereas for the $0-$ definition, $\mathbf{L}[\delta(t)] = 1$. In the case when no impulses or higher derivatives of impulses are involved, then $f(0-) = f(0+)$ for continuous function. Therefore, all of the "strong results" resulting from a rigorous treatment of Laplace transforms* obtained by using $t = 0+$ as a lower limit also apply for the $0-$ definition.

In order for a function to possess a Laplace transform, it must obey the condition,

$$\int_{0-}^{\infty} |f(t)| e^{-\sigma t} dt < \infty \tag{9.14}$$

for a real, positive σ. Note that, for a function to have a Fourier transform, it must obey the condition,

$$\int_{-\infty}^{\infty} |f(t)| \, dt < \infty \tag{9.15}$$

As a result, a ramp function or a step function will not possess a Fourier transform† but will have a Laplace transform because of the added convergence factor $e^{-\sigma t}$. However, the function e^{t^2} will not even have a Laplace transform. In transient problems, the Laplace transform is preferred to the Fourier transform, not only because a larger class of waveforms have Laplace transforms, but also because the Laplace transform takes directly into account initial conditions at $t = 0-$ due to the lower limit of integration in the Laplace transform. In contrast, the Fourier transform has limits of integration $(-\infty, \infty)$ and, in order to take into account initial conditions due to a switch closing at $t = 0$, the forcing function must take a form as $f(t)\,u(t) + x(0-)\delta(t)$, where $x(0-)$ represents the initial condition.

The inverse transform, $\mathbf{L}^{-1}[F(s)]$ is

$$f(t) = \frac{1}{2\pi j} \int_{\sigma_1 - j\infty}^{\sigma_1 + j\infty} F(s)\, e^{st} ds \tag{9.16}$$

where σ_1 is a real positive quantity which is greater than the σ convergence factor in Eq. 9.14. Note that the inverse transform as defined involves a complex integration known as a *contour integration*.‡ Since it is

*D. V. Widder, *The Laplace Transform*, Princeton University Press, Princeton, 1941.

†These functions do not possess Fourier transforms in the strict sense, but may possess a *generalized* Fourier transform containing impulses in frequency; see M. J. Lighthill, *Fourier Analysis and Generalized Functions*, Cambridge University Press, New York, 1959.

‡For a lucid treatment, see S. Goldman, *Transformation Calculus and Electrical Transients*, Prentice-Hall, Englewood Cliffs, N.J., 1949, Chapter 7.

beyond the intended scope of this book to cover contour integration, we use a partial fraction expansion procedure to obtain the inverse transform. To find inverse transforms by recognition, we must remember certain basic transform pairs and also use a table of Laplace transforms. Two of the most basic transform pairs are now given. Consider first the transform of a unit step function, $u(t)$.

Example 9.1.1 $f(t) = u(t)$:

$$F(s) = \int_{0-}^{\infty} u(t)\, e^{-st}\, dt = -\frac{e^{-st}}{s}\bigg|_{0-}^{\infty} = 0 - \left(-\frac{1}{s}\right) = \frac{1}{s} \tag{9.17}$$

Next, let us find the transform of an exponential function of time.

Example 9.1.2 $f(t) = e^{at}u(t)$:

$$F(s) = \int_{0-}^{\infty} e^{at} e^{-st}\, dt = -\frac{e^{-(s-a)t}}{s-a}\bigg|_{0-}^{\infty} = \frac{1}{s-a} \tag{9.18}$$

We will now discuss a number of important properties of Laplace transforms. Using these properties we will build up a table of transforms. To facilitate this task, we illustrate each property by considering the transforms of important signal waveforms—first, the *linearity* property.

Linearity

The transform of a finite sum of time functions is the sum of the transforms of the individual functions, that is,

$$\mathbf{L}\left[\sum_i f_i(t)\right] = \sum_i \mathbf{L}[f_i(t)] \tag{9.19}$$

This property follows readily from the definition of the Laplace transform.

Example 9.1.3 $f(t) = \sin \omega t$. Expanding $\sin \omega t$ by Euler's identity, we have

$$f(t) = \frac{1}{2j}\left(e^{j\omega t} - e^{-j\omega t}\right)$$

The Laplace transform of $f(t)$ is the sum of the transforms of the individual cisoidal $e^{\pm j\omega t}$ terms. Thus

$$\mathbf{L}[\sin \omega t] = \frac{1}{2j}\left(\frac{1}{s-j\omega} - \frac{1}{s+j\omega}\right) = \frac{\omega}{s^2+\omega^2} \tag{9.20}$$

Real Differentiation

Given that $\mathbf{L}[f(t)] = F(s)$, then

$$\mathbf{L}\left[\frac{df}{dt}\right] = sF(s) - f(0-) \tag{9.21}$$

where $f(0-)$ is the value of $f(t)$ at $t = 0-$.

Proof. By definition,

$$\mathbf{L}[f'(t)] = \int_{0-}^{\infty} e^{-st} f'(t)\,dt \tag{9.22}$$

Integrating Eq. 9.22 by parts, we have

$$\mathbf{L}[f'(t)] = e^{-st} f(t)\big|_{0-}^{\infty} + s\int_{0-}^{\infty} f(t)\,e^{-st}\,dt \tag{9.23}$$

Since $e^{-st} \to 0$ as $t \to \infty$, and because the integral on the right-hand side is $\mathbf{L}[f(t)] = F(s)$, we have

$$\mathbf{L}[f'(t)] = sF(s) - f(0-) \tag{9.24}$$

Similarly, we can show for the nth derivative,

$$\mathbf{L}\left[\frac{d^n f(t)}{dt^n}\right] = s^n F(s) - s^{n-1} f(0-) - s^{n-2} f'(0-) - \cdots - f^{(n-1)}(0-) \tag{9.25}$$

where $f^{(n-1)}(0-)$ is the $(n-1)$st derivative of $f(t)$ at $t = 0-$. We thus see that differentiating by t in the time domain is equivalent to multiplying by s in the complex frequency domain. In addition, the initial conditions are taken into account by the terms $f^{(i)}(0-)$. It is this property that transforms differential equations in the time domain to algebraic equations in the frequency domain.

Example 9.1.4 $f(t) = u(t)$. Let us find $\mathbf{L}[f'(t)]$, which is the transform of the unit impulse. We know that

$$\mathbf{L}[u(t)] = \frac{1}{s} \tag{9.26}$$

Then,

$$\mathbf{L}[\delta(t)] = s\left(\frac{1}{s}\right) = 1 \tag{9.27}$$

since $u(0-) = 0$.

Real Integration

If $\mathbf{L}[f(t)] = F(s)$, then the Laplace transform of the integral of $f(t)$ is $F(s)$ divided by s, that is,

$$\mathbf{L}\left[\int_{0-}^{t} f(\tau)\, d\tau\right] = \frac{F(s)}{s} \tag{9.28}$$

Proof. By definition

$$\mathbf{L}\left[\int_{0-}^{t} f(\tau)\, d\tau\right] = \int_{0-}^{\infty} e^{-st}\left[\int_{0-}^{t} f(\tau)\, d\tau\right] dt \tag{9.29}$$

Integrating by parts, we obtain

$$\mathbf{L}\left[\int_{0-}^{t} f(\tau)\, d\tau\right] = -\frac{e^{-st}}{s}\int_{0-}^{t} f(\tau)\, d\tau\bigg|_{0-}^{\infty} + \frac{1}{s}\int_{0-}^{\infty} e^{-st} f(t)\, dt \tag{9.30}$$

Since $e^{-st} \to 0$ as $t \to \infty$, and since

$$\int_{0-}^{t} f(\tau)\, d\tau\bigg|_{t=0-} = 0$$

the above proposition follows.

Example 9.1.5 $f(t) = \sin \omega t$. Let us find

$$\mathbf{L}[\cos \omega t] = \mathbf{L}\left[\frac{1}{\omega}\frac{d}{dt}\sin \omega t\right] \tag{9.31}$$

By the real differentiation property, we have

$$\mathbf{L}\left[\frac{d}{dt}\sin \omega t\right] = s\left(\frac{\omega}{s^2 + \omega^2}\right) \tag{9.32}$$

so that

$$\mathbf{L}[\cos \omega t] = \frac{s}{\omega}\left(\frac{\omega}{s^2 + \omega^2}\right) = \frac{s}{s^2 + \omega^2} \tag{9.33}$$

In this example, note that $\sin \omega(0-) = 0$. We then have

$$\mathbf{L}\left[\int_{0-}^{t} f(\tau)\, d\tau\right] = \frac{F(s)}{s} \tag{9.34}$$

Example 9.1.6 Let us find the transform of the unit ramp function, $\rho(t) = tu(t)$. We know that

$$\int_{0-}^{t} u(\tau)\, d\tau = \rho(t) \tag{9.35}$$

Since

$$\mathbf{L}[u(t)] = \frac{1}{s} \tag{9.36}$$

then

$$\mathbf{L}[\rho(t)] = \frac{L[u(t)]}{s} = \frac{1}{s^2} \tag{9.37}$$

Differentiation by *s*

Differentiation by s in the complex frequency domain corresponds to multiplication by t in the time domain, that is,

$$\mathbf{L}[tf(t)] = -\frac{dF(s)}{ds} \tag{9.38}$$

Proof. From the definition of the Laplace transform, we see that

$$\frac{dF(s)}{ds} = \int_{0-}^{\infty} f(t)\frac{d}{ds}e^{-st}dt = -\int_{0-}^{\infty} tf(t)e^{-st}dt = -\mathbf{L}[tf(t)] \tag{9.39}$$

Example 9.1.7 Given $f(t) = e^{-\alpha t}$, whose transform is

$$F(s) = \frac{1}{s+\alpha} \tag{9.40}$$

let us find $\mathbf{L}[te^{-\alpha t}]$. By the above theorem

$$\mathbf{L}[te^{-\alpha t}] = -\frac{d}{ds}\left(\frac{1}{s+\alpha}\right) = \frac{1}{(s+\alpha)^2} \tag{9.41}$$

Similarly, we can show that

$$\mathbf{L}[t^n e^{-\alpha t}] = \frac{n!}{(s+\alpha)^{n+1}} \tag{9.42}$$

where n is a positive integer.

Complex Translation

By the complex translation property, if $F(s) = \mathbf{L}[f(t)]$, then,

$$F(s-a) = \mathbf{L}[e^{at}f(t)] \tag{9.43}$$

where a is a complex number.

Proof. By definition,

$$\mathbf{L}[e^{at}f(t)] = \int_{0-}^{\infty} e^{at}f(t)e^{-st}dt = \int_{0-}^{\infty} e^{-(s-a)t}f(t)\,dt = F(s-a) \tag{9.44}$$

Example 9.1.8 Given $f(t) = \sin \omega t$, find $\mathbf{L}[e^{-at} \sin \omega t]$. Since

$$\mathbf{L}[\sin \omega t] = \frac{\omega}{s^2+\omega^2} \tag{9.45}$$

by the above theorem,

$$\mathbf{L}[e^{-at} \sin \omega t] = \frac{\omega}{(s+a)^2+\omega^2} \tag{9.46}$$

Similarly, we can show that

$$\mathbf{L}[e^{-at} \cos \omega t] = \frac{s+a}{(s+a)^2+\omega^2} \tag{9.47}$$

Real Translation (Shifting Theorem)

Here we consider the very important concept of the transform of a shifted or delayed function of time. If $\mathbf{L}[f(t)] = F(s)$, then the transform of the function delayed by time a is

$$\mathbf{L}[f(t-a)u(t-a)] = e^{-as}F(s) \tag{9.48}$$

Proof. By definition,

$$\mathbf{L}[f(t-a)u(t-a)] = \int_a^{\infty} e^{-st} f(t-a)\, dt \tag{9.49}$$

Introducing a new dummy variable, $\tau = t-a$, we have

$$\mathbf{L}[f(\tau)u(\tau)] = \int_{0-}^{\infty} e^{-s(\tau+a)} f(\tau)\, d\tau = e^{-as} \int_{0-}^{\infty} f(\tau) e^{-s\tau}\, d\tau = e^{-as}F(s) \tag{9.50}$$

In Eq. 9.50, $f(\tau)u(\tau)$ is the shifted or delayed time function; therefore, the theorem is proved.

It is important to recognize that the term e^{-as} is a time-delay operator. If we are given the function, $e^{-as}G(s)$, and were required to find $\mathbf{L}^{-1}[e^{-as}G(s)] = g_1(t)$, we can discard, for the moment, e^{-as}, find the inverse transform $\mathbf{L}^{-1}[G(s)] = g(t)$, and then take into account the time delay by setting

$$g(t-a)u(t-a) = g_1(t) \tag{9.51}$$

Example 9.1.9 Given the square pulse $f(t)$ in Fig. 9.1, let us first find its transform $F(s)$. Then, let us determine the inverse transform of the square of $F(s)$, that is, let us find

$$f_1(t) = \mathbf{L}^{-1}[F^2(s)] \tag{9.52}$$

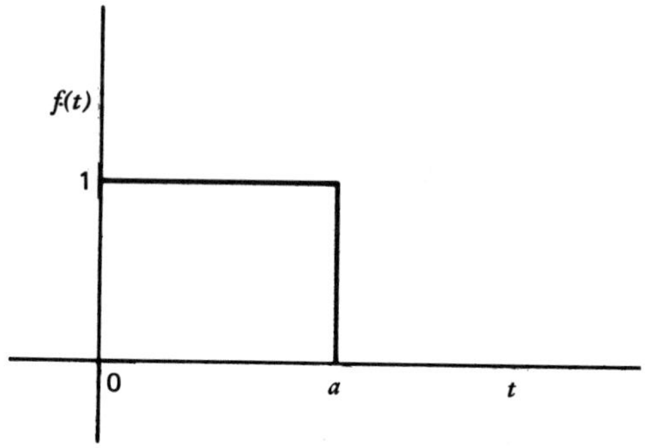

Fig. 9.1 A square pulse.

Solution. The square pulse is given in terms of step functions as

$$f(t) = u(t) - u(t-a) \tag{9.53}$$

Its Laplace transform is then

$$F(s) = \frac{1}{s}(1 - e^{-as}) \tag{9.54}$$

Squaring $F(s)$, we obtain

$$F^2(s) = \frac{1}{s^2}(1 - 2e^{-as} + e^{-2as}) \tag{9.55}$$

To find the inverse transform of $F^2(s)$, we need only to determine the inverse transform of the term with zero delay, which is

$$\mathbf{L}^{-1}\left[\frac{1}{s^2}\right] = tu(t) \tag{9.56}$$

Then

$$\mathbf{L}^{-1}[F^2(s)] = tu(t) - 2(t-a)u(t-a) + (t-2a)u(t-2a) \tag{9.57}$$

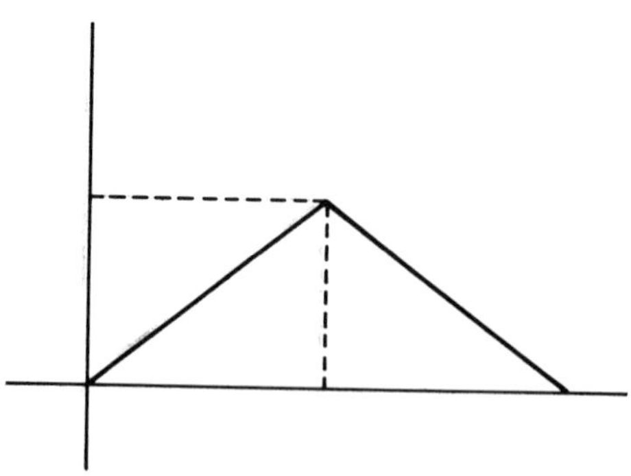

Fig. 9.2 A triangular pulse.

so that the resulting waveform is shown in Fig. 9.2. From this example, we see that the square of a transformed function *does not* correspond to the square of its inverse transform.

Example 9.1.10 In Fig. 9.3 we have a train of impulses

$$f(t) = K_0\delta(t) + K_1\delta(t-T_1) + \cdots + K_n\delta(t-nT_1) \tag{9.58}$$

The Laplace transform of this impulse train is

$$\mathbf{L}[f(t)] = K_0 + K_1e^{-sT_1} + K_2e^{-2sT_1} + \cdots + K_ne^{-nsT_1} \tag{9.59}$$

In dealing with sampled signals, the substitution $z = e^{sT_1}$ is often used. We can represent the transform of the impulse train as

$$\mathbf{L}[f(t)] = K_0 + \frac{K_1}{z} + \frac{K_2}{z^2} + \cdots + \frac{K_n}{z^n} \tag{9.60}$$

The transform in Eq. 9.60 is called the *z-transform* of $f(t)$. This transform is widely used in connection with sampled-data control systems.

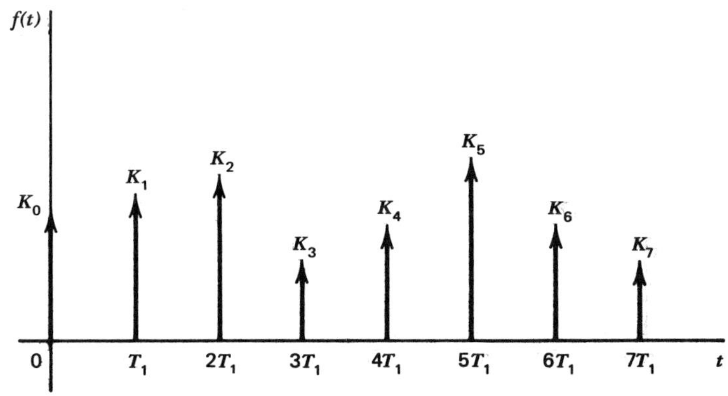

Fig. 9.3 An impulse train.

Periodic Waveforms

The Laplace transform of a periodic waveform can be obtained in two ways:

(a) through summation of an infinite series as illustrated in Example 9.1.12.

(b) through the formula derived below.

$$\mathbf{L}[f(t)] = \int_{0-}^{T} f(t)e^{-st}\,dt + \int_{T}^{2T} f(t)e^{-st}\,dt + \cdots + \int_{nT}^{(n+1)T} f(t)e^{-st}\,dt + \cdots \tag{9.61}$$

Since $f(t)$ is periodic, Eq. 9.61 reduces to

$$\mathbf{L}[f(t)] = \int_{0-}^{T} f(t)e^{-st}\,dt + e^{-sT}\int_{0-}^{T} f(t)e^{-st}\,dt + \cdots + e^{-snT}\int_{0-}^{T} f(t)e^{-st}\,dt + \cdots$$

$$= (1 + e^{-sT} + e^{-s2T} + \cdots)\int_{0-}^{T} f(t)e^{-st}\,dt$$

$$= \frac{1}{1-e^{-sT}}\int_{0-}^{T} f(t)e^{-st}\,dt \tag{9.62}$$

Example 9.1.11 Given the periodic pulse train in Fig. 9.4 let us use Eq. 9.62 to determine its Laplace transform.

$$F(s) = \frac{1}{1-e^{-sT}}\int_{0-}^{a} e^{-st}\,dt$$

$$= \frac{-1}{s(1-e^{-sT})}\, e^{-st}\bigg|_{0-}^{a}$$

$$= \frac{1}{s}\frac{1-e^{-as}}{1-e^{-sT}} \tag{9.63}$$

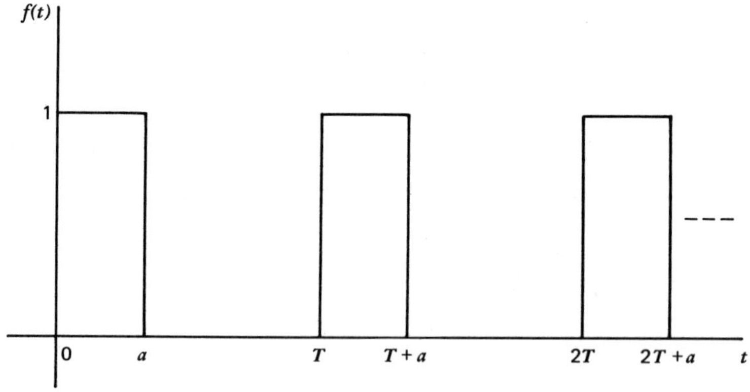

Fig. 9.4 A periodic pulse train.

Example 9.1.12 In this example we will calculate the Laplace transform of $f(t)$ in Fig. 9.4 using summation of an infinite series. The periodic pulse train can be represented as

$$f(t) = u(t) - u(t-a) + u(t-T) - u[t-(T+a)] + u(t-2T) - u[t-(2T+a)] + \cdots \tag{9.64}$$

Its Laplace transform is

$$F(s) = \frac{1}{s}\,(1 - e^{-as} + e^{-sT} - e^{-(T+a)s} + \cdots)$$

$$= \frac{1}{s}\,[1 - e^{-as} + e^{-sT}(1 - e^{-as}) + e^{-2sT}(1 - e^{-as}) + \cdots] \qquad (9.65)$$

which simplifies further to give

$$F(s) = \frac{1}{s}\,(1 - e^{-as})\,(1 + e^{-sT} + e^{-2sT} + \cdots) \qquad (9.66)$$

We then see that $F(s)$ can be given in the closed form

$$F(s) = \frac{1}{s}\,\frac{1 - e^{-as}}{1 - e^{-sT}} \qquad (9.67)$$

Other periodic pulse trains can also be given in closed form. The reader is referred to the problems at the end of this chapter.

Also at the end of the chapter is a table of Laplace transforms. Most of the entries are obtained through simple applications of the properties just discussed. It is important to keep these properties in mind, because many transform pairs that are not given in the table can be obtained by using these properties. For example, let us find the inverse transform of

$$F(s) = \frac{\omega s}{(s + \sigma)^2 + \omega^2} \qquad (9.68)$$

Since the s in the numerator implies differentiation in the time domain, we can write

$$F(s) = s\,\frac{\omega}{(s + \sigma)^2 + \omega^2} \qquad (9.69)$$

Using the differentiation property, we obtain,

$$\mathbf{L}^{-1}[F(s)] = \frac{d}{dt}\,(e^{-\sigma t} \sin \omega t)$$

$$= (-\sigma \sin \omega t + \omega \cos \omega t)e^{-\sigma t} \qquad (9.70)$$

Note that $e^{-\sigma t} \sin \omega t$ at $t = 0-$ is zero.

9.2 PARTIAL-FRACTION EXPANSIONS

In the previous section we saw that the real differentiation and real integration properties of the Laplace transform change differential equations in time to algebraic equations in frequency. Let us consider some examples of its applications.

Example 9.2.1 Let us solve the differential equation,

$$x''(t) + 3x'(t) + 2x(t) = 4e^t \tag{9.71}$$

given the initial conditions, $x(0-) = 1, x'(0-) = -1$.

Solution. We first proceed by taking the Laplace transform of the differential equation that then becomes

$$[s^2X(s) - sx(0-) - x'(0-)] + 3[sX(s) - x(0-)] + 2X(s) = \frac{4}{s-1} \tag{9.72}$$

Substituting the initial conditions into Eq. 9.72 and simplifying, we have

$$(s^2 + 3s + 2)X(s) = \frac{4}{s-1} + s + 2 \tag{9.73}$$

We then obtain $X(s)$ explicitly as

$$X(s) = \frac{s^2 + s + 2}{(s-1)(s+1)(s+2)} \tag{9.74}$$

To find the inverse transform, $x(t) = \mathbf{L}^{-1}[X(s)]$, we expand $X(s)$ into partial fractions,

$$X(s) = \frac{K_{-1}}{s-1} + \frac{K_1}{s+1} + \frac{K_2}{s+2} \tag{9.75}$$

Solving for K_{-1}, K_1, and K_2 algebraically, we obtain

$$K_{-1} = \tfrac{2}{3} \qquad K_1 = -1 \qquad K_2 = \tfrac{4}{3}$$

The final solution is the inverse transform of $X(s)$ or

$$x(t) = \tfrac{2}{3}e^t - e^{-t} + \tfrac{4}{3}e^{-2t} \tag{9.76}$$

Example 9.2.2 Given the set of simultaneous differential equations,

$$\begin{aligned} 2x'(t) + 4x(t) + y'(t) + 7y(t) &= 5u(t) \\ x'(t) + x(t) + y'(t) + 3y(t) &= 5\delta(t) \end{aligned} \tag{9.77}$$

with the initial conditions $x(0-) = y(0-) = 0$, let us find $x(t)$ and $y(t)$.

Solution. Transforming the set of equations, we obtain

$$\begin{aligned} 2(s+2)X(s) + (s+7)Y(s) &= \frac{5}{s} \\ (s+1)X(s) + (s+3)Y(s) &= 5 \end{aligned} \tag{9.78}$$

Solving for $X(s)$ and $Y(s)$ simultaneously, we have

$$X(s) = \frac{(5/s)\Delta_{11} + 5\Delta_{21}}{\Delta} \tag{9.79}$$

$$Y(s) = \frac{(5/s)\Delta_{12} + 5\Delta_{22}}{\Delta} \tag{9.80}$$

where Δ is the determinant of the set of equations in Eq. 9.78 and Δ_{ij} is the ijth cofactor of Δ. More explicitly, $X(s)$ is

$$X(s) = \frac{-5s^2 - 30s + 15}{s(s^2 + 2s + 5)} \tag{9.81}$$

Expanding $X(s)$ in partial fractions, we have

$$X(s) = \frac{K_1}{s} + \frac{K_2 s + K_3}{s^2 + 2s + 5} \tag{9.82}$$

Multiplying both sides of Eq. 9.82 by s and letting $s = 0$, we find

$$K_1 = sX(s)|_{s=0} = 3 \tag{9.83}$$

K_2 and K_3 are then obtained from the equation

$$X(s) - \frac{3}{s} = \frac{-8s - 36}{s^2 + 2s + 5} \tag{9.84}$$

A further simplification occurs by completing the square of the denominator of $X(s)$, that is,

$$s^2 + 2s + 5 = (s+1)^2 + 4 \tag{9.85}$$

As a result of Eq. 9.85 we can rewrite $X(s)$ as

$$X(s) = \frac{3}{s} - \frac{8(s+1) + 14(2)}{(s+1)^2 + (2)^2} \tag{9.86}$$

so that the inverse transform is

$$x(t) = (3 - 8e^{-t}\cos 2t - 14e^{-t}\sin 2t)\,u(t) \tag{9.87}$$

In similar fashion, we obtain $Y(s)$ to be

$$Y(s) = -\frac{1}{s} + \frac{11s + 17}{s^2 + 2s + 5} = -\frac{1}{s} + \frac{11(s+1) + 3(2)}{(s+1)^2 + (2)^2} \tag{9.88}$$

The inverse transform is then seen to be

$$y(t) = (-1 + 11e^{-t}\cos 2t + 3e^{-t}\sin 2t)u(t) \tag{9.89}$$

As we have seen, the ease with which we use transform methods depends upon how quickly we are able to obtain the partial-fraction expansion of a given transform function. In this section, we will elaborate on some simple and effective methods for partial-fraction expansions.

It should be recalled that, if the degree of the numerator is greater or equal to the degree of the denominator, we can divide the numerator by the denominator such that the remainder can be expanded more easily into partial fractions. Consider the following example,

$$F(s) = \frac{N(s)}{D(s)} = \frac{s^3 + 3s^2 + 3s + 2}{s^2 + 2s + 2} \tag{9.90}$$

Since the degree of $N(s)$ is greater than the degree of $D(s)$, we divide $D(s)$ into $N(s)$ to give

$$F(s) = s + 1 - \frac{s}{s^2 + 2s + 2} \tag{9.91}$$

Here we see the remainder term can be easily expanded into partial fractions. However, there is no real need at this point because the denominator $s^2 + 2s + 2$ can be written as

$$s^2 + 2s + 2 = (s+1)^2 + 1 \tag{9.92}$$

We can then write $F(s)$ as

$$F(s) = s + 1 - \frac{(s+1) - 1}{(s+1)^2 + 1} \tag{9.93}$$

so that the inverse transform can be obtained directly from the transform tables, namely

$$\mathbf{L}^{-1}[F(s)] = \delta'(t) + \delta(t) + e^{-t}(\sin t - \cos t) \tag{9.94}$$

From this example we see that intuition and a knowledge of the transform table can often save considerable work. Consider some further examples in which intuition plays a dominant role.

Example 9.2.3 Find the partial-fraction expansion of

$$F(s) = \frac{2s + 3}{(s+1)(s+2)} \tag{9.95}$$

If we see that $F(s)$ can also be written as

$$F(s) = \frac{(s+1) + (s+2)}{(s+1)(s+2)} \tag{9.96}$$

then the partial-fraction expansion is trivially,

$$F(s) = \frac{1}{s+1} + \frac{1}{s+2} \tag{9.97}$$

Example 9.2.4 Find the partial-fraction expansion of

$$F(s) = \frac{s+5}{(s+2)^2} \tag{9.98}$$

We see that $F(s)$ can be rewritten as

$$F(s) = \frac{(s+2)+3}{(s+2)^2} = \frac{1}{(s+2)} + \frac{3}{(s+2)^2} \tag{9.99}$$

Now, let us discuss some formal methods for partial-fraction expansions.

Denominator Having Real Roots

Consider the function

$$F(s) = \frac{N(s)}{(s-s_0)(s-s_1)(s-s_2)} \tag{9.100}$$

where s_0, s_1, and s_2 are distinct, real roots, and the degree of $N(s) < 3$. Expanding $F(s)$ we have,

$$F(s) = \frac{K_0}{s-s_0} + \frac{K_1}{s-s_1} + \frac{K_2}{s-s_2} \tag{9.101}$$

Let us first obtain the constant K_0. We proceed by multiplying both sides of the equation by $(s-s_0)$ to give

$$(s-s_0)F(s) = K_0 + \frac{(s-s_0)K_1}{s-s_1} + \frac{(s-s_0)K_2}{s-s_2} \tag{9.102}$$

If we let $s = s_0$ in the last equation, we obtain

$$K_0 = (s-s_0)F(s)|_{s=s_0} \tag{9.103}$$

Similarly, we see that the other constants can be evaluated through the general relation,

$$K_i = (s-s_i)F(s)|_{s=s_i} \tag{9.104}$$

Example 9.2.5 Let us find the partial-fraction expansion for

$$F(s) = \frac{s^2+2s-2}{s(s+2)(s-3)} = \frac{K_0}{s} + \frac{K_1}{s+2} + \frac{K_2}{s-3} \tag{9.105}$$

Using Eq. 9.104 we find

$$\begin{aligned} K_0 &= sF(s)|_{s=0} \\ &= \left.\frac{s^2+2s-2}{(s+2)(s-3)}\right|_{s=0} = \frac{1}{3} \end{aligned} \tag{9.106}$$

$$K_1 = \left.\frac{s^2+2s-2}{s(s-3)}\right|_{s=-2} = -\frac{1}{5} \tag{9.107}$$

$$K_2 = \left.\frac{s^2+2s-2}{s(s+2)}\right|_{s=3} = \frac{13}{15} \tag{9.108}$$

Denominator Having Complex Roots

Eq. 9.104 is also applicable to a function with complex roots in its denominator. Suppose $F(s)$ is given by

$$F(s) = \frac{N(s)}{D_1(s)(s-\alpha-j\beta)(s-\alpha+j\beta)}$$

$$= \frac{K_1}{s-\alpha-j\beta} + \frac{K_2}{s-\alpha+j\beta} + \frac{N_1(s)}{D_1(s)} \tag{9.109}$$

where N_1/D_1 is the remainder term. Using Eq. 9.104, we have

$$K_1 = \frac{N(\alpha+j\beta)}{2j\beta D_1(\alpha+j\beta)} \tag{9.110}$$

$$K_2 = \frac{N(\alpha-j\beta)}{-2j\beta D_1(\alpha-j\beta)} \tag{9.111}$$

where we assume that $s = \alpha \pm j\beta$ are not zeros of $D_1(s)$.

It can be shown that the constants K_1 and K_2 associated with conjugate roots are themselves conjugate. Therefore, if we denote K_1 as

$$K_1 = A + jB \tag{9.112}$$

then

$$K_2 = A - jB = K_1^* \tag{9.113}$$

If we denote the inverse transform of the complex conjugate terms as $f_1(t)$, we see that

$$f_1(t) = \mathbf{L}^{-1}\left[\frac{K_1}{s-\alpha-j\beta} + \frac{K_1^*}{s-\alpha+j\beta}\right]$$

$$= e^{\alpha t}(K_1 e^{j\beta t} + K_1^* e^{-j\beta t})$$

$$= 2e^{\alpha t}(A\cos\beta t - B\sin\beta t) \tag{9.114}$$

Example 9.2.6 Let us find the inverse transform of

$$F(s) = \frac{s^2+3}{(s^2+2s+5)(s+2)} \tag{9.115}$$

For the simple root, $s = -2$, the constant K is

$$K = (s+2)F(s)\big|_{s=-2} = \frac{7}{5} \tag{9.116}$$

For the complex conjugate roots

$$s^2 + 2s + 5 = (s+1+j2)(s+1-j2) \tag{9.117}$$

We see that $\alpha = -1$, $\beta = 2$, so

$$K_1 = \left.\frac{s^2+3}{2j\beta(s+2)}\right|_{s=-1+j2}$$

$$= \frac{-1+j2}{5} \tag{9.118}$$

Thus $A = -1/5$ and $B = 2/5$ so that

$$f(t) = -2e^{-t}(\tfrac{1}{5}\cos 2t + \tfrac{2}{5}\sin 2t) + \tfrac{7}{5}e^{-2t} \tag{9.119}$$

Denominator Having Multiple Roots

Next let us consider the case when the partial fraction involves repeated or multiple roots. We will consider two methods. The first requires differentiation, the second does not.

Method A. Suppose we are given the function

$$F(s) = \frac{N(s)}{(s-s_0)^n D_1(s)} \tag{9.120}$$

with multiple roots of degree n at $s = s_0$. The partial fraction expansion of $F(s)$ is

$$F(s) = \frac{K_0}{(s-s_0)^n} + \frac{K_1}{(s-s_0)^{n-1}} + \frac{K_2}{(s-s_0)^{n-2}} + \cdots + \frac{K_{n-1}}{s-s_0} + \frac{N_1(s)}{D_1(s)} \tag{9.121}$$

where $N_1(s)/D_1(s)$ represents the remaining terms of the expansion. The problem is to obtain $K_0, K_1, \ldots, K_{n-1}$. For the K_0 term we use the method cited earlier for simple roots

$$K_0 = (s-s_0)^n F(s)|_{s=s_0} \tag{9.122}$$

However, if we were to use the same formula to obtain the factors $K_1, K_2, \ldots, K_{n-1}$ we invariably arrive at the indeterminate 0/0 condition. Instead, let us multiply both sides of Eq. 9.121 by $(s-s_0)^n$ and define

$$F_1(s) \equiv (s-s_0)^n F(s) \tag{9.123}$$

Thus

$$F_1(s) = K_0 + K_1(s-s_0) + \cdots + K_{n-1}(s-s_0)^{n-1} + R(s)(s-s_0)^n \tag{9.124}$$

where $R(s)$ indicates the remaining terms. If we differentiate Eq. 9.124 by s, we obtain

$$\frac{d}{ds}F_1(s) = K_1 + 2K_2(s-s_0) + \cdots + K_{n-1}(n-1)(s-s_0)^{n-2} + \cdots \tag{9.125}$$

It is evident that

$$K_1 = \frac{d}{ds} F_1(s)\Big|_{s=s_0} \tag{9.126}$$

On the same basis,

$$K_2 = \frac{1}{2}\frac{d^2}{ds^2} F_1(s)\Big|_{s=s_0} \tag{9.127}$$

and in general,

$$K_j = \frac{1}{j!}\frac{d^j}{ds^j} F_1(s)\Big|_{s=s_0}; \qquad j = 0, 1, 2, \ldots, n-1 \tag{9.128}$$

Example 9.2.7 Consider the function,

$$F(s) = \frac{s-2}{s(s+1)^3} \tag{9.129}$$

which we represent in expanded form as

$$F(s) = \frac{K_0}{(s+1)^3} + \frac{K_1}{(s+1)^2} + \frac{K_2}{s+1} + \frac{A}{s} \tag{9.130}$$

The constant A for the simple root at $s = 0$ is

$$A = sF(s)|_{s=0} = -2 \tag{9.131}$$

To obtain the constants for the multiple roots we first get $F_1(s)$,

$$F_1(s) = (s+1)^3 F(s) = \frac{s-2}{s} \tag{9.132}$$

Using the general formula for the multiple root expansion, we obtain

$$K_0 = \frac{1}{0!}\frac{d^0}{ds^0}\left(\frac{s-2}{s}\right)\Big|_{s=-1} = 3 \tag{9.133}$$

$$K_1 = \frac{1}{1!}\frac{d}{ds}\left(\frac{s-2}{s}\right)\Big|_{s=-1} = \frac{2}{s^2}\Big|_{s=-1} = 2 \tag{9.134}$$

$$K_2 = \frac{1}{2!}\frac{d}{ds}\left(\frac{2}{s^2}\right)\Big|_{s=-1} = \left(-\frac{2}{s^3}\right)\Big|_{s=-1} = 2 \tag{9.135}$$

so that

$$F(s) = \frac{3}{(s+1)^3} + \frac{2}{(s+1)^2} + \frac{2}{(s+1)} - \frac{2}{s} \tag{9.136}$$

Method B. The second method for arriving at the partial fraction expansion for multiple roots requires no differentiation. It involves a modified power series expansion. Let us consider $F(s)$ and $F_1(s)$ defined in Eqs. 9.120 and 9.123 in Method A. We define a new variable p such that

$$p = s - s_0 \tag{9.137}$$

Then we can write $F_1(s)$ as

$$F_1(p + s_0) = \frac{N(p + s_0)}{D_1(p + s_0)} \tag{9.138}$$

Dividing $N(p + s_0)$ by $D_1(p + s_0)$, with both polynomials written in *ascending* powers of p, we obtain

$$F_1(p + s_0) = K_0 + K_1 p + K_2 p^2 + \cdots + K_{n-1} p^{n-1} + \frac{K_n p^n}{D_1(p + s_0)} \tag{9.139}$$

The original function $F(s)$ is related to $F_1(p + s_0)$ by the equation

$$F(p + s_0) = \frac{F_1(p + s_0)}{p^n} = \frac{K_0}{p^n} + \frac{K_1}{p^{n-1}} + \cdots + \frac{K_{n-1}}{p} + \frac{K_n}{D_1(p + s_0)} \tag{9.140}$$

Substituting $s - s_0 = p$ in the above equation, we obtain

$$F(s) = \frac{K_0}{(s - s_0)^n} + \frac{K_1}{(s - s_0)^{n-1}} + \cdots + \frac{K_{n-1}}{(s - s_0)} + \frac{K_n}{D_1(s)} \tag{9.141}$$

We have thus found the partial fraction expansion for the multiple-root terms. The remaining terms $K_n/[D_1(s)]$ still must be expanded into partial fractions. Consider the following example.

Example 9.2.8

$$F(s) = \frac{2}{(s + 1)^3 (s + 2)} \tag{9.142}$$

Using the method given above, $F_1(s)$ is

$$F_1(s) = (s + 1)^3 F(s) = \frac{2}{s + 2} \tag{9.143}$$

Setting $p = s + 1$, we then have

$$F_1(p - 1) = \frac{2}{p + 1} \tag{9.144}$$

The expansion of $F_1(p - 1)$ into a series as given in Eq. 9.139 requires dividing the numerator 2 by the denominator $1 + p$, with both numerator and denominator arranged in ascending power of p. The division here is

$$
\begin{array}{r l}
 & 2-2p+2p^2 \\
1+p) & 2 \\
 & \underline{2+2p} \\
 & -2p \\
 & \underline{-2p-2p^2} \\
 & 2p^2 \\
 & \underline{2p^2+2p^3} \\
 & -2p^3
\end{array}
$$

Since the multiplicity of the roots is $N=3$, we terminate the division after we have three terms in the quotient. We then have

$$F_1(p-1) = 2-2p+2p^2-\frac{2p^3}{p+1} \tag{9.145}$$

The original function $F(p-1)$ is

$$F(p-1) = \frac{F_1(p-1)}{p^3} = \frac{2}{p^3}-\frac{2}{p^2}+\frac{2}{p}-\frac{2}{p+1} \tag{9.146}$$

Substituting $s+1=p$, we have

$$F(s) = \frac{2}{(s+1)^3}-\frac{2}{(s+1)^2}+\frac{2}{(s+1)}-\frac{2}{s+2} \tag{9.147}$$

Example 9.2.9 As a second example, consider the function

$$F(s) = \frac{s+2}{s^2(s+1)^2} \tag{9.148}$$

Since we have two sets of multiple roots here—at $s=0$ and $s=-1$—we have a choice of expanding $F_1(s)$ about $s=0$ or $s=-1$. Let us arbitrarily choose to expand about $s=0$; since $p=s$ here, we do not have to make any substitutions. $F_1(s)$ is then

$$F_1(s) = \frac{s+2}{(s+1)^2} = \frac{2+s}{1+2s+s^2} \tag{9.149}$$

Expanding $F_1(s)$, we obtain

$$F_1(s) = 2-3s+\frac{s^2(3s+4)}{(s+1)^2} \tag{9.150}$$

$F(s)$ is then

$$F(s) = \frac{2}{s^2}-\frac{3}{s}+\frac{3s+4}{(s+1)^2} \tag{9.151}$$

We must now repeat this process for the term

$$\frac{3s+4}{(s+1)^2}$$

Fortunately we see that the term can be written as

$$\frac{3s+4}{(s+1)^2} = \frac{3(s+1)+1}{(s+1)^2} = \frac{3}{(s+1)} + \frac{1}{(s+1)^2} \tag{9.152}$$

The final answer is then

$$F(s) = \frac{2}{s^2} - \frac{3}{s} + \frac{3}{s+1} + \frac{1}{(s+1)^2} \tag{9.153}$$

9.3 ANALYSIS OF CIRCUITS USING LAPLACE TRANSFORMS

With the help of a table of Laplace transforms and the technique of partial fraction expansion, we are now ready to analyze any arbitrary linear time invariant circuit, with or without initial energy storage. Naturally we will start with basic elements like R, L, and C and combine them later on for analysis of more complicated circuits. For any signal $v(t)$ or $i(t)$, we have, if $v(t)$ and $i(t)$ are Laplace transformable

$$\begin{aligned} \mathbf{L}[v(t)] &= V(s) \\ \mathbf{L}[i(t)] &= I(s) \end{aligned} \tag{9.154}$$

Resistor

Now the resistor R is defined by the v-i relationship

$$v(t) = Ri(t) \tag{9.155}$$

Hence we have, after taking the Laplace transform of both sides of Eq. 9.155

$$V(s) = RI(s) \tag{9.156}$$

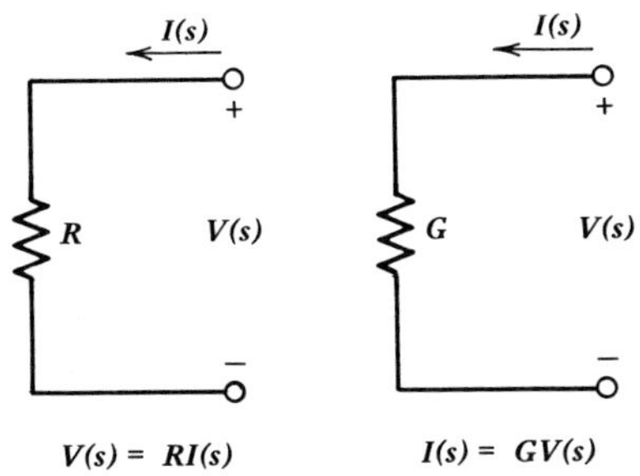

Fig. 9.5 A resistor.

In Eq. 9.156 the resistor is considered as an element with impedance equal to R. On the other hand, if we rewrite Eq. 9.156 as

$$I(s) = \frac{1}{R} V(s) = GV(s) \tag{9.156a}$$

then the resistor is considered as an element with admittance equal to G. The circuit representation for a resistor is shown in Fig. 9.5.

Inductor

For an inductor the voltage-current relationship in the time domain is given as

$$v(t) = L\frac{di}{dt} \tag{9.157}$$

Taking transforms on both sides and remembering the transform for a derivative, we obtain

$$V(s) = Ls\,I(s) - Li(0-) \tag{9.158}$$

In Eq. 9.158 we see that the terminal voltage of an inductor could be considered as a sum of two terms: the first one is a voltage drop caused by its inherent impedance Ls, the second term is a voltage rise due to its initial current. The equivalent circuit is shown in Fig. 9.6. Of course we could also rewrite Eq. 9.158 as

$$I(s) = \frac{1}{sL} V(s) + \frac{i(0-)}{s} \tag{9.159}$$

Then the inductor is considered a current source in parallel with an admittance. Eq. 9.159 is depicted in Fig. 9.7.

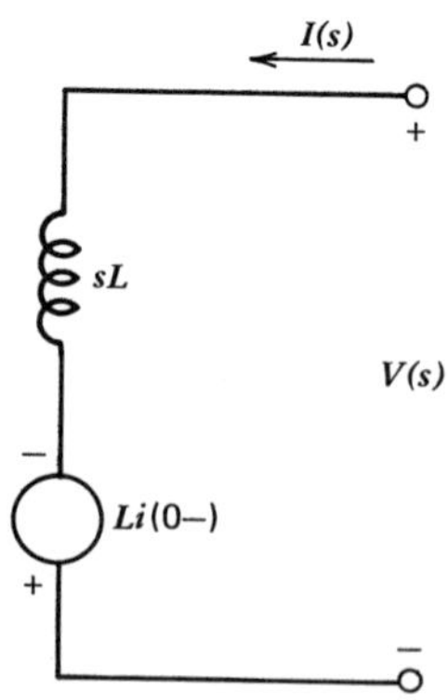

Fig. 9.6 An equivalent circuit for an inductor.

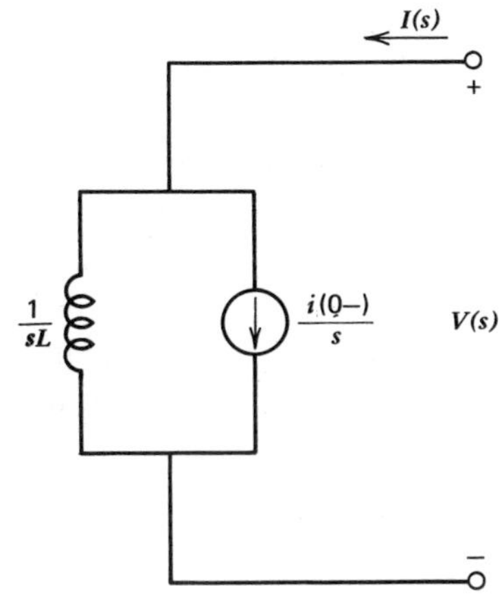

Fig. 9.7 An equivalent circuit for an inductor.

Capacitor

A set of similar equations could be found for a capacitor. We have

$$i(t) = C\,\frac{dv(t)}{dt} \tag{9.160}$$

Taking Laplace transforms, we have

$$I(s) = Cs\,V(s) - Cv(0-) \tag{9.161}$$

Rewriting Eq. 9.161, we have

$$V(s) = \frac{1}{Cs}I(s) + \frac{v(0-)}{s} \tag{9.162}$$

Eqs. 9.161 and 9.162 are shown in Figs. 9.8 and 9.9 respectively.

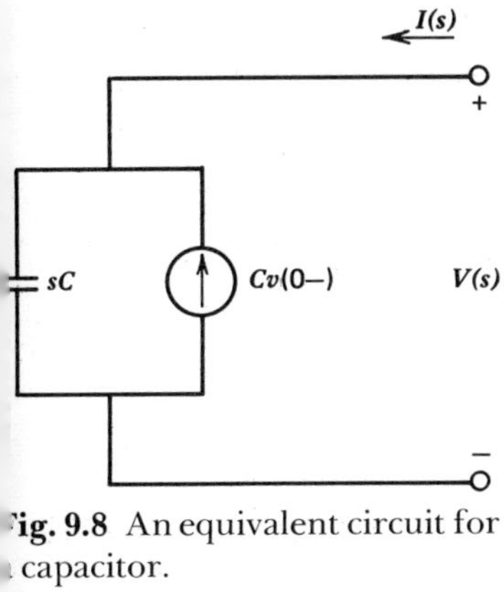

Fig. 9.8 An equivalent circuit for a capacitor.

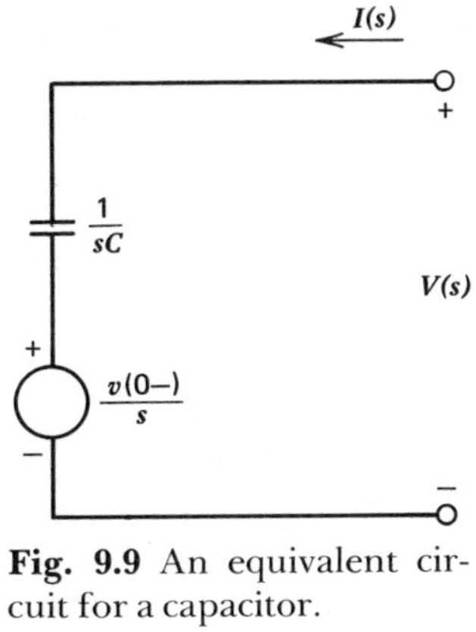

Fig. 9.9 An equivalent circuit for a capacitor.

From the previous analysis we see that in the frequency-domain representation the network elements can be represented as *impedances* and *admittances* in series or parallel with energy sources. Working from a transformed circuit diagram, we can therefore write mesh or node equations on an impedance or admittance basis directly. Let us look at one example.

Example 9.3.1 Consider the circuit shown in Fig. 9.10. The switch is thrown from position 1 to 2 at time $t = 0$. Just before the switch is thrown the initial conditions are $i_L(0-) = 2$ amp, $v_C(0-) = 2V$. Let us find the current $i(t)$ after the switching action. Since the circuit elements are in series, we choose the series representation for both L and C. The circuit is now redrawn in Fig. 9.11 as a transformed circuit. Note that since the switch is closed at $t = 0$, the 5-V battery is transformed as $5/s$. The loop equation for the circuit in Fig. 9.11 is

$$\frac{5}{s} = \left(3 + s + \frac{2}{s}\right) I(s) - 2 + \frac{2}{s} \tag{9.163}$$

Solving for $I(s)$, we have

$$I(s) = \frac{2s+3}{s^2+3s+2} = \frac{1}{s+1} + \frac{1}{s+2} \tag{9.164}$$

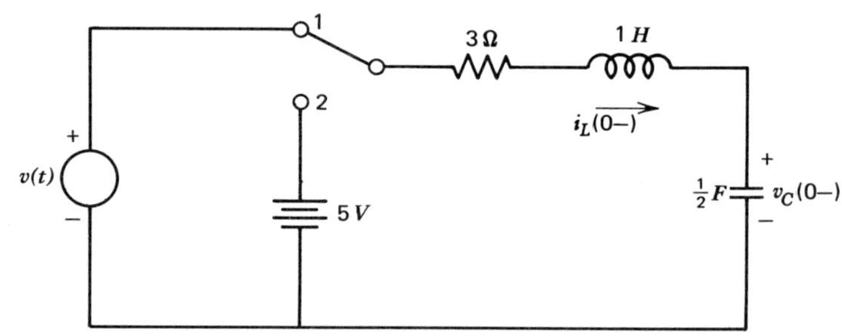

Fig. 9.10 A simple circuit.

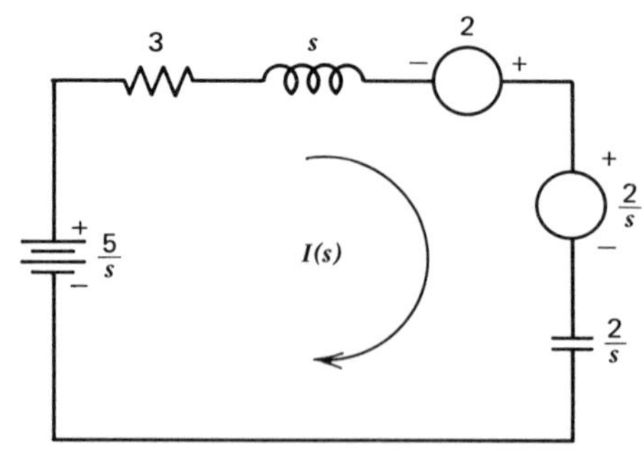

Fig. 9.11 The transformed circuit.

Hence the current $i(t)$ is found as

$$i(t) = \mathbf{L}^{-1}[I(s)] = e^{-t} + e^{-2t} \tag{9.165}$$

Often we are required to determine a *single* branch current through a given element or a *single* voltage across an element. In such cases we need Thévenin's or Norton's theorem. We have discussed the time-domain version in Chapter 7. An analogous version of the theorems is also available in the frequency domain. We will state only the results without proofs.

Thévenin's Theorem

To determine the current $I(s)$ through an element of impedance $Z_1(s)$, shown in Fig. 9.12 the rest of the network N can be replaced by an equivalent impedance $Z_e(s)$ in series with an equivalent voltage source $V_e(s)$, as depicted in Fig. 9.13. The equivalent impedance $Z_e(s)$ is the impedance "looking into" N from the terminals of $Z_1(s)$ when all voltage sources in N are short-circuited and all current sources are open-circuited. The

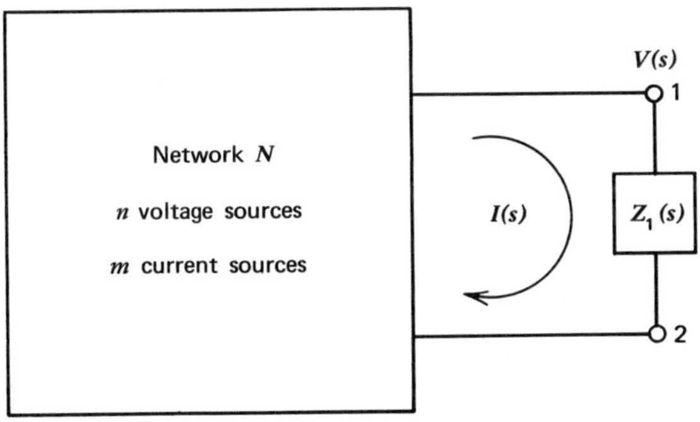

Fig. 9.12 An arbitrary network.

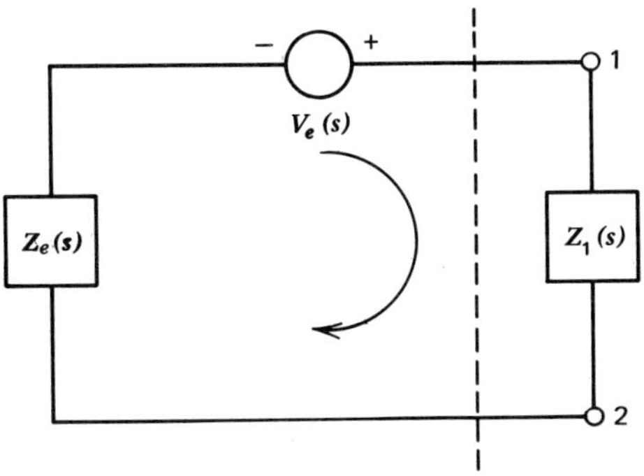

Fig. 9.13 Thévenin's equivalent circuit.

equivalent voltage source $V_e(s)$ is the voltage that appears between the terminals 1 and 2 in Fig. 9.12 when the element $Z_1(s)$ is removed or open-circuited.

Norton's Theorem

To determine the voltage $V(s)$ across an element whose admittance is $Y_1(s)$, the rest of the network can be represented as an equivalent admittance $Y_e(s)$ in parallel with an equivalent current source $I_e(s)$, as shown in Fig. 9.14. The admittance $Y_e(s)$ is the reciprocal of the Thévenin impedance. The current $I_e(s)$ is the current that flows through a short circuit across $Y_1(s)$.

For both theorems, the element whose current or voltage is to be determined must not be magnetically coupled to any element in the rest of the network. It is easily seen from Figs. 9.13 and 9.14 that we have

$$I(s) = \frac{V_e(s)}{Z_1(s) + Z_e(s)} \tag{9.166}$$

$$V(s) = \frac{I_e(s)}{Y_1(s) + Y_e(s)} \tag{9.167}$$

Furthermore, since $Y_1(s) = 1/Z_1(s)$ and $Y_e(s) = 1/Z_e(s)$, we have

$$I_e(s) = \frac{V_e(s)}{Z_e(s)} \tag{9.168}$$

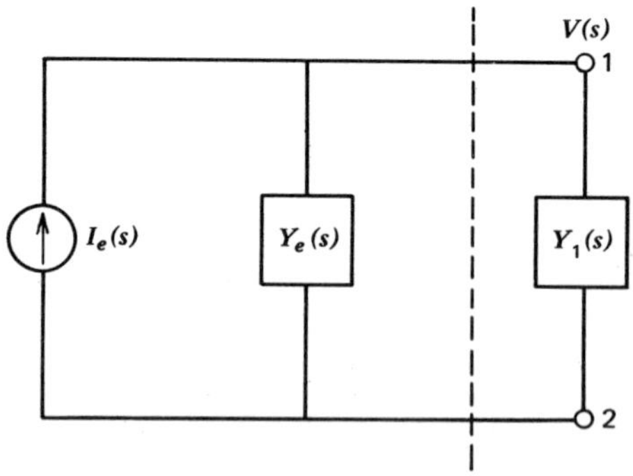

Fig. 9.14 Norton's equivalent circuit.

Example 9.3.2 Let us determine by Thévenin's theorem the current $I_1(s)$ flowing through the capacitor in the network shown in Fig. 9.15. First, let us obtain $Z_e(s)$ by opening all current sources and short-circuit-

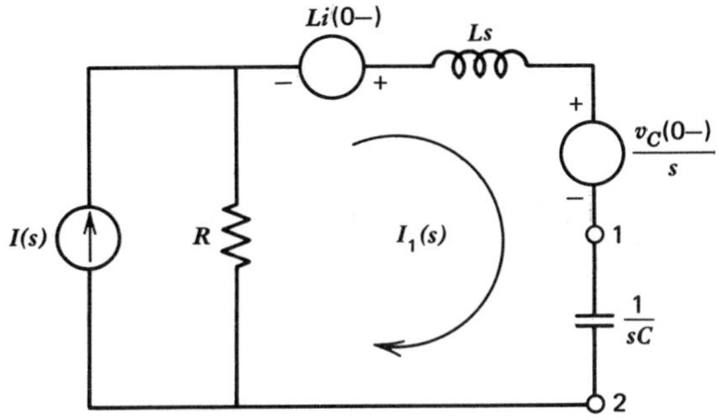

Fig. 9.15 An *RLC* circuit.

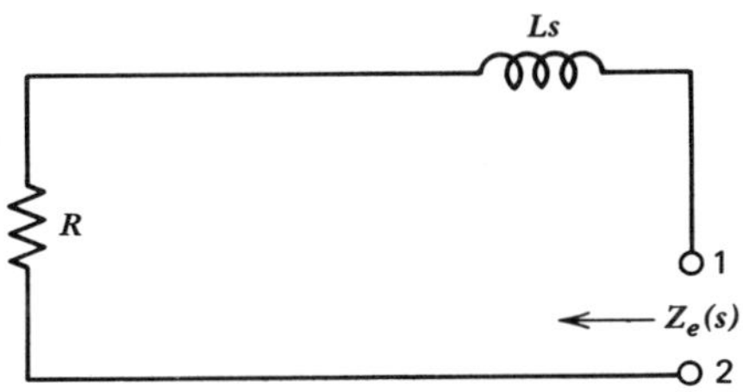

Fig. 9.16 Circuit in finding $Z_e(s)$.

ing all voltage sources; this is shown in Fig. 9.16. We have

$$Z_e(s) = R + Ls \tag{9.169}$$

Next we find $V_e(s)$ by removing the capacitor so that the open-circuit voltage between the terminals 1 and 2 is $V_e(s)$, as shown in Fig. 9.17. We have

$$V_e(s) = RI(s) + Li(0-) - \frac{v_C(0-)}{s} \tag{9.170}$$

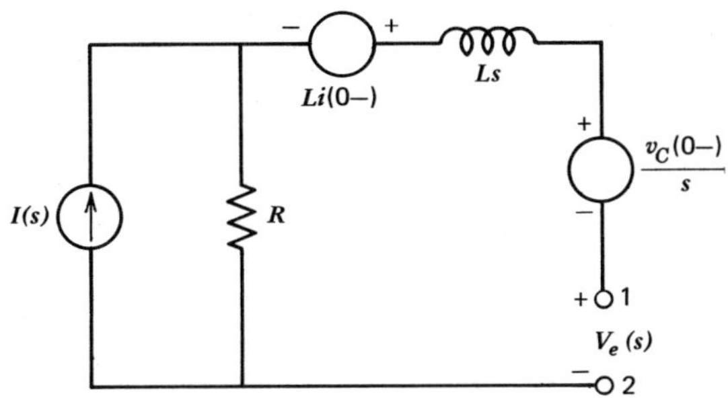

Fig. 9.17 Circuit in finding $V_e(s)$.

By Thévenin's theorem we then have

$$I_1(s) = \frac{V_e(s)}{Z_e(s) + Z(s)} = \frac{RI(s) + Li(0-) - v_C(0-)/s}{R + sL + \dfrac{1}{sC}} \tag{9.171}$$

Example 9.3.3 In the network shown in Fig. 9.18 the switch closes at $t = 0$. It is given that $v(t) = 0.1e^{-5t}$ and all initial currents and voltages are zero. Let us find the current $i_2(t)$ by Norton's theorem.

The transformed circuit is given in Fig. 9.19. To find the Norton equivalent current source, we short circuit points 1 and 2 in the network shown. Then $I_e(s)$ is the current flowing in the short-circuit, or

$$I_e(s) = \frac{V(s)}{R_1 + sL} = \frac{0.1/(s+5)}{R_1 + sL} = \frac{1}{(s+5)(s+10)} \tag{9.172}$$

The equivalent admittance of the circuit as viewed from points 1 and 2 is

$$Y_e(s) = sC + \frac{1}{R_1 + sL} = \frac{0.5s^2 + 5s + 10}{s + 10} \tag{9.173}$$

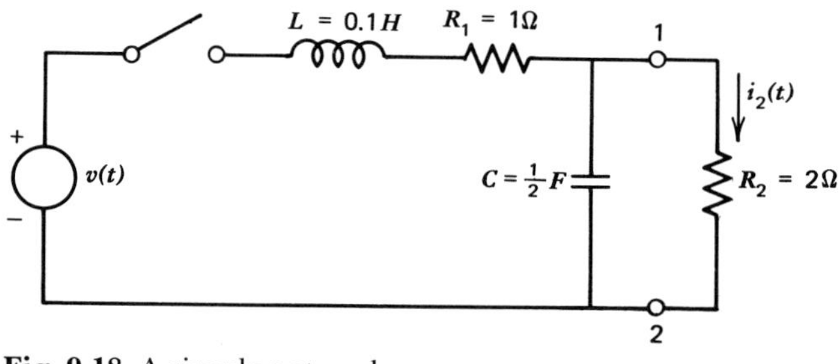

Fig. 9.18 A simple network.

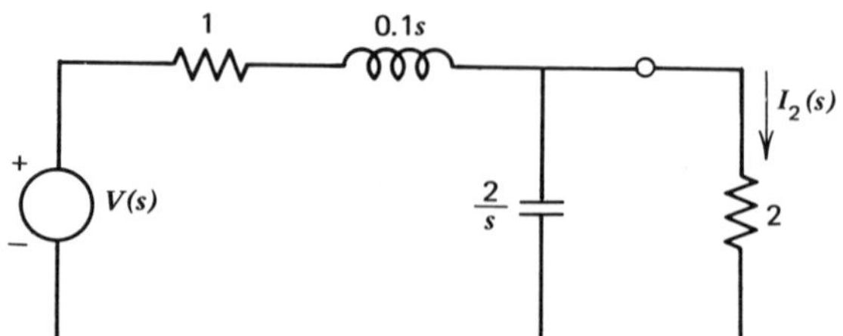

Fig. 9.19 The transformed circuit.

Hence $I_2(s)$ is then equal to

$$I_2(s) = \frac{I_e(s)}{R_2[Y_e(s) + G_2]}$$

$$= \frac{1}{(s+5)^2(s+6)} \tag{9.174}$$

Expanding Eq. 9.174 by partial fractions, we have

$$I_2(s) = \frac{1}{(s+5)^2} - \frac{1}{s+5} + \frac{1}{s+6} \tag{9.175}$$

Hence we have

$$i_2(t) = \mathbf{L}^{-1}[I_2(s)] = [te^{-5t} - e^{-5t} + e^{-6t}]u(t) \tag{9.176}$$

9.4 SYSTEM FUNCTIONS

In Chapter 7 we discussed how the impedance concept is used to find the frequency response of a given circuit subject to an exponential input with frequency ω. Thus the impedances of a resistor R, an inductor L, and a capacitor C are, respectively,

$$\begin{aligned} Z_R &= R \\ Z_L &= j\omega L \\ Z_C &= \frac{1}{j\omega C} \end{aligned} \tag{9.177}$$

And when these components are connected in series or parallel, the resultant impedance could be found by using KCL or KVL as is discussed in Chapter 7.

Incidentally, if the inductor and capacitor we considered in Section 9.3 are free of initial energy, in other words, if we have zero-state circuit elements R, L and C, we have a similar result by forming the ratio of the voltage response transform to the current excitation transform. Indeed from Eqs. 9.156, 9.158, and 9.162 we have, after setting $i_L(0-) = v_C(0-) = 0$,

$$\begin{aligned} \frac{V_R(s)}{I_R(s)} &= R \\ \frac{V_L(s)}{I_L(s)} &= Ls \\ \frac{V_C(s)}{I_C(s)} &= \frac{1}{Cs} \end{aligned} \tag{9.178}$$

where $s = \sigma + j\omega$ is a complex variable. Hence the impedance functions we derived in Chapter 7 are special cases of the ratio of the Laplace transformed response and excitation for zero-state network elements. We now define the system function as such that when the system is in zero-state, the transformed excitation $E(s)$ and the transformed response $R(s)$ of the system are related by the system function $H(s)$ as

$$R(s) = E(s)H(s) \tag{9.179}$$

In other words, the system function $H(s)$ is the response of the network per unit of excitation, or, it is the ratio of output to input as

$$\frac{R(s)}{E(s)} \triangleq H(s) \tag{9.180}$$

If one of these quantities is a voltage and the other a current, then the response function is an impedance or an admittance; while if excitation and response are both currents or both voltages, then the system function is dimensionless. Thus for a resistor, an inductor, or a capacitor the system function is given in Eq. 9.178 if voltages are the responses and currents are the excitations. Since they are the response-excitation ratios taken at same terminals, they are called the driving-point impedances following the network functions definition. Similarly we have the transfer system functions if excitation and response are at different pairs of terminals. We will discuss transfer functions in the next chapter.

For series or parallel circuits, we can get an equivalent impedance or admittance function by simply following the methods used in Chapter 7 in replacing all $j\omega$ by s in the final result. For more complicated networks, we could write down the Kirchhoff's laws in the transform domain and solve them for the appropriate voltage or current variables we are interested in and then form the appropriate system functions. Let us now look at one example.

Example 9.4.1 Assume that Fig. 9.20 is a zero-state circuit driven by a current source $I(s)$ (in time domain the current is $i(t)$).

The system function for current excitation is the impedance. The voltage $v(t)$ is the impedance times the current excitation. When we transform both the current $i(t)$ and the voltage $v(t)$ to the frequency domain, the admittance in the frequency domain is

$$Y(s) = Cs + \frac{1}{Ls} + \frac{1}{R} \tag{9.181}$$

Since the impedance function is the reciprocal of the admittance func-

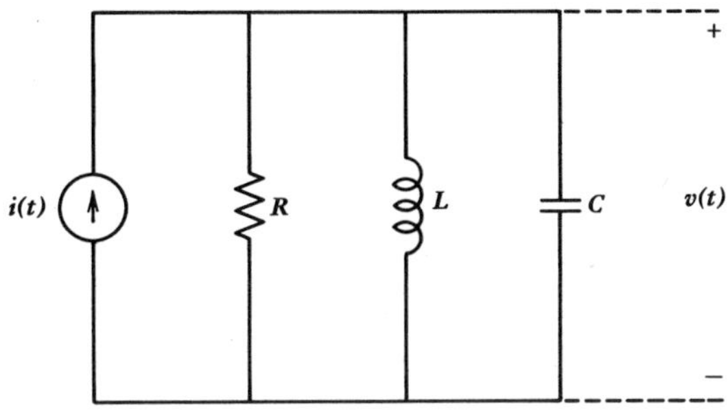

Fig. 9.20 An *RLC* circuit.

tion, it can be written as

$$Z(s) = \frac{s}{C\left[s^2 + \frac{1}{RC}s + \frac{1}{LC}\right]} \tag{9.182}$$

This can be simplified to read as

$$Z(s) = \frac{a_0 s}{(s+s_1)(s+s_2)} \tag{9.183}$$

where a_0 is a constant.

Let $i(t)$ be a unit step function. Then

$$I(s) = \frac{1}{s}$$

$$V(s) = I(s) \cdot Z(s)$$

$$\therefore \quad V(s) = \frac{1}{s} \cdot \frac{a_0 s}{(s+s_1)(s+s_2)} \tag{9.184}$$

Using partial-fraction expansion

$$V(s) = \frac{k_1}{s+s_1} + \frac{k_2}{s+s_2} + \frac{k_3}{s} \tag{9.185}$$

It can easily be shown that $k_3 = 0$; hence we have

$$v(t) = k_1 e^{-s_1 t} + k_2 e^{-s_2 t} \tag{9.186}$$

The relation $R(s) = E(s)H(s)$ can also be obtained directly from taking the Laplace transform of the convolution integral

$$r(t) = \int_0^t e(\tau) h(t-\tau)\, d\tau \tag{9.187}$$

where $r(t)$ is the response function, $e(t)$ is the excitation and $h(t)$ is the impulse response of the system. The Laplace transform of $r(t)$ is

$$\mathbf{L}[r(t)] = \int_0^\infty e^{-st}\left[\int_0^t e(\tau)h(t-\tau)\,d\tau\right]dt$$

$$= \int_0^\infty e^{-s(t-\tau)}e^{-s\tau}\left[\int_0^t e(\tau)h(t-\tau)\,d\tau\right]dt \qquad (9.188)$$

From the definition of the shifted step function

$$u(t-\tau) = 1 \qquad \tau \leqslant t \qquad (9.189)$$
$$= 0 \qquad \tau > t$$

we have the identity

$$\int_0^t e(\tau)h(t-\tau)\,d\tau = \int_0^\infty e(\tau)u(t-\tau)h(t-\tau)\,d\tau \qquad (9.190)$$

Substituting Eq. 9.190 into Eq. 9.188, we have

$$\mathbf{L}[r(t)] = \int_0^\infty\left[\int_0^\infty e(\tau)h(t-\tau)u(t-\tau)\,d\tau\right]e^{-s(t-\tau)}e^{-s\tau}dt \qquad (9.191)$$

If we let $x = t-\tau$ so that $dx = dt$, then we have

$$\mathbf{L}[r(t)] = \int_0^\infty\left[\int_0^\infty e(\tau)h(x)u(x)\,d\tau\right]e^{-sx}e^{-s\tau}dx \qquad (9.192)$$

Interchanging the order of integration, we have

$$\mathbf{L}[r(t)] = \int_0^\infty\left[\int_0^\infty h(x)u(x)\,e^{-sx}dx\right]e(\tau)\,e^{-s\tau}d\tau$$

or

$$R(s) = H(s)\int_0^\infty e(\tau)\,e^{-s\tau}d\tau = H(s)E(s) \qquad (9.193)$$

The result of Equation 9.193 in transforming a time-domain convolution to a multiplication in the frequency domain is similar to the results in using Fourier transform in Chapter 8. This is of course what we have expected as Laplace transform is a special form of Fourier transform.

Now we can relate the system function $H(s)$ with the impulse response of the system. Suppose the excitation is given as an impulse function $\delta(t)$, then we have

$$E(s) = 1 \qquad (9.194)$$

So the response function

$$R(s) = E(s) \cdot H(s) = H(s) \tag{9.195}$$

Thus $H(s)$ is the response transform to an impulse excitation. So $r(t)$, which is the inverse transform of the response function, is equal to the impulse response $h(t)$. Or, in other words,

$$\mathbf{L}[h(t)] = H(s) \tag{9.196}$$

The system function $H(s)$ and the impulse response form a Laplace transform pair. To calculate the impulse response, we have seen earlier that we have to solve differential equations. A much easier way is to get the system function and take its inverse transform to get the impulse response.

Example 9.4.2 Consider the circuit in Fig. 9.21. The system function relating $V_2(s)$ to $V_1(s)$ is

$$H(s) = \frac{V_2(s)}{V_1(s)} = \frac{\frac{1}{sC}}{R + \frac{1}{RC}} = \frac{\frac{1}{RC}}{s + \frac{1}{RC}}$$

The impulse response is therefore obtained as

$$\mathbf{L}^{-1}[H(s)] = h(t) = \frac{1}{RC} \cdot e^{-t/RC}$$

In the time-domain analysis we have come across quite frequently the step response. In the Laplace theory this becomes very easy. Since the system function $H(s)$ is the transform of the impulse response and since the step response is the integral of the impulse response, we have to divide by s and take the inverse transform to obtain the step response.

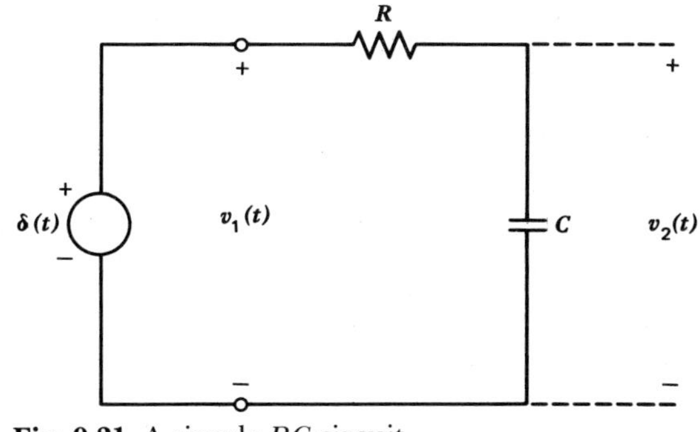

Fig. 9.21 A simple *RC* circuit.

$$\text{Step response} = \mathbf{L}^{-1}\left[\frac{1}{s}H(s)\right] = \alpha(t)$$

In other words, the transform of the step response is equal to the system function times the transform of unit step, which is equal to $1/s$.

Example 9.4.3 Let us find the current step response of the *RL* circuit in Fig. 9.22. Since $I(s)$ is the response and $V_g(s)$ is the excitation, the system function is

$$H(s) = \frac{I(s)}{V_g(s)} = \frac{1}{R+sL} \tag{9.197}$$

Therefore $H(s)/s$ is

$$\frac{H(s)}{s} = \frac{1}{s(R+sL)} = \frac{1}{R}\left(\frac{1}{s} - \frac{1}{s+R/L}\right) \tag{9.198}$$

The step response $\alpha(t)$ is now obtained as

$$\alpha(t) = \mathbf{L}^{-1}\left[\frac{1}{s}H(s)\right] = \frac{1}{R}\left(1 - e^{-(R/L)t}\right)u(t) \tag{9.199}$$

It is readily seen that if we know the impulse response of a zero-state linear system, we can determine the response of the system due to any other excitation. In other words, the impulse response alone is sufficient to characterize the system from the stand point of excitation and response.

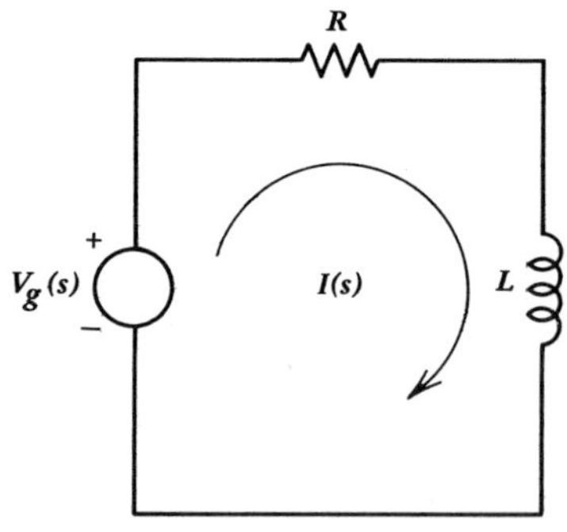

Fig. 9.22 An *RL* circuit.

9.5 POLES AND ZEROES OF RATIONAL FUNCTIONS

We have seen that for zero-state networks, an arbitrary excitation gives forth a response function that is equal to the system function times the excitation function. Or we have

$$R(s) = H(s) \cdot E(s) \tag{9.200}$$

The inverse of $R(s)$ can be found easily by the help of partial-fraction expansion and a table of Laplace transforms. In general $R(s)$ is a rational function of s. In other words, we have

$$R(s) = \frac{q(s)}{p(s)} \tag{9.201}$$

Now the roots of the denominator polynomial equation $p(s) = 0$ play an important role in determining the partial-fraction expansions of $R(s)$. These roots are called the poles of $R(s)$ as whenever the complex variable s is substituted for these roots the value of $R(s)$ approaches infinity. Similarly the roots of the polynomial equation $q(s) = 0$ are called the zeroes of $R(s)$. We will discuss some implications of the poles and zeroes of a given rational function upon its time domain behavior. First let us denote, for a given rational function, poles by small crosses and zeroes by small circles in a complex s-plane. Thus, for the function

$$F(s) = \frac{s(s-1+j1)(s-1-j1)}{(s+1)^2(s+j2)(s-j2)} \tag{9.202}$$

the poles are at $s = -1$ (double), $s = -j2$, $s = +j2$ and the zeroes are at $s = 0$, $s = 1+j1$, $s = 1-j1$, $s = \infty$. These are shown in Fig. 9.23.

We can now consider some pole-zero diagrams corresponding to standard signals. First we have an exponential signal $e^{-\sigma_0 t}$, where $\sigma_0 > 0$,

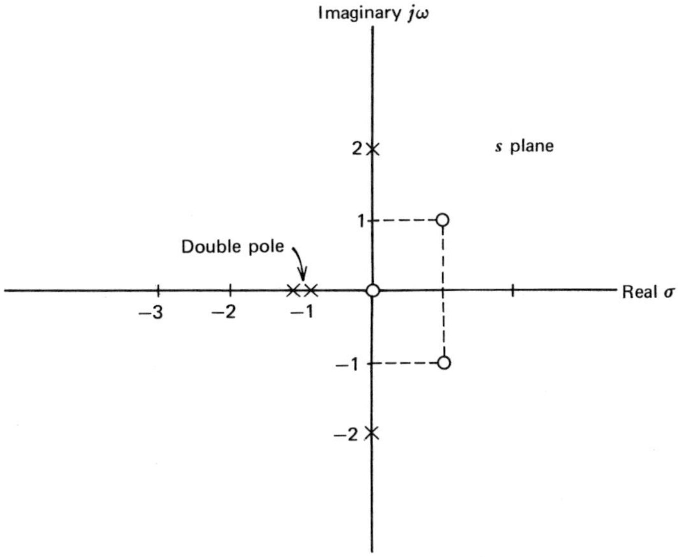

Fig. 9.23 Pole-zero diagram of $F(s)$.

whose transform is

$$\mathbf{L}[e^{-\sigma_0 t}] = \frac{1}{s+\sigma_0} \tag{9.203}$$

The pole is at $s = -\sigma_0$, as is indicated in Fig. 9.24. We observe from Fig. 9.24 that if we increase the value of σ_0, the pole is moving further to the left of the origin. But a larger σ_0 value implies a faster decaying exponential; therefore we can conclude that *the further the pole is from the origin on the $-\sigma$ axis, the more rapid the exponential decay.*

Let us now consider a damped sine wave whose transform is

$$\mathbf{L}[e^{-\sigma_0 t} \sin \omega_0 t] = \frac{1}{(s+\sigma_0)^2 + \omega_0^2}$$

The poles are at $s_1 = -\sigma_0 + j\omega_0$ and $s_2 = -\sigma_0 - j\omega_0$ and are shown in Fig. 9.25. From Fig. 9.25 we see again the factor σ_0 determines how fast the function is decaying. The larger is σ_0, the faster decays the function. Now the imaginary part ω_0 represents the frequency of oscillation; therefore from Fig. 9.25 we conclude that *the greater the magnitude of the imaginary part of the pole, the higher the frequency of oscillation.*

In case $\sigma_0 > 0$, then we have an exponentially increasing function. The real part of the poles is positive in this case. Hence, if a function whose transform has poles on the right-half of the s-plane, it is unstable. Another unbounded situation arises if there is a pair of *double poles* on the $j\omega$ axis. For example, this is true for the transform function

$$F(s) = \frac{s}{(s^2+\omega_0^2)^2} \tag{9.204}$$

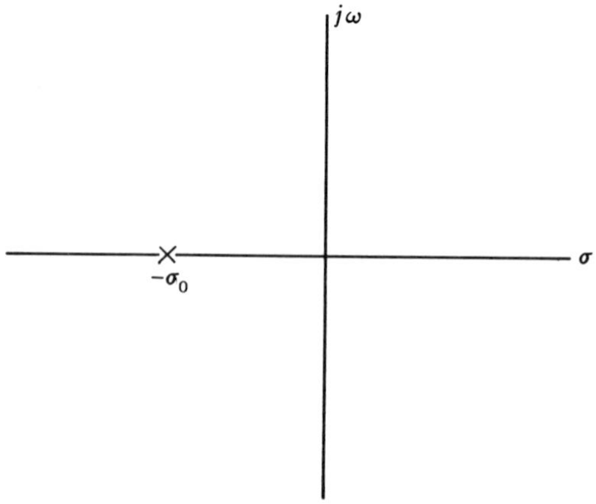

Fig. 9.24 The pole-zero diagram of $1/s+\sigma_0$.

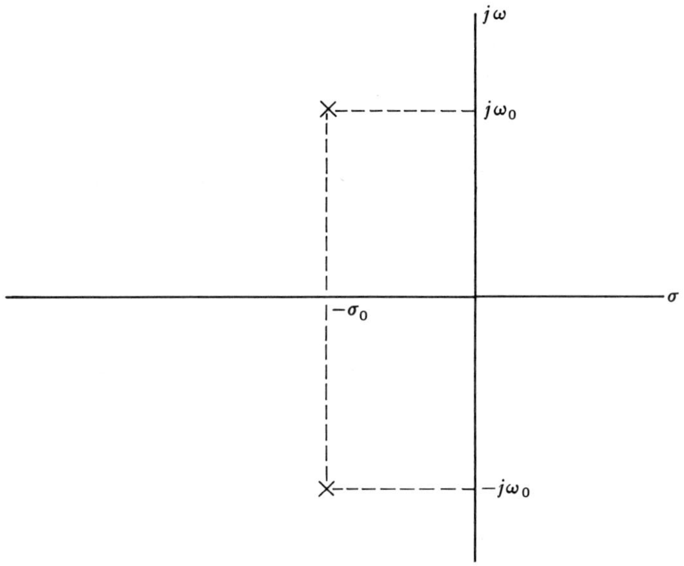

Fig. 9.25 The pole-zero diagram of $1/(s+\sigma_0)^2+\omega_0^2$.

whose pole-zero diagram is shown in Fig. 9.26. This is readily seen from its inverse transform

$$f(t) = \frac{t}{2\omega_0}(\sin \omega_0 t)u(t) \tag{9.205}$$

which is an increasing function of t. It is apparent that a stable system function cannot have multiple poles on the $j\omega$ axis.

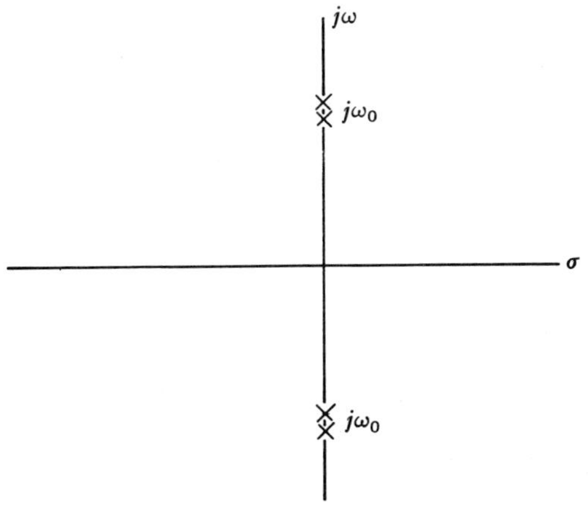

Fig. 9.26 The pole-zero diagram of an unstable function.

So far we have seen that knowing the positions of the poles of a rational function will tell us qualitatively how the function will behave in the time domain. To determine it quantitatively the coefficients of its partial-fraction expansion have to be evaluated. These are of course affected by both the poles and zeroes of the transformed function. We will discuss methods for their evaluations in the next section.

Now if we go back to Eq. 9.200 and assuming the poles and zeroes of the excitation $E(s)$ and the system function $H(s)$ are given, then it is clear that $R(s)$ will have the poles and zeroes of both $H(s)$ and $E(s)$ except in the case where a pole-zero cancellation occurs. So we can write Eq. 9.200 in partial-fraction form as

$$\begin{aligned} R(s) &= H(s) \cdot E(s) \\ &= \sum_{i=1}^{n} \frac{A_i}{s - s_i} + \sum_{j=1}^{m} \frac{B_j}{s - s_j} \end{aligned} \tag{9.206}$$

where s_i are the poles of $H(s)$ and s_j are the poles of $E(s)$. Taking the inverse transform, we have

$$r(t) = \sum_{i=1}^{n} A_i e^{s_i t} + \sum_{j=1}^{m} B_j e^{s_j t} \tag{9.207}$$

From Eq. 9.206 we see that the poles of $H(s)$ are independent of the external excitation. Hence the first term on the right-hand side of Eq. 9.207 corresponds to the natural or transient response of the system and the second term is the so-called forced or driven response. We have seen in the previous chapters how the theory of differential equations is used to find the transient and the driven responses of a system. Now by Laplace transform we can achieve the same with less effort if we know the poles of the response function. Again we should emphasize that the determination of poles is not trivial if the degree of the denominator polynomial is very large. And for systems with multiexcitation and multiresponse it would be easier to solve the original differential equations of the system by using numerical methods discussed in Chapters 2 and 3 than by taking the Laplace transform approach.

9.6 DIGITAL COMPUTATION IX: ROOT-FINDING METHODS AND PARTIAL-FRACTION EXPANSIONS

The Laplace transform method enables us to transform the differential equations describing the circuits into algebraic equations. Then the desired response in transformed form is expressed in terms of the transformed excitation function and system function. Thus if we have an *RLC* series circuit in Fig. 9.27:

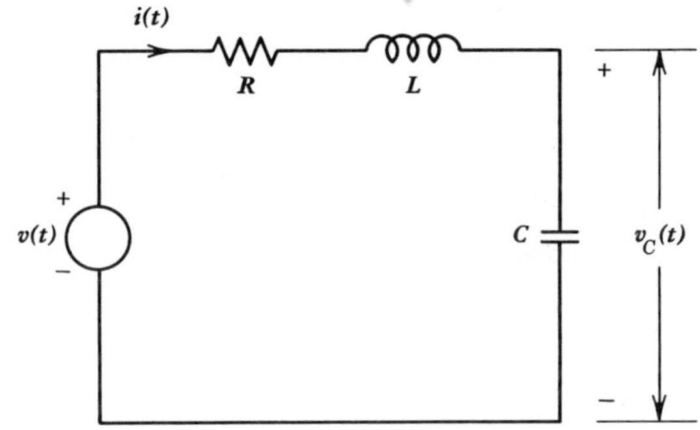

Fig. 9.27 An *RLC* circuit.

and the voltage across the capacitor is the desired response, then we have, after applying Laplace transform,

$$\frac{V_C(s)}{V(s)} = \frac{\frac{1}{Cs}}{R + Ls + \frac{1}{Cs}} = \frac{1}{LCs^2 + RCs + 1} \tag{9.208}$$

Hence the voltage across the capacitor can be expressed as

$$V_C(s) = V(s) \cdot \left[\frac{1}{LC_s^2 + RCs + 1}\right] = V(s) \cdot H(s) \tag{9.209}$$

for whatever $V(s)$ is.

In general, the transformed response is expressed as

$$R(s) = E(s)H(s) \tag{9.210}$$

where $R(s)$ is the desired response, $E(s)$ is the system excitation and $H(s)$ is the system function.

In order to find $r(t)$, the inverse Laplace transform is used and we have

$$r(t) = \mathbf{L}^{-1}[R(s)] = \frac{1}{2\pi j} \int_{\sigma_1 - j\infty}^{\sigma_1 + j\infty} R(s)e^{st} ds \tag{9.211}$$

where σ_1 is a certain positive number. This inversion process is seldom used practically and the most commonly used method is by partial-fraction expansions of the unknown $R(s)$ as is discussed in Section 9.4. If

$R(s)$ is expressed as a ratio of two polynomials of s, then we have

$$R(s) = \frac{q(s)}{p(s)} \tag{9.212}$$

where the degree of $p(s)$ is usually less than the degree of $q(s)$. Thus we are required to find the roots of the equation

$$p(s) = 0 \tag{9.213}$$

in order to find the partial-fraction expansion of $R(s)$.

We will first present some methods for finding roots of a polynomial equation and then discuss some methods for partial-fraction expansions.

Root-Finding Methods

Our first equation is a quadratic equation. Thus we have

$$p(s) = a_2 s^2 + a_1 s + a_0 = 0 \tag{9.214}$$

The roots of Eq. 9.213 are easily evaluated by the quadratic formula

$$s_1 = \frac{-a_1 + \sqrt{a_1{}^2 - 4a_0a_2}}{2a_2} \qquad s_2 = \frac{-a_1 - \sqrt{a_1{}^2 - 4a_0a_2}}{2a_2} \tag{9.215}$$

Assuming that $a_1 > 0$, then s_1 is a smaller root. Suppose further that $a_1{}^2 \gg 4a_0a_2$, then s_1 is the result of subtraction of two nearly equal numbers a_1 and $\sqrt{a_1{}^2 - 4a_0a_2}$. As is indicated in Section 1.9, the significance error introduced by subtracting two nearly equal numbers reduces the accuracy seriously. Hence in this case we should use the following formula for s_1:

$$s_1 = \frac{-2a_0}{a_1 + \sqrt{a_1{}^2 - 4a_0a_2}} \tag{9.216}$$

If a_1 is negative, then the sign before the radical should also be minus.

For an nth degree polynomial, there is usually no analytical formula that will enable us to find all of the roots of this polynomial. Hence in actual computation we often use iterative method, in which a new approximation is computed based on the result of a previous approximation, to find a certain root. We will describe two simple iterative methods based on the approximation of the Taylor's expansion around the neighborhood of the unknown root.

Let us designate one of the unknown root of $p(s)$ as σ. Thus we have

$$p(\sigma) = 0 \tag{9.217}$$

Furthermore we assume that we have obtained an approximation s_i that is very close to σ such that we have

$$|\sigma - s_i| \ll 1 \tag{9.218}$$

Now the Taylor's expansion around σ could be expressed as

$$0 = p(\sigma) = p(s_i) + (\sigma - s_i)p'(s_i) + \frac{(\sigma - s_i)^2}{2} p''(s_i) + \cdots \tag{9.219}$$

Since we have Eq. 9.218, the higher order terms on the right-hand side of Eq. 9.219 are negligible compared to the low order terms. Hence if we discard all terms higher than or equal to the second order term, we have

$$0 \approx p(s_i) + (\sigma - s_i)p'(s_i) \tag{9.220}$$

The real root σ could therefore be estimated from Eq. 9.220 as

$$\sigma \approx s_i - \frac{p(s_i)}{p'(s_i)} \tag{9.221}$$

Based on Eq. 9.221, we have the following method.

Newton's Method. Let s_0 be an initial approximation to a zero $s = \sigma$ of $p(s)$. The successive approximations can be generated from the following iterative formula:

$$s_{i+1} = s_i - \frac{p(s_i)}{p'(s_i)} \tag{9.222}$$

If σ is real, then a geometric interpretation of Newton's method is shown in Fig. 9.28. Here s_i is an approximation to the root $s = \sigma$. Then

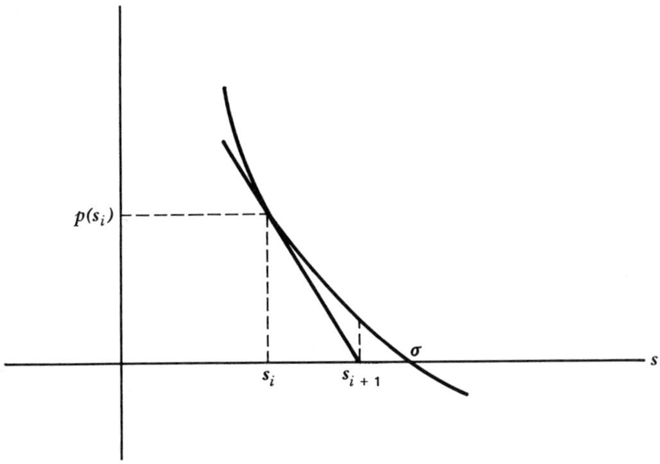

Fig. 9.28 Newton's method.

$p'(s_i)$ is the slope of the tangent to the curve $y = p(s)$ at $s = s_i$, and s_{i+1} is the intersection of this tangent with the s axis.

Let us now consider the error of the estimate s_{i+1}. Define $e_{i+1} = \sigma - s_{i+1}$, then from Eq. 9.222 we have

$$e_{i+1} = \sigma - s_{i+1} = \sigma - s_i + \frac{p(s_i)}{p'(s_i)} \tag{9.223}$$

$$= \frac{(\sigma - s_i)p'(s_i) + p(s_i)}{p'(s_i)}$$

Now from Taylor's expansion for $p(\sigma)$ we have

$$0 = p(\sigma) = p(s_i) + (\sigma - s_i)p'(s_i) + \frac{(\sigma - s_i)^2}{2!}p''(s_i) + \cdots$$

If s_i is sufficiently close to σ, then the following approximation is also true:

$$0 \approx p(s_i) + (\sigma - s_i)p'(s_i) + \frac{(\sigma - s_i)^2}{2!}p''(s_i) \tag{9.224}$$

Substituting this approximation into Eq. 9.224, we have

$$e_{i+1} = -\frac{(\sigma - s_i)^2 p''(s_i)}{2p'(s_i)} = -e_i^2 \frac{p''(s_i)}{2p'(s_i)} \tag{9.225}$$

Hence if σ is a simple root of $p(s) = 0$ and if $\lim_{i \to \infty} s_i = \sigma$, then

$$\lim_{i \to \infty} \frac{e_{i+1}}{(e_i)^2} = -\frac{p''(\sigma)}{2p'(\sigma)} \tag{9.226}$$

This means that once s_i is near σ, the error in the next step is proportional to the square of the error in s_i. An iteration that exhibits this behavior is said to *converge quadratically*. Let us now look at some examples:

Suppose we have

$$p(s) = s^2 - 2 = 0 \tag{9.227}$$

Then

$$p'(s) = 2s \tag{9.228}$$

Substituting Eqs. 9.227 and 9.228 into Eq. 9.222, we obtain the iteration

$$s_{i+1} = s_i - \frac{s_i^2 - 2}{2s_i} = \tfrac{1}{2}\left(s_i + \frac{2}{s_i}\right) \tag{9.229}$$

In particular, let us choose $s_0 = 1.4$. Then on applying Eq. 9.229, we obtain the sequence

$$s_0 = 1.4$$

$$s_1 = \tfrac{1}{2}\left(1.4 + \frac{2}{1.4}\right) = 1.4142857$$

$$s_2 = \tfrac{1}{2}\left(s_1 + \frac{2}{s_1}\right) = 1.4142136$$

$$s_3 = \tfrac{1}{2}\left(s_2 + \frac{2}{s_2}\right) = 1.4142136$$

The sequence is obviously converging quite rapidly. Since s_2 and s_3 are identical, we can conclude that this result is correct to at least seven significant figures.

For complex roots, let us consider another example. Let the polynomial equation be given as

$$p(s) = s^2 + 1 = 0 \tag{9.230}$$

Again we have

$$p'(s) = 2s \tag{9.231}$$

Hence we have the iteration

$$s_{i+1} = s_i - \frac{s_i^2 - 1}{2s_i} = \tfrac{1}{2}\left(s_i - \frac{1}{s_i}\right) \tag{9.232}$$

Starting with an initial $s_0 = 1 + j1$, we have the following sequence from Eq. 9.232.

$$s_0 = 1 + j1$$

$$s_1 = \tfrac{1}{2}\left(1 + j1 - \frac{1}{1 + j1}\right) = \tfrac{1}{4} + \tfrac{3}{4}j = 0.25 + 0.75j$$

$$s_2 = \tfrac{1}{2}\left(s_1 - \frac{1}{s_1}\right) = -\tfrac{3}{40} + \tfrac{39}{40}j = -0.075 + 0.975j$$

$$s_3 = \tfrac{1}{2}\left(s_2 - \frac{1}{s_2}\right) = 0.001948 + 1.000321j$$

$$s_4 = \tfrac{1}{2}\left(s_3 - \frac{1}{s_3}\right) = 0.000001 + 0.998537j$$

Apparently the iteration will converge to one of the true root $\sigma = j1$.

Newton's method makes use of the first two terms of the Taylor's expansion of $p(\sigma)$ about the point s_i. If we use the first three terms of the Taylor's expansion of $p(\sigma)$, then we have

$$0 \approx p(s_i) + (\sigma - s_i)p'(s_i) + \frac{(\sigma - s_i)^2}{2}p''(s_i) \tag{9.233}$$

Therefore we have

$$(\sigma - s_i)\left[p'(s_i) + \frac{(\sigma - s_i)}{2}p''(s_i)\right] \approx -p(s_i) \tag{9.234}$$

Now we can estimate the factor $(\sigma - s_i)$ inside the bracket of Eq. 9.234 by using the approximation

$$\sigma - s_i \approx -\frac{p(s_i)}{p'(s_i)} \tag{9.235}$$

Substituting Eq. 9.235 into Eq. 9.234, we have

$$\begin{aligned}\sigma - s_i &\approx \frac{-p(s_i)}{p'(s_i) - \dfrac{p(s_i)p''(s_i)}{2p'(s_i)}} \\ &= -\frac{2p(s_i)p'(s_i)}{2[p'(s_i)]^2 - p(s_i)p''(s_i)}\end{aligned} \tag{9.236}$$

This leads to the following method.

Modified Newton's Method. Let s_0 be an initial approximation to a zero $s = \sigma$ of $p(s)$. The successive approximations can be generated from the following iterative formula:

$$s_{i+1} = s_i - \frac{2p(s_i)p'(s_i)}{2[p'(s_i)]^2 - p(s_i)p''(s_i)} \tag{9.237}$$

It can be shown that if σ is a simple root of $p(s) = 0$ and if $\lim_{i\to\infty} s_i = \sigma$, then we have

$$\lim_{i\to\infty} \frac{e_{i+1}}{(e_i)^3} = \frac{3[p''(\sigma)] - 2p'''(\sigma)}{12[p'(\sigma)]^2} \tag{9.238}$$

This means that once s_i is near σ, the error in the next step is proportional to the cubic of the error in s_i. Hence it converges more rapidly than the Newton's method. As an example, let us again find the square root of 2 using this method. We have

$$p(s) = s^2 - 2 = 0 \tag{9.239}$$

$$p'(s) = 2s \tag{9.240}$$

$$p''(s) = 2 \tag{9.241}$$

Substituting Eqs. 9.239, 9.240, 9.241 into Eq. 9.237, we obtain the iteration

$$s_{i+1} = s_i - \frac{2s_i(s_i^2 - 2)}{3s_i^2 + 2} \tag{9.242}$$

Using $s_0 = 1.4$, then we have the following results:

$$s_0 = 1.4$$

$$s_1 = 1.4 - \frac{2 \times 1.4(1.4^2 - 2)}{3 \times 1.4^2 + 2} = 1.4142132$$

$$s_2 = s_1 - \frac{2s_1(s_1{}^2 - 2)}{3s_1{}^2 + 2} = 1.4142136$$

We see the first iteration has already produced a result s_1 that is correct to six significant figures of the exact answer.

We should note that for multiple roots, the Newton's method may fail since $p'(s_i)$, which is the denominator of Eq. 9.222, is also approaching zero. However, for a root σ with multiplicity m, the $(m-1)$th derivative of the polynomial has exactly only one root σ. Hence we can set up a criterion such that whenever $p(s_i)$ and $p'(s_i)$ satisfy the following condition simultaneously,

$$\begin{aligned} |p(s_i)| &\leqslant \epsilon_1 \\ |p'(s_i)| &\leqslant \epsilon_2 \end{aligned} \tag{9.243}$$

where ϵ_1 and ϵ_2 are small positive numbers, then a multiple root is suspected; hence derivatives of the polynomial should be used to find this root.

To set up the criterion of Eq. 9.243, let us look at a polynomial $p(s)$ with a root σ_1 of multiplicity $m_1 \geqslant 2$. We could write

$$p(s) = (s - \sigma_1)^{m_1} p_1(s) \tag{9.244}$$

where $p_1(s)$ is the other factor of $p(s)$. Taking the first and second derivatives of Eq. 9.244, we have

$$\begin{aligned} p'(s) &= m_1(s-\sigma_1)^{(m_1-1)} p_1(s) + (s-\sigma_1)^{m_1} p_1'(s) \\ p''(s) &= m_1(m_1-1)(s-\sigma_1)^{(m_1-2)} p_1(s) + 2m_1(s-\sigma_1)^{(m_1-1)} p_1'(s) \\ &\quad + (s-\sigma_1)^{m_1} p_1''(s) \end{aligned} \tag{9.245}$$

Now if s_i is sufficiently close to σ_1, then we have

$$\begin{aligned} p(s_i) &= (s_i - \sigma_1)^{m_1} p_1(s_i) \\ p'(s_i) &\approx m_1(s-\sigma_1)^{(m_1-1)} p_1(s_i) \\ p''(s_i) &\approx m_1(m_1-1)(s-\sigma_1)^{(m_1-2)} p_1(s_i) \end{aligned} \tag{9.246}$$

by neglecting the higher order terms of $s_i - \sigma_1$. From Eq. 9.246, we have the following relation:

$$\frac{p(s_i)}{p'(s_i)} \approx \frac{1}{m_1}(s_i - \sigma_1)$$

$$\frac{p'(s_i)}{p''(s_i)} \approx \frac{1}{m_1 - 1}(s_i - \sigma_1)$$

Hence we could estimate the ratio of $p(s_i)/p'(s_i)$ by using the available ratio $p'(s_i)/p''(s_i)$. Or we have

$$\left|\frac{p(s_i)}{p'(s_i)}\right| \approx \left|\frac{m_1-1}{m_1}\right| \cdot \left|\frac{p'(s_i)}{p''(s_i)}\right| \geqslant \frac{1}{2}\left|\frac{p'(s_i)}{p''(s_i)}\right| \tag{9.248}$$

which reduces to

$$p'(s_i) \leqslant \sqrt{2p(s_i)p''(s_i)} \tag{9.249}$$

Now if s_i satisfies the criterion that

$$|p(s_i)| \ll \epsilon_1 \tag{9.250}$$

then from Eq. 9.249 we have

$$|p'(s_i)| \leqslant \sqrt{2\epsilon_1 p''(s_i)} = \epsilon_2 \tag{9.251}$$

Hence Eq. 9.250 and Eq. 9.251 could be used to indicate the possibility of having multiple roots.

If $m_1 = 1$, then it can easily be seen from Eq. 9.245 that we have

$$\begin{aligned} p'(s_i) &\approx p_1(s_i) \\ p''(s_i) &\approx 2p_1'(s_i) \end{aligned} \tag{9.252}$$

which indicates that the magnitudes of $p'(s_i)$ and $p''(s_i)$ are some fairly large constants. Hence Eq. 9.251 will not be satisfied for s_i sufficiently close to the root σ_1. Thus we don't have to worry about having multiple roots in this case.

Once we have found a root σ_1 of the polynomial equation $p(s) = 0$, then we can find another polynomial $\gamma(s)$ such that

$$p(s) = (s-\sigma_1)\gamma(s)$$

by using the Horner's rule. Specifically, if we have

$$\begin{aligned} p(s) &= a_n s^n + a_{n-1}s^{n-1} + \cdots + a_1 s + a_0 \\ \gamma(s) &= c_{n-1}s^{n-1} + c_{n-2}s^{n-2} + \cdots + c_1 s + c_0 \end{aligned}$$

then we have

$$\begin{aligned} p(s) &= (s-\sigma_1)(c_{n-1}s^{n-1} + c_{n-2}s^{n-2} + \cdots + c_1 s + c_0) \\ &= c_{n-1}s^n + (c_{n-2} - \sigma_1 c_{n-1})s^{n-1} + \cdots + (c_0 - \sigma_1 c_1)s - \sigma_1 c_0 \end{aligned}$$

On equating these coefficients to those of $p(s)$ and rearranging, we obtain the following recursion formula:

$$\begin{aligned} c_{n-1} &= +a_n \\ c_{n-k} &= +a_{n-k+1} + \sigma_1 c_{n-k+1} \qquad k = 2, 3, \ldots, n \\ -c_0\sigma_1 &= a_0 \end{aligned} \tag{9.253}$$

where the last equality $-c_0\sigma_1 = a_0$ could be used to check the accuracy of the c_i's. Once $\gamma(s)$ has been determined, the Newton's method or the modified one can then be used to determine the next zero of the new polynomial equation $\gamma(s) = 0$ and so on.

Partial-Fraction Expansion

We now turn to the question of determining the coefficients in the partial-fraction expansion of the transfer function $H(s)$.

First let us look at some examples. If the transfer function $H(s)$ is given as

$$H(s) = \frac{2}{(s+2)(s+1)^3} \tag{9.254}$$

Then the partial fraction is expressed as

$$H(s) = \frac{K_1}{(s+2)} + \frac{K_{21}}{(s+1)} + \frac{K_{22}}{(s+1)^2} + \frac{K_{23}}{(s+1)^3} \tag{9.255}$$

If we multiply both sides of Eq. 9.255 with $(s+2)(s+1)^3$, we have the following result:

$$2 = (s+1)^3 K_1 + (s+1)^2(s+2)K_{21} + (s+2)(s+1)K_{22} + (s+2)K_{23} \tag{9.256}$$

Now we have four unknowns in Eq. 9.256; hence we need four independent equations to find these unknowns. Observe that we have two distinct roots for the denominator polynomial of $H(s)$. If we set $s = -2$ and -1 respectively, then we have two simple equations in two unknowns,

$$\begin{aligned} 2 &= -K_1 \\ 2 &= +K_{23} \end{aligned} \tag{9.257}$$

We need two more equations. These are obtained by substituting s with two distinct values other than the two values used. Thus, for example, if we set $s = 0$ and 1 respectively, then we have the following two additional equations:

$$\begin{aligned} K_1 + 2K_{21} + 2K_{22} + 2K_{23} &= 2 \\ 8K_1 + 12K_{21} + 6K_{22} + 3K_{23} &= 2 \end{aligned} \tag{9.258}$$

Eq. 9.258 together with Eq. 9.257 forms four linear equations with four unknowns which can then be solved by calling the Gaussian elimination subroutine CGE(N, A, B) where N should be 4 and A is the coefficient matrix of the system, or

$$A = \begin{bmatrix} -1 & 0 & 0 & 0 \\ 0 & 0 & 0 & +1 \\ 1 & 2 & 2 & 2 \\ 8 & 12 & 6 & 3 \end{bmatrix} \tag{9.259}$$

and B is the 4-vector

$$B = \begin{bmatrix} 2 \\ 2 \\ 2 \\ 2 \end{bmatrix} \tag{9.260}$$

The solution can be shown as $K_1 = -2, K_{21} = 2, K_{22} = -2, K_{23} = 2$.

In general $H(s)$ is given as

$$H(s) = \frac{q(s)}{p(s)} \tag{9.261}$$

where $p(s)$ is in factored form as $(s-\sigma_1)^{n_1}(s-\sigma_2)^{n_2}\cdots(s-\sigma_j)^{n_j}$. In other words there are j distinct roots for $p(s)=0$; the multiplicity of each distinct root $+\sigma_i$ is n_i with $i=1,2,\cdots,j$. Then the partial-fraction expansion takes the foilowing format:

$$\begin{aligned} H(s) = {} & \frac{k_{11}}{s-\sigma_1} + \frac{k_{12}}{(s-\sigma_1)^2} + \cdots + \frac{k_{1n_1}}{(s-\sigma_1)^{n_1}} \\ & + \frac{k_{21}}{(s-\sigma_2)} + \frac{k_{22}}{(s-\sigma_2)^2} + \cdots + \frac{k_{2n_2}}{(s-\sigma_2)^{n_2}} \\ & + \cdots + \\ & \frac{k_{j1}}{(s-\sigma_j)} + \frac{k_{j2}}{(s-\sigma_j)^2} + \cdots + \frac{k_{jn_j}}{(s-\sigma_j)^{n_j}} \end{aligned} \tag{9.262}$$

Hence we need an $\ell = n_1 + n_2 + \cdots + n_j$ linear equations to determine these coefficients. If we multiply Eq. 9.262 by $p(s)$, then we have

$$\begin{aligned} q(s) = {} & p_{11}(s)k_{11} + p_{12}(s)k_{12} + \cdots + p_{1n_1}(s)k_{1n_1} \\ & + p_{21}(s)k_{21} + p_{22}(s)k_{22} + \cdots + p_{2n_2}(s)k_{2n_2} \\ & + \cdots + \\ & p_{j1}(s)k_{j1} + p_{j2}(s)k_{j2} + \cdots + p_{jn_j}(s)k_{jn_j} \end{aligned} \tag{9.263}$$

where

$$p_{rt}(s) = p(s)\cdot\frac{1}{(s-\sigma_r)^t} \qquad \begin{aligned} r &= 1, 2, \cdots, j \\ t &= 1, 2, \cdots, n_r \end{aligned}$$

Now we can form $\ell = n_1 + n_2 + \cdots + n_j$ linear equations by substituting ℓ distinct values for s into Eq. 9.263. To avoid a zero denominator in evaluating $p_{rt}(s)$, it is preferable not to choose s equal to any one of the σ_i's. This practice will not give us simple equations such as Eq. 9.257 in our simple example. However, it is simpler to evaluate $p_{rt}(s)$ without having to worry about a zero denominator. Thus we have ℓ linear equations in ℓ unknowns as follows,

$$\begin{aligned}
&p_{11}(s_1)k_{11}+\cdots+p_{1n_1}(s_1)k_{1n_1}+p_{21}(s_1)k_{21}+\cdots+p_{2n_2}(s_1)k_{2n_2}\\
&\qquad+\cdots+p_{j1}(s_1)k_{j1}+\cdots+p_{jn_j}(s_1)k_{jn_j}=q(s_1)\\
&p_{11}(s_2)k_{11}+\cdots+p_{1n_1}(s_2)k_{1n_1}+p_{21}(s_2)k_{21}+\cdots+p_{2n_2}(s_2)k_{2n_2}\\
&\cdots\qquad+\cdots+p_{j1}(s_2)k_{j1}+\cdots+p_{jn_j}(s_2)k_{jn_j}=q(s_2)\\
&p_{11}(s_\ell)k_{11}+\cdots+p_{1n_1}(s_\ell)k_{1n_1}+p_{21}(s_\ell)k_{21}+\cdots+p_{2n_2}(s_\ell)k_{2n_2}\\
&\qquad+\cdots+p_{j1}(s_\ell)k_{j1}+\cdots+p_{jn_j}(s_\ell)k_{jn_j}=q(s_\ell)
\end{aligned} \tag{9.264}$$

Once we have Eq. 9.264, then the Gaussian elimination subroutine CGE(N, A, B) could be used to determine these unknowns.

Computer Program for Finding Roots

To implement the modified Newton's method in finding the roots of the given polynomial $p(s)$, we need first the following information:

1. The degree n of the polynomial $p(s)$.
2. The coefficients of $p(s)$.

The coefficients are usually stored as an $n+1$ vector ordered from the coefficient of the smallest power of s to the coefficient of the largest power of s. Thus if we have

$$p(s)=3+2s+1s^2,$$

then we need a 3-vector $P(I)$ with $I=1, 2, 3$ to store the corresponding coefficient and we have

$$P(1)=3$$

$$P(2)=2$$

$$P(3)=1$$

The evaluation of $p(s)$, $p'(s)$ and $p''(s)$ for a specific s can all be done using Horner's rule. To start the iteration, we use $j1$ as the initial approximation, this choice will lead to a real root if the roots are all real and complex root if there is any complex root. For multiple roots, we also need the numbers ϵ_1 and ϵ_2 that appear in Eq. 9.243. These are given in Eq. 9.250 and Eq. 9.251. We first choose an appropriate small number as ϵ_1, then Eq. 9.251 is used to set ϵ_2. If for a certain s_i, only Eq. 9.250 is satisfied, then s_i is a good approximation of this particular root. If both Eq. 9.250 and Eq. 9.251 are satisfied, then we will use the derivative polynomial of $p(s)$ to find this "multiple root."

After one root σ is found by using the said process, then the original $p(s)$ is divided by the factor $s-\sigma$ and the quotient polynomial $\gamma(s)$

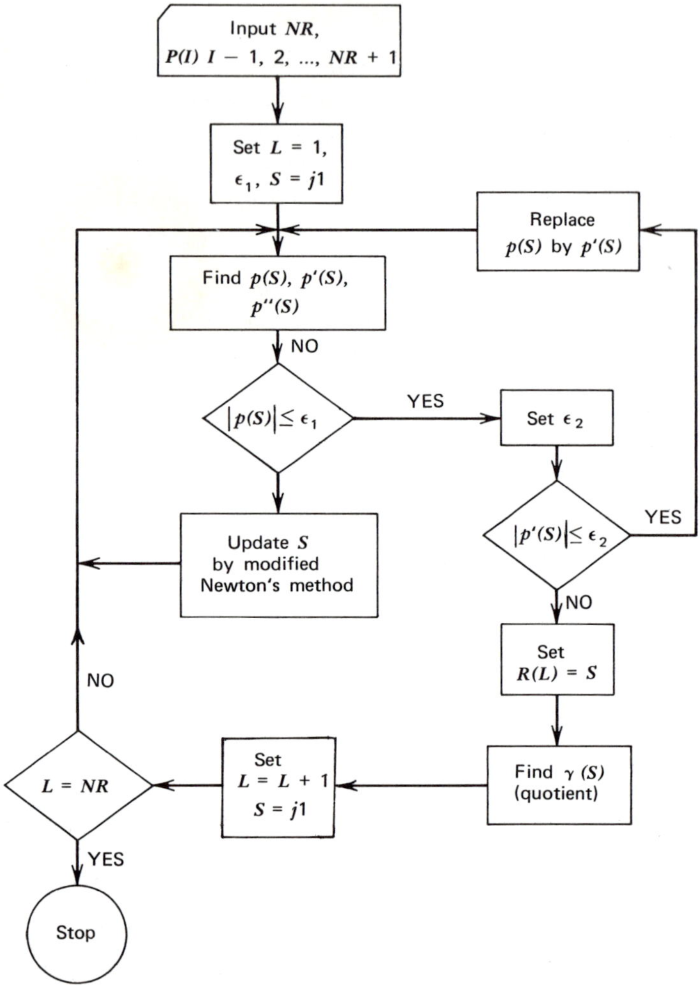

Fig. 9.29 Flowchart for finding roots of a polynomial.

found by using Horner's rule. This quotient polynomial will now be used to find the rest of the roots of $p(s)$.

A flowchart of this process is shown in Fig. 9.29. A subroutine ROOT (NR,P,R) is written based on this flowchart to find all roots of an NRth degree polynomial whose coefficients are stored in $P(I)$ where $I = 1, 2, \cdots, NR+1$. The results are stored in $R(J)$ where $J = 1, 2, \cdots, NR$.

The degree of the polynomial $p(s)$ is restricted to less than or equal to 20. For NR > 20, the dimensions of the vectors P and R are to be modified accordingly.

Example 9.6.1 Find all the roots of the polynomial

$$p(s) = (s+1)^5 = 1+5s+10s^2+10s^3+5s^4+s^5$$

by using the subroutine ROOT.

Solution. The FORTRAN program, the subroutine ROOT and the computer output are shown in Figs. 9.30 through 9.32. The parameters are the following:

P	Vector for storing coefficients of $p(s)$
R	Vector for storing roots of $p(s)$
NR	Degree of $p(s)$
PP	Working vector
S	Complex root (for iteration)
PZ	Value of $p(s)$
$P1$	Value of $p'(s)$
$P2$	Value of $p''(s)$
EPS	ϵ_1
$EPS1$	ϵ_2
L	Counter for the Lth root
ITR	Number of iterations for a root

```
      DIMENSION P(21),R(20)
      COMPLEX P,R
   10 FORMAT(10X,2E14.5)
      NR=5
      P(1)=1.
      P(2)=5.
      P(3)=10.
      P(4)=10.
      P(5)=5.
      P(6)=1.
      CALL ROOT(NR,P,R)
      DO 1 I=1,NR
         PRINT 10, R(I)
    1 CONTINUE
      END
```

Fig. 9.30 Program for finding roots of a polynomial using subroutine ROOT.

```
      SUBROUTINE ROOT(NR,P,R)
      DIMENSION PP(21),P(21),R(20)
      COMPLEX S,PZ,P1,P2,PP,P,R
      EPS=1.0E-06
      NP=NR+1
      DO 1 I=1,NP
        PP(I)=P(I)
    1 CONTINUE
      DO 500 L=1,NR
        ITR=0
        NN=NR-L+1
        S=(0.0,1.0)
  101   NZ=NN+1
        N1=NN-1
        N2=NN-2
        PZ=P(NZ)
        DO 10 I=1,NN
          PZ=PZ*S+P(NZ-I)
   10   CONTINUE
        P1=NN*P(NZ)
        IF(N1-1) 20,30,30
   30   DO 12 I=1,N1
          P1=P1*S+(NN-I)*P(NZ-I)
   12   CONTINUE
   20   CONTINUE
        IF(L-NR) 102,16,16
   16   P2=0.
        GO TO 104
  102   P2=NN*(NN-1)*P(NZ)
        IF(N2) 104,104,103
  103   DO 15 I=1,N2
          P2=P2*S+(NN-I)*(NN-I-1)*P(NZ-I)
   15   CONTINUE
  104   CONTINUE
```

Fig. 9.31 (*a*) First part of the subroutine ROOT.

```
        IF(CABS(PZ)-EPS) 17,28,28
   17   EPS1=SQRT(2.*EPS*CABS(P2))
        IF(CABS(P1)-EPS1) 18,200,200
   18   DO 19 I=1,NN
          P(I)=P(I+1)*I
   19   CONTINUE
        NN=NN-1
        GO TO 101
   28   S=S-PZ/(P1-PZ*P2/(P1*2))
        ITR=ITR+1
        GO TO 101
  200   R(L)=S
        NL=NR-L+1
        IF(NL-1) 240,240,230
  230   P(NL)=PP(NL+1)
        NK=NL-1
        DO 22 I=1,NK
          P(NL-I)=PP(NL-I+1)+S*P(NL-I+1)
   22   CONTINUE
        DO 23 I=1,NL
          PP(I)=P(I)
   23   CONTINUE
  240   CONTINUE
  500 CONTINUE
      RETURN
      END
```

Fig. 9.31 (*b*) Second part of subroutine ROOT.

Re(σ)	Im(σ)
-.10000E+01	.43368E-18
-.10000E+01	0.
-.10000E+01	.86736E-18
-.10000E+01	0.
-.10000E+01	0.

Fig. 9.32 Sample computer output.

Computer Program for Partial-Fraction Expansion

Once the roots and their multiplicities are known, then we can find the coefficients of the partial-fraction expansion of a given transfer function $H(s)$ by using the linear equation solver *CGE*. Specifically, we have

$$H(s) = \frac{q(s)}{p(s)}$$

where $q(s)$ is an unfactored polynomial of s and $p(s)$ is given in factored form as

$$p(s) = (s-\sigma_1)^{n_1}(s-\sigma_2)^{n_2}\cdots(s-\sigma_j)^{n_j}$$

and we have j distinct roots each with multiplicity n_k where $k = 1, 2, \ldots, j$. Hence $\ell = n_1 + n_2 + \cdots + n_j$ is known and we need ℓ linear equations to find our partial-fraction expansion coefficients. We use the following procedure to find ℓ distinct s values for evaluating the $p_{rt}(s)$ and $q(s)$ in Eq. 9.265. First we order the σ_i as

$$\sigma_1 > \sigma_2 > \cdots > \sigma_j$$

then we select

$$s_I = \sigma_j + I(\sigma_j - \sigma_1) \qquad I = 1, 2, \cdots, \ell \tag{9.265}$$

With this choice for the s_I, ℓ linear algebraic equations are formed which could be solved by using the subroutine CGE.

A flowchart for finding the partial-fraction expansion coefficients using the linear equation approach is shown in Fig. 9.33. A subroutine PAFRA(*R,Q,XK,NR,M,MQ*) is written based on this flowchart. Here *R* is an *NR*-vector storing the distinct roots. *Q* is a vector for storing the coefficients of the unfactored polynomial $q(s)$. *XK* is a vector storing the results of $q(s_I)$. $M(J)$ is the multiplicity of the *J*th root. *MQ* is the degree of the polynomial $q(s)$. The resultant coefficients are stored in *XK*. We need, of course, the subroutine CGE in order to use PAFRA. The total number of roots in $p(s)$ is restricted to 20. For a polynomial with more than 20 roots, the dimension statements in all related programs are to be modified to the desired number of roots.

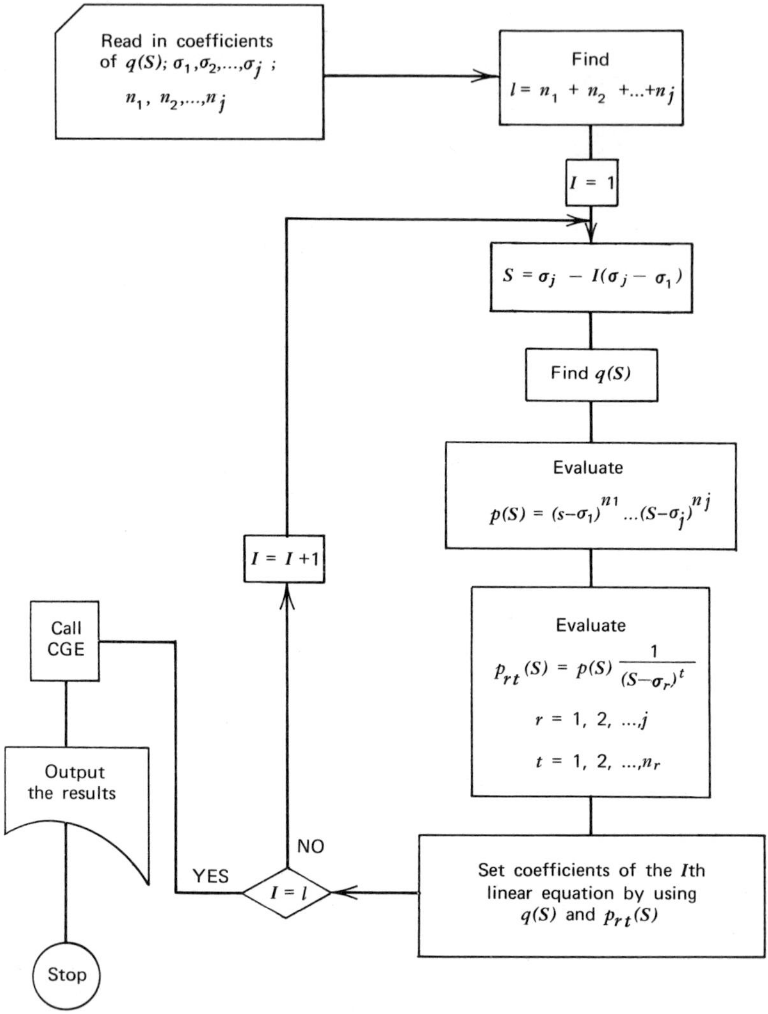

Fig. 9.33 Flowchart for evaluating partial-fraction coefficients.

Example 9.6.2 Find the partial-fraction expansion of the following transfer function by using PAFRA.

$$H(s) = \frac{2}{(s+1)^3(s+2)}$$

The exact expansion is

$$H(s) = \frac{2}{s+1} - \frac{2}{(s+1)^2} + \frac{2}{(s+1)^3} + \frac{2}{s+2}$$

Solution. The FORTRAN program, the subroutines PAFRA and CGE, the computer output are shown in Figs. 9.34 to 9.37. The parameters are the following:

- R Storage for distinct roots
- Q Storage for coefficients of $q(s)$
- M Storage for multiplicity of roots
- NR Number of distinct roots
- MQ Degree of $q(s)$
- S Value selected for forming one linear equation
- XK Value of $q(s)$, also the results for the coefficients of the partial fraction expansion.
- P Value of $p(s)$
- A Coefficient matrix of the resultant system of linear equations.
- L Number of linear equations formed

Table 9.6.1 Laplace transforms

$f(t)$	$F(s)$
1. $(d/dt)(f(t))$	$sF(s)-f(0-)$
2. $(d^n/dt^n)(f(t))$	$s^nF(s)-\sum_{j=1}^{n} s^{n-j}f^{(j-1)}(0-)$
3. $\int_{0-}^{t} f(\tau)\,d\tau$	$F(s)/s$
4. $(-t)^n f(t)$	$(d^n/ds^n)F(s)$
5. $e^{at}f(t)$	$F(s-a)$
6. $f(t-a)u(t-a)$	$e^{-as}F(s)$
7. $u(t)$	$1/s$
8. $\delta(t)$	1
9. t	$1/s^2$
10. e^{-at}	$1/(s+a)$
11. $t^n e^{-at}$	$n!/(s+a)^{n+1}$
12. $\sin bt$	$b/(s^2+b^2)$
13. $\cos bt$	$s/(s^2+b^2)$
14. $\sin h\, at$	$a/(s^2-a^2)$
15. $\cos h\, at$	$s/(s^2-a^2)$
16. $e^{-at}\sin bt$	$b/((s+a)^2+b^2)$
17. $e^{-at}\cos bt$	$(s+a)/((s+a)^2+b^2)$
18. $e^{-at}t^n/n!$	$1/(s+a)^{n+1}$
19. $(e^{-at}-e^{-bt})/(b-a)$	$1/((s+a)(s+b))$
20. $(t\sin bt)/2b$	$s/(s^2+b^2)^2$

```
      DIMENSION R(20),Q(20),M(20),XK(20)
      COMPLEX R,Q,XK
    2 FORMAT(10X,2E14.5)
      R(1)=-1.
      R(2)=-2.
      M(1)=3
      M(2)=1
      Q(1)=2.
      NR=2
      MQ=0
      CALL PAFRA(R,Q,XK,NR,M,MQ)
      L=0
      DO 1 I=1,NR
        L=L+M(I)
    1 CONTINUE
      PRINT 2, (XK(I), I=1,L)
      END
```

Fig. 9.34 Program for finding partial-fraction expansion coefficients.

```
      SUBROUTINE CGE(N,A,B)
      DIMENSION A(20,20),B(20)
      COMPLEX A,B,XMAX,SAVE
      N1=N-1
      DO 300 I=1,N1
        XMAX=0.
        DO 40 K=I,N
          IF(CABS(XMAX)-CABS(A(K,I))) 50,40,40
   50     XMAX=A(K,I)
          L=K
   40   CONTINUE
        IF(CABS(XMAX)-0.0) 60,60,70
   60   PRINT 80
   80   FORMAT(10X,19H SYSTEM IS SINGULAR)
        RETURN
   70   DO 90 J=I,N
          SAVE=A(I,J)
          A(I,J)=A(L,J)
          A(L,J)=SAVE
   90   CONTINUE
        SAVE=B(I)
        B(I)=B(L)
        B(L)=SAVE
        KK=I+1
        DO 100 K=KK,N
          A(K,I)=A(K,I)/XMAX
  100   CONTINUE
        DO 120 K=KK,N
          DO 110 JK=KK,N
            A(K,JK)=A(K,JK)-A(K,I)*A(I,JK)
  110     CONTINUE
          B(K)=B(K)-A(K,I)*B(I)
  120   CONTINUE
  300 CONTINUE
      B(N)=B(N)/A(N,N)
      DO 510 I=1,N1
        IB=N-I
        IC=N
        DO 500 K=1,I
          B(IB)=B(IB)-A(IB,IC)*B(IC)
          IC=IC-1
  500   CONTINUE
        B(IB)=B(IB)/A(IB,IB)
  510 CONTINUE
      RETURN
      END
```

Fig. 9.35 The subroutine CGE.

```
      SUBROUTINE PAFRA(R,C,XK,NR,M,MQ)
      DIMENSION R(20),Q(20),M(20),A(20,20),XK(20)
      COMPLEX S,R,XK,C,P,A
      L=0
      DO 1 I=1,NR
        L=L+M(I)
    1 CONTINUE
      DO 10 I=1,L
        S=R(NR)+I*(R(NR)-R(1))
        MMQ=MQ+1
        XK(I)=C(MMQ)
        IF(MMQ-1) 30,30,2
    2   DO 3 J=1,MMQ
          XK(I)=S*XK(I)+C(MMC-J)
    3   CONTINUE
   30   P=1
        DO 4 J=1,NR
          P=P*((S-R(J))**M(J))
    4   CONTINUE
        LK=0
        DO 5 J=1,NR
          MJ=M(J)
          DO 6 K=1,MJ
            LK=LK+1
            A(I,LK)=P/((S-R(J))**K)
    6     CONTINUE
    5   CONTINUE
   10 CONTINUE
      CALL CGE(L,A,XK)
      RETURN
      END
```

Fig. 9.36 The subroutine PAFRA.

```
    Real          Imaginary

 0.20001E 01      0.0
-0.20000E 01      0.0
 0.20001E 01      0.0
-0.20001E 01      0.0
```

Fig. 9.37 The computer output.

PROBLEMS

9-1. We have seen that the unit-step function $u(t)$ is not Fourier transformable. However, $e^{-\sigma t}u(t)$ is Fourier transformable since

$$\int_{-\infty}^{\infty} e^{-\sigma t}u(t)\,dt < \infty \qquad \text{for } \sigma > 0$$

The function $e^{-\sigma t}$ where $\sigma > 0$ is the convergence factor for $u(t)$. Determine the convergence factors for the following functions:

(i) $e^{at}u(t)$ where $a > 0$
(ii) $\cos wt\,u(t)$
(iii) $e^{jwt}u(t)$
(iv) $t^2u(t)$

9-1a. Show that the Laplace transform of sinh at determined from Eq. 9.13 is

$$\mathbf{L}(\sinh at) = \frac{a}{s^2 - a^2}$$

9-1b. Repeat 9-1a for the following transformation:

$$\mathbf{L}(\cosh at) = \frac{s}{s^2 - a^2}$$

9-1c. Repeat 9-1a for the following transformations:

$$\text{(i) } \mathbf{L}(e^{-\alpha t} \sin wt) = \frac{w}{(s+\alpha)^2 + w^2}, \quad \alpha > 0.$$

$$\text{(ii) } \mathbf{L}(e^{-\alpha t} \cos wt) = \frac{s+\alpha}{(s+\alpha)^2 + w^2}, \quad \alpha > 0.$$

$$\text{(iii) } \mathbf{L}\left(\frac{t}{2w} \sin wt\right) = \frac{s}{(s^2 + w^2)}$$

9-1d. Repeat 9-1a for the following transformations:

(i) $\mathbf{L}(t^n/n!) = 1/s^{n+1}$ (ii) $\mathbf{L}(e^{-\alpha t}t^n/n) = 1/(s+\alpha)^{n+1}, \alpha > 0$

$$\text{(iii) } \mathbf{L}\left[\frac{1}{\beta - \alpha}(e^{-\alpha t} - e^{-\beta t})\right] = 1/(s+\alpha)(s+\beta)$$

9-1e. Show that $\mathbf{L}[f(t)/t] = \int_s^{\infty} F(s)\, ds$.

9-1f. Repeat 9-1 for the functions of 9-1a and 9-1b.
9-1g. Repeat 9-1 for the functions of 9-1c.
9-1h. Repeat 9-1 for the functions of 9-1d.
9-2. Find the Laplace transforms of the following functions:
(i) $f(t) = \sin(wt + \theta)$
(ii) $f(t) = e^{-(t+a)} \cos(wt + \theta)$
(iii) $f(t) = (t^3 + 1)e^{-at}$
(iv) $f(t) = K^t, K > 1$
(v) $f(t) = t\delta'(t)$
(vi) $f(t) = (t^2 + ae^{-t})e^{-bt}$
9-3. Find the Laplace transforms of the following functions:
(i) $f(t) = \cos(t - \pi/2)\, u(t - \pi/4)$
(ii) $f(t) = \cos(t - \pi/4)\, u(t)$
(iii) $f(t) = u(t^3 - 4t)$
9-4. Find the Laplace transforms for the functions shown in Fig. P9-4.

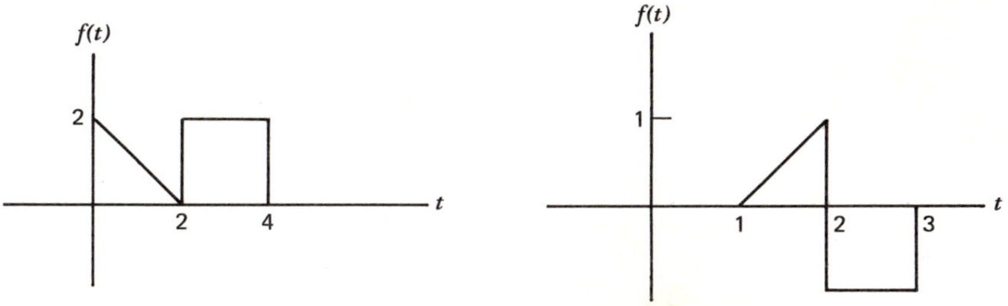

Fig. P9-4

9-5. Find the Laplace transforms for the derivatives of the functions of problem 9-4.

9-6. Find the Laplace transforms of the functions shown in Fig. P9-6.

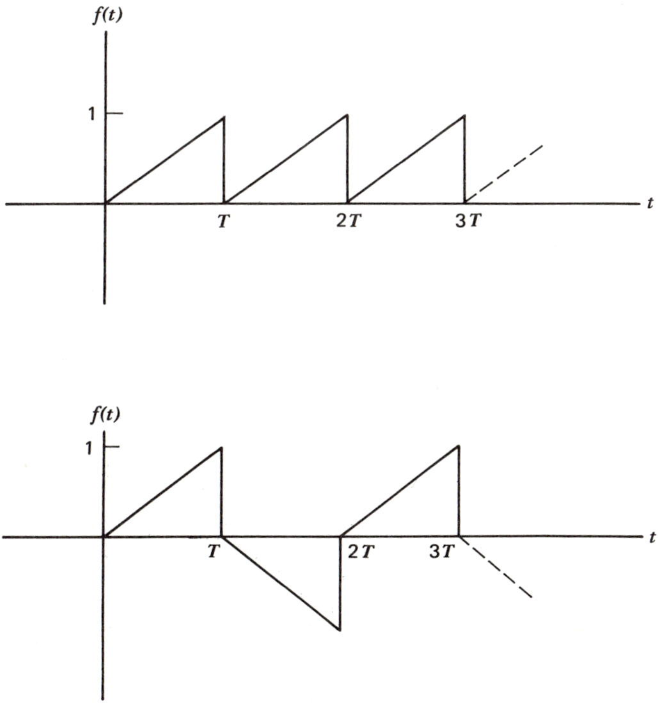

Fig. P9-6

9-6a. Expand the following functions as partial fractions:

(i) $H(s) = \dfrac{(s+1)(s+3)}{s(s+2)(s+4)}$ (ii) $H(s) = \dfrac{(s^2+1)(s^2+3)}{s(s^2+2)(s^2+4)}$

(iii) $H(s) = \dfrac{1}{1+2s+2s^2+s^3}$ (iv) $H(s) = \dfrac{s+4}{(s+1)^2(s+3)^2}$

9-7. Solve the following differential equations using Laplace transforms.

(i) $x''(t)+6x'(t)+9x(t) = \cos 2t$
(ii) $x''(t)+3x'(t)+2x(t) = \delta(t)$
(iii) $x''(t)+5x'(t)+4x(t) = (e^{-t}+e^{-4t})u(t)$

It is given that $x(0-) = x'(0-) = 0$ for all the equations.

9-8. Using Laplace transforms, solve the following sets of simultaneous equations:

(i) $2x'(t)+4x(t)+y'(t)+2y(t) = \delta(t)$
$x'(t)+3x(t)+y'(t)+2y(t) = 0$
(ii) $2x'(t)+x(t)+y'(t)+4y(t) = 2e^{-t}$
$x'(t)+y'(t)+8y(t) = \delta(t)$

The initial conditions are all zero at $t = 0-$.

9-9. Find the inverse transforms for the following transforms:

(i) $F(s) = \dfrac{s}{(s^2+9)(s+3)}$ (ii) $F(s) = \dfrac{s^3+3s+1}{s^2+s}$

(iii) $F(s) = \dfrac{(s+1)^4}{s^4}$ (iv) $F(s) = \dfrac{3s^3-s^2-3s+2}{s^2(s-1)^2}$

9-10. Find the inverse transforms for the following transforms:

(i) $F(s) = \dfrac{1+e^{-2s}}{s^2(s+2)}$

(ii) $F(s) = \dfrac{se^{-3s}-e^{-5s}}{s^2+3s+2}$

(iii) $F(s) = \dfrac{se^{-as}}{s+b}$

9-10a. Find

$$\mathbf{L}^{-1}\left[\frac{n!}{s(s+1)(s+2)\cdots(s+n)}\right]$$

9-11. Using Laplace transform, solve problem 2-1. by partial-fraction expansion.
9-11a. Repeat 9-11 for problem 2-5.
9-12. Repeat 9-11 for problem 2-7.
9-12a. Repeat 9-11 for problem 2-8.
9-12b. Repeat 9-11 for problem 2-9.
9-12c. Repeat 9-11 for problem 2-10.
9-13. Repeat 9-11 for problem 2-14.

9-13a. Repeat 9-11 for problem 2-17a.
9-13b. Repeat 9-11 for problem 2-17d.
9-13c. Repeat 9-11 for problem 2-17e.
9-13d. Repeat 9-11 for problem 2-17f.
9-13e. Repeat 9-11 for problem 3-4.
9-13f. Repeat 9-11 for problem 3-5.
9-13g. Repeat 9-11 for problem 3-5a.
9-14. Repeat 9-11 for problem 3-6.
9-15. Repeat 9-11 for problem 3-7.
9-15a. Repeat 9-11 for problem 3-7a.
9-15b. Repeat 9-11 for problem 3-8(iv).
9-15c. Repeat 9-11 for problem 3-9(iv).
9-15d. Repeat 9-11 for problem 3-20f.
9-15e. Repeat 9-11 for problem 4-8.
9-16. Solve the following integral equations for $x(t)$ using Laplace transform:

(i) $x(t) + \int_0^t (t-\tau)x(\tau)\,d\tau = 1$

(ii) $\sin t = \int_0^t x(\tau)e^{-(t-\tau)}\,d\tau$

9-17. The current generator of the circuit of Fig. P9-17 delivers a constant current of 4 amperes. At $t = 0$ the switch S is opened and the resulting voltage across the terminals 1-1′ is $v(t) = 6e^{-4t} + 12$ volts.
(i) Find $Z(s)$ looking into the terminals 1-1′.
(ii) Find a network realization for $Z(s)$.

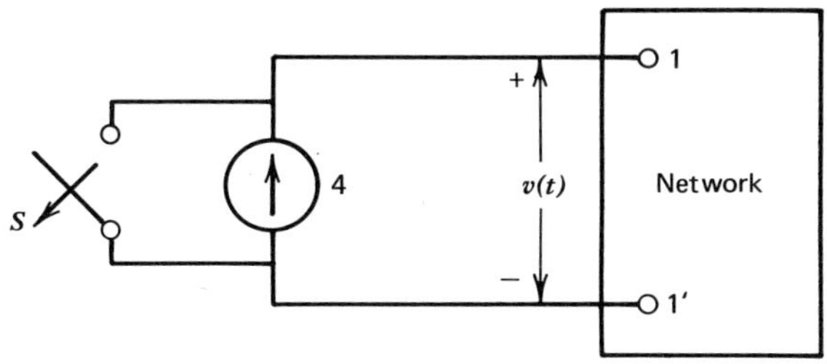

Fig. P9-17

9-17a. Plot the poles and zeroes in the s plane for the following transfer functions:

(i) $H(s) = \dfrac{s^2 + 3s + 2}{s^2 + 3s + 4}$

(ii) $H(s) = \dfrac{(s^2+4)(s+1)}{(s^2+1)(s^2+2s+5)}$

(iii) $H(s) = \dfrac{(s^2-1)}{(s-2)(s^2+2s+2)}$

(iv) $H(s) = \dfrac{(s^2-2s+5)(s^2-4s+4)}{(s^2+2s+5)(s^2+4s+4)}$

9-17b. Assume that each of the transfer functions given in 9-17a is the impulse response of a certain network. Determine which network will have unbounded response if a unit-step excitation is applied to each of the networks.

9-18. Given the pole-zero diagrams shown in Fig. P9-18, write the rational functions $F_1(s)$ and $F_2(s)$ as quotients of polynomials. What can you say about the relationship between the poles in the right-half plane and the signs of the coefficients of the denominator polynomials?

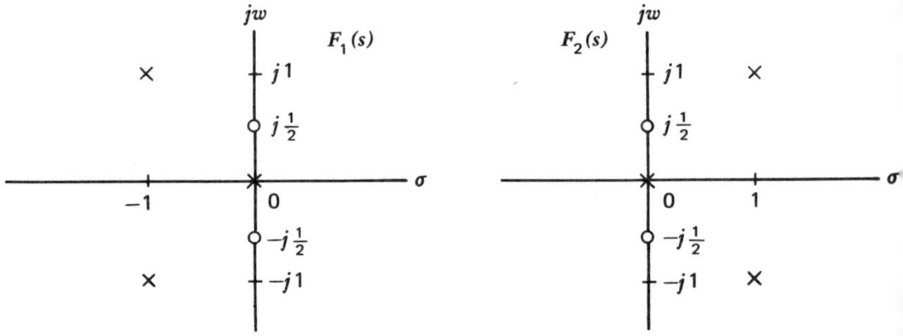

Fig. P9-18

9-19. The pole-zero plots of certain functions $F(s)$ are shown in Fig. P9-19.

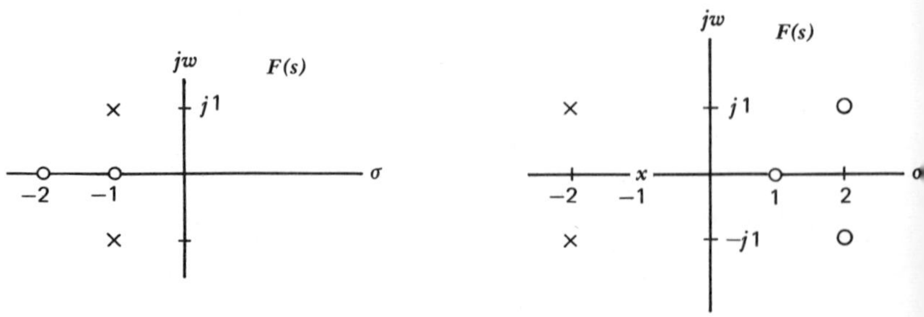

Fig. P9-19

(i) Find simple rational equations as quotients of polynomials for $F(s)$.
(ii) Find the partial-fraction coefficients for those $F(s)$ of (i).
(iii) If these $F(s)$ are the transfer functions of filters, how would you classify them—stable or unstable?

9-20. Two response transforms, $R_1(s)$ and $R_2(s)$, have the pole-zero plots shown in Fig. P9-20. In addition, it is known that $r_1(0+) = 4$ and $r_2(\infty) = 4$. Find $r_1(t)$ and $r_2(t)$.

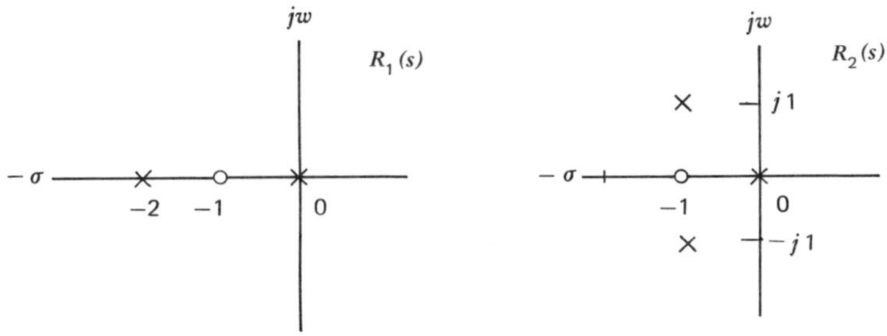

Fig. P9-20

9-21. For the simple function $p(s) = s^2 - 2$:
(i) Find the positive root of $p(s)$ by using Newton's method with 5-digit mantissa chopped floating-point arithmetic. Set the initial approximation $s_0 = 10$ with 10 iterations.
(ii) Calculate $e_i = \sigma - s_i$ for each iteration. When is the Newton's method convergent quadratically?

9-22. The method of false position for finding the real root of $p(s) = 0$ can be described as follows. Referring to Fig. P9-22, let $(t, p(t))$ be a fixed point on $y = p(s)$. Draw a chord through this point and the point $(s_i, p(s_i))$ so that it intersects the s axis at a point $(s_{i+1}, 0)$. Then we have the following recursive equation:

$$s_{i+1} = \frac{t\,p(s_i) - s_i\,p(t)}{p(s_i) - p(t)}$$

This method can also be used to find complex roots if t and $p(t)$ are given as complex numbers.
(i) Solve 9-21(i) by the method of false position. Use $t = 11$ and $s_0 = 10$ with 10 iterations.
(ii) It can be shown that for the method of false position we have

$$e_{i+1} - e_i = \frac{(t - s_i)}{p(t) - p(s_i)}\,p(s_i)$$

Calculate e_i for each iteration and verify the above equation.

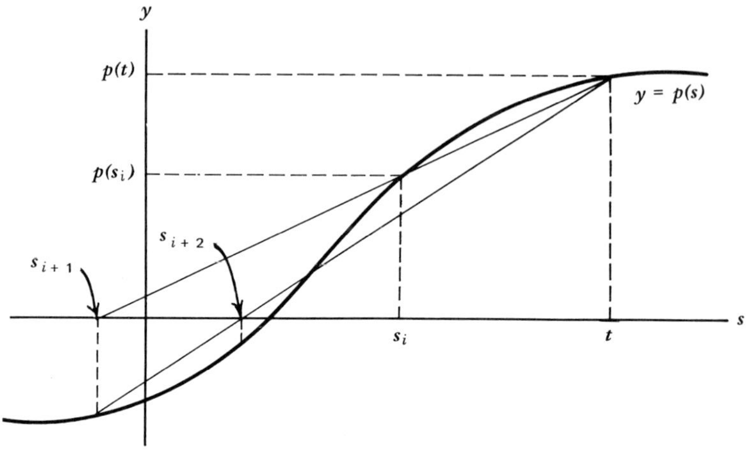

Fig. P9-22

9-22a. Using the subroutine ROOT, find the roots of the following polynomial equations:

(i) $p(s) = 15 - 28s + 14s^2 - 2s^3 = 0$
(ii) $p(s) = 5.75 - 11s + 6s^2 - s^3 = 0$
(iii) $p(s) = 4 + 4s - 4s^2 - 2s^3 + s^4$

Ans: (i) 0.860, 2.230, 3.910
(ii) 0.893, 2.270, 2.834
(iii) 0.732, 1.414, −1.414, −2.732

9-22b. Repeat 9-22a for the following polynomial equations:

(i) $p(s) = 18 - 6s - 9s^2 + s^3 + s^4 = 0$
(ii) $p(s) = 1 + 2s - 5s^2 - 6s^3 + 9s^4 = 0$
(iii) $p(s) = 187 - 57s - 110s^2 + 36s^3 + 16s^4 - 5s^5 = 0$

Ans: (i) −2.45, −2.30, 1.30, 2.45
(ii) −0.434, −0.434, 0.768, 0.768
(iii) −1.949, −1.852, 1.783, 2.052, 3.166.

9-22c. Repeat 9-22a for the following polynomial equations:

(i) $p(s) = 18 + 12s - s^2 - 4s^3 + 2s^4 = 0$
(ii) $p(s) = 70 - 62s + 11s^2 + 10s^3 - 6s^3 - 6s^4 + s^5 = 0$
(iii) $p(s) = 196 - 28s + 4s^2 - 2s^3 + s^4 = 0$

Ans: (i) $2 \pm j1.414, -1 \pm j0.707$;
(ii) $-2.096, 1.500 \pm j1.658, 2.548 \pm j0.424$;
(iii) $-2 \pm j3.162, 3 \pm j2.236$.

9-22d. Solve 9-6a using the subroutine PAFRA.

9-22e. Repeat 9-22d for 9-10a with n set as 10.

9-22f. Find the partial-fraction expansion of each of the transfer functions given in 9-17a using the subroutine PAFRA.

9-23. Find the Laplace transforms of the capacitor voltages of the circuit of Fig. P3-18. Reduce them into simple rational functions as quotients of polynomials.

9-24. Find the partial-fraction expansions of the capacitor voltages $V_{C1}(s)$, $V_{C2}(s)$, $V_{C3}(s)$ of problem 9-23 by using the subroutines ROOT and PAFRA. Assume $v_1(t) = v_2(t) = 0$, $v_{C1}(0-) = -0.2078$ V, $v_{C2}(0-) = 0.5609\ V$, and $v_{C3}(0-) = -0.8015\ V$.

9-25. From the results of 9-24, find the time-domain expressions for the capacitor voltages. Check your answers with problem 3-23 at some discrete time instants.

9-26. Repeat 9-24 for $v_1(t) = 2\ V$ and $v_2(t) = 4\ V$ with all initial states set to zero. Check your answers by finding the final states of the capacitor voltages by using methods of Chapter 5.

CHAPTER 10
Two-Port Network Characterization

In electric network theory, the word *port* has a special meaning. A port may be regarded as a pair of terminals in which the current into one terminal is equal to the current out of the other terminal. In our discussions in the previous chapters, we have most often assumed that we have access to all elements and all nodes in the network under study. This is true in the early stages of design of a circuit so that it is possible to measure the voltages or currents and to adjust the components. However, once the preliminary design is complete, the network is then transferred to a chassis or perhaps it is encased in plastic. Then we are dealing with networks to which access is possible only at the network ports. Thus we are facing the problem of *characterizing the specific network in terms of the excitation-response measurements at the network ports*.

For easier presentation, we will only consider in this chapter the two-port network characterization problem. The methods discussed in this chapter can easily be extended to n-port cases.

10.1 TWO-PORT MODELS

As we have discussed earlier, a port may be regarded as a pair of terminals in which the current into one terminal equals the current out of the other. For the one-port network shown in Fig. 10.1, $I = I'$. A one-port network is completely specified when the voltage-current relationship at the terminals of the port are given. For example, if $V = 10\ V$ and $I = 2$ amp, then we know that the *input or driving-point* impedance of the one-port is given as

$$Z_{in} = \frac{V}{I} = 5\Omega \tag{10.1}$$

Whether the one-port is actually a single 5-Ω resistor, or two 2.5-Ω

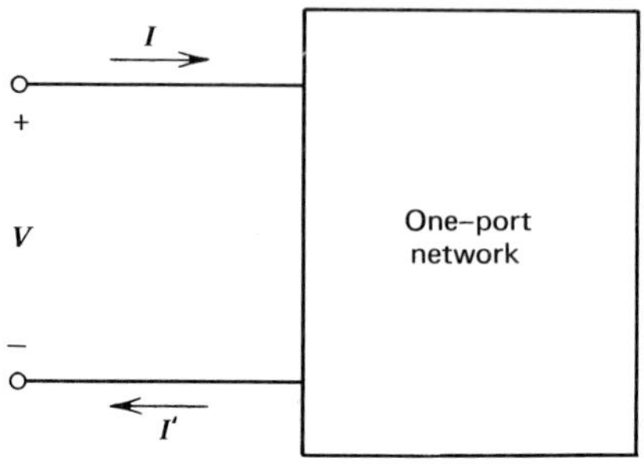

Fig. 10.1 One-port network.

resistors in series, or two 10-Ω resistors in parallel is of little importance because the primary concern is the current-voltage relationship at the port. Consider the example when $I = 2s + 3$ and $V = 1$; then the input admittance of the one port is

$$Y_{\text{in}} = \frac{I}{V} = 2s + 3 \tag{10.2}$$

that corresponds to a 2-*fd* capacitor in parallel with a 1/3-Ω resistor in its simplest case (Fig. 10.2).

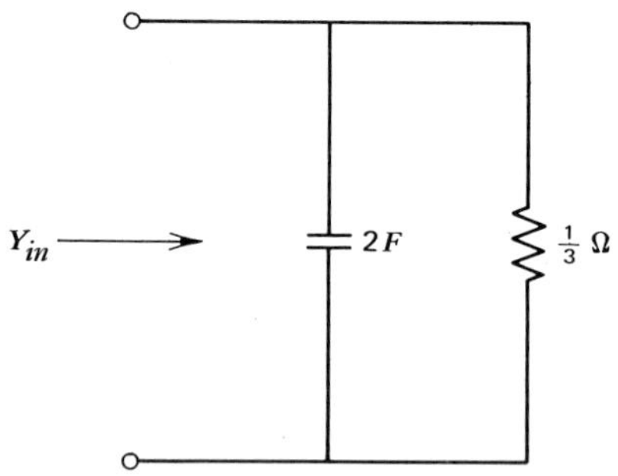

Fig. 10.2 A simple realization of $Y_{\text{in}} = 2s + 3$.

Two-Port Parameters

A general *two-port* network, shown in Fig. 10.3 has two pairs of voltage-current relationships. The variables are V_1, V_2, I_1, I_2. Two of these are *dependent* variables, the other two, *independent* variables. The number of possible combinations generated by four variables taken two at a time is six. Thus there are six possible sets of equations describing a two-port. We will discuss the four most useful descriptions here.

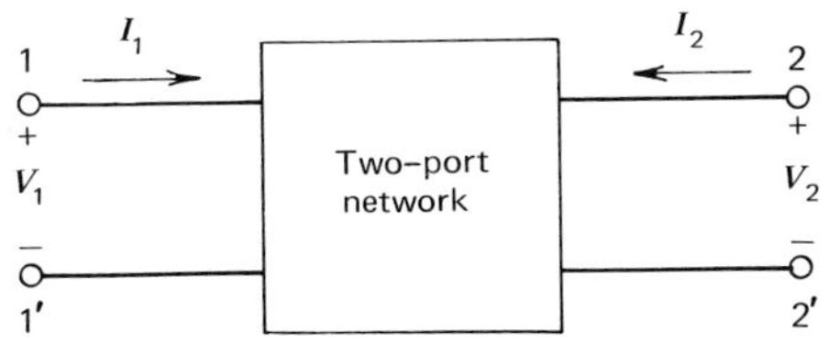

Fig. 10.3 A general two-port network.

The z Parameters

A particular set of equations that describe a two-port network are the z parameter equations

$$\begin{aligned} V_1 &= z_{11}I_1 + z_{12}I_2 \\ V_2 &= z_{21}I_1 + z_{22}I_2 \end{aligned} \tag{10.3}$$

In these equations, the variables V_1 and V_2 are dependent and I_1, I_2 are independent. The individual z parameters are defined by

$$\begin{aligned} z_{11} &= \frac{V_1}{I_1}\bigg|_{I_2=0} \qquad & z_{12} &= \frac{V_1}{I_2}\bigg|_{I_1=0} \\ z_{21} &= \frac{V_2}{I_1}\bigg|_{I_2=0} \qquad & z_{22} &= \frac{V_2}{I_2}\bigg|_{I_1=0} \end{aligned} \tag{10.4}$$

It is observed that all the z parameters have the dimensions of impedance. Moreover, the individual parameters are specified only as the current in one of the ports is set to zero. This corresponds to one of the ports being *open-circuited* from which the z parameters also derive the name *open-circuit* parameters. Note that z_{11} relates the current and voltage in the 1-1′ port only whereas z_{22} gives the current voltage relationship for the 2-2′ port. Such parameters are called open-circuit driving-point impedances. On the other hand, the parameters z_{12} and z_{21} relate the voltage in one port to the current in the other. These are known as (open-circuit) transfer impedances. As an example, let us find the open-circuit parameters for the T circuit in Fig. 10.4. We obtain the z parameters by inspection

$$\begin{aligned} z_{11} &= \frac{V_1}{I_1}\bigg|_{I_2=0} = Z_a + Z_b \\ z_{22} &= \frac{V_2}{I_2}\bigg|_{I_1=0} = Z_b + Z_c \\ z_{12} &= \frac{V_1}{I_2}\bigg|_{I_1=0} = Z_b \\ z_{21} &= \frac{V_2}{I_1}\bigg|_{I_2=0} = Z_b \end{aligned} \tag{10.5}$$

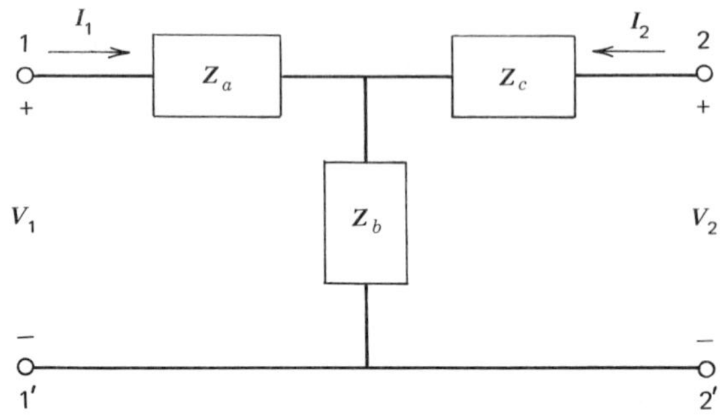

Fig. 10.4 A T-network.

Observe that $z_{12} = z_{21}$. When the open-circuit transfer impedances of a two-port network are equal, the network is *reciprocal.*

Most two-port networks, whether passive or active, can be characterized by a set of open-circuit parameters. Usually, the network is sufficiently complicated so that we cannot obtain the z parameters by inspection, as we did for the T circuit in Fig. 10.4. The question is now: how do we obtain the z parameters for *any* circuit in general? The procedure is as follows. We write a set of node equations with the voltages at the ports, V_1 and V_2, and other node voltages within the two ports, $V_3, V_4, \ldots, V_k$ as the dependent variables. The independent variables are the currents I_1 and I_2, which we will take to be current sources. We then proceed to write a set of node equations,

$$\begin{aligned}
I_1 &= n_{11}V_1 + n_{12}V_2 + n_{13}V_3 + \cdots + n_{1k}V_k \\
I_2 &= n_{21}V_1 + n_{22}V_2 + \cdots \qquad \cdots + n_{2k}V_k \\
0 &= n_{31}V_1 + \cdots \quad + \cdots \qquad\quad + n_{3k}V_k \\
&\cdots\cdots\cdots\cdots\cdots\cdots\cdots\cdots\cdots\cdots \\
0 &= n_{k1}V_1 + \cdots \quad + \cdots \qquad\quad + n_{kk}V_k
\end{aligned} \tag{10.6}$$

where n_{ij} represents the admittance between the ith and jth nodes, that is,

$$n_{ij} = G_{ij} + sC_{ij} + \frac{1}{sL_{ij}}$$

If the circuit is made up of RLC elements only, then it is clear that $n_{ij} = n_{ji}$. As a result, the ijth cofactor of the determinant of the node equations, Δ_{ij}, must be equal to the jith cofactor, Δ_{ji}, that is,

$$\Delta_{ij} = \Delta_{ji}$$

This result leads directly to the reciprocity condition $z_{21} = z_{12}$ as we shall see.

Returning to the set of node equations in Eq. 10.6, let us solve for V_1

and V_2. We obtain

$$V_1 = \frac{\Delta_{11}}{\Delta} I_1 + \frac{\Delta_{21}}{\Delta} I_2$$
$$V_2 = \frac{\Delta_{12}}{\Delta} I_1 + \frac{\Delta_{22}}{\Delta} I_2 \tag{10.7}$$

Relating this last set of equations to the defining equations for the z parameters, it is clear that

$$z_{11} = \frac{\Delta_{11}}{\Delta} \qquad z_{12} = \frac{\Delta_{21}}{\Delta}$$
$$z_{21} = \frac{\Delta_{12}}{\Delta} \qquad z_{22} = \frac{\Delta_{22}}{\Delta} \tag{10.8}$$

Since for a passive network $\Delta_{21} = \Delta_{12}$, it follows that $z_{21} = z_{12}$; the network is then reciprocal.

As an example, let us find the z parameters of the *Pi* circuit in Fig. 10.5. First, the node equations are:

$$I_1 = (Y_A + Y_C)V_1 - Y_C V_2$$
$$I_2 = -Y_C V_1 + (Y_B + Y_C)V_2 \tag{10.9}$$

The determinant for this set of equations is

$$\Delta Y = Y_A Y_B + Y_A Y_C + Y_B Y_C$$

In terms of ΔY, the open-circuit parameters for the *Pi* circuit are:

$$z_{11} = \frac{Y_B + Y_C}{\Delta Y} \qquad z_{21} = \frac{Y_C}{\Delta Y}$$
$$z_{12} = \frac{Y_C}{\Delta Y} \qquad z_{22} = \frac{Y_A + Y_C}{\Delta Y} \tag{10.10}$$

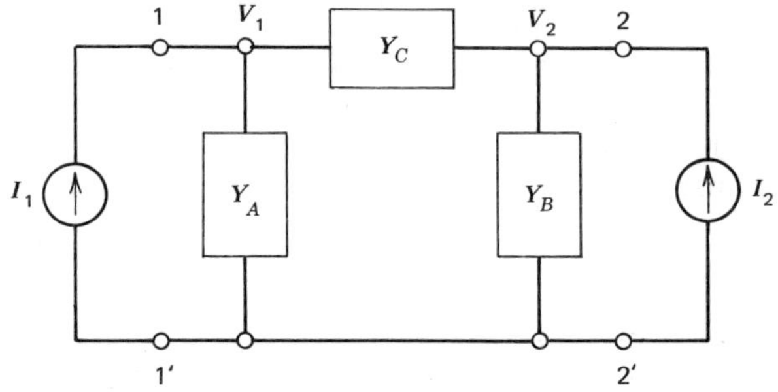

Fig. 10.5 A π-network.

Now, let us perform a delta-wye transformation for the circuits in Figs. 10.4 and 10.5. In other words, let us find relationships between the immittances of the two circuits so that they both have the same z parameters. We readily obtain

$$\begin{aligned} z_{12} &= Z_b = \frac{Y_C}{\Delta Y} \\ z_{22} &= Z_b + Z_c = \frac{Y_A + Y_C}{\Delta Y} \\ z_{11} &= Z_a + Z_b = \frac{Y_B + Y_C}{\Delta Y} \end{aligned} \tag{10.11}$$

We then find

$$\begin{aligned} Z_a &= \frac{Y_B}{\Delta Y} \\ Z_c &= \frac{Y_A}{\Delta Y} \end{aligned} \tag{10.12}$$

The y Parameters

Suppose we were to write a set of mesh equations for the two-port in Fig. 10.3. Then the voltages V_1 and V_2 would become independent sources and the currents I_1 and I_2 would be just two of the dependent mesh currents. Consider the general set of mesh equations

$$\begin{aligned} V_1 &= m_{11}I_1 + m_{12}I_2 + \cdots + m_{1k}I_k \\ V_2 &= m_{21}I_1 + m_{22}I_2 + \cdots + m_{2k}I_k \\ 0 &= m_{31}I_1 + \quad \cdot \quad \cdots + m_{3k}I_k \\ &\cdots\cdots\cdots\cdots\cdots\cdots\cdots\cdots \\ 0 &= m_{k1}I_1 + \quad \cdot \quad \cdots + m_{kk}I_k \end{aligned} \tag{10.13}$$

where m_{ii} represents the sum of the impedances in the ith mesh and m_{ij} is the common impedance between mesh i and mesh j. We note here again that, for an RLC network, $m_{ij} = m_{ji}$ for all i and j. Thus reciprocity holds.

Solving the set of mesh equations for I_1 and I_2, we obtain the following equations:

$$\begin{aligned} I_1 &= \frac{\Delta_{11}}{\Delta} V_1 + \frac{\Delta_{21}}{\Delta} V_2 \\ I_2 &= \frac{\Delta_{12}}{\Delta} V_1 + \frac{\Delta_{22}}{\Delta} V_2 \end{aligned} \tag{10.14}$$

The above equations define the *short-circuit admittance parameters* as

$$I_1 = y_{11}V_1 + y_{12}V_2$$
$$I_2 = y_{21}V_1 + y_{22}V_2 \tag{10.15}$$

where $y_{ij} = \Delta_{ji}/\Delta$ for all i and j.

Let us find the y parameters for the bridged-T circuit given in Fig. 10.6. The mesh equations for the circuit are

$$V_1 = \left(\frac{1}{s}+1\right)I_1 + I_2 - \frac{1}{s}I_3$$
$$V_2 = I_1 + \left(\frac{1}{s}+1\right)I_2 + \frac{1}{s}I_3 \tag{10.16}$$
$$0 = -\frac{1}{s}I_1 + \frac{1}{s}I_2 + 2\left(\frac{1}{s}+1\right)I_3$$

In straightforward fashion we obtain

$$\Delta = \frac{2(2s+1)}{s^2}$$
$$\Delta_{11} = \Delta_{22} = \frac{2s^2+4s+1}{s^2} \tag{10.17}$$
$$\Delta_{12} = \Delta_{21} = \frac{2s^2+2s+1}{s^2}$$

The short-circuit parameters are then

$$y_{11} = y_{22} = \frac{2s^2+4s+1}{2(2s+1)}$$
$$y_{21} = y_{12} = -\frac{2s^2+2s+1}{2(2s+1)}$$

When $y_{11} = y_{22}$ or $z_{11} = z_{22}$, the network is symmetrical.

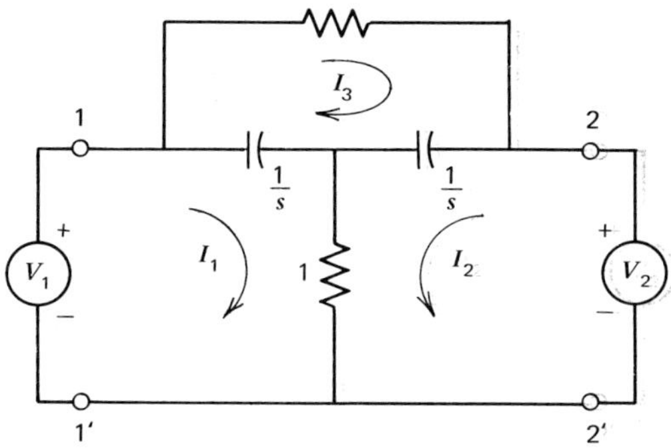

Fig. 10.6 A bridged-T network.

Returning to Eq. 10.15 that defines the y parameters, we see that the y parameters are expressed explicitly as

$$
\begin{aligned}
y_{11} &= \left.\frac{I_1}{V_1}\right|_{V_2=0} \\
y_{12} &= \left.\frac{I_1}{V_2}\right|_{V_1=0} \\
y_{21} &= \left.\frac{I_2}{V_1}\right|_{V_2=0} \\
y_{22} &= \left.\frac{I_2}{V_2}\right|_{V_1=0}
\end{aligned}
\tag{10.19}
$$

The reason why the y parameters are also called *short-circuit admittance* parameters is now apparent. In obtaining y_{11} and y_{21}, the 2-2′ port must be short-circuited, and when we find y_{22} and y_{12}, the 1-1′ port must be short-circuited, as shown in Figs. 10.7*a* and 10.7*b*.

As a second example, let us obtain the y parameters of the *Pi* circuit in Fig. 10.5. To obtain y_{11} and y_{21}, we short-circuit terminals 2-2′. We then have

$$
\begin{aligned}
y_{11} &= Y_A + Y_C \\
y_{21} &= -Y_C
\end{aligned}
\tag{10.20}
$$

We next short-circuit terminals 1-1′ to obtain

$$
\begin{aligned}
y_{22} &= Y_B + Y_C \\
y_{21} &= -Y_C
\end{aligned}
\tag{10.21}
$$

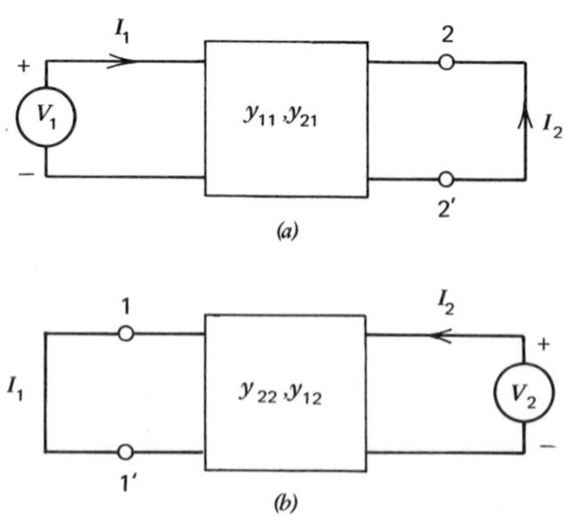

Fig. 10.7 Circuits for Y-parameter determination.

The h Parameters

A set of parameters that are extremely useful in describing transistor circuits are the h parameters given by the equations

$$\begin{aligned} V_1 &= h_{11}I_1 + h_{12}V_2 \\ I_2 &= h_{21}I_1 + h_{22}V_2 \end{aligned} \tag{10.22}$$

The individual parameters are defined by the relationships

$$\begin{aligned} h_{11} &= \left.\frac{V_1}{I_1}\right|_{V_2=0} \qquad & h_{12} &= \left.\frac{V_1}{V_2}\right|_{I_1=0} \\ h_{21} &= \left.\frac{I_2}{I_1}\right|_{V_2=0} \qquad & h_{22} &= \left.\frac{I_2}{V_2}\right|_{I_1=0} \end{aligned} \tag{10.23}$$

We see that h_{11} and h_{21} are short-circuit type parameters and h_{12} and h_{22} are open-circuit type parameters. The parameter h_{11} can be interpreted as the input impedance at port 1 with port 2 short-circuited. It is easily seen that h_{11} is merely the reciprocal of y_{11},

$$h_{11} = \frac{1}{y_{11}} \tag{10.24}$$

The parameter h_{22} is an open-circuit admittance parameter and is related to z_{22} by

$$h_{22} = \frac{1}{z_{22}} \tag{10.25}$$

Both the remaining h parameters are transfer functions—h_{21} is a short-circuit current ratio and h_{12} is an open-circuit voltage ratio. Their relationships to the z and y parameters will be discussed later.

For the *Pi* circuit in Fig. 10.5 the h parameters are:

$$\begin{aligned} h_{11} &= \frac{1}{Y_A + Y_C} \\ h_{12} &= \frac{Y_C}{Y_A + Y_C} \\ h_{21} &= \frac{-Y_C}{Y_A + Y_C} \\ h_{22} &= Y_B + \frac{Y_A Y_C}{Y_A + Y_C} \end{aligned} \tag{10.26}$$

Observe that for the *Pi* circuit, $h_{21} = -h_{12}$. This is the reciprocity condition for the h parameters and can be derived from their relationships to either the z or y parameters.

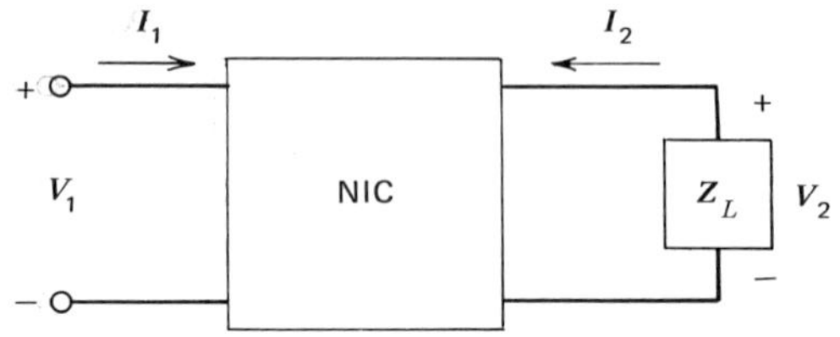

Fig. 10.8 NIC with load impedance.

Next let us consider the h parameters of an ideal device called the *negative impedance converter* (NIC) which converts a positive load impedance into a negative impedance at its input port.* Consider the NIC with a load impedance Z_L shown in Fig. 10.8. Its input impedance is

$$Z_{1N} = -Z_L \tag{10.27}$$

which can be rewritten as

$$\frac{V_1}{I_1} = \frac{V_2}{I_2} \tag{10.28}$$

The following voltage-current relationships hold for the NIC,

$$\begin{aligned} V_1 &= kV_2 \\ I_1 &= kI_2 \end{aligned} \tag{10.29}$$

If we interpret Eq. 9.154 using h parameters, we arrive at the following conditions,

$$\begin{aligned} h_{11} &= h_{22} = 0 \\ h_{12} &= \frac{1}{h_{21}} = k \end{aligned} \tag{10.30}$$

We see that since $h_{12} \neq -h_{21}$, the NIC is nonreciprocal.

In matrix notation, the h matrix of the NIC is

$$\begin{bmatrix} h_{11} & h_{12} \\ h_{21} & h_{22} \end{bmatrix} = \begin{bmatrix} 0 & k \\ \dfrac{1}{k} & 0 \end{bmatrix} \tag{10.31}$$

The NIC is a convenient device in the modeling of active circuits. It is not however, a device that exists only in our imagination. Practica

*For a lucid discussion of the properties of the NIC, see L. P. Huelsman, *Circuits, Matrices and Linear Vector Spaces,* McGraw-Hill, 1963, Chapter 4.

realizations of NIC's have been achieved using transistors. Some of these are described in an article by Larky.*

The ABCD Parameters

Let us take as the dependent variables, the voltage and current at the port 1, and define the following equation:

$$\begin{bmatrix} V_1 \\ I_1 \end{bmatrix} = \begin{bmatrix} A & B \\ C & D \end{bmatrix} \begin{bmatrix} V_2 \\ -I_2 \end{bmatrix} \tag{10.32}$$

This matrix equation defines the A, B, C, D parameters, whose matrix is known as the *transmission* matrix because it relates the voltage and current at the input port to their corresponding quantities at the output. The reason the current I_2 carries a negative sign is that most transmission engineers like to regard their output current as coming *out* of the output port instead of going into the port as per standard usage.

In explicit form the $ABCD$ parameters can be expressed as

$$\begin{aligned} A &= \left.\frac{V_1}{V_2}\right|_{I_2=0} & B &= -\left.\frac{V_1}{I_2}\right|_{V_2=0} \\ C &= \left.\frac{I_1}{V_2}\right|_{I_2=0} & D &= -\left.\frac{I_1}{I_2}\right|_{V_2=0} \end{aligned} \tag{10.33}$$

From these relations, we see that A represents an open-circuit voltage transfer function; B is a short-circuit transfer impedance; C is an open-circuit transfer admittance; and D is a short-circuit current ratio. Note that all four parameters are transfer functions so that the term *transmission matrix* is a very appropriate one. Let us describe the short-circuit transfer functions B and D in terms of y parameters, and the open-circuit transfer functions A and C in terms of z parameters. Using straightforward algebraic operations, we obtain

$$\begin{aligned} A &= \frac{z_{11}}{z_{21}} & B &= -\frac{1}{y_{21}} \\ C &= \frac{1}{z_{21}} & D &= -\frac{y_{11}}{y_{21}} \end{aligned} \tag{10.34}$$

For the $ABCD$ parameters, the reciprocity condition is expressed by the equation,

$$\det \begin{bmatrix} A & B \\ C & D \end{bmatrix} = AD - BC = 1 \tag{10.35}$$

*A. I. Larky, "Negative-Impedance Converters," Trans., *IRE on Circuit Theory*, Vol. CT-4, No. 3, September 1957, pp. 124–131.

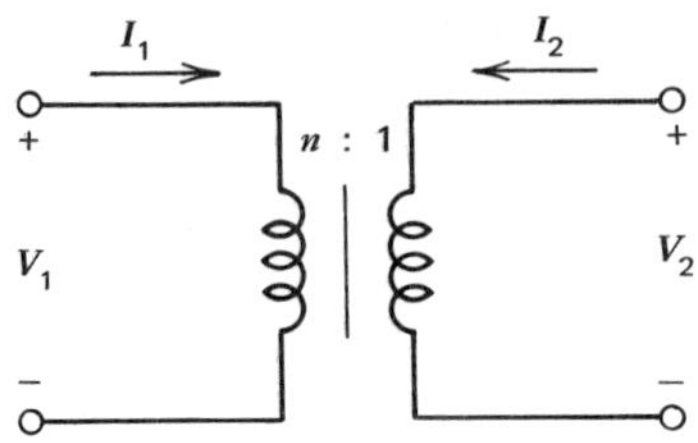

Fig. 10.9 An ideal transformer.

Let us find, as an example, the *ABCD* parameter for the *ideal transformer* in Fig. 10.9, whose defining equations are:

$$\begin{aligned} V_1 &= nV_2 \\ I_1 &= \frac{1}{n}(-I_2) \end{aligned} \tag{10.36}$$

If we express Eq. 10.36 in matrix form, we have

$$\begin{bmatrix} V_1 \\ I_1 \end{bmatrix} = \begin{bmatrix} n & 0 \\ 0 & \frac{1}{n} \end{bmatrix} \begin{bmatrix} V_2 \\ -I_2 \end{bmatrix} \tag{10.37}$$

so that the transmission matrix of the ideal transformer is

$$\begin{bmatrix} A & B \\ C & D \end{bmatrix} = \begin{bmatrix} n & 0 \\ 0 & \frac{1}{n} \end{bmatrix} \tag{10.38}$$

Note, incidentally, that the ideal transformer does *not* possess an impedance or admittance matrix because the self and mutual inductances are infinite.*

For the ideal transformer terminated in a load impedance shown in Fig. 10.10(a), the following equations apply:

$$\begin{aligned} V_1 &= nV_2 \\ I_1 &= \frac{V_2}{nZ_L} \end{aligned} \tag{10.39}$$

Taking the ratio of V_1 to I_1 we find the input impedance at port 1 to be

$$Z_1 = \frac{V_1}{I_1} = n^2 Z_L \tag{10.40}$$

*For a detailed discussion concerning ideal transformers, see M. E. Van Valkenburg *Modern Network Synthesis*, John Wiley, New York, 1960.

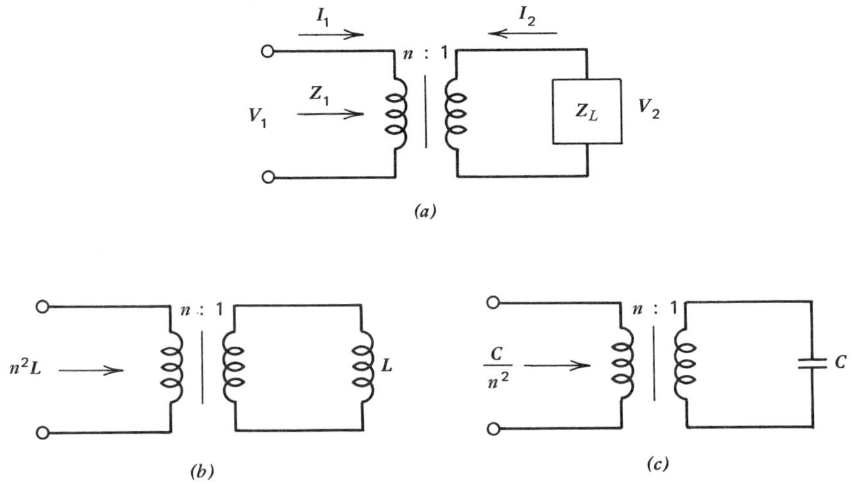

Fig. 10.10 Ideal transformer as an impedance transformer.

Thus we see that an ideal transformer is an *impedance transformer*. If the load element were an inductor L, (Fig. 10.10*b*) at port 1 we would see an equivalent inductor of value n^2L. Similarly a capacitor C at the load would appear as a capacitor value C/n^2 at port 1 (Fig. 10.10*c*).

As a second example concerning the use of the transmission matrix in network analysis, consider the *ABCD* parameters of the *Pi* circuit in Fig. 10.5

$$\begin{aligned} A &= \frac{Y_B + Y_C}{Y_C} \\ B &= \frac{1}{Y_C} \\ C &= \frac{Y_A Y_B + Y_B Y_C + Y_A Y_C}{Y_C} \\ D &= \frac{Y_A + Y_C}{Y_C} \end{aligned} \tag{10.41}$$

If we check for reciprocity from Eq. 10.41, we see that

$$\begin{aligned} AD - BC &= \frac{(Y_A + Y_C)(Y_B + Y_C) - (Y_A Y_B + Y_B Y_C + Y_A Y_C)}{Y_C{}^2} \\ &= \frac{Y_C{}^2}{Y_C{}^2} = 1 \end{aligned} \tag{10.42}$$

10.2 RELATIONSHIPS BETWEEN TWO-PORT PARAMETERS

The relationships between two-port parameters are quite simply obtained because of the simple algebraic nature of the two-port equations. For example, we have seen that $h_{11} = 1/y_{11}$ and $h_{22} = 1/z_{22}$. To derive h_{12} in terms of open-circuit parameters consider the z parameter equations when port 1 is open-circuited ($I_1 = 0$),

$$\begin{aligned} V_1 &= z_{12} I_2 \\ V_2 &= z_{22} I_2 \end{aligned} \tag{10.43}$$

Therefore we have

$$h_{12} = \left.\frac{V_1}{V_2}\right|_{I_1=0} = \frac{z_{12}}{z_{22}} \tag{10.44}$$

Similarly, since h_{21} is defined as a short-circuit type parameter, we can derive h_{21} in terms of y parameters as

$$h_{21} = \frac{y_{21}}{y_{11}} \tag{10.45}$$

We can express all the h parameters as functions of only the z parameters or y parameters alone. An easy way to accomplish this task is to find out the relationships between the z and y parameters themselves. Certainly, by their very nature, the z and y parameters are not simply reciprocals of each other, as the novice might guess, since one set of parameters is defined for open-circuit conditions, the other for short-circuit.

The z and y relationships can be obtained very easily by using matrix notation. We define the z matrix as

$$[Z] = \begin{bmatrix} z_{11} & z_{12} \\ z_{21} & z_{22} \end{bmatrix} \tag{10.46}$$

and the y matrix as

$$[Y] = \begin{bmatrix} y_{11} & y_{12} \\ y_{21} & y_{22} \end{bmatrix} \tag{10.47}$$

In simplified notation we can write the two sets of equations as

$$[V] = [Z][I] \tag{10.48}$$

and

$$[I] = [Y][V] \tag{10.49}$$

Replacing $[I]$ in Eq. 10.48 by $[Y][V]$, we obtain,

$$[V] = [Z][Y][V] \tag{10.50}$$

so that the product $[Z][Y]$ must yield the unit matrix $[U]$. The matrices $[Y]$ and $[Z]$ must therefore be inverses of each other, that is,

$$[Z]^{-1} = [Y] \text{ and } [Y]^{-1} = [Z] \tag{10.51}$$

From the relationship we can find the relations between the individual z and y parameters.

$$\begin{aligned} y_{11} &= \frac{z_{22}}{\Delta_z} \qquad & y_{22} &= \frac{z_{11}}{\Delta_z} \\ y_{12} &= -\frac{z_{12}}{\Delta_z} \qquad & y_{21} &= -\frac{z_{21}}{\Delta_z} \end{aligned} \tag{10.52}$$

Table 10.2.1 Matrix Conversion Table

$$\Delta_x = x_{11}x_{22} - x_{12}x_{21}$$

	$[z]$		$[y]$		$[h]$		$[T]$	
$[z]$	z_{11}	z_{12}	$\frac{y_{22}}{\Delta_y}$	$-\frac{y_{12}}{\Delta_y}$	$\frac{\Delta_h}{h_{22}}$	$\frac{h_{12}}{h_{22}}$	$\frac{A}{C}$	$\frac{\Delta_T}{C}$
	z_{21}	z_{22}	$-\frac{y_{21}}{\Delta_y}$	$\frac{y_{11}}{\Delta_y}$	$-\frac{h_{21}}{h_{22}}$	$\frac{1}{h_{22}}$	$\frac{1}{C}$	$\frac{D}{C}$
$[y]$	$\frac{z_{22}}{\Delta_z}$	$-\frac{z_{12}}{\Delta_z}$	y_{11}	y_{12}	$\frac{1}{h_{11}}$	$-\frac{h_{12}}{h_{11}}$	$\frac{D}{B}$	$-\frac{\Delta_T}{B}$
	$-\frac{z_{21}}{\Delta_z}$	$\frac{z_{11}}{\Delta_z}$	y_{21}	y_{22}	$\frac{h_{21}}{h_{11}}$	$\frac{\Delta_h}{h_{11}}$	$-\frac{1}{B}$	$\frac{A}{B}$
$[h]$	$\frac{\Delta_z}{z_{22}}$	$\frac{z_{12}}{z_{22}}$	$\frac{1}{y_{11}}$	$-\frac{y_{12}}{y_{11}}$	h_{11}	h_{12}	$\frac{B}{D}$	$\frac{\Delta_T}{D}$
	$-\frac{z_{21}}{z_{22}}$	$\frac{1}{z_{22}}$	$\frac{y_{21}}{y_{11}}$	$\frac{\Delta_y}{y_{11}}$	h_{21}	h_{22}	$-\frac{1}{D}$	$\frac{C}{D}$
$[T]$	$\frac{z_{11}}{z_{21}}$	$\frac{\Delta_z}{z_{21}}$	$-\frac{y_{22}}{y_{21}}$	$-\frac{1}{y_{21}}$	$-\frac{\Delta_h}{h_{21}}$	$-\frac{h_{11}}{h_{21}}$	A	B
	$\frac{1}{z_{21}}$	$\frac{z_{22}}{z_{21}}$	$-\frac{\Delta_y}{y_{21}}$	$-\frac{y_{11}}{y_{21}}$	$-\frac{h_{22}}{h_{21}}$	$-\frac{1}{h_{21}}$	C	D

where $\Delta_z = z_{11}z_{22} - z_{12}z_{21}$; and

$$z_{11} = \frac{y_{22}}{\Delta_y} \qquad y_{22} = \frac{y_{11}}{\Delta_y}$$
$$z_{12} = -\frac{y_{12}}{\Delta_y} \qquad y_{21} = -\frac{y_{21}}{\Delta_y} \tag{10.53}$$

where $\Delta_y = y_{11}y_{22} - y_{12}y_{21}$. Using these identities we can derive the h or $ABCD$ parameters in terms of either the z or y parameters. A conversion table is given in Table 10.2.1 to facilitate the process. Note that in the table,

$$\Delta_T = AD - BC$$

10.3 SYSTEM FUNCTIONS USING TWO-PORT PARAMETERS

In this section we will examine how we determine driving-point and transfer functions of a two-port using two-port parameters. These functions fall into two broad categories. The first applies to two-ports without load and source impedances. These transfer functions can be described by means of z or y parameters alone. For example, let us derive the expressions for the *open-circuit voltage ratio* V_2/V_1, using z parameters first and y parameters next. Consider the z parameter equations for the two-port when port 2 is open-circuited.

$$V_2 = z_{21}I_1$$
$$V_1 = z_{11}I_1 \tag{10.54}$$

If we take the ratio of V_2 to V_1, we obtain

$$\frac{V_2}{V_1} = \frac{z_{21}}{z_{11}} \tag{10.55}$$

By letting I_2 of the second y parameter equation go to zero we derive the open-circuit voltage ratio to be:

$$\frac{V_2}{V_1} = -\frac{y_{21}}{y_{22}} \tag{10.56}$$

In similar manner we can derive the *short-circuit current ratio* of a two-port as

$$\frac{I_2}{I_1} = \frac{y_{21}}{y_{11}} \tag{10.57}$$

and

$$\frac{I_2}{I_1} = -\frac{z_{21}}{z_{22}} \tag{10.58}$$

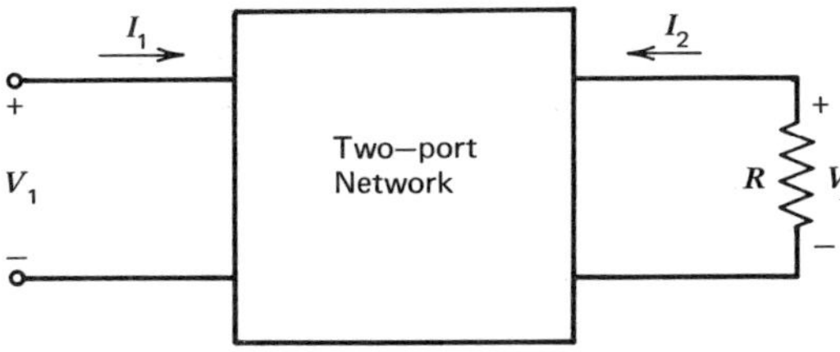

Fig. 10.11 A two-port network.

The open and short-circuit transfer functions are not ones we usually deal with in practice since there are usually source and load impedances to account for. The second category of two-port transfer functions are those including source or load impedances. These transfer functions are functions of the two-port parameters z, h, or y *and* the source and/or load impedance. For example, let us derive the transfer admittance I_2/V_1 of a two-port network that is terminated in a resistor of R ohms, as given in Fig. 10.11. For this two-port network, the following equations apply:

$$\begin{aligned} I_2 &= y_{21}V_1 + y_{22}V_2 \\ V_2 &= -I_2R \end{aligned} \tag{10.59}$$

By eliminating the variable V_2, we obtain

$$Y_{21} = \frac{I_2}{V_1} = \frac{y_{21}/R}{y_{22} + 1/R} \tag{10.60}$$

Note that Y_{21} and y_{21} are not the same. Y_{21} is the transfer admittance of the two-port network terminated in a resistor R, and y_{21} is the transfer admittance when port 2 is short-circuited. We must be careful to make this distinction in other cases of a similar nature.

In order to solve for transfer functions of two-ports terminated at either port by an impedance Z_L, it is convenient to use the equivalent circuit of the two-port network given in terms of its z parameters (Fig. 10.12) or y parameters (Fig. 10.13). The equivalent voltage sources $z_{12}I_2$ and $z_{21}I_1$ in Fig. 10.12 are called *controlled sources* because they depend upon a current somewhere in the network.* Similarly, the current sources $y_{12}V_2$ and $y_{21}V_1$ are also controlled sources. For the circuit in Fig. 10.12, let us find the transfer impedance

$$z_{21} = \frac{V_2}{I_1} \tag{10.61}$$

*For a lucid treatment of controlled sources, see E. J. Angelo, *Electronic Circuits*, McGraw-Hill, New York, 1958.

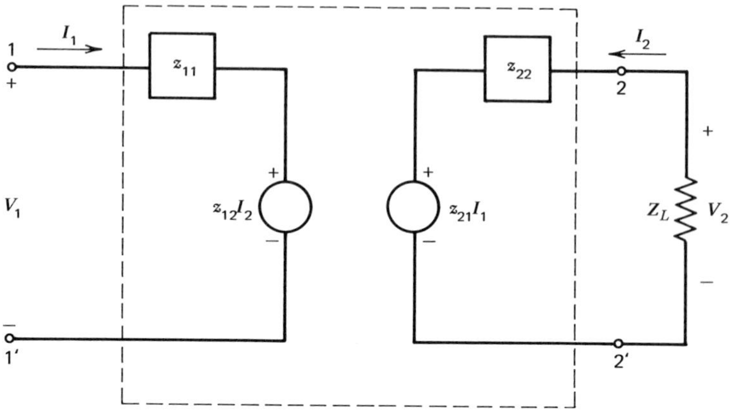

Fig. 10.12 Two-port equivalent.

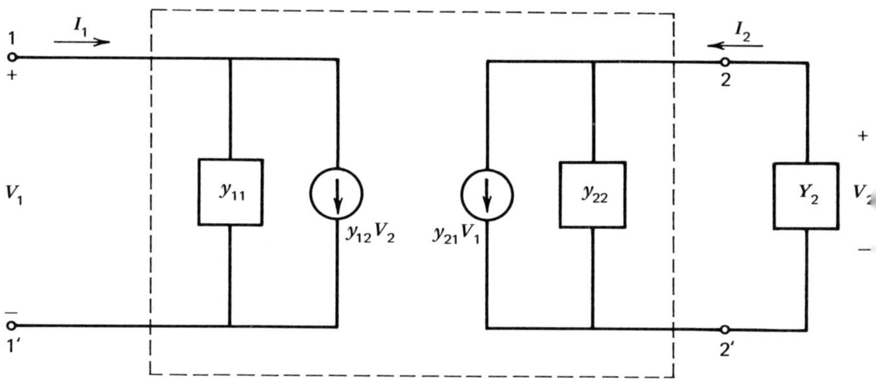

Fig. 10.13 Two-port equivalent.

with port 2 terminated in a load impedance Z_L. If we write the mesh equation for the I_2 mesh, we have

$$-z_{21}I_1 = (z_{22} + Z_L)I_2 \tag{10.62}$$

Since $V_2 = -I_2Z_L$, we readily obtain

$$z_{21} = \frac{V_2}{I_1} = \frac{z_{21}Z_L}{z_{22} + Z_L} \tag{10.63}$$

Also, it is clear that the current-ratio transfer function for the terminated two-port network is

$$\frac{I_2}{I_1} = \frac{-z_{21}}{z_{22} + Z_L} \tag{10.64}$$

In similar fashion, we obtain the voltage-ratio transfer function for the circuit represented in Fig. 10.13 as

$$\frac{V_2}{V_1} = -\frac{y_{21}}{Y_2 + y_{22}} \tag{10.65}$$

Next, suppose we are required to find the transfer function V_2/V_g for the two-port network terminated at both ends, as shown in Fig. 10.14. We first write the two mesh equations,

$$\begin{aligned} V_g &= (R_1 + z_{11})I_1 + z_{12}I_2 \\ 0 &= z_{21}I_1 + (z_{22} + R_2)I_2 \end{aligned} \tag{10.66}$$

Next, we solve for I_2 to give

$$I_2 = \frac{V_g z_{21}}{(R_1 + z_{11})(R_2 + z_{22}) - z_{12}z_{21}} \tag{10.67}$$

From the equation $V_2 = -R_2 I_2$, we may now arrive at the following solution,

$$\frac{V_2}{V_g} = -\frac{R_2 I_2}{V_g} = \frac{z_{21}R_2}{(R_1 + z_{11})(R_2 + z_{22}) - z_{21}z_{12}} \tag{10.68}$$

Note that the equivalent circuits of the two-ports in Figs. 10.12 and 10.13 are not unique. Two other examples are given in Figs. 10.15 and 10.16. Observe that the controlled sources are nonzero in these equivalent circuits only if the circuit is nonreciprocal.

Finally, let us consider the hybrid equivalent circuit shown in Fig. 10.17. Observe the voltage controlled source, $h_{12}V_2$ at port 1 and the current controlled source $h_{21}I_1$ at port 2. Let us find the input impedance Z_{in}. The pertinent equations are:

$$V_1 = h_{11}I_1 + h_{12}V_2 \tag{10.69}$$

and

$$V_2 = -Z_L I_2 = -(h_{21}I_1 + h_{22}V_2)Z_L \tag{10.70}$$

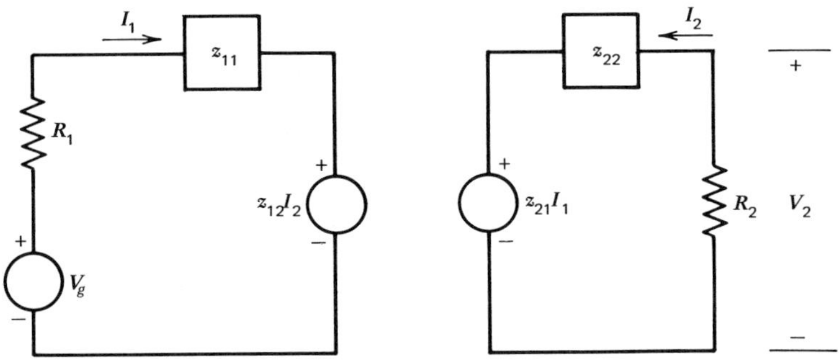

Fig. 10.14 Network with source impedance.

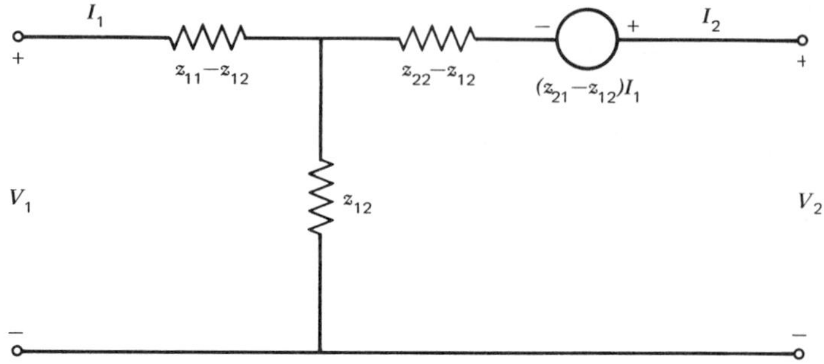

Fig. 10.15 Two-port equivalent circuit with one controlled voltage source.

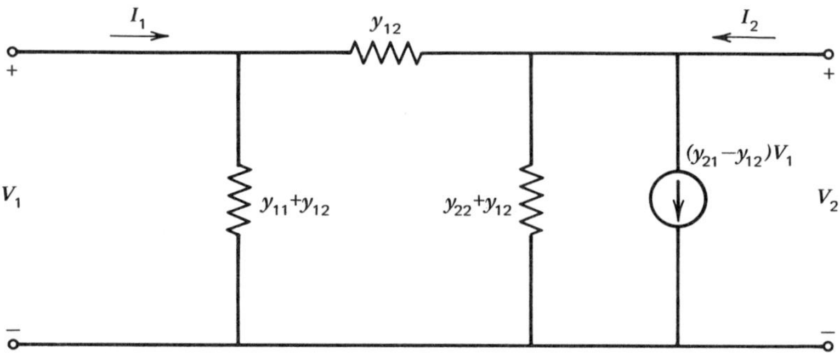

Fig. 10.16 Two-port equivalent circuit with one controlled current source.

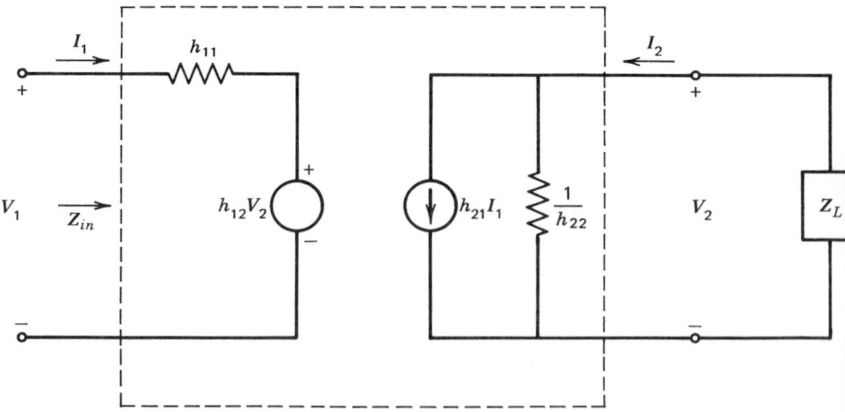

Fig. 10.17 Hybrid equivalent circuit.

Solving Eq. 10.70 for V_2 we find

$$V_2 = -\frac{h_{21}Z_L I_1}{1+h_{22}Z_L} \tag{10.71}$$

Substituting V_2 in Eq. 10.71 into Eq. 10.69, we get,

$$V_1 = \left(h_{11} - \frac{h_{12}h_{21}Z_L}{1+h_{22}Z_L}\right)I_1 \tag{10.72}$$

so that

$$Z_{in} = \frac{V_1}{I_1} = h_{11} - \frac{h_{12}h_{21}Z_L}{1+h_{22}Z_L} \tag{10.73}$$

We can easily check to see that Eq. 10.73 is dimensionally correct, since h_{11} has the dimensions of impedance, h_{22} is an admittance and h_{12}, h_{21} are dimensionless since they represent voltage and current ratios respectively.

10.4 INTERCONNECTION OF TWO-PORTS

In this section we will consider various interconnections of two-ports. We will see that when a pair of two-ports are cascaded, the overall transmission matrix is equal to the product of the individual transmission matrices of the two-ports. When two two-ports are connected in series, their z matrices add; when they are connected in parallel, their y matrices add. First let us consider the case when we connect a pair of two-ports, N_a and N_b in cascade or in tandem, as shown in Fig. 10.18. We see that

$$\begin{bmatrix} V_{2a} \\ -I_{2a} \end{bmatrix} = \begin{bmatrix} V_{1b} \\ I_{1b} \end{bmatrix} \tag{10.74}$$

The transmission matrix equation for N_a is:

$$\begin{bmatrix} V_1 \\ I_1 \end{bmatrix} = \begin{bmatrix} A_a & B_a \\ C_a & D_a \end{bmatrix} \begin{bmatrix} V_{2a} \\ -I_{2a} \end{bmatrix} \tag{10.75}$$

Correspondingly for N_b we have

$$\begin{bmatrix} V_{2a} \\ -I_{2a} \end{bmatrix} = \begin{bmatrix} V_{1b} \\ I_{1b} \end{bmatrix} = \begin{bmatrix} A_b & B_b \\ C_b & D_b \end{bmatrix} \begin{bmatrix} V_2 \\ -I_2 \end{bmatrix} \tag{10.76}$$

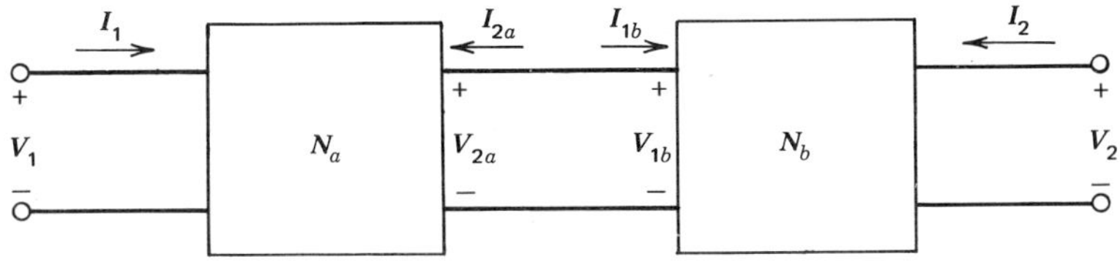

Fig. 10.18 Cascade connection of two ports.

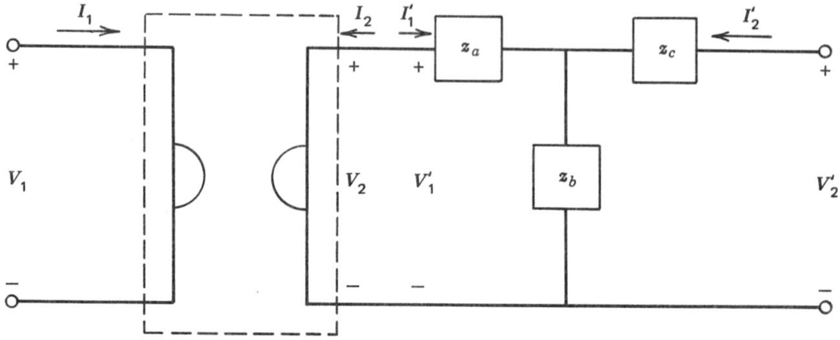

Fig. 10.19 Gyrator in tandem with T circuit.

Substituting the second matrix into the first we obtain

$$\begin{bmatrix} V_1 \\ I_1 \end{bmatrix} = \begin{bmatrix} A_a & B_a \\ C_a & D_a \end{bmatrix} \cdot \begin{bmatrix} A_b & B_b \\ C_b & D_b \end{bmatrix} \begin{bmatrix} V_2 \\ -I_2 \end{bmatrix}$$

We see that the transmission matrix of the overall two-port network is simply the product of the transmission matrices of the individual two-ports.

As an example, let us calculate the overall transmission matrix of a *gyrator** in tandem with a T network shown in Fig. 10.19. The ideal gyrator is an impedance inversion device whose input impedance Z_{in} is related to its load impedance Z_L by

$$\begin{aligned} Z_{in} &= a^2 Y_L \\ &= \frac{a^2}{Z_L} \end{aligned} \tag{10.77}$$

The constant a in Eq. 10.77 is defined as the *gyration resistance*. If we regard the gyrator as a two-port, its defining equations are:

$$\begin{aligned} V_1 &= a(-I_2) \\ I_1 &= \frac{1}{a} V_2 \end{aligned} \tag{10.78}$$

so that the transmission matrix of the gyrator is

$$\begin{bmatrix} 0 & a \\ \frac{1}{a} & 0 \end{bmatrix}$$

*B. D. H. Tellegen, "The Gyrator, a New Electric Network Element," *Phillips Research Repts.*, Vol. 3, April 1948, pp. 81–101, also see the book by L. P. Huelsman, *op. cit*, pp. 140–148.

We see that for the gyrator

$$AD - BC = -1 \tag{10.79}$$

Therefore the gyrator is a nonreciprocal device, although it is passive.*

The overall transmission matrix of the configuration in Fig. 10.19 is obtained by the product of the individual transmission matrices,

$$\begin{bmatrix} A & B \\ C & D \end{bmatrix} = \begin{bmatrix} 0 & a \\ \dfrac{1}{a} & 0 \end{bmatrix} \begin{bmatrix} \dfrac{z_a + z_b}{z_b} & \dfrac{z_a z_b + z_b z_c + z_a z_c}{z_b} \\ \dfrac{1}{z_b} & \dfrac{z_b + z_c}{z_b} \end{bmatrix}$$

$$= \begin{bmatrix} \dfrac{a}{z_b} & \dfrac{a(z_b + z_c)}{z_b} \\ \dfrac{z_a + z_b}{a z_b} & \dfrac{z_a z_b + z_b z_c + z_a z_c}{a z_b} \end{bmatrix} \tag{10.80}$$

If we check the configuration in Fig. 10.19 for reciprocity, we see that for the transmission matrix in Eq. 10.80

$$AD - BC = -1 \tag{10.81}$$

We thus see that any reciprocal network connected in tandem with a gyrator yields a configuration that is nonreciprocal.

Next consider the situation where a pair of two-ports N_a and N_b are connected in parallel, as shown in Fig. 10.20. Let us find the y parameters

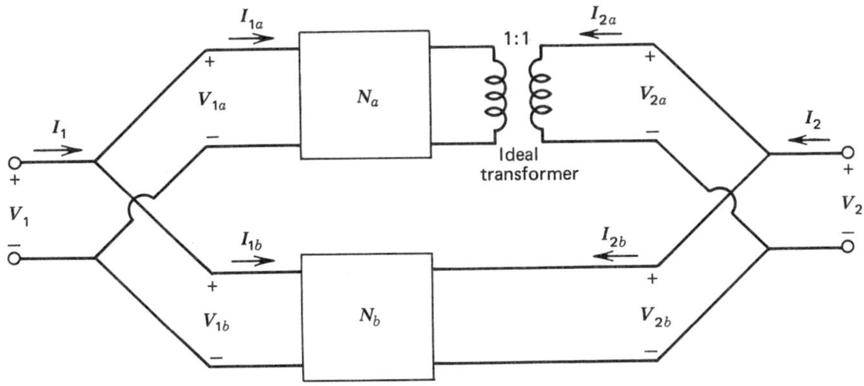

Fig. 10.20 Parallel connection of two ports.

*Most gyrators are microwave devices that depend upon the *Hall-effect* in ferrites.

for the overall two-port network. The matrix equations for the individual two-ports are:

$$\begin{bmatrix} I_{1a} \\ I_{2a} \end{bmatrix} = \begin{bmatrix} y_{11a} & y_{12a} \\ y_{21a} & y_{22a} \end{bmatrix} \begin{bmatrix} V_{1a} \\ V_{2a} \end{bmatrix} \tag{10.82}$$

and

$$\begin{bmatrix} I_{1b} \\ I_{2b} \end{bmatrix} = \begin{bmatrix} y_{11b} & y_{12b} \\ y_{21b} & y_{22b} \end{bmatrix} \begin{bmatrix} V_{1b} \\ V_{2b} \end{bmatrix} \tag{10.83}$$

From Fig. 10.20, we see that the following equations must hold:

$$\begin{aligned} V_1 &= V_{1a} = V_{1b} \qquad & V_2 &= V_{2a} = V_{2b} \\ I_1 &= I_{1a} + I_{1b} \qquad & I_2 &= I_{2a} + I_{2b} \end{aligned} \tag{10.84}$$

In connecting two-ports in series or in parallel, we must be careful lest the individual character of a two-port network is altered when connected in series or parallel with another two-port. For example, when we connect the two-ports in Fig. 10.21 in parallel, the impedances Z_7 and Z_8 will be short-circuited. Therefore, to ensure that a two-port network does not alter in any extent the character of the other, ideal transformers are used to provide the necessary isolation. In matrix notation, the sum of the individual $[I_i]$ matrices of the two-ports in

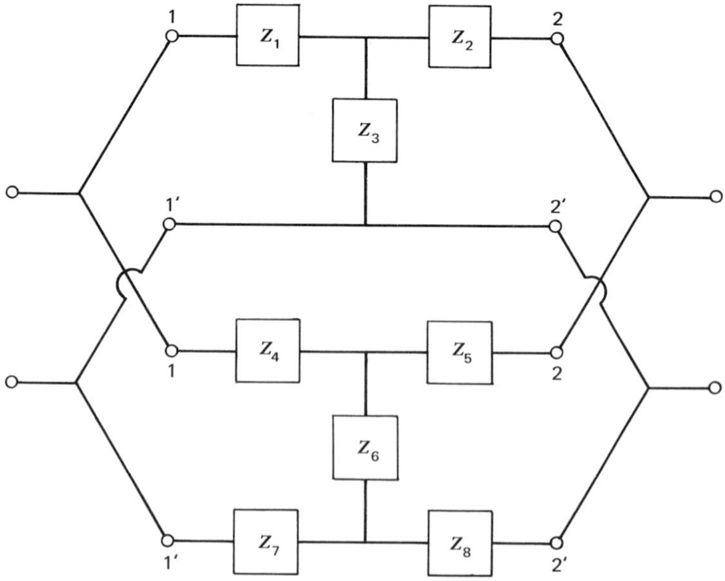

Fig. 10.21 A two-port network.

parallel must equal the $[I]$ matrix of the overall two-port network. Thus

$$\begin{bmatrix} I_1 \\ I_2 \end{bmatrix} = \begin{bmatrix} I_{1a} \\ I_{2a} \end{bmatrix} + \begin{bmatrix} I_{1b} \\ I_{2b} \end{bmatrix}$$

$$= \left[\begin{bmatrix} y_{11a} & y_{12a} \\ y_{21a} & y_{22a} \end{bmatrix} + \begin{bmatrix} y_{11b} & y_{12b} \\ y_{21b} & y_{22b} \end{bmatrix} \right] \begin{bmatrix} V_1 \\ V_2 \end{bmatrix} \quad (10.85)$$

so that the y parameters of the overall two-port network can be expressed in terms of the y parameters of the individual two-ports as

$$\begin{bmatrix} y_{11} & y_{12} \\ y_{21} & y_{22} \end{bmatrix} = \begin{bmatrix} y_{11a} + y_{11b} & y_{12a} + y_{12b} \\ y_{21a} + y_{21b} & y_{22a} + y_{22b} \end{bmatrix} \quad (10.86)$$

If we connect two-ports in series, as shown in Fig. 10.22, we can express the z parameters of the overall two-port network in terms of the z parameters of the individual two-ports as

$$\begin{bmatrix} z_{11} & z_{12} \\ z_{21} & z_{22} \end{bmatrix} = \begin{bmatrix} z_{11a} + z_{11b} & z_{12a} + z_{12b} \\ z_{21a} + z_{21b} & z_{22a} + z_{22b} \end{bmatrix} \quad (10.87)$$

In summary, it can be said that

1. When two-ports are connected in parallel, find the y parameters first and, from the y parameters, derive the other two-port parameters.
2. When two-ports are connected in series, it is usually easiest to find the z parameters.

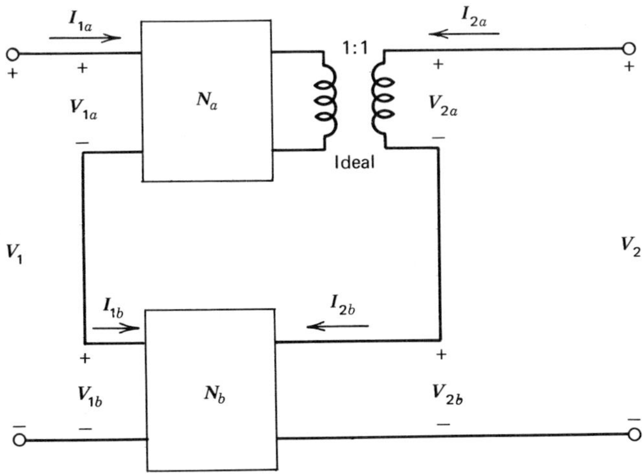

Fig. 10.22 Series connection of two ports.

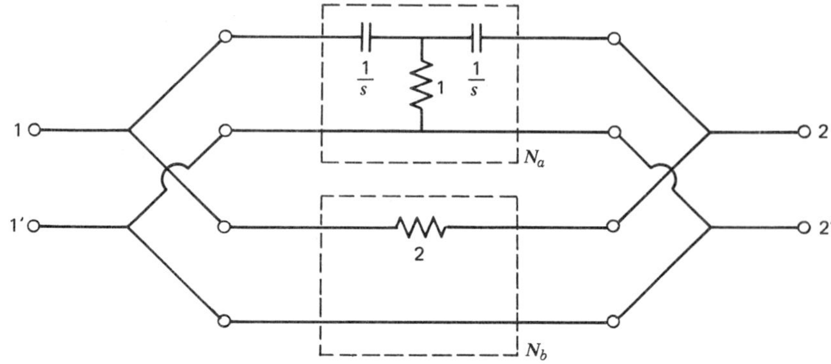

Fig. 10.23 A parallel T network.

3. When two-ports are connected in tandem, the transmission matrix is generally easier to obtain.

As a final example, let us find the y parameters of the bridged-T circuit in Fig. 10.6. We see that the bridged-T circuit would be decomposed into a parallel connection of two-ports, as shown in Fig. 10.23. Our task is to first find the y parameters of the two-ports N_a and N_b. The y parameters of N_b are obtained by inspection to be

$$y_{12b} = y_{21b} = -\tfrac{1}{2} \qquad y_{11b} = y_{22b} = \tfrac{1}{2}$$

N_a is a T circuit so that the z parameters can be obtained by inspection. These are

$$\begin{aligned} z_{11a} &= z_{22a} = \frac{s+1}{s} \\ z_{12a} &= z_{21a} = 1 \end{aligned} \tag{10.88}$$

We then find the y parameters from the equations,

$$\begin{aligned} y_{11a} &= y_{22a} = \frac{z_{22a}}{\Delta z_a} = \frac{s(s+1)}{2s+1} \\ y_{12a} &= y_{21a} = \frac{z_{12a}}{\Delta z_a} = -\frac{s^2}{2s+1} \end{aligned} \tag{10.89}$$

Since both N_a and N_b are symmetrical two ports, we know that $y_{11} = y_{22}$ for the overall bridged-T circuit. The y parameters for the bridged-T circuit are now obtained as

$$\begin{aligned} y_{11} &= y_{11a} + y_{11b} \\ &= \frac{s(s+1)}{2s+1} + \frac{1}{2} = \frac{2s^2+4s+1}{2(2s+1)} \\ y_{12} &= y_{12a} + y_{12b} \\ &= -\frac{s^2}{2s+1} - \frac{1}{2} = -\frac{2s^2+2s+1}{2(2s+1)} \end{aligned} \tag{10.90}$$

10.5 ANALYSIS OF LADDER NETWORKS

In this section, we will consider a simple method to obtain the network functions of a ladder network in a single operation.* This method depends only upon relationships that exist between the branch currents and node voltages of the ladder. Consider the network shown in Fig. 10.24 where all the series branches are given as impedances and all the parallel branches are given as admittances. If the v denote node voltages and i denote branch currents, then the following relationships apply.

$$\begin{aligned} v_{j+1} &= i_{j+2}Z_{j+2} + v_{j+3} \\ i_j &= v_{j+1}Y_{j+1} + i_{j+2} \end{aligned} \tag{10.91}$$

These equations form the basis of the method we will discuss here.

To illustrate this method, consider the network in Fig. 10.25 for which

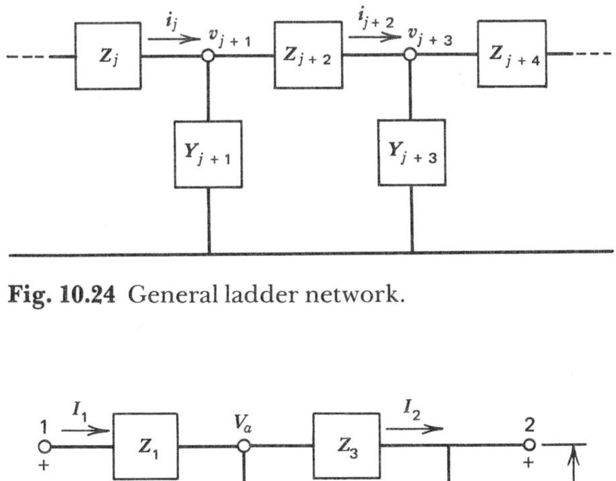

Fig. 10.24 General ladder network.

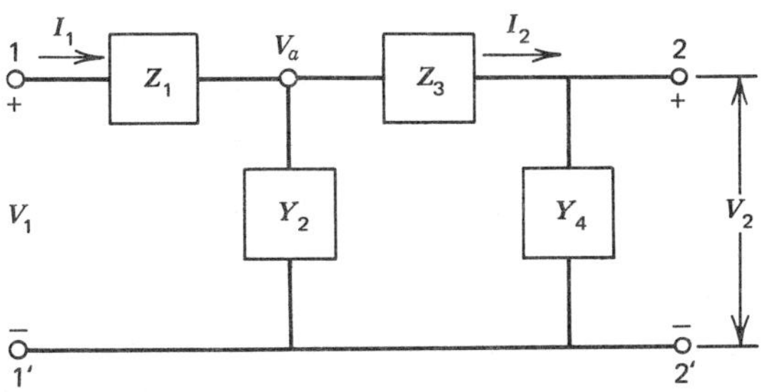

Fig. 10.25 A simple ladder network.

*F. F. Kuo and G. H. Leichner, "An Iterative Method for Determining Ladder Network Functions," *Proc. IRE*, 47, No. 10 (Oct. 1959), 1782-1783.

the following node voltage and branch current relationships apply.

$$
\begin{aligned}
V_2(s) &= V_2(s) \\
I_2(s) &= Y_4(s)V_2(s) \\
V_a(s) &= I_2(s)Z_3(s) + V_2(s) = [1 + Z_3(s)Y_4(s)]V_2(s) \\
I_1(s) &= Y_2(s)V_a(s) + I_2(s) \\
&= \{Y_2(s)[1 + Z_3(s)Y_4(s)] + Y_4(s)\}V_2(s) \\
V_1(s) &= I_1(s)Z_1(s) + V_a(s) \\
&= \{Z_1[Y_2(1 + Z_3Y_4) + Y_4] + (1 + Z_3Y_4)\}V_2(s)
\end{aligned}
\tag{10.92}
$$

We see upon examining the set of equations given above that each equation depends upon the two previous equations only. The first equation, $V_2 = V_2$ is, of course, unnecessary. But, as we shall see later, it is helpful as a starting point. In writing these equations, we begin at the 2-2′ port of the ladder and work towards the 1-1′ port. Each succeeding equation takes into account one new immittance. We see, further, that with the exception of the first two equations, each subsequent equation is obtained by multiplying the equation just preceding it by the immittance next down the line, and then adding to this product the equation twice preceding it. For example, we see that $I_2(s)$ is obtained by multiplying the preceding equation $V_2(s)$ by the admittance $Y_4(s)$. The next immittance is $Z_3(s)$. We obtain $V_a(s)$ by multiplying the previous equation $I_2(s)$ by $Z_3(s)$ to obtain I_2Z_3; then we add to this product the equation twice preceding it, $V_2(s)$, to obtain $V_a = I_2Z_3 + V_2$. The process is then easily *mechanized* according to the following algorithm: (1) alternate writing node voltage and branch current equations; (2) the next equation is obtained by multiplying the present equation by the next immittance (as we work from one port to the other), and adding to this product the results of the previous equation.

Using this set of equations, we obtain the input impedance $z_{11}(s)$ by dividing the equation for $V_1(s)$ by the equation for $I_1(s)$. We obtain the voltage-ratio transfer function $V_2(s)/V_1(s)$ by dividing the first equation $V_2(s)$ by the last equation $V_1(s)$. We obtain other network functions such as transfer immittances and current ratios in similar manner. Note that every equation contains $V_2(s)$ as a factor. In taking ratios of these equations, the $V_2(s)$ term is canceled. Therefore, our analysis can be simplified if we let $V_2(s) = 1$.

If the first equation of the set were a current variable $I_i(s)$ instead of the voltage $V_2(s)$, the subsequent equations would contain the current variable $I_i(s)$ as a factor that we could also normalize to $I_i(s) = 1$. An example where the first equation is a current rather than a voltage equation occurs in determining the y parameters of a two-port network.

Before we embark upon some numerical examples, it is important to note that we must represent the series branches as impedances and the shunt branches as admittances. Suppose the series branch consisted of a resistor $R = 1\Omega$ in parallel with a capacitor $C = 1/2F$. Then the impedance of the branch is

$$Z(s) = \frac{R(1/sC)}{1/sC + R} = \frac{2}{s+2} \tag{10.93}$$

and must be considered as a single entity in writing the equations for the ladder. Similarly, if a shunt branch consisted of a resistor $R_1 = 2\Omega$ and an inductor $L_1 = 1\ h$, then the admittance of the branch is

$$Y(s) = \frac{1}{2+s} \tag{10.94}$$

The key point in this discussion is that we must use the *total* impedance or admittance of a branch in writing the equations for the ladder.

Example 10.5.1 Let us find the voltage ratio V_2/V_1, the current ratio I_2/I_1, the input impedance $Z_1 = V_1/I_1$, and the transfer impedance $Z_{21} = V_2/I_1$ for the network in Fig. 10.26. First, we must represent the series branches as impedances and the shunt branches as admittances, as shown in Fig. 10.27. The branch current and node voltage equations for the network are:

$$\begin{aligned} V_2(s) &= 1 \\ I_2(s) &= 2s \\ V_a(s) &= 3s(2s) + 1 = 6s^2 + 1 \\ I_1(s) &= \frac{2}{s}(6s^2+1) + 2s = 14s + \frac{2}{s} \\ V_1(s) &= \frac{4}{s}\left(14s + \frac{2}{s}\right) + 6s^2 + 1 = 6s^2 + 57 + \frac{8}{s^2} \end{aligned} \tag{10.95}$$

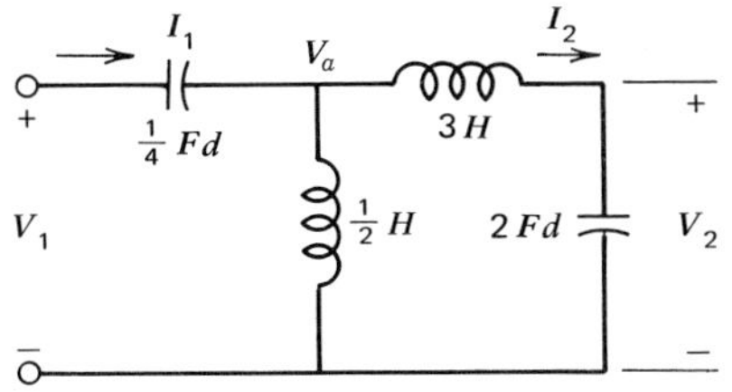

Fig. 10.26 A simple ladder network.

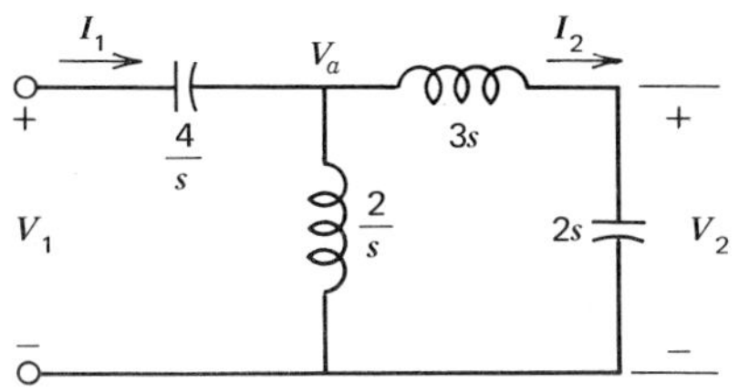

Fig. 10.27 A simple ladder network.

The various network functions are then obtained:

$$\text{(a)} \qquad Z_{\text{in}} = \frac{V_1}{I_1} = \frac{6s^4+57s^2+8}{14s^3+2s}$$

$$\text{(b)} \qquad \frac{V_2}{V_1} = \frac{s^2}{6s^4+57s^2+8}$$

$$\text{(c)} \qquad \frac{I_2}{I_1} = \frac{2s^2}{14s^2+2} \tag{10.96}$$

$$\text{(d)} \qquad Z_{21} = \frac{V_2}{I_1} = \frac{s}{14s^2+2}$$

Example 10.5.2 Find the short-circuit admittance functions y_{11} and y_2 for the network in Fig. 10.28. To obtain the short-circuit functions, we must short-circuit the 2-2′ port. Then we represent the series branches as impedances and the shunt branches as admittances so that the resulting network is given in Fig. 10.29.

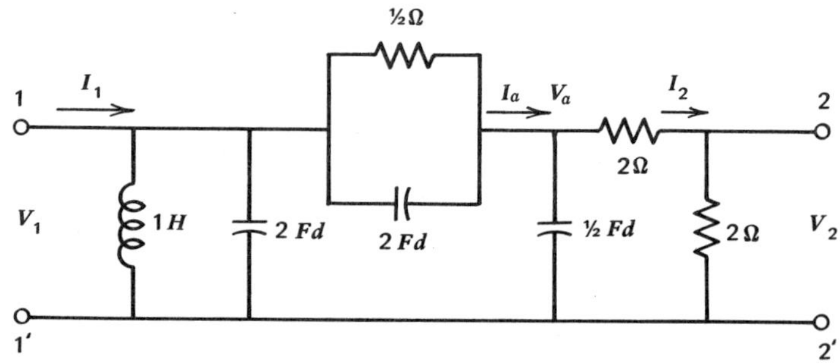

Fig. 10.28 A ladder network.

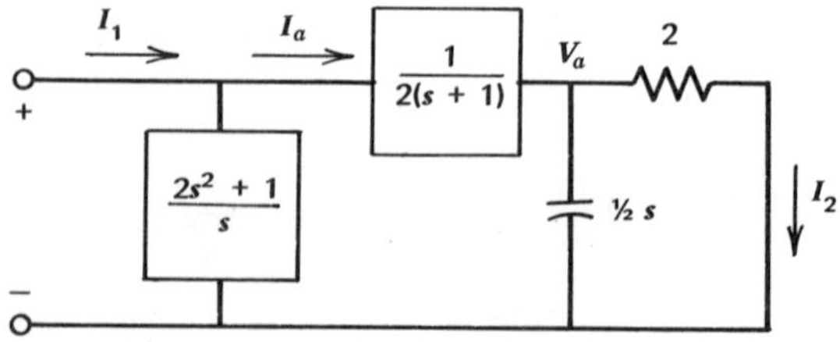

Fig. 10.29 A ladder network.

The pertinent equations are:

$$I_2 = 1 \qquad V_a = 2$$
$$I_a = \tfrac{1}{2}s(2) + 1 = s + 1$$
$$V_1 = \frac{1}{2(s+1)}(s+1) + 2 = \frac{5}{2} \tag{10.97}$$
$$I_1 = \frac{2s^2+1}{s}\left(\frac{5}{2}\right) + (s+1) = 6s + 1 + \frac{5/2}{s}$$

We now obtain our short-circuit functions as

$$y_{11} = \frac{I_1}{V_1} = \tfrac{12}{5}s + \tfrac{2}{5} + \frac{1}{s} \qquad y_{21} = \frac{-I_2}{V_1} = -\tfrac{2}{5} \tag{10.98}$$

PROBLEMS

10-1. Find the z, y, h, and T parameters of the networks shown in Fig. P10-1. Some of the parameters may not be defined for particular circuit configurations.

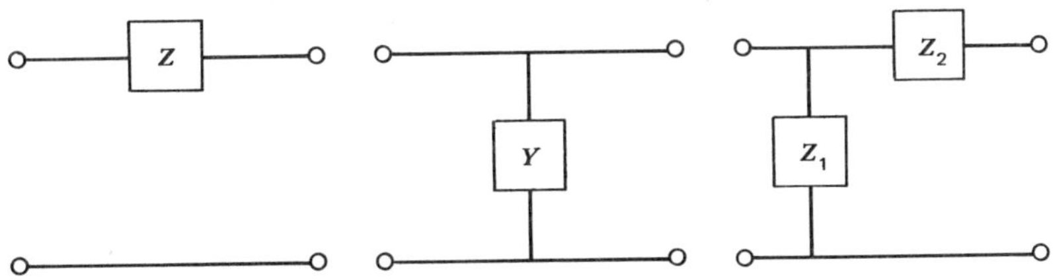

Fig. P10-1

10-1a. For the circuit shown in Fig. P10-1a,
(i) Find the z and y parameters.
(ii) Show that $z_{11} = z_{22}$ if $R_1 = R_2$ and $L_1 = L_2$.

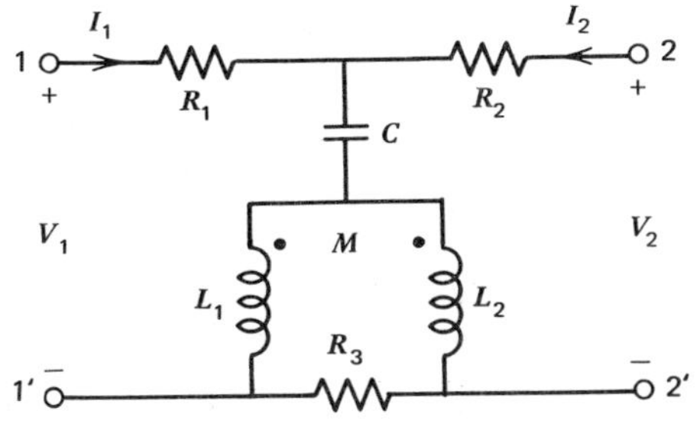

Fig. P10-1a

10-2. Find the z, y, h, and T parameters for the networks shown in Fig. P10-2.

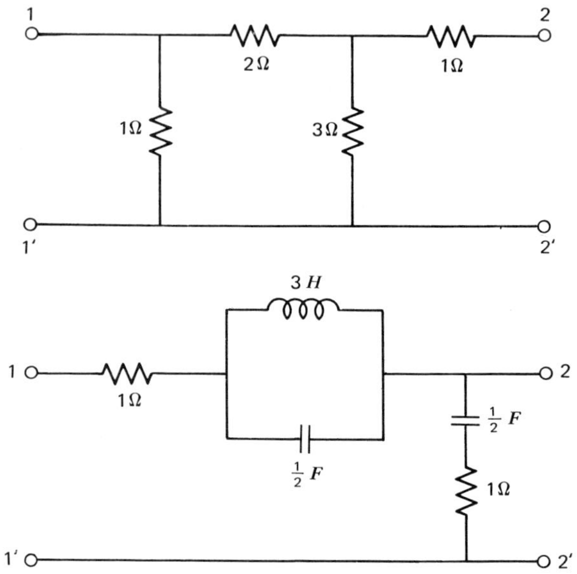

Fig. P10-2

10-3. Find the z parameters for the lattice and the bridge circuits shown in Fig. P10-3. (The results should be identical.)

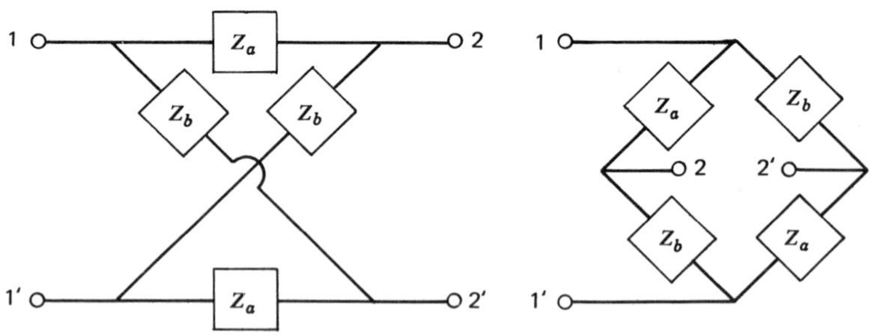

Fig. P10-3

10-4. For the circuits of Fig. P10-3, find the y parameters in terms of the admittances $Y_a = 1/Z_a$ and $Y_b = 1/Z_b$.

10-4a. The circuits shown in Fig. P10-4a have equivalent z parameters, find Z_a, Z_b, and Z_c.

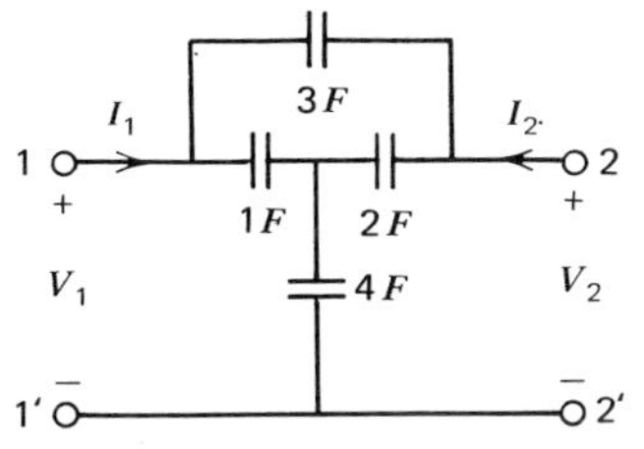

Fig. P10-4*a*

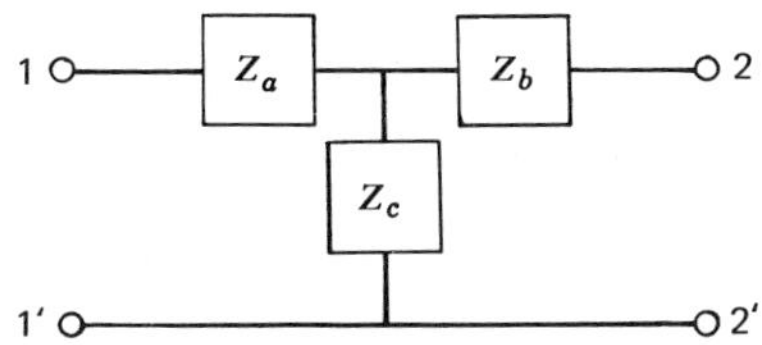

10-5. For the circuit shown, find the voltage-ratio transfer function V_2/V_1 and the input impedance $Z_1 = V_1/I_1$ in terms of the z parameters of the two-port network N and the load resistor R_L.

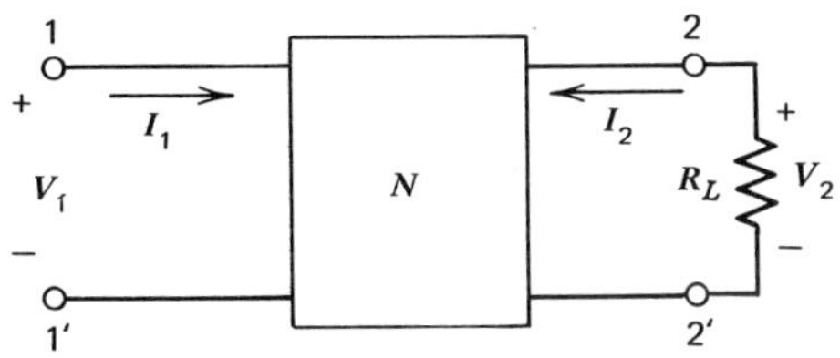

Fig. P10-5

10-6. For the cascade connection of two-ports depicted in the figure, show that the transfer impedance z_{12} of the overall circuit is given in terms of the z parameters of the individual two-ports by the equation

$$z_{12} = \frac{z_{12a}z_{12b}}{z_{12a} + z_{12b}}$$

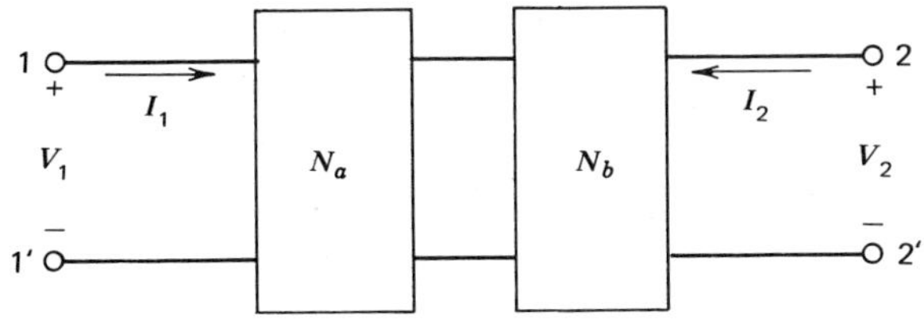

Fig. P10-6

10-7. Find the z and y parameters of the nonideal transformer shown in Fig. P10-7. Determine the T and π equivalent circuits for the transformer in terms of L_1, L_2, and M. (Hint: Use the z parameters for the T equivalent circuit and the y parameters for the π equivalent circuit.)

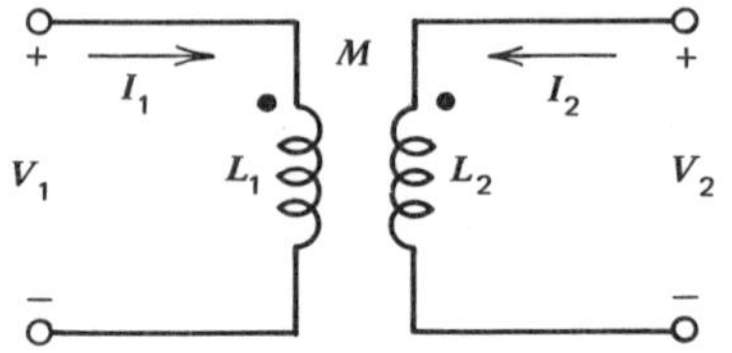

Fig. P10-7

10-7a. Find the h parameters of the circuit shown in Fig. P10-7a.

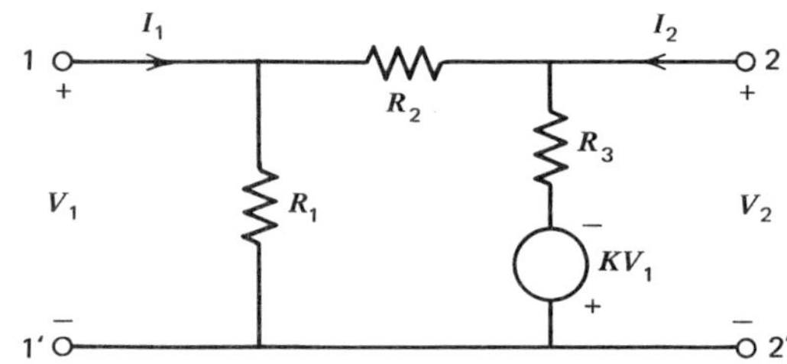

Fig. P10-7a

10-7b. Repeat 10-7a for the circuit shown in Fig. P10-7b.

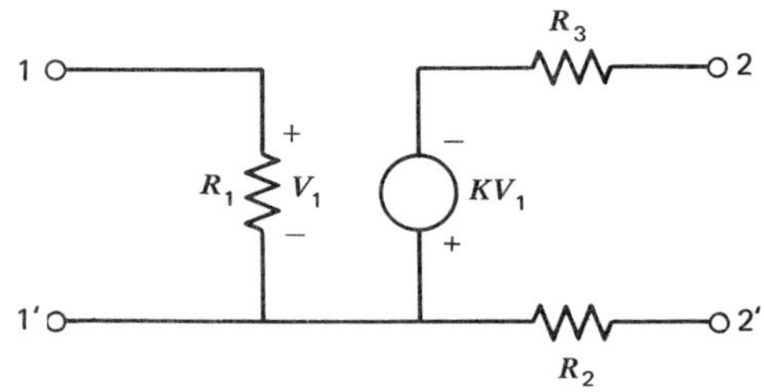

Fig. P10-7b

10-7c. Repeat 10-7a for the circuit shown in Fig. P10-7c.

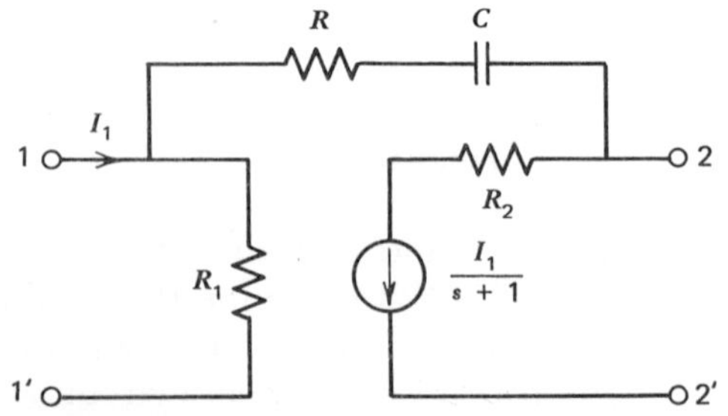

Fig. P10-7c

10-8. The *inverse hybrid* parameters of a two-port network are defined by the equation

$$\begin{bmatrix} I_1 \\ V_2 \end{bmatrix} = \begin{bmatrix} g_{11} & g_{12} \\ g_{21} & g_{22} \end{bmatrix} \begin{bmatrix} V_1 \\ I_2 \end{bmatrix}$$

Express the g parameters in terms of either the z or y parameters and give a physical interpretation of the meaning of these parameters; that is, say whether a parameter is open or short-circuit parameter, and whether it is a driving point or transfer function. Finally, derive the conditions of reciprocity for the g parameters.

10-8a. Find the T parameters for the circuit shown in Fig. P10-8a.

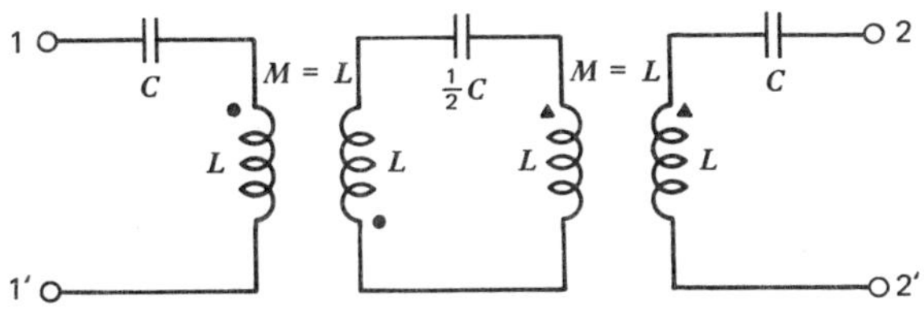

Fig. P10-8*a*

10-9. Prove that for a passive reciprocal network the T parameters satisfy $AD - BC = 1$.

10-10. Find the z and h parameters of the common emitter transistor represented by its T circuit model.

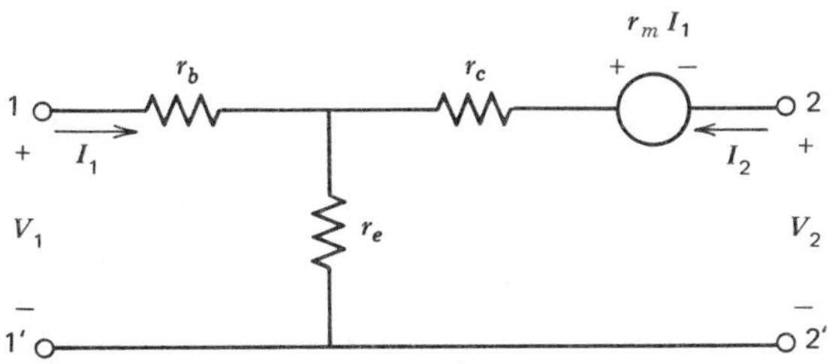

Fig. P10-10

10-11. The circuit in part (*a*) of the figure is to be described by an *equivalent input circuit* shown in part (*b*). Determine Z_{eq} in (*b*) as a function of the elements and voltages in (*a*).

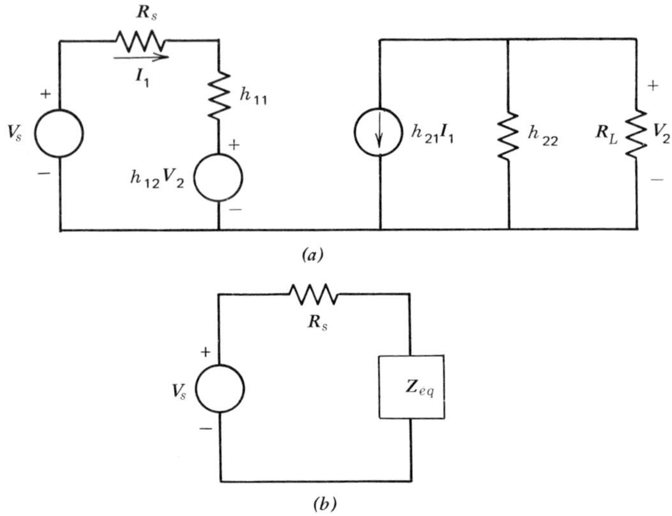

Fig. P10-11

10-12. Find the T parameters of the circuit shown in Fig. P10-12.

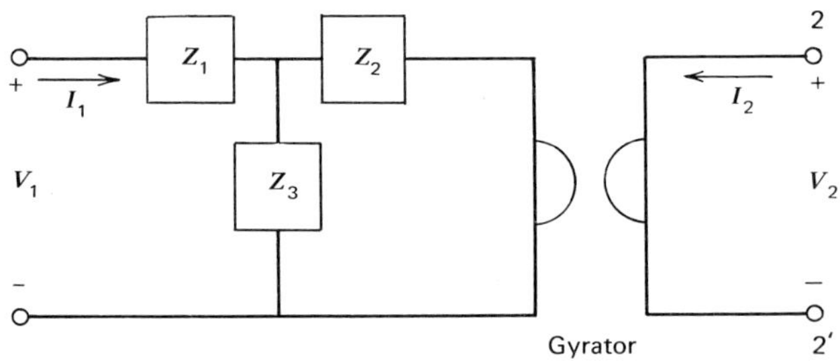

Fig. P10-12

10-13. Find the y parameters of the twin-T circuit shown in Fig. P10-1

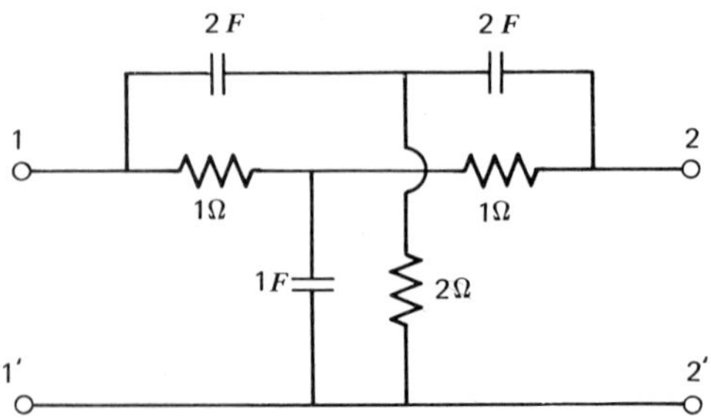

Fig. P10-13

by considering the circuit to be made up of two T circuits in parallel.

10-14. Find the z parameters of the circuit shown in Fig. P10-14.

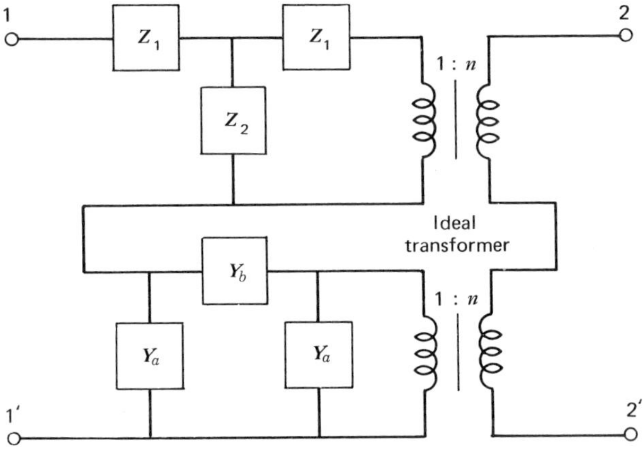

Fig. P10-14

10-15. Repeat 10-14 for the circuit of Fig. P10-15.

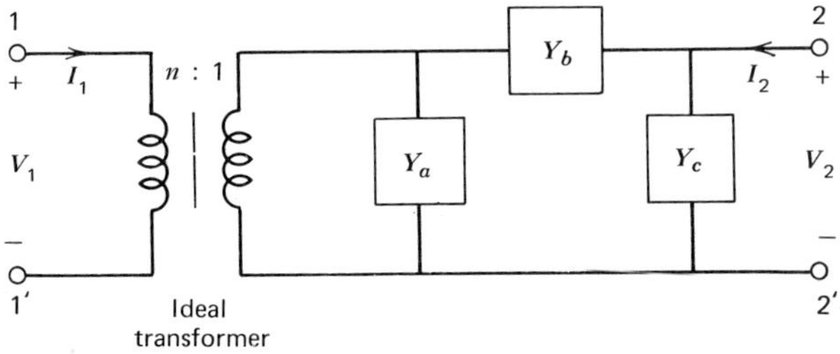

Fig. P10-15

10-16. Repeat 10-14 for the circuit of Fig. P10-16.

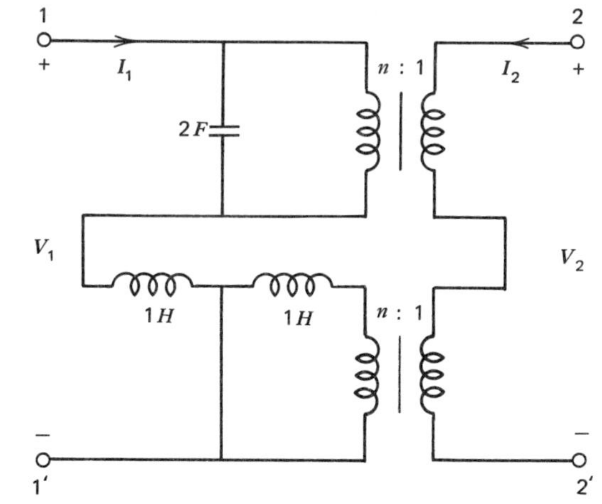

Fig. P10-16

10-16a. Find the z and y parameters for the series-parallel circuit shown in Fig. P10-16*a*.

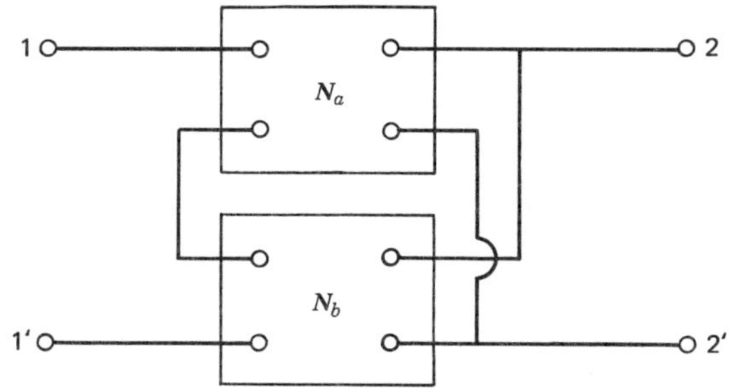

Fig. P10-16*a*

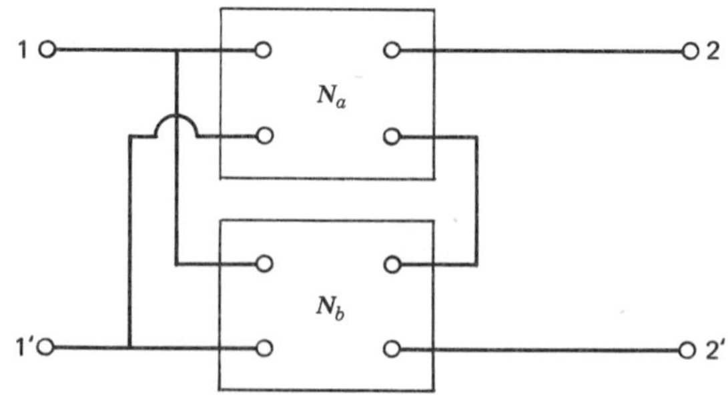

Fig. P10-16*b*

10-16b. Find the z and y parameters for the parallel series circuit shown in Fig. P10-16*b*.

10-17. Find the transfer impedance V_2/I_1 for the circuit shown in Fig. P10-17. Plot the poles and zeros for the transfer impedance obtained.

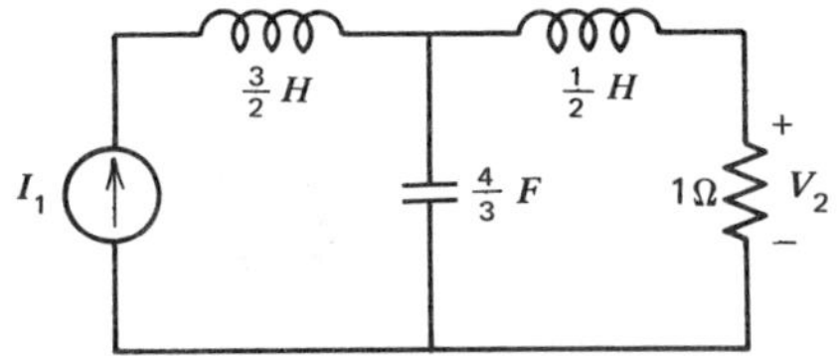

Fig. P10-17

10-18. Find the voltage ratio V_2/V_1 for the circuit shown in Fig. P10-18.

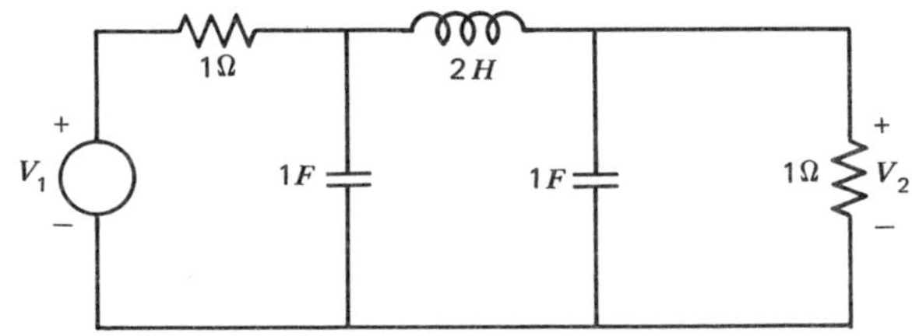

Fig. P10-18

10-19. Find the short-circuit parameters for the ladder network shown in Fig. P10-19.

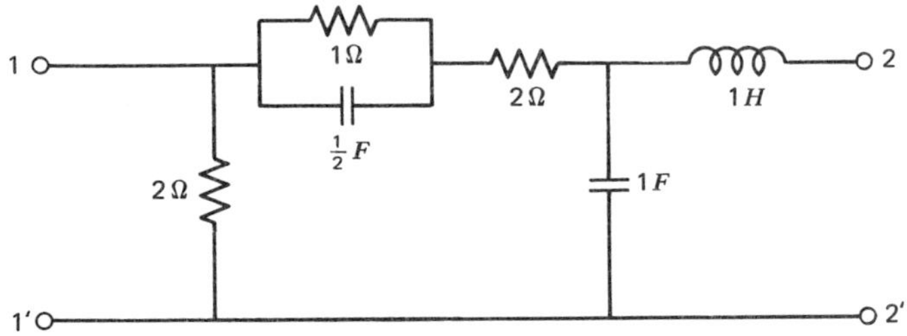

Fig. P10-19

10-20. Determine the voltage ratio V_2/V_1, the current ratio I_2/I_1, the transfer impedance V_2/I_1, and the driving point impedance V_1/I_1 for the network shown in Fig. P10-20.

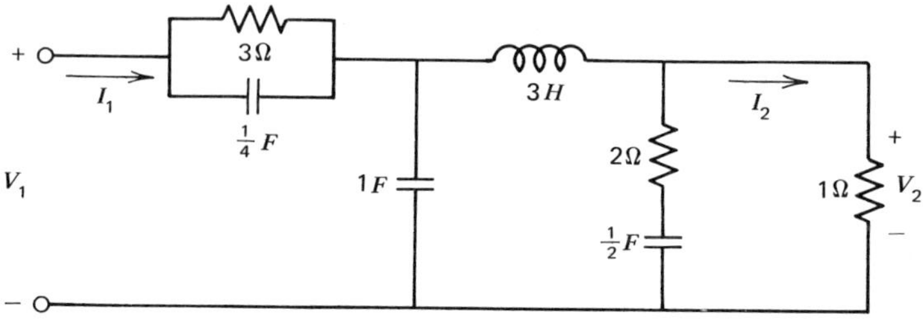

Fig. P10-20

10-20a. For the circuit of Fig. P10-1*a*, find the currents I_1 and I_2 if voltage sources $V_1 = V_2 = 1/s$ are applied to the circuit.

10-20b. For the circuit of Fig. P10-1*a*, find V_1 and V_2 if current sources $I_1 = I_2 = 1/s$ are applied to the circuit.

10-20c. Repeat 10-20a for the circuit of Fig. P10-4*a*.

10-20d. Repeat 10-20b for the circuit of Fig. P10-4*a*.

10-20e. Find I_1 and V_2 of the circuit of Fig. P10-7*a* if the excitations are $I_2 = 1/s$ and $V_1 = 1/s$.

10-20f. Find I_2 and V_1 if the excitations to the circuit of Fig. P10-7*a* are changed to $I_1 = 1/s$ and $V_2 = 1/s$.

10-20g. For the circuit of Fig. P10-7*a*, find the transfer function V_2/I_1 if a resistor R_L is connected across 2-2′ and a current source I_1 is connected to 1-1′.

10-20h. Repeat 10-20g if a voltage source V_1 is applied to the terminal pair 1-1′.

10-20i. Apply the results of 10-8 to find the inverse hybrid parameters of the circuit of Fig. P10-7*b*.

10-20j. Show that $[h][g] = [I]$, where $[I]$ is the identity matrix, for the circuit of Fig. P10-7*b*.

10-20k. Find the transfer ratio V_2/V_1 of the circuit of Fig. P10-16*a* in terms of the z parameters of N_a and N_b if a resistor R_L is connected across 2-2′.

10-20l. Repeat 10-20k for the circuit of Fig. P10-16*b*.

APPENDIX

Introduction to Matrix Algebra

A.1 FUNDAMENTAL OPERATIONS

Matrix notation is merely a shorthand method of algebraic symbolism, enabling one to carry out the algebraic operations more quickly.

The theory of matrices originated primarily from the need (1) to solve simultaneous linear equations and (2) to have a compact notation for linear transformations from one set of variables to another.

As an example of (2), consider the set of simultaneous linear equations:

$$\begin{array}{c} a_{11}x_1 + a_{12}x_2 + \cdots + a_{1n}x_n = y_1 \\ a_{21}x_1 + a_{22}x_2 + \cdots + a_{2n}x_n = y_2 \\ \vdots \qquad\qquad\qquad \vdots \\ a_{m1}x_1 + a_{m2}x_2 + \cdots + a_{mn}x_n = y_m \end{array} \tag{A.1}$$

These may express a general linear transformation from the x_i to the y_i. In general, $m \neq n$. An example where the numbers of variables in the two sets are unequal is that of representing a 3-dimensional object in 2 dimensions (in a perspective drawing). Here, $m = 2$ and $n = 3$.

Definition

A matrix is an ordered rectangular array of numbers, generally the coefficients of a linear transformation. The matrix is defined by giving all its elements, and the location of each.

The matrix of the equations in Eq. A.1, written as

$$\begin{bmatrix} a_{11} & a_{12} & \cdots & a_{1n} \\ a_{21} & a_{22} & \cdots & a_{2n} \\ \vdots & \vdots & & \vdots \\ a_{2m} & a_{m2} & \cdots & a_{mn} \end{bmatrix},$$

is of *order* $m \times n$. (The 1st number here is the number of rows; the 2nd is the number of columns.) The matrix may be denoted by a single capital letter A or by $[a_{ij}]$.

A matrix is a single complete entity, like a position in chess. Two matrices A and B are equal only if all corresponding elements are the same, $a_{ij} = b_{ij}$ for all i and j.

A matrix may consist of a single row or single column. The complete matrix notation applied to Eq. A.1 is:

$$\begin{bmatrix} a_{11} & a_{12} & \cdots & a_{1n} \\ a_{21} & a_{22} & \cdots & a_{2n} \\ \cdot & \cdot & & \cdot \\ \cdot & \cdot & & \cdot \\ \cdot & \cdot & & \cdot \\ a_{m1} & a_{m2} & \cdots & a_{mn} \end{bmatrix} \begin{bmatrix} x_1 \\ \\ \cdot \\ \cdot \\ \cdot \\ x_n \end{bmatrix} = \begin{bmatrix} y_1 \\ y_2 \\ \cdot \\ \cdot \\ \cdot \\ y_m \end{bmatrix} \tag{A.2}$$

which says, if put crudely, that

$$(A) \text{ operates on } (x_i) \text{ to yield} \rightarrow (y_1)$$

This emphasizes the similarity between a matrix and a transformation. The **x**- and **y**-matrices are column matrices.

Row and column matrices are called *vectors* (specially, row vectors and column vectors) and their similarity to the more usual type of vector will appear later. Vectors will be written as small letters, such as **x** or **y**. Note that the elements of vectors need only one subscript, while elements of matrices need two.

A.2 ELEMENTARY CONCEPTS

Square Matrix

A square matrix has the same number of rows as columns, that is, a matrix of order (nn). The Y matrix is an example of a square matrix

$$Y = \begin{bmatrix} y_{11} & y_{12} \\ y_{21} & y_{22} \end{bmatrix} \tag{A.3}$$

Diagonal Matrix

A diagonal matrix is a square matrix whose elements off the main diagonal are zero, that is, one in which $\alpha_{ij} = 0$ for $i \neq j$. The following matrix is diagonal.

$$\begin{bmatrix} 1 & 0 & 0 \\ 0 & -2 & 0 \\ 0 & 0 & 3 \end{bmatrix} \tag{A.4}$$

Unit Matrix

A unit matrix is a diagonal matrix for which $a_{ij} = 1$ for $i = j$, and is denoted as I. For example, I is

$$I = \begin{bmatrix} 1 & 0 & 0 \\ 0 & 1 & 0 \\ 0 & 0 & 1 \end{bmatrix} \tag{A.5}$$

Equality

Two matrices are equal if they have the same number of rows and columns and if the corresponding elements are equal. Suppose $y_1 = 2$, $y_2 = -3$, and $y_3 = -6$. If we write this set of equations in matrix form, we have

$$\begin{bmatrix} y_1 \\ y_2 \\ y_3 \end{bmatrix} = \begin{bmatrix} 2 \\ -3 \\ -6 \end{bmatrix} \tag{A.6}$$

Transpose

The transpose of a matrix A denoted as A^T, is the matrix formed by interchanging the rows and columns of A. Thus, if

$$A = \begin{bmatrix} 1 & 0 \\ -6 & 5 \\ 3 & 2 \end{bmatrix} \tag{A.7}$$

then

$$A^T = \begin{bmatrix} 1 & -6 & 3 \\ 0 & 5 & 2 \end{bmatrix} \tag{A.8}$$

Determinant of a Matrix

The determinant of a matrix is defined only for square matrices. For example,

$$\det \begin{bmatrix} 1 & 2 \\ -5 & 4 \end{bmatrix} = \begin{vmatrix} 1 & 2 \\ -5 & 4 \end{vmatrix} = 14 \tag{A.9}$$

Note that the determinant of a matrix has a particular value, whereas the matrix itself is merely an array of quantities.

Cofactor

The cofactor A_{ij} of the element a_{ij} of the matrix A is defined as the product of $(-1)^{i+j}$ with the determinant formed by deleting the ith row and jth column of A. Thus, A_{21} of 3 of the matrix

$$A = \begin{bmatrix} 2 & 6 \\ 3 & 4 \end{bmatrix} \tag{A.10}$$

is

$$A_{21} = (-1)^{2+1} \times 6 = -6 \tag{A.11}$$

Adjoint Matrix

The adjoint matrix of a square matrix A is formed by replacing each element of A by its cofactor and transposing. For example, for the matrix in Eq. A.10.

$$\begin{aligned} \operatorname{adj} A &= \begin{bmatrix} 4 & -3 \\ -6 & 2 \end{bmatrix}^T \\ &= \begin{bmatrix} 4 & -6 \\ -3 & 2 \end{bmatrix} \end{aligned} \tag{A.12}$$

Singular and Nonsingular Matrices

A singular matrix is a square matrix A for which $\det A = 0$. A nonsingular matrix is one for which $\det A \neq 0$.

A.3 OPERATIONS ON MATRICES

Addition

Two matrices may be added if both matrices are of the same order. Each element of the first matrix is added to the element of the second matrix whose row and column orientation is the same. An example of matrix addition is shown below.

$$\begin{bmatrix} -2 & 4 \\ 3 & 0 \end{bmatrix} + \begin{bmatrix} 3 & 6 \\ -7 & -2 \end{bmatrix} = \begin{bmatrix} 1 & 10 \\ -4 & -2 \end{bmatrix} \tag{A.13}$$

Thus if matrices $A, B, C, \ldots, K$ are all of the same order, then

$$A + B + \cdots + K = [a_{ij} + b_{ij} + \cdots + k_{ij}] \tag{A.14}$$

The *associative* and *commutative* laws apply.

$$\begin{aligned} &\textit{Associative:} \quad A + (B + C) = (A + B) + C \\ &\textit{Commutative:} \; A + B = B + A \end{aligned} \tag{A.15}$$

Multiplication by a Scalar

We define multiplication by a scalar as

$$\lambda A = \lambda[a_{ij}] = [\lambda a_{ij}] \tag{A.16}$$

Thus to multiply a matrix by a scalar, multiply each of its elements by the scalar.

Example A.1

$$3\begin{bmatrix} 2 & 0 \\ -1 & 1 \\ -3 & 2 \end{bmatrix} = \begin{bmatrix} 6 & 0 \\ -3 & 3 \\ -9 & 6 \end{bmatrix} \tag{A.17}$$

Linear Combination of Matrices

If the rules of addition and multiplication by a scalar are combined, we have, for two matrices of the same order,

$$\alpha A + \beta B = [\alpha a_{ij} + \beta b_{ij}] \tag{A.18}$$

where α and β are scalars.

Multiplication

In order for matrix multiplication AB to be possible, the number of columns of the first matrix A must equal the number of rows of the second matrix B. The product C will have the number of rows of the first and the number of columns of the second matrix. In other words, if A has m rows and n columns and B has n rows and p columns, then the product C will have m rows and p columns. The individual elements of C are given by

$$c_{ij} = \sum_{k=1}^{n} a_{ik} b_{kj} \tag{A.19}$$

Example A.2

$$\begin{bmatrix} 2 & -1 & 0 \\ -1 & 0 & 3 \end{bmatrix} \begin{bmatrix} 4 & 1 \\ 2 & 0 \\ -3 & 2 \end{bmatrix} = \begin{bmatrix} 6 & 2 \\ -13 & 5 \end{bmatrix} \tag{A.20}$$

Example A.3 The system of equations

$$\begin{aligned} z_{11}I_1 + z_{12}I_2 &= V_1 \\ z_{21}I_1 + z_{22}I_2 &= V_2 \end{aligned} \tag{A.21}$$

can be written in matrix notation as

$$\begin{bmatrix} z_{11} & z_{12} \\ z_{21} & z_{22} \end{bmatrix} \begin{bmatrix} I_1 \\ I_2 \end{bmatrix} = \begin{bmatrix} V_1 \\ V_2 \end{bmatrix} \tag{A.22}$$

We see that systems of equations can be very conveniently written in matrix notation.

Matrix multiplication is not generally commutative, that is,

$$A_{mn}B_{np} \neq B_{np}A_{mn} \tag{A.23}$$

Observe that the product BA is not defined unless $p = m$. Even a product of square matrices is generally not commutative as seen by the following example:

$$\begin{bmatrix} 1 & 0 \\ -1 & 2 \end{bmatrix} \begin{bmatrix} -1 & 0 \\ 0 & 2 \end{bmatrix} = \begin{bmatrix} -1 & 0 \\ 1 & 4 \end{bmatrix} \tag{A.24}$$

If we interchange the order of multiplication, we obtain

$$\begin{bmatrix} -1 & 0 \\ 0 & 2 \end{bmatrix} \begin{bmatrix} 1 & 0 \\ -1 & 2 \end{bmatrix} = \begin{bmatrix} -1 & 0 \\ -2 & 4 \end{bmatrix} \tag{A.25}$$

Because of the noncommutative nature of matrix multiplication, we must distinguish between premultiplication and postmultiplication. In BA, A is premultiplied by B, B is postmultiplied by A.

For matrix multiplication, the associative and distributive laws apply.

$$\begin{aligned} &\textit{Associative:} \quad A(BC) = (AB)C = ABC \\ &\textit{Distributive:} \quad A(B+C) = AB + AC \end{aligned} \tag{A.26}$$

Transpose of a Product

The transpose of a product AB is equal to the product, in reverse order, of the transpose of the individual matrices, A and B, that is,

$$(AB) = B^T A^T \tag{A.27}$$

and

$$(ABC)^T = C^T(AB)^T = C^T B^T A^T \tag{A.28}$$

The product $\mathbf{x}^T\mathbf{x}$, if $\mathbf{x}$ is a column vector, is a scalar number equal to the sum of the squares of the elements of $\mathbf{x}$. Thus

$$\mathbf{x}^T\mathbf{x} = [x_1 x_2 \cdots x_n] \begin{bmatrix} x_1 \\ x_2 \\ \cdot \\ \cdot \\ \cdot \\ x_n \end{bmatrix} \tag{A.29}$$

$$= x_1^2 + x_2^2 + \cdots + x_n^2$$

The product $\mathbf{x}\mathbf{x}^T$ is a square matrix C such that $C = C^T$.

Common Expressions in Simple Notation

(a) The sum of products

$$a_1 b_1 + a_2 b_2 + \cdots + a_n b_n$$

which are the typical element of a product matrix, may be written as

$$\mathbf{a}^T\mathbf{b} \qquad \text{or} \qquad \mathbf{b}^T\mathbf{a}$$

where $\mathbf{a}$ and $\mathbf{b}$ are column vectors.

(b) The sum of squares

$$x_1^2 + x_2^2 + \cdots + x_n^2$$

is thus $\mathbf{x}^T\mathbf{x}$ ($\mathbf{x}$, column vector). It is also $\mathbf{x}\mathbf{x}^T$ if $\mathbf{x}$ is a row vector.

(c) The expression

$$a_{11}x_1^2 + a_{22}x_2^2 + \cdots + a_{nn}x_n^2$$

is $\mathbf{x}^T A\mathbf{x}$ ($\mathbf{x}$, column vector) where A is a *diagonal* matrix with elements a_{jj}. Similarly

$$a_{11}x_1y_1 + a_{22}x_2y_2 + \cdots + a_{nn}x_ny_n$$

is $\mathbf{x}^T A\mathbf{y}$ or $\mathbf{y}^T A\mathbf{x}$.

(d) An expression such as

$$\sum_i \sum_j a_{ij}x_ix_j \qquad a_{ij} = a_{ji}$$

is a *quadratic form*. This is

$$\begin{aligned} a_{11}x_1^2 + 2a_{12}x_1x_2 + \cdots \qquad \cdots + 2a_{1n}x_1x_n \\ + a_{22}x_2^2 + 2a_{23}x_2x_3 + \cdots + 2a_{2n}x_2x_n \\ + \cdots + a_{nn}x_n^2 = \mathbf{x}^T A\mathbf{x} \end{aligned} \tag{A.30}$$

Inverse

Division is not defined in matrix algebra. The analogous operation i that of obtaining the *inverse* of a square matrix. The inverse A^{-1} of matrix A is defined by the relation

$$A^{-1}A = AA^{-1} = I \tag{A.31}$$

To obtain $A^{-1}A$ we first obtain the adjoint of A, adj A. Then we obtai the determinant of A. The inverse A^{-1} is equal to adj A divided b $|A|$, that is,

$$A^{-1} = \frac{1}{|A|} \operatorname{adj} A \tag{A.32}$$

Example A.4 let A be given as

$$A = \begin{bmatrix} 2 & 1 \\ -1 & 1 \end{bmatrix} \tag{A.33}$$

Its determinant is

$$|A| = 3 \tag{A.34}$$

and the cofactors are

$$\begin{aligned} A_{11} &= 1 \qquad & A_{12} &= 1 \\ A_{21} &= -1 \qquad & A_{22} &= 2 \end{aligned} \tag{A.35}$$

The adjoint matrix is

$$\begin{aligned} \operatorname{adj} A &= \begin{bmatrix} 1 & 1 \\ -1 & 2 \end{bmatrix}^T \\ &= \begin{bmatrix} 1 & -1 \\ 1 & 2 \end{bmatrix} \end{aligned} \tag{A.36}$$

so that $\mathbf{A}^{-1}$ is

$$\begin{aligned} A^{-1} &= \tfrac{1}{3} \begin{bmatrix} 1 & -1 \\ 1 & 2 \end{bmatrix} \\ &= \begin{bmatrix} \frac{1}{3} & -\frac{1}{3} \\ \frac{1}{3} & \frac{2}{3} \end{bmatrix} \end{aligned} \tag{A.37}$$

As a check we see that

$$\begin{aligned} \begin{bmatrix} \frac{1}{3} & -\frac{1}{3} \\ \frac{1}{3} & \frac{2}{3} \end{bmatrix} \begin{bmatrix} 2 & 1 \\ -1 & 1 \end{bmatrix} &= \begin{bmatrix} 1 & 0 \\ 0 & 1 \end{bmatrix} \\ \begin{bmatrix} 2 & 1 \\ -1 & 1 \end{bmatrix} \begin{bmatrix} \frac{1}{3} & -\frac{1}{3} \\ \frac{1}{3} & \frac{2}{3} \end{bmatrix} &= \begin{bmatrix} 1 & 0 \\ 0 & 1 \end{bmatrix} \end{aligned} \tag{A.38}$$

If the determinant of the matrix is zero, then the inverse is not defined In other words, only nonsingular square matrices have inverses.

A.4 SOLUTIONS OF LINEAR EQUATIONS

Consider a set of linear algebraic equations, to be solved simultaneously:

$$\begin{aligned} a_{11}x_1 + a_{12}x_2 + \cdots + a_{1n}x_n &= h_1 \\ a_{21}x_1 + a_{22}x_2 + \cdots + a_{2n}x_n &= h_2 \\ \vdots \qquad\qquad \vdots \qquad\qquad & \quad \vdots \\ a_{n1}x_1 + a_{n2}x_2 + \cdots + a_{nn}x_n &= h_n \end{aligned} \tag{A.39}$$

The h_1 are constants. It is desired to solve for the x_j. In matrix notation,

$$A\mathbf{x} = \mathbf{h} \tag{A.40}$$

Premultiplying by A^{-1},

$$\mathbf{x} = \mathbf{A}^{-1}\mathbf{h} \tag{A.41}$$

In expanded form, this is

$$\begin{bmatrix} x_1 \\ x_2 \\ x_3 \end{bmatrix} = \frac{1}{|A|} \begin{bmatrix} A_{11} & A_{21} & A_{31} \\ A_{12} & A_{22} & A_{32} \\ A_{13} & A_{23} & A_{33} \end{bmatrix} \begin{bmatrix} h_1 \\ h_2 \\ h_3 \end{bmatrix} \tag{A.42}$$

Thus

$$x_1 = \frac{h_1A_{11} + h_2A_{21} + h_3A_{31}}{|A|}$$

$$= \frac{\begin{vmatrix} h_1 & a_{12} & a_{13} \\ h_2 & a_{22} & a_{23} \\ h_3 & a_{32} & a_{33} \end{vmatrix}}{|A|}$$

and similarly for x_2 and x_3. This is the familiar *Cramer's rule* for solving such equations.

Example A.5. Solve for x,y,z:

$$\begin{aligned} x - y + z &= 2 \\ 2x + y &= 1 \\ -x + 3y + z &= -1 \end{aligned} \tag{A.44}$$

In matrix form these equations are written as $A\mathbf{x} = \mathbf{h}$ where

$$A = \begin{bmatrix} 1 & -1 & 1 \\ 2 & 1 & 0 \\ -1 & 3 & 1 \end{bmatrix} \qquad \mathbf{x} = \begin{bmatrix} x \\ y \\ z \end{bmatrix} \qquad \mathbf{h} = \begin{bmatrix} 2 \\ 1 \\ -1 \end{bmatrix} \tag{A.45}$$

Also, $|A| = 10$.

Thus

$$\mathbf{x} = A^{-1}\mathbf{h}$$

$$= \frac{1}{10}\begin{bmatrix} 1 & 4 & -1 \\ -2 & 2 & 2 \\ 7 & -2 & 3 \end{bmatrix}\begin{bmatrix} 2 \\ 1 \\ -1 \end{bmatrix} = \frac{1}{10}\begin{bmatrix} 7 \\ -4 \\ 9 \end{bmatrix} \tag{A.46}$$

$\therefore \quad x = 0.7 \qquad y = -0.4 \qquad z = 0.9$

A.5 REFERENCES ON MATRIX ALGEBRA

Below is a short list of books on matrices.

Aitken, A. C. *Determinants and Matrices,* 9th Ed., Interscience, New York, 1956.

Bellman, R. *Introduction to Matrix Analysis,* McGraw-Hill, New York, 1960.

Eisenman, R. L. *Matrix Vector Analysis,* McGraw-Hill, New York, 1963.

Faddeev, D. K., and V. N. Faddeeva, *Computational Methods of Linear Algebra,* Freeman, San Francisco, 1963.

Gantmacher, F. R. *Applications of the Theory of Matrices,* Interscience, New York, 1959.

Hohn, F. E. *Elementary Matrix Algebra,* Macmillan, New York, 1964.

Huelsman, L. P. *Circuits, Matrices, and Linear Vector Spaces,* McGraw-Hill, New York, 1963.

LeCorbeiller, P. *Matrix Analysis of Electric Networks,* Wiley, New York, 1950.

Marcus, M. and H. Minc, *Survey of Matrix Theory and Matrix Inequalities,* Allyn and Bacon, Boston, 1964.

Nering, E. D. *Linear Algebra and Matrix Theory,* Wiley, New York, 1964.

Perlis, S. *Theory of Matrices,* Addison Wesley, Reading, Mass., 1952.

Pipes, L. A. *Matrix Methods for Engineering,* Prentice-Hall, Englewood Cliffs, New York, 1963.

Tropper, A. M. *Matrix Theory for Electrical Engineering Students,* Harrop, London, 1962.

von Weiss, A. *Matrix Analysis for Electrical Engineers,* Van Nostrand, Princeton, 1964.

Index